Coal-burning Type of Endemic Fluorosis

Zhi-Zhong Guan
Editor

Coal-burning Type of Endemic Fluorosis

Pathophysiology and Clinical Treatments

Springer

Editor
Zhi-Zhong Guan
Department of Pathology, Affiliated
Hospital and Key Laboratory of Endemic
and Ethnic Diseases of the Ministry of
Education
Guizhou Medical University
Guiyang, Guizhou, People's Republic of China

ISBN 978-981-16-1497-2 ISBN 978-981-16-1498-9 (eBook)
https://doi.org/10.1007/978-981-16-1498-9

This Springer imprint is published by the registered company Springer Nature Singapore Pte Ltd.
The registered company address is: 152 Beach Road, #21-01/04 Gateway East, Singapore 189721, Singapore

Preface

Endemic fluorosis is a geochemical disease. In addition to the typically clinic manifestations of skeletal and dental fluorosis, it also causes systemic damages. In the natural environment, excessive fluoride intake in the human body through drinking water, air, food, tea, and other sources results in systematacially chronic toxic lesions. In international community, endemic fluorosis affects more than 20 countries, including China, India, Bangladesh, Sri Lanka, Vietnam, Thailand, Iran, Argentina, the United States, Canada, Mexico, Morocco, Algeria, Tunisia, Egypt, South Africa, and so on. Among these countries, China and India are worst afflicted by endemic fluorosis with the most widespread range. In China, most provinces and regions have areas with endemic fluorosis, in which three types are found, i.e., drinking water, coal-burning, and brick tea drinking type of endemic fluorosis. There were as many as 1308 counties with such areas in China with a population of 116 million affected by the disease based on past surveys.

The concept regarding excessive fluoride intake from food instead of water contamination as previously thought was first proposed in the 1970s by a team, leading by the well-known professor Zan-Dao Wei, from Guizhou Medical University. Later, it was confirmed by the scientists in the Center for Disease Control and Prevention in Guizhou and Guiyang Institute of Geochemistry at the Chinese Academy of Sciences. The fact is that local residents in these fluorosis areas burned the coal containing high amount of fluoride indoors and used stoves without exhaust systems to cook, bake (grains, peppers, and vegetables), and keep warm in the winter. Such results lead to chronic poisoning caused by excessive fluoride intake, which is termed as coal-burning type of endemic fluorosis. At present, there are 13 provinces (Guizhou, Yunnan, Sichuan, Chongqing, Hubei, and Hunan with worst condition of such endemic fluorosis) with a population of 34 million at risk of coal-burning type of endemic fluorosis.

For a long time, Chinese scientists and preventative workers have done a lot of basic and on-site prevention and control research on coal-burning type of endemic fluorosis and achieved remarkable results.

This book covers these aspects concerning origin, relevant epidemiology, pathology, molecular mechanism, biological markers of early diagnosis, clinical treatment and prevention, and evaluation of comprehensive control on the coal-burning type of endemic fluorosis.

I have been working on fluorosis research since the 1970s. The study of endemic fluorosis could be daunting with hardships and difficulties, but when we discover effective research to address the sufferings of the population living in the disease areas, the joy is unparalleled. I believe that this book could be a good reference and helpful for researchers, graduate students, and undergraduates in this field to raise their awareness of this disease and take effectively preventive measures for the goal to completely eliminate the disease.

Guiyang, Guizhou, People's Republic of China Zhi-Zhong Guan

Acknowledgments

I would like to express my heartfelt gratitude to my colleagues and friends who have contributed in many ways toward the completion of the book.

First and foremost, I wish to express my utmost appreciation to the key contributing authors in China for their commitment to this book. They are Zi-Gui Zhang (at Zhejiang Normal University), Hui Xu (at Jilin University), Jian-Cao Bian (at Shandong Institute of Prevention and Control for Endemic Disease), Fa Sun (Health Commission of Guizhou Province), Fu-Chen Li (at Liupanshui Center for Disease Control and Prevention), and Di-Dong Lou (at Guizhou University of Traditional Chinese Medicine); my colleagues in Guizhou Medical University: Chang-Xue Wu, Yi Zhao, Xiao-Lan Qi, Wen-Feng Yu, Ri-Guang Liu, Yong Wang, Xu-Guang Chen, Qing-Hong Duan, Jian Liao, Ling Mao, Kai-ling Zhang, Yan-Jie Liu, Na Wei, Jie Deng, and Yang-Ting Dong; also the coauthors (young scientists and PhD students): Feng-Yang Guo, Zhong-Yuan Zhang, Man Zhang, Ji Bian, Zhi Huang, Yue-Hai Xiao, Long-Yan Ran, Jie Xiang, Xiao-Xiao Zeng, Xiao Xiao, Jian Zou, and Yan-Long Yu. These authors have years of experience in the research and prevention of endemic fluorosis and achieved valuable results in the field. The research teams of several authors are also first-tier ones in the field of endemic fluorosis research. The authors have profound understanding concerning the progress of chapters in relevant international scientific societies, the prevention and control of the disease, and the needs of the reader, which serve as guarantee for the quality of the book.

In particular, I want to acknowledge my supervisors, Professors Jia-Liu Liu and Zan-Dao Wei, of Guizhou Medical University, who guided me to conduct the research on endemic fluorosis when I was a young graduate student, and then conducted useful discussion on the contents of this book when I planned to compose it many years ago. Without their guidance and help, there would be no such book. In addition, I would like to express my sincere appreciation to Professor Nan-Shan Zhong, the internationally renowned expert and academician of the Chinese Academy of Engineering, who gave me and my colleagues long-term support and sent his research group to participate in the clinical research in the areas of coal-burning type of endemic fluorosis. It was such effort and assistance that made us far ahead in the field of coal-burning type of endemic fluorosis research. The valuable results of the research were also included in the book.

Importantly, this book is one of the implementations of these funds: *National Natural Science Foundation of China* (U1812403, 81760571) and the *Projects of Department of Science and Technology of Guizhou Province and Guizhou Medical University*. Therefore, I would like to express my sincere thanks to these foundations for their financial support and assistance in publishing this book.

Zhi-Zhong Guan

Contents

Part I Discovery and Epidemiology of Endemic Fluorosis

1 Discovery and Concept of Endemic Fluorosis 3
Zhi Zhong Guan

2 Epidemic Status of Coal-burning Type of Endemic Fluorosis 13
Zhi-Zhong Guan, Fu-Cheng Li, Yi Zhao, and Chang-Xue Wu

Part II Pathological Changes of Endemic (Chronic) Fluorosis

3 Skeletal Fluorosis 31
Jie Xiang, Xiao-Xiao Zeng, and Zhi-Zhong Guan

4 Medical Imaging Manifestation of Skeletal Fluorosis 63
Qing-Hong Duan, Xu-Guang Chen, and Zhi-Zhong Guan

5 Epidemic Characteristic and Pathogenesis of Dental Fluorosis 85
Jian Liao

6 Damage in Nervous System 105
Na Wei and Zhi-Zhong Guan

7 Modification of Digestive System 125
Jian Zou and Zhi-Zhong Guan

8 Modification of Urinary System 137
Xiao Xiao and Zhi-Zhong Guan

9 Modification of Cardiovascular System 151
Jian-Chao Bian, Man Zhang, and Ji Bian

10 Damage of Endocrine System 161
Xiao-Xiao Zeng, Jie Xiang, and Zhi-Zhong Guan

11 **Damages to Reproductive System by Coal-burning Type of Endemic Fluorosis** 173
Yue-Hai Xiao and Fa Sun

Part III Molecular Pathogenesis of Endemic (Chronic) Fluorosis

12 **Enhanced Oxidative Stress** 195
Long-Yan Ran and Zhi-Zhong Guan

13 **High Level of Inflammation** 211
Wen-Feng Yu and Zhi Huang

14 **Effect of Fluoride on Signal Transduction Pathways** 225
Xiao-Lan Qi

15 **Changes of Receptors** 251
Yang-Ting Dong and Zhi-Zhong Guan

16 **Alteration of Endoplasmic Reticulum Stress** 269
Hui Xu, Feng-Yang Guo, and Zhong-Yuan Zhang

17 **Mitochondrial Changes** 283
Di-Dong Lou and Zhi-Zhong Guan

18 **Calcium Metabolism Disorder in Chronic Fluorosis** 297
Zi-Gui Zhang

19 **Altering of Carbohydrate Metabolism** 307
Kai-Lin Zhang and Zhi-Zhong Guan

20 **Changes in Gene Expression** 319
Jie Deng and Zhi-Zhong Guan

Part IV Clinical Diagnosis and Treatments of Coal-Burning Type of Endemic Fluorosis

21 **Biological Indicators for Early Diagnosis of Endemic Fluorosis** 337
Yan-Jie Liu and Zhi-Zhong Guan

22 **Comprehensive Prevention and Control of Coal-burning Type of Endemic Fluorosis** 355
Fu-Cheng Li and Chang-Xue Wu

23 **Surgical Treatment of Joint Damage in Skeletal Fluorosis** 377
Ri-Guang Liu

24 **Clinical Prevention and Treatment of Coal-burning Type of Endemic Fluorosis** 401
Ling Mao and Yong Wang

25 Drug Development 421
Zhi-Zhong Guan, Yi Zhao, and Chang-Xue Wu

26 Improvement of the Health Condition of Local Residents After Comprehensive Treatments of Coal-Burning Type of Endemic Fluorosis 441
Zhi-Zhong Guan, Yan-Long Yu, and Yi Zhao

List of Abbreviations 455

Contributors

Ji Bian Shandong Normal University, Jinan, People's Republic of China

Jian-Chao Bian Shandong Institute of Prevention and Control for Endemic Disease, Jinan, People's Republic of China

Xu-Guang Chen Department of Radiology at the Affiliated Hospital of Guizhou Medical University, Guiyang, Guizhou, People's Republic of China

Jie Deng The Key Laboratory of Endemic and Ethnic Diseases (Guizhou Medical University) of the Ministry of Education, Guiyang, Guizhou, People's Republic of China
The Provincial Key Laboratory of Medical Molecular Biology, Guiyang, Guizhou, People's Republic of China

Yang-Ting Dong The Key Laboratory of Endemic and Ethnic Diseases (Guizhou Medical University), Ministry of Education, Guiyang, Guizhou, People's Republic of China
The Provincial Key Laboratory of Medical Molecular Biology, Guiyang, Guizhou, People's Republic of China

Qing-Hong Duan Department of Radiology at the Affiliated Cancer Hospital of Guizhou Medical University, Guiyang, Guizhou, People's Republic of China

Zhi-Zhong Guan The Key Laboratory of Endemic and Ethnic Diseases (Guizhou Medical University), Ministry of Education, Guiyang, Guizhou, People's Republic of China
Department of Pathology at the Affiliated Hospital of Guizhou Medical University, Guiyang, Guizhou, People's Republic of China
The Provincial Key Laboratory of Medical Molecular Biology, Guiyang, Guizhou, People's Republic of China

Feng-Yang Guo School of Pharmaceutical Sciences, Jilin University, Changchun, Jilin, People's Republic of China

Zhi Huang Department of Interventional Medicine at the Affiliated Hospital of Guizhou Medical University, Guiyang, Guizhou, People's Republic of China

Jian Liao The Key Laboratory of Head and Neck Cancer Research, Hospital and School of Stomatology, Guizhou Medical University, Guiyang, Guizhou, People's Republic of China

Fu-Cheng Li The Key Laboratory of Endemic and Ethnic Diseases (Guizhou Medical University) of the Ministry of Education, Guiyang, People's Republic of China
Liupanshui Center for Disease Control and Prevention, Liupanshui, People's Republic of China

Ri-Guang Liu Department of Orthopedics at the Affiliated Hospital of Guizhou Medical University, Guiyang, Guizhou, People's Republic of China

Yan-Jie Liu Department of Pathology at the Affiliated Hospital of Guizhou Medical University, Guiyang, Guizhou, People's Republic of China

Di-Dong Lou School of Basic Medicine Sciences, Guizhou University of Traditional Chinese Medicine, Guian, People's Republic of China

Ling Mao The Key Laboratory of Head and Neck Cancer Research, Hospital and School of Stomatology, Guizhou Medical University, Guiyang, Guizhou, People's Republic of China

Xiao-Lan Qi The Key Laboratory of Endemic and Ethnic Diseases (Guizhou Medical University) of the Ministry of Education, Guiyang, Guizhou, People's Republic of China
The Provincial Key Laboratory of Medical Molecular Biology, Guiyang, Guizhou, People's Republic of China

Long-Yan Ran The Key Laboratory of Endemic and Ethnic Diseases (Guizhou Medical University), Ministry of Education, Guiyang, Guizhou, People's Republic of China
Department of Pathology at the Affiliated Hospital of Guizhou Medical University, Guiyang, Guizhou, People's Republic of China

Fa Sun Department of Urology at the Affiliated Hospital of Guizhou Medical University, Guiyang, Guizhou, People's Republic of China
Health Commission of Guizhou Province, Guiyang, Guizhou, People's Republic of China

Yong Wang The Key Laboratory of Head and Neck Cancer Research, Hospital and School of Stomatology, Guizhou Medical University, Guiyang, Guizhou, People's Republic of China

Na Wei Department of Pathology at the Affiliated Hospital of Guizhou Medical University, Guiyang, Guizhou, People's Republic of China

Chang-Xue Wu The Key Laboratory of Endemic and Ethnic Diseases (Guizhou Medical University), Ministry of Education, Guiyang, Guizhou, People's Republic of China
The Provincial Key Laboratory of Medical Molecular Biology, Guiyang, Guizhou, People's Republic of China

Jie Xiang The Key Laboratory of Endemic and Ethnic Diseases (Guizhou Medical University), Ministry of Education, Guiyang, Guizhou, People's Republic of China
Department of Pathology at the Affiliated Hospital of Guizhou Medical University, Guiyang, Guizhou, People's Republic of China

Xiao Xiao The Key Laboratory of Endemic and Ethnic Diseases (Guizhou Medical University), Ministry of Education, Guiyang, Guizhou, People's Republic of China
Department of Pathology at the Affiliated Hospital of Guizhou Medical University, Guiyang, Guizhou, People's Republic of China

Yue-Hai Xiao Department of Urology at the Affiliated Hospital of Guizhou Medical University, Guiyang, Guizhou, People's Republic of China

Hui Xu School of Pharmaceutical Sciences, Jilin University, Changchun, Jilin, People's Republic of China

Wen-Feng Yu Department of Biochemistry and Molecular Biology, School of Basic Medical Science, Guizhou Medical University, Guiyang, Guizhou, People's Republic of China
The Key Laboratory of Endemic and Ethnic Diseases (Guizhou Medical University), Ministry of Education, Guiyang, Guizhou, People's Republic of China
The Provincial Key Laboratory of Medical Molecular Biology, Guiyang, Guizhou, People's Republic of China

Yan-Long Yu The Key Laboratory of Endemic and Ethnic Diseases (Guizhou Medical University), Ministry of Education, Guiyang, Guizhou, People's Republic of China
The Provincial Key Laboratory of Medical Molecular Biology, Guiyang, Guizhou, People's Republic of China

Xiao-Xiao Zeng The Key Laboratory of Endemic and Ethnic Diseases (Guizhou Medical University), Ministry of Education, Guiyang, Guizhou, People's Republic of China
Department of Pathology at the Affiliated Hospital of Guizhou Medical University, Guiyang, Guizhou, People's Republic of China

Kai-Lin Zhang Basic Medical College, Guizhou Medical University, Guiyang, Guizhou, People's Republic of China

Man Zhang Shandong Institute of Prevention and Control for Endemic Disease, Jinan, People's Republic of China

Zhong-Yuan Zhang School of Pharmaceutical Sciences, Jilin University, Changchun, Jilin, People's Republic of China

Zi-Gui Zhang College of Chemistry and Life Science, Zhejiang Normal University, Jinhua, Zhejiang, People's Republic of China

Yi Zhao College of Foreign Languages, Guizhou Medical University, Guiyang, Guizhou, People's Republic of China

Jian Zou The Key Laboratory of Endemic and Ethnic Diseases (Guizhou Medical University) of the Ministry of Education, Guiyang, Guizhou, People's Republic of China
Department of Pathology at the Affiliated Hospital of Guizhou Medical University, Guiyang, Guizhou, People's Republic of China

Part I
Discovery and Epidemiology of Endemic Fluorosis

Chapter 1
Discovery and Concept of Endemic Fluorosis

Zhi-Zhong Guan

Abstract Endemic fluorosis refers to the systemically chronic toxic lesions caused by exposure to excessive fluoride through drinking water, air, food, tea, and other means in a given natural environment. Currently, more than 20 countries in the world are affected by this disease, of which China and India have the most widespread areas of endemic fluorosis as well as the most severe damage to patients. Three main types of endemic fluorosis in most provinces and regions of China are caused by drinking water, burning coal, or drinking brick-tea. The coal-burning type of endemic fluorosis was first discovered in the Guizhou province of China in the 1970s and later recognized in other regions of southwestern China as well. In these regions, the residents get used to dry food (such as corn, chili, and vegetables) by burning coal with high level of fluoride in open stoves without a chimney during the autumn harvest or heating in the winter time; thereby, contaminating the food and air with fluoride. Long-term exposure to fluoride in this manner can give rise to chronic fluorosis, with dental and skeletal fluorosis and damages to most organs. To control and prevent this disease, much effort has been put into understanding pathogenesis, clinical treatments and improving stoves with remarkable results. However, a better understanding of the pathogenesis and a sustainable strategy to control this disease are still required.

Keywords Clinic treatment · Epidemiology · Pathogenesis · Prevention and control · Types of Endemic fluorosis

---

Z.-Z. Guan (✉)
The Key Laboratory of Endemic and Ethnic Diseases (Guizhou Medical University), Ministry of Education, Guiyang, Guizhou, People's Republic of China

Department of Pathology at the Affiliated Hospital of Guizhou Medical University, Guiyang, Guizhou, People's Republic of China

The Provincial Key Laboratory of Medical Molecular Biology, Guiyang, Guizhou, People's Republic of China
e-mail: zzg@gmc.edu.cn

Z.-Z. Guan (ed.), *Coal-burning Type of Endemic Fluorosis*,
https://doi.org/10.1007/978-981-16-1498-9_1

General Background about Fluorine

Fluorine, a halogen and the most reactive one of the non-metallic elements, is considered as an essential trace element for human beings despite remaining a controversy. Although the main natural component of tooth enamel is hydroxyapatite, fluoride is indispensable to prevent dental caries (cavities) by converting hydroxyapatite into more resistant fluorapatite. In addition, fluoride in small amounts contributes to the normal development of bones, especially preventing osteoporosis.

In many countries, fluoride is added to drinking water (water fluoridation) and toothpaste to prevent dental caries. In addition, fluoridated mouth rinses, lozenges, gels, foams, and varnishes are used as topical therapy. This therapy may repair instead of preventing damage to the teeth by causing the mineral fluorapatite to be incorporated into damaged enamel. Many people believe appropriately fluoridation of water protects the teeth without posing risks to health.

However, others have very different opinions about water fluoridation, which is not surprising in light of the numerous contradictory findings concerning the benefits and toxicity of systemic fluoride therapy (Spittle 2019; Fawell et al. 2006). This debate began roughly 70 years ago, when communities first began to use water fluoridation to prevent tooth decay. Activists to opposed water fluoridation emphasize that this is a form of involuntary medication that causes millions of children to develop dental fluorosis, a discoloration of the teeth that can lead to embarrassment and anxiety. In addition, fluoridated water can cause severe bone disorder and arthritis, damage the developing brain, impair thyroid function, and possibly result in osteosarcoma (bone cancer). Moreover, certain studies have not observed any consistent or meaningful difference in the prevalence of cavities in fluoridated and non-fluoridated regions.

On the other hand, in extensive areas of the world excess amounts of fluoride are present naturally. In such places, the body can gradually take up much fluoride through drinking water, air, food, tea, and other means, which can in the long ran lead to chronic systemic poisoning, which is referred to endemic fluorosis.

The Discovery of Endemic Fluorosis

About 110 years have passed since the discovery of mottled teeth (now called dental fluorosis) by Greene Vardiman Black, a renowned dental researcher and dean of the Northwestern University Dental School of Chicago, USA, in 1909 (Smith et al. 1931). Since then, our understanding of such disease has advanced considerably. The earliest pathological signs of endemic fluorosis usually appear in the bones (skeletal fluorosis) and/or teeth (dental fluorosis). At present, the common understanding is that such disease is accompanied by pathological changes in multiple systems and organs.

Shortly after Black's discovery of mottled teeth, he and Mckay found in 1916 that the prevalence of dental fluorosis is connected to drinking water (Black and McKay 1916). In 1931, Churchill, the chief chemist at Aluminum Company of America in Pennsylvania, first reported that the water in areas with a high prevalence of this syndrome contained high concentrations of fluoride (Churchill 1931). One year later, in 1932, Moller and Gudjonsson coined the term fluorosis to describe osteosclerosis (skeletal fluorosis) induced by industrial fluoride intoxication (Moller and Gudjonsson 1967). In 1937, Shortt and colleagues reported 10 cases of endemic fluorosis in Madras, India (Shortt et al. 1937) and Roholm documented 68 cases of industrial fluorosis (Roholm 1937); both these investigators examined organs other than the skeletal system of patients with fluorosis and thereby deepened our general understanding of fluorosis as systemic changes. In 1976, Dr. Waldbott, the Chief-Editor of *Fluoride*, an international journal, reported the research findings on chronic non-skeletal fluorosis at an international conference (Waldbott 1988).

In China, there is still early evidence for endemic fluorosis (Guan et al. 2017; Anderson and Stevenson 1930). The 100,000-year-old human incisor and skull fossils unearthed by Chinese archaeologists in Hsu Chia Yao village, Yanggao county of Shanxi province bear the traces of both dental and skeletal fluorosis. Furthermore, there are indications of dental fluorosis in the fossil of "Dingcun man" (also 100,000 years old) found in Xiangfen county of Shanxi province, as well as in the remains of an patient suspected with skeletal fluorosis, living at the time of the Xia dynasty (4000 years ago) in Dadianzi Village, Aohan Banner, Ju UD City of Inner Mongolia. Moreover, during the time of the Three Kingdoms, Ji Kang living in the Wei Kingdom (225–262 AD) mentioned "inhabitants in Shanxi with yellow colored teeth" in his writings on Regimen Theory (about maintaining good health), which is considered to be the earliest known description of a connection between mottled teeth and environment.

In 1930, Anderson and colleagues described 398 patients with dental fluorosis living close to Beijing and 109 cases near Taiyuan in China (Anderson and Stevenson 1930). In 1935, Dr. Kiborn reported mottled enamel and joint stiffness in residents of Shimenkan in the western region of Guizhou, as well as the observation that the well water there contained 12 ppm fluoride, and suspected these individuals were suffering from fluorosis (Lyth 1945). One year later, Lyth reported four cases of skeletal fluorosis in adults and 134 children with dental fluorosis among members of Miao minority all living in the same village of Weining Autonomous county, in Guizhou in *Lancet* (Lyth 1946), in which the situation was later proven to be the result of food contaminated by fluoride that originated from burning coal. From 1930s to 1960s, Chinese researchers discovered that the drinking water in northeast and northwest China contains high level of fluoride, which correlates with the prevalence of dental fluorosis in these regions (Guan et al. 2017).

Today, fluorosis is known to be largely endemic in over 20 countries, with China and India being the worst stricken, along with Bangladesh, Sri Lanka, Vietnam, Iran, Thailand, Argentina, the USA, Canada, Mexico, Morocco, Algeria, Tunisia, Egypt, South Africa, and others.

Types of Endemic Fluorosis

Three different types of endemic fluorosis (Guan et al. 2017) are due to drinking water, brick-tea, and coal-burning, with the former being most common internationally. In China, endemic fluorosis caused by drinking water is found mainly in northern regions, coal-burning type of endemic fluorosis in southern regions, and brick-tea type of endemic fluorosis in minority areas (Tibetan nationality). By 2014, the total number of counties where fluorosis was endemic was estimated to be 1544, with 126 million residents, including more than 39.5 million with dental fluorosis and 2.88 million with skeletal fluorosis.

Drinking Water Type of Endemic Fluorosis

Drinking water type of endemic fluorosis can be ultimately induced by long-term excessive intake of fluoride by drinking subterranean water. The water which Chinese citizens drink is mainly groundwater and underground water. Accordingly, its fluoride content and origin are closely related to natural causes such as geological formations, weathering, and leaching, the chemical composition of runoff water, precipitation, and evaporation.

The fluoride concentration in natural groundwater fluctuates widely, being determined primarily by the composition of rocks in riverbeds, the fluorine contained in soil and the solubility of fluoride. In arid and semi-arid areas, fluorine-rich materials or runoffs accumulate in groundwater in low-lying flat terrain, where occurs extensive evaporation of water and pre-concentration of fluoride in ground water and then forms the endemic areas.

The World Health Organization recommends that the content of fluoride in drinking water should be below 1.5 ppm, and the average level in China is below 1.2 ppm. However, this content in areas of the drinking water type of endemic fluorosis normally exceeds 2 ppm and is sometimes as high as 20 ppm, primarily due to the exceedingly high fluorine content of local rock and soil. In such areas, the incidence of dental fluorosis reached 100% and that of skeletal fluorosis 29.8%. In the northern part of China, 1055 counties with a total population of 61.8 million occurred the drinking water type of endemic fluorosis, exhibiting 18.16 million residents with dental fluorosis and 1.27 million skeletal fluorosis.

Coal-Burning Type of Endemic Fluorosis

Coal-burning type of endemic fluorosis, also unique to China, was first discovered in Guizhou province in the 1970s (Wei et al. 1981). Building on the basic finding of chronic fluorosis in this province by Kiborn in 1934 (Lyth 1945) and Lyth in 1945

(Lyth 1946), after a large number of epidemiological surveys the eminent professor Zan-Dao Wei at Guizhou Medical University first pinpointed that this was due to the excessive fluoride content of food rather than drinking water (Wei et al. 1981). Subsequently, food was shown to be contaminated by burning coal by researchers in Hubei and Guizhou provinces of China.

The local people burn coal indoors with no exhaust system to dry foods (corn, chili, and vegetables) for long-term retention after the autumn harvest season or for heating in winter, thereby polluting the air and food with fluoride (Zheng et al. 2007). Currently, 13 provinces or regions, mainly located in southwestern in China, are known to be affected by coal-burning type of endemic fluorosis. Altogether, 173 counties (cities or districts) are affected by this type of endemic fluorosis with 3.2 million people at risk and in a previous report 14.6 million patients were suffering from dental fluorosis, and 1.9 million from skeletal fluorosis.

Brick-Tea Type of Endemic Fluorosis

Brick-tea type of endemic fluorosis, which is unique to China, occurs due to the excessive intake for a long-term brick-, milk-, and oil-tea which contain high level of fluoride. Brick-tea is fermented and aged by using tea leaves and stalks containing higher concentrations of fluoride (300–1000 ppm). Due to environmental conditions, access to vegetables and fruits is traditionally very limited or even non-existent in major areas of Tibet and Mongolia. As a tradition, the residents resort to brick-tea for its refreshed taste, an aid in digestion and a vitamin supplement as they consume more beef and mutton.

Brick-tea type of endemic fluorosis was first discovered in an area of Sichuan province inhabited by Tibetan nationals in the 1980s. The endemic areas in western China, where Tibetan and Mongolian nationals live, hold 316 counties with a total population of 31 million.

The occurrence of endemic fluorosis is profoundly influenced by economic conditions, culture, level of education and personal hygiene, as well as diet. This condition is prevalent primarily in areas with an underdeveloped economy, low living standards, and widespread malnutrition. In addition, poor knowledge about health and inadequate self-health and medical care all contribute to the popularity of the disease. Moreover, insufficient intake of energy, protein, calcium, and vitamins lower resistance to chronic fluorosis. Since pregnant women and children require more nutrients, they are more prone to malnutrition and symptoms of chronic fluorosis.

Effect of Coal-Burning Type of Endemic Fluorosis on the Human Body

Skeletal and dental fluorosis are the main symptoms of endemic fluorosis. Furthermore, osteogenic tissues are clearly affected by fluoride. The pathological skeletal changes associated with chronic fluorosis are varied, including osteosclerosis, osteoporosis, osteomalacia and secondary bone disease induced by primary hyperparathyroidism. Osteogenic lesions, the most common of these, involve disrupted turnover of bone due to changes in calcium metabolism. In connection with ossification or the osteoclastic process, metabolism of calcium phosphorus, enzyme activity in bone, cellular regulation, microelements and nutritional status play control roles and are closely related to the manifestation of skeletal fluorosis. The involvement of growth factor derived from bone, osteonectin, osteocalcin, and cytokine in bone metabolism is the focus of current research in this area (Chen et al. 2001).

In addition, the nervous, urinary, digestive, endocrine, and cardiovascular systems are major targets of chronic fluorosis and the focus of research (Zahvoronkov and Strochkova 1981; Guan et al. 2000; Gui et al. 2010; Wu et al. 2018). With respect to the central nervous system, it is now known that fluoride can across the blood–brain barrier and induce injury, including neuropathological changes in brain morphology, the cholinergic nervous system, biological membranes, receptors for neurotransmitters and neuronal signal transductions, as well as mental retardation. In clinical practice, cognitive impairment in adults and mental retardation (the lower intelligence quotient) among children have been found to be more prevalent in areas of coal-burning type of endemic fluorosis.

Furthermore, there are indications that chronic fluorosis changes the morphology and function of the kidney, i.e., the ultrastructure of the renal tubular epithelial cells of experimental animals exposed to high level of fluoride is abnormal and renal functions of patients with fluorosis altered. Moreover, liver, the main organ of detoxification and key regulator of energy metabolism, is altered by chronic fluorosis, including fewer clear boundaries between hepatocytes, an abnormal structure in histology. Hepatic functions are also disturbed, as revealed by elevated activity of transaminase and lowered capacity for detoxification observed in experimental animals and patients with chronic fluorosis. In addition, the reduced levels of thyroid hormones T_3 and T_4, and a compensatory increase in thyroid-stimulating hormone (TSH) are observed in the blood of adults and children living in areas of coal-burning type of endemic fluorosis. The thyroid cell follicles are enlarged and contain irregular amounts of colloid in experimental animals with chronic fluorosis. With respect to the heart, ventricular conductivity and electrocardiogram are abnormal and lipid metabolism is disturbed in patients with this condition. Finally, experimental animal with chronic fluorosis also develops atherosclerosis.

Currently, endemic fluorosis is considered as a disease with systemic damages, in which dental and skeletal fluorosis is a typically clinical manifestation and simultaneously multiple organs or systems are also affected.

Molecular Mechanisms underlying the Pathological Changes Caused by Chronic Fluorosis

The pathogenesis of the multiple deleterious effects of endemic fluorosis is quite complex. Many attempts to explain the underlying mechanism(s) based on the alterations in nucleic acids and proteins, metabolic enzymes, collagen, and the balance between major and trace elements have not yet succeeded fully. At the beginning of the 1990s, we proposed that oxidative stress can explain the systemic pathological injuries resulting from endemic fluorosis, i.e., more production of free radicals and an attenuated capacity of antioxidant enzymes and/or compounds may be a central cause in pathogenesis of the disease (Guan et al. 1989). At present, most researchers in this area agree that the oxidative stress provides a relatively plausible explanation (Lou et al. 2013; Dong et al. 2015).

At the same time, certain issues concerning the systemic injuries associated with chronic fluorosis remain to be resolved. For example, how does excessive accumulation of fluoride affect interactions between trace elements, hormones, enzymes, and cytokines related to bone metabolism? How sensitive are various organs to long-term exposure to fluoride? What is the relationship between skeletal and non-skeletal effects? Is the nervous system a specific target or simply affected by the general toxicity of excessive fluoride? What role does genetic susceptibility play? More importantly, successful therapeutic drug has not yet been developed and the patients with chronic fluorosis continue to suffer. All of these may be interesting aspects to encourage researchers for further study.

Clinical Treatments of Coal-Burning Type of Endemic Fluorosis

Over the past decades, a great deal of effort has been devoted to the prevention and treatment of endemic fluorosis and significant reduction in the intake of fluoride achieved (Guan et al. 2017).

The main and obvious preventive measure for endemic fluorosis due to drinking water is to reduce the fluoride content in drinking water by changing the sources and by defluoridation. In regions with brick-tea type of endemic fluorosis, the fundamental measures are to reduce the fluoride content of tea and, at the same time, make people aware of this problem and encourage them changing their tea-drinking habits.

The principle means of prevention for fluorosis caused by burning coal is to replace traditional open stoves with stoves that have a chimney or run an electricity or methane instead of coal to be used as an energy power (Guan et al. 2018). In addition, a great deal of effort has been put into educating local residents about the harmful effects of fluoride pollution from burning coal in a stove without a chimney.

Long-term experience has made it abundantly clear that the occurrence of coal-burning type of endemic disease is related not only to behavior and lifestyle, but also to the social-economic development in an area, in which it is difficult to attain sustainable prevention and control simply by changing and improving traditional stoves. Therefore, it is important to carry out comprehensive prevention and control on the disease.

In the 1980s, researchers from Guizhou initiated prevention and treatment of coal-burning type of endemic fluorosis. They were the first to propose a comprehensive concept, i.e., "Health education as the foundation, improved stoves as the main measure and other forms as assistance," involving cooperation between government and medical institutes, as well as the participation of local residents. This approach has been successful in areas with coal-burning type of endemic fluorosis (Guan et al. 2017).

Importantly, stomatologists and surgeons in Guizhou actively treat patients with dental and skeletal fluorosis in area of coal-burning type of endemic fluorosis (Guan et al. 2018). In addition, drugs designed to remove fluoride and reduce free radical damage are being developed with some success.

Elimination of the coal-burning type of endemic fluorosis still presents many challenges. This situation is most prevalent in poor areas of China with aged cultural notion, an underdeveloped economy and large population, and therefore a long-term sustainable strategy for the prevention and control of this disease is extremely important.

It is important continually to identify relevant epidemic factors, clarify the underlying molecular pathogenesis, expand clinical treatments, develop effective drugs, and improve medical evaluation. The complete elimination of endemic fluorosis caused by burning coal is our ultimate goal.

Conclusion

Endemic fluorosis is caused by exposure to excessive fluoride through drinking water, air, food, tea, and other means in a given natural environment. Three main types of endemic fluorosis in most provinces and regions of China are caused by drinking water, burning coal, or drinking brick-tea. The coal-burning type of endemic fluorosis was first discovered in the Guizhou province of China in the 1970s and later recognized in other regions of southwestern China as well. In these regions, the residents get used to dry food (such as corn, chili, and vegetables) by burning coal with high level of fluoride in open stoves without a chimney during the autumn harvest or heating in the winter time; thereby, contaminating the food and air with fluoride. Long-term exposure to fluoride in this manner can give rise to chronic fluorosis, with dental and skeletal fluorosis and damages to most organs. To control and prevent this disease, much effort has been put into understanding pathogenesis, clinical treatments and improving stoves with remarkable results.

References

Anderson BG, Stevenson MA. The occurrence of mottled enamel among the Chinese. J Dent Res. 1930;10:233–8.

Black GV, McKay FS. Mottled teeth: an endemic developmental imperfection of the enamel of the teeth heretofore unknown in the literature of dentistry. Dent Cosmosl. 1916;1916:129.

Chen XG, Ke QP, Sun MH, Wei ZD. Skeletal radiographs, hydroxyproline, alkaline phosphatase, and other biochemical parameters in endemic fluorosis. Fluoride. 2001;34:202–3.

Churchill HV. The occurrence of fluorides in some waters of the United States. J Ind Eng Chem. 1931;23:996–8.

Dong YT, Wang Y, Wei N, Zhang QF, Guan ZZ. Deficit in learning and memory of rats with chronic fluorosis correlates with the decreased expressions of M1 and M3 muscarinic acetylcholine receptors. Arch Toxicol. 2015;89:1981–91.

Fawell J, Bailey K, Chilton J, Dahi E, Fewtrell L, Magara Y. Fluoride in drinking-water. London: World Health Organization; 2006.

Guan ZZ, Yang PS, Pan S, Yu LD. An experimental study of blood biochemical diagnostic indices for chronic fluorosis. Fluoride. 1989;3:112–5.

Guan ZZ, Xiao KQ, Zeng XY, Long YG, Cheng YH, et al. Changed cellular membrane lipid composition and lipid peroxidation of kidney in rats with chronic fluorosis. Arch Toxicol. 2000;74:602–8.

Guan ZZ, Wang LH, Sun DJ. Endemic fluorosis. In: Sun DJ, editor. Endemiology. Beijing: People's Medical Publishing House of China; 2017.

Guan ZZ, Wang Y, Duan QH, Liu RG, Li FC, Xu SX, et al. Basic investigation and clinic treatment for the coal-burning type of endemic fluorosis in Guizhou, China. Fluoride. 2018;52:83.

Gui CZ, Ran LY, Li JP, Guan ZZ. Changes of learning and memory ability and brain nicotinic receptors of rat offspring with coal burning fluorosis. Neurotoxicol Teratol. 2010;32:536–41.

Lou DD, Guan ZZ, Liu YJ, Liu YF, Zhang KL, Pan JG, et al. The influence of chronic fluorosis on mitochondrial dynamics morphology and distribution in cortical neurons of the rat brain. Arch Toxicol. 2013;87:449–57.

Lyth O. Endemic fluorosis in Western Kweichow. Ind Med Gz. 1945;80:370.

Lyth O. Endemic fluorosis in Kweichow, China. Lancet. 1946;247:233–7.

Moller PF, Gudjonsson SV. Massive fluorosis of bones and ligaments. Clin Orthop Relat Res. 1967;55:5–15.

Roholm K. Fluorine intoxication: a clinical hygienic study with a review of the literature and some experimental investigation. London: HK Lewis; 1937.

Shortt HE, Pandit CG, Raghavachari TN. Endemic fluorosis in the Madras presidency. Indian J Med Res. 1937;5:553–68.

Smith MC, Lantz EM, Smith HV. The cause of mottled enamel. Science. 1931;74:244.

Spittle B. Publishing on the harmful effects of fluoride exposure. Fluoride. 2019;52:193–8.

Waldbott GL. The pre skeletal phase of chronic fluoride intoxication. Fluoride. 1988;31:13–20.

Wei ZD, Zhou LY, Bao RC. Endemic food-borne fluorosis in Guizhou, China. Fluoride. 1981;14:91–3.

Wu CX, Wang YH, Li Y, Guan ZZ, Qi XL. Changes of DNA repair gene methylation in blood of chronic fluorosis patients and rats. J Trace Elem Med Biol. 2018;50:223–8.

Zahvoronkov AA, Strochkova LS. Fluorosis: geographical pathology and some experimental findings. Fluoride. 1981;14:183–91.

Zheng B, Wu D, Wang B, Liu X, Wang M, et al. Fluorosis caused by indoor coal combustion in China: discovery and progress. Environ Geochem Health. 2007;29:103–8.

Chapter 2
Epidemic Status of Coal-burning Type of Endemic Fluorosis

Zhi-Zhong Guan, Fu-Cheng Li, Yi Zhao, and Chang-Xue Wu

Abstract Fluoride gains its access to human body through digestive and respiratory tracts and then is excreted from kidney and widely found in hard tissues like bones and teeth, which are rich in calcium and phosphorus. Fluoride integrates with hydroxyapatite and then forms fluorapatite, which can enhance the mechanical strength of bones and play a partial role in resistance to tooth decay. In natural conditions, the excessive intake of fluoride by drinking, eating, and breathing ultimately results in endemic fluorosis. Major manifestations of this disease are skeletal and dental fluorosis, and fluoride can also trigger pathological damages on multiple systems and organs. Endemic fluorosis is classified into three types: drinking water type, coal-burning and brick-tea types of endemic fluorosis. Drinking water type of endemic fluorosis, which is widely found in international community, is mainly discovered in the northern part of China. Coal-burning and brick-tea types of endemic fluorosis are unique to China; the former is mainly found in southern China, while the latter in Tibetan and Mongolian living areas of China. At the present, Chinese government has formulated national standards for the prevention and control of endemic fluorosis in light of China's specific conditions. Coal-burning

Z.-Z. Guan (✉) · C.-X. Wu
The Key Laboratory of Endemic and Ethnic Diseases (Guizhou Medical University), Ministry of Education, Guiyang, Guizhou, People's Republic of China

Department of Pathology at the Affiliated Hospital of Guizhou Medical University, Guiyang, Guizhou, People's Republic of China

The Provincial Key Laboratory of Medical Molecular Biology, Guiyang, Guizhou, People's Republic of China
e-mail: zzg@gmc.edu.cn

F.-C. Li
The Key Laboratory of Endemic and Ethnic Diseases (Guizhou Medical University) of the Ministry of Education, Guiyang, People's Republic of China

Liupanshui Center for Disease Control and Prevention, Liupanshui, People's Republic of China

Y. Zhao
College of Foreign Languages, Guizhou Medical University, Guiyang, Guizhou, People's Republic of China

Z.-Z. Guan (ed.), *Coal-burning Type of Endemic Fluorosis*,
https://doi.org/10.1007/978-981-16-1498-9_2

type of endemic fluorosis was first discovered by scientists from Guizhou province of China: residents of the endemic areas tried to dry grains, chili, and vegetables, cook foods and heat for keeping warm in winter by burning coal containing high content of fluoride in open stoves without exhaust system; therefore, the foods and air contaminated by fluoride in-door lead to hundreds of times more fluoride than normal intake over the health standard and then trigger chronic fluorosis. The current control and prevention of coal-burning type of endemic fluorosis lies in replacing old-style stove, health education, improving economy, and prevention ability.

Keywords Coal-burning type of endemic fluorosis · Division of endemic area · Fluoride metabolism · Prevention and treatment · Toxic effect

Introduction

Endemic fluorosis is a disease of inflicting great damage on human body, including affecting people with skeletal and dental fluorosis, and causing injuries of non-bone systems and organs (Guan et al. 2017). Endemic fluorosis prevails in over 20 countries, while China and India hold the most numbers of cases. It is estimated that half a billion population are exposed to the environment with a source of high fluoride content and 200 million of people suffered from chronic fluorosis in the world. Three types of endemic fluorosis, i.e., drinking water, coal-burning and brick-tea types, are all found in China. Coal-burning type of endemic fluorosis, which is unique to China, is a chronic fluorosis caused by excessive fluoride intake through digestive and respiratory tracts with a couple to hundreds of times higher than the standard of fluoride intake in China. This is resulted from local routine of drying food (corn, chili, and vegetables), cooking and heating on open stoves in-door without exhaust system and then contaminating food and air by fluoride. This disease inflicts tremendous pain on society, family, and individuals by depriving patients' ability to work and even causing disability (Wei et al. 1981).

Guizhou province is the worst hit by the coal-burning type of endemic fluorosis with the most numbers of cases and the widest distribution of endemic areas in the world. Of all the 88 counties and cities in Guizhou, 37 with a total population of 19 million were found to have such areas of coal-burning type of endemic fluorosis (Guan et al. 2017). For years, Chinese scientists and professionals of prevention and control for endemic disease have made great breakthrough on this specific fluorosis from studies on epidemiology, cause of disease, pathological changes, and pathogenesis. At the same time, effective measures have been taken to reduce number of new patients and slow down the progression for old patients, including replacing the stoves with new one with exhaust system, enhancing health education, promoting economic development of endemic area, improving living standard of the local residents, upgrading health cares and carrying out clinical treatments.

Fluoride Metabolism

Fluoride Ingestion

Fluoride enters the human body through digestive and respiratory tracts and after ingestion the ion from various sources immediately enters the blood circulation system (Whitford 1990). Most of soluble fluoride can be absorbed by the body quickly and its concentration in plasma is obviously increased within minutes. The half-absorption period of fluoride is about 30 min and the concentration of fluoride reaches the peak from 30 to 60 min. NaF can be completely absorbed as it is soluble, while CaF_2, MgF_2, and AlF_3 cannot be absorbed quickly or completely because of their low solubility.

Fluoride, once it enters digestive tract, is mainly absorbed by stomach with a simple process of passive diffusion. The absorption amount of fluoride is closely related with pH value of gastric fluid. Hyperacidity forms non-ionizing hydrofluoric acid which can enable fluoride to penetrate cell wall and to be absorbed in great amount in stomach. The rest fluoride can be absorbed by small intestine regardless pH conditions. Fluoride absorption from food is largely dependent on solubility level of inorganic fluoride and calcium content in diet. Normally, fluoride absorbed from diet is about 80%, and the absorption rate can be lowered by 50% if calcium or aluminum compound is added to diet. In addition, respiratory tract, skin, and oral mucosa can also absorb fluoride. In addition, Fluoride can enter the blood system of fetus through placental barrier or brain tissues through blood–brain barrier.

Free inorganic fluoride and combined organic fluoride are found in human blood. Seventy-five percent of the fluoride is present in plasma, while the rest is in erythrocytes. Fluoride in plasma is basically all ionic with the half-life period of 4–10 h, which do not integrate with plasma proteins or other components. Fluoride contents in milk and saliva are very low.

In body, about 99% of fluoride is deposited in bone and dental tissues. Fluoride integrates with bone crystals in the form of fluorapatite or hydroxy fluorapatite. Fluoride deposited in cancellous tissue can be released into blood and functioned as a fluoride reservoir. Fluoride enters the dental tissue and accumulates in the surface of enamel layer in the process of dentification, mineralization, and post-mineralization. During the tooth development, drinking water in moderate fluoride concentration can guarantee the average fluoride concentration of permanent tooth enamel to be 150 ppm, and the surface enamel 2000 ppm.

Fluoride Excretion

As urine discharged 75% of the total excreted fluoride, kidney is the lead organ to excrete fluoride in body. Urinary excretion of fluoride is closely related to glomerular filtration rate, and 10–90% filtered fluoride can be absorbed again in renal tubules. Long-term intake of a certain amount of fluoride could ensure the balance between urinary fluoride excretion and fluoride content in bone. The daily excretion of urinary fluoride is an index to reflect the total intake of fluoride. Hence, urinary fluoride level is widely accepted as a major index to monitor fluoride intake in population (most of population shows a urinary fluoride concentration less than 1 ppm). Excrement of kidney is responsible for 12.6–19.5% discharge of fluoride; sweat gland 7–10%; tears, hair, and nails can excrete trace amount of fluoride.

Physiological Function of Fluoride

The common knowledge believes that fluoride shoulders certain physiological function as it is an essential trace element in human body and widely distributed in hard tissues like bones and teeth which are rich in calcium and phosphorous (Dhar and Bhatnagar 2009). Fluoride is prone to integrate with hydroxyapatite from hard tissues and replace hydroxyl group with fluorapatite. Fluorapatite can improve the mechanic strength and acid resistant ability of bones and teeth, and enhance the stability of calcium and phosphorus in these tissues. A certain amount of fluoride intake can maintain the tooth health and enhance the susceptibility of tooth decay (especially for children).

In certain European countries and the USA, fluoride is added to drinking water, tooth paste, and various drinks to prevent tooth decay (Peckham and Awofeso 2014). However, fluoride reveals its nature as a double-sided effect to the surprise of scientists: dental fluorosis is on the rise in many children who are being treated with water fluorination to reduce tooth decay. There are also reports to indicate that the effect of fluoride against dental decay is not ideal as previous prediction.

Epidemic Types and Features of Endemic Fluorosis

In many developing countries, the negative effects of fluoride are more prominent. As people from these countries take excessive fluoride from drinking water, food or air, or drinking tea in natural conditions, they thus lead to chronic accumulated intoxication of fluoride, or termed as endemic fluorosis (Waldbott 1988). Existing studies consider that endemic fluorosis triggers comprehensive pathological damages on multiple systems and organs.

Drinking Water Type of Endemic Fluorosis

Drinking water type of endemic fluorosis is resulted from drinking water with high fluoride content (>1.2 ppm), which originates from the rock and soil with excessive content of fluoride. Drinking water type of endemic fluorosis is found in these countries like China, India, Bangladesh, Sri Lanka, Vietnam, Thailand, Iran, Argentina, USA, Canada, Mexico, Morocco, South Africa, Tanzania, Algeria, Tunisia, Egypt, and so on (Srivastava and Flora 2020).

China is one of the obviously affected countries by this type of endemic fluorosis in terms of endemic area and extent of damage. Endemic areas are distributed over 28 provinces (cities and districts), and the following provinces (cities and districts) hold a wide range of endemic areas: Liaoning, Jilin, Heilongjiao, Tianjin, Hebei, Shanxi, Inner Mongolia, Shandong, Henan, Shaanxi, Gansu, Qinghai, Ningxia, and Xinjiang.

Epidemic features of drinking water type of endemic fluorosis include both natural and social factors. Natural factors causing this endemic disease include geological, geographical, climate and hydrological conditions. Fluoride content in rock is the basic reason of fluoride content in water. The sources of fluoride come from surface and underground water. The source and content of fluoride in water are largely determined by natural factors like geologic formation, weathering and leaching degree, runoff living soil chemical composition, precipitation, and evaporation.

The relationship between fluoride concentration in natural drinking water and fluorosis epidemiological features is also determined by weather, climate, and water chemical composition. Drought and heat compel an increase in water consumption, then naturally increase fluoride concentration for intake. The water hardness and basicity degree of water chemical composition pose a certain influence on fluoride concentration in water and its biological effect.

Social factors serve as complementary conditions for endemic fluorosis. Such factors (economy, culture, education, and medical service) affect residents in endemic areas on their health and nutrition status, and behaviors. Endemic fluorosis usually prevails in poverty stricken or underdeveloped areas of China. Residents of these areas suffer from malnutrition due to low living standard in the form of insufficient intake in calorie, proteins, and vitamins, as well as low proportion of animal proteins in particular, thus lead to low resistance against the toxicity of excessive fluoride. Since pregnant women and children require more nutrition, they are more likely to suffer from malnutrition and be affected by chronic fluorosis.

Furthermore, individual health care is also an influencing factor. Generally, several facts in endemic areas contribute to the prevailing of chronic fluorosis, including underdeveloped education, poor health status, absence of individual health care awareness and competence, and high prevalence rate of endemic and other chronic diseases.

Coal-Burning Type of Endemic Fluorosis

Coal-burning type of endemic fluorosis was first identified in Guizhou in 1970s. This endemic fluorosis mainly prevails in remote area and mountainous regions of southern China. High concentration of fluoride in rocks, soil, and coal is the cause of endemic area, and give birth to an endemic fluorosis that is unique to China (Wei et al. 1981; Guan et al. 2017). Thirteen provinces were found with these endemic fluorosis areas: Guizhou, Yunnan, Sichuan, Chongqing, Hubei, Hunan, Shaanxi, Henan, Jiangxi, Shanxi, Guangxi, Zhejiang, and Liaoning. Guizhou was the first province to be identified with such endemic fluorosis with large areas of endemic and a substantial number of patients in critical condition. The severe fluoride contamination on in-door air, food, and chili originated from long-term use of burning coal containing high fluoride content in open stoves (with no exhaust systems) or coal-mixed clay for drying food or heating. For example, fresh corn contains a fluoride content of 1.2–1.6 ppm, coal-burning dried corn shows the fluoride content of 4.2–300 ppm; fresh chili has a fluoride content of 17.7–18.5 ppm, but open-stove dried one has 167–1267 ppm.

Epidemiological features of coal-burning type of endemic fluorosis also include intwined natural, environmental, social and economic factors, as well as behavioral and habitual factors. The climate condition in endemic area features with the lack of sunshine in harvest season, which forces residents to dry the corn (main agricultural products) on stoves in-door and then store in the open-air; in addition, residents of mountainous regions get used to use chili as a major condiment, which is also dried in the same fashion due to the lack of sunshine. Furthermore, heating by burning coal in open stove and discharging the smoke directly in-door, since there is absence of exhaust system on the stove, all deteriorates the fluoride contamination. In the end, the mixture of coal and clay (with more high content of fluoride) for mixed burning severely contaminates food and air in-door (Zheng et al. 2007)

Endemic areas of this disease share similarities in inconvenient transportation, underdeveloped economic and social welfare conditions, low residential density, low-level of education, poor health care awareness, fragile self-care and medical care competence together with malnutrition. All these features expose residents to the harm of fluoride and other harmful substance. In the process of improving traditional stoves together with other prevention measures, change of habit and behaviors poses the biggest challenge as residents are weak in self-protection and reluctant to accept health education.

Brick-Tea Type of Endemic Fluorosis

Brick-tea type of endemic fluorosis was first identified in Ngawa area of Sichuan province in 1984. This endemic type was also unique to China, and triggered by drinking brick-, milk-, and oil-tea. The fluoride content in such teas ranges from 52.5 to 1175 ppm. Long-term consumption of these teas in large quantity would accumulate fluoride in body and occur chronic fluorosis. Endemic areas of this disease distribute in these provinces or autonomous regions, such as Tibet, Sichuan, Gansu, Inner Mongolia, Qinhai, Xinjiang, Yunnan, Ningxia and Tibet and Sichuan, in which the dominant ethnic groups are Tibetan and Mongolian.

Living habit is the primarily influencing factor of brick-tea type of endemic fluorosis. The endemic areas in question distributed in plateau at high altitude. Residents in these areas live on husbandry and highland barley growing, which is the root cause of diet without diversity and habitual brick-tea consumption for residents. Since Tibetans take highland barley as staple food, they prefer eating it with decocted brick-tea and adults consume 4–5 l of tea daily. Meanwhile, residents have a diet with little vegetables invariably which lead to lack of diet nutrients, like vitamin C and E. All these factors make Tibetans the worst affected group by this endemic.

National Standards for Division of Endemic Fluorosis Area and Control of Endemic Fluorosis in China

Since endemic fluorosis is rampant in large areas of Chinese territory, establishing national standards for division of fluorosis area and control of endemic fluorosis are necessary and beneficial to detain its damage with coordinated resources and manpower. However, since the endemic types and influence factors in China are complex, it is difficult to take account of the complicated realities of various areas. So, the standards for determining the division of fluorosis areas and the control of endemic fluorosis are in constant improvement.

Division Standard of Endemic Fluorosis Areas

The national division standard of endemic fluorosis area (GB17018-2011) was determined in 2011 and defines the differences among drinking water, coal-burning and brick-tea types of endemic fluorosis.

One classification of the division standard of endemic fluorosis areas is based on administratively registered or natural villages and endemic areas should meet the following indicators.

Drinking water type of endemic fluorosis area: fluoride content in drinking water is>1.2 ppm and locally born and bred children from 8 to 12 years old show a prevalence rate of dental fluorosis>30%.

Coal-burning type of endemic fluorosis area: residents get used to burning coal in open stoves (without exhaust systems), and locally born and bred children (8–12 years old) show a prevalence rate of dental fluorosis >30%.

Brick-tea type of endemic fluorosis area: residents (over 16 years old) consume fluoridated tea >3.5 mg daily per capita, and X-ray confirms the patients presented skeletal fluorosis.

Another classification of the division standard of endemic fluorosis areas is based on the disease severity of population.

Mild endemic area: locally born and bred children (8–12 years old) in drinking water or coal-burning type of endemic fluorosis area show moderate and worse symptoms with dental fluorosis prevalence rate$\leq$20%, or X-ray confirms mild skeletal fluorosis patients but no moderate or worse symptoms; residents (36–45 years old) of brick-tea type of endemic fluorosis area show no moderate or worse symptom of skeletal fluorosis complications after X-ray check.

Moderate endemic area: locally born and bred children (8–12 years old) in drinking water or coal-burning type of endemic fluorosis area show moderate and worse symptoms with dental fluorosis prevalence rate of 20–40%, or X-ray confirms mild or worse skeletal fluorosis, but severe skeletal fluorosis prevalence rate $\leq$2%; residents (36–45 years old) of brick-tea type of endemic fluorosis area show moderate or worse symptom with skeletal fluorosis prevalence rate$\leq$10%.

Severe endemic area: locally born and bred children (8–12 years old) in drinking water or coal-burning type of endemic fluorosis area show moderate and worse symptoms with dental fluorosis prevalence rate over 40%, or X-ray confirms severe skeletal fluorosis patients with prevalence rate>2%; residents (36–45 years old) of brick-tea type of endemic fluorosis area show moderate or worse symptom with skeletal fluorosis prevalence rate>10%.

Control Criteria for Endemic Fluorosis

The existing effective national control standard for endemic fluorosis area (GB 17017-2010) is applicable to the endemic fluorosis areas defined as drinking water, coal-burning and brick-tea types, as well as the mixed fluoride source in administratively registered or natural villages.

Drinking water type of endemic fluorosis areas must meet these two control indexes: (1) large scale central water supply in countryside (daily water supply $\geq$1000 m^3, or to the population $\geq$10,000) with fluoride content $\leq$1.0 ppm, or small scale central water supply (daily water supply <1000 m^3, or to the population <10,000) with fluoride content $\leq$1.2 ppm; (2) Locally born and bred children (8–12 years old) have dental fluorosis prevalence rate $\leq$30%.

Coal-burning type of endemic fluorosis areas must meet these two control indexes: (1) both qualified improved stove (including adopting clean energy, such as electricity, liquid gas or methane) and correct stove handling rates above 90%; (2) locally born and bred children (8–12 years old) have dental fluorosis prevalence rate $\leq$30%.

Brick-tea type of endemic fluorosis areas must meet these three control indexes: (1) fluoride content in brick-tea $\leq$300 ppm; (2) local residents (30–60 years old) skeletal fluorosis prevalence rate is lowered three years in a row, and X-ray confirms no new moderate and worse skeletal fluorosis patients; (3) locally born and bred children (8–12 years old) have dental fluorosis prevalence rate $\leq$30%.

A History of Discovering Coal-Burning Type of Endemic Fluorosis

Before 1970s, scientific community only identified drinking water type of endemic fluorosis which was caused by high fluoride content in drinking water. In 1946, British scientist Olive Lyth published a paper named as *Endemic Fluorosis in Kweichow (Guizhou), China* in Lancet, in which he stated that the chronic fluorosis was caused by drinking water containing high fluoride content (5.9–6.3 ppm) (Lyth 1946). Then, there was a period of time saw no any relevant report or study on endemic fluorosis in Guizhou. In the beginning of 1970s, a report was indicated that many villagers in the rural areas of Bijie city, Guizhou had "tawny teeth, impaired mobility on extremities and joints, even kyphosis stoop and walking difficulty, and getting off bed for the severe cases, which seriously impacted ability of their agricultural work and quality of life."

In the middle of 1970s, Professor Zan-Dao Wei of Guizhou Medical University (then Guiyang Medical College), China rallied an endemic fluorosis research group to study those cases in Guizhou (Wei et al. 1981). Later, the Endemic Fluorosis Research Team (consisted of Guizhou Medical University, Guizhou Provincial Disease Control and Prevention Center [then Guizhou Health and Epidemic Prevention Station], Institute of Geochemistry of Chinese Academy of Sciences) was established to investigate the endemic fluorosis in Guizhou province. The findings revealed that fluoride content in water from endemic areas was significantly lower (less than 0.5 ppm of fluoride) than national standard (1 ppm of fluoride), which therefore could not support the conclusion of endemic fluorosis as a water type of according to tradition. Furthermore, there were two opposing views on whether or not there was endemic fluorosis in Guizhou since clinical findings and skeletal imaging showed similar results to endemic fluorosis. Thanks to the reverse thinking pattern proposed by the renovating scientists from Guizhou, which aimed to prove whether or not there was excessive fluoride in patients body from metabolism in vivo. The tested urine samples served as compelling evidence for diagnosis as tested fluoride concentrations (average fluoride content reached 6.4 ppm and maximum concentration reached 20.4 ppm) far exceeded normal fluoride excretion. In addition, the positive rate of dental fluorosis reached 98.2%, and moderate symptom patients accounted for 97.6%; while the positive rate of skeletal fluorosis 43.6%. Ultimately, the research team concluded that this was a new type of endemic fluorosis rather than water sourced one since tested water fluoride content was lower than national health standard. The fluoride intake originated from food with high fluoride content in endemic areas, thus it was termed Guizhou foodborne endemic fluorosis. This finding was published on an international journal, *Fluoride*, by Professor Zan-Dao Wei in 1980. Then the journal also published editorial on this article (Editorial 1981).

Chinese scientists established systematic assay to provide comprehensive and reliable data on fluoride contents of environmental samples (Zheng et al. 2007). They collected thousands of samples (rocks, coal, soil, water from various sources,

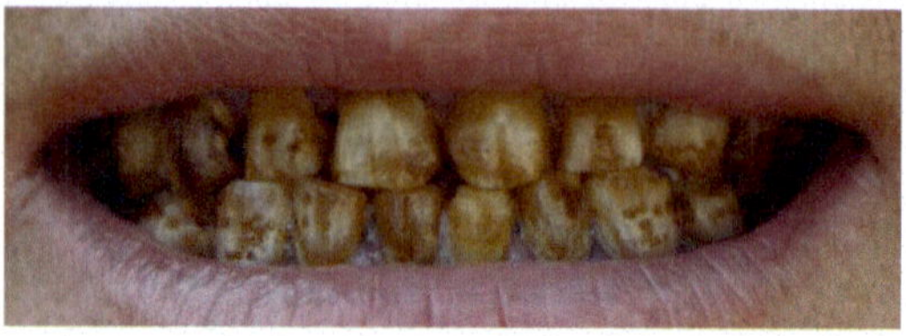

Fig. 2.1 Dental fluorosis from a patient with endemic fluorosis. Dental fluorosis from a patient who lives in the area of coal-burning type of endemic fluorosis, which exhibits more obviously dark pigmented patches in the surface and defects in the dentin

and corn and pepper as well as other crops) in both Guizhou and other areas of China to analyze fluorine and other elements. The following investigations revealed more facts about this type of endemic fluorosis: normally found in mountainous region at high altitude with high humidity; harvested food (basically corn and chili) in autumn could not be dried completely by sunlight so people baked them with stoves by burning coal containing high amount of fluoride. The fluoride contents of dried food (corn, chili, and vegetables) obviously increased and served as the main source of fluoride contamination. Therefore, such endemic was termed as coal-burning type of endemic fluorosis. The studies found that residents of the endemic fluorosis areas heavily relied on coals containing 76–2637 ppm of fluoride. Further investigations discovered that coal-mixed clay for economical combustion had fluoride content as high as 440–7000 ppm. Such clay could release the fluoride while burning and play as major source of contamination on food and air. Since the long-term habit of cooking, drying, and heating by open stoves with no exhaust system, food, and in-door air were severely contaminated by fluoride in these areas. Naturally, people took in fluoride with dozens or hundreds of times higher than health standard level and lead to chronic fluorosis, exhibiting more severe dental (Fig. 2.1) and skeletal (Fig. 2.2) fluorosis than other type of endemic fluorosis.

Prevention and Control Measures of Coal-Burning Type of Endemic Fluorosis

The international communities established their own prevention and control measures based on individual conditions in different countries to endemic fluorosis (Susheela and Toteja 2018; Guan et al. 2017). Such measures against coal-burning type of endemic fluorosis in China focus on several aspects to detain the prevalence of endemic: improving stoves, health education, boosting local economy, and so on.

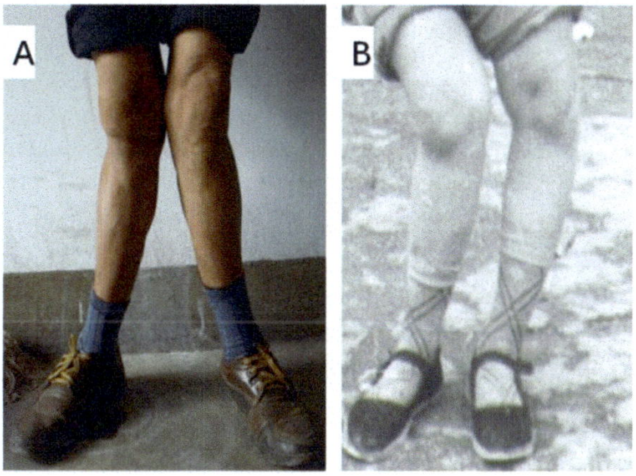

Fig. 2.2 Skeletal fluorosis from a patient with endemic fluorosis. Skeletal fluorosis from the patients who lives in the area of coal-burning type of endemic fluorosis, which exhibits more severe bone deformities. (**a**) The knee-joints of the bones are significantly deformed, presenting an X-shaped leg. (**b**) Leg deformation due to severe osteomalacia

Improving Stoves

The measure aims to discharge the smoke to the outdoor to avoid in door coal burning contamination, including replacing stoves, chimney or specially designed stoves.

For better-off areas, large quantity of finished stove products can be purchased in way of bidding, then being checked and accepted by professional. After-sale service stores can be set up with manufacturers in those areas, which can better solve practical problems of users. In other cases, clean energy like methane, liquid gas, and electricity should be encouraged to use if local conditions are fit for such development. Reducing using coals is an efficient way to alleviate in-door contamination.

For less better-off areas, improving stoves must be the primary focus. Such project is to replace the clay built open stove into lid-covered stove with exhaust system. In addition, the new exhaust outlet must be higher than the eave. Inside the stove, the chamber must be equipped with grate for better ventilation to improve combustion efficiency. A viewport should be saved around the corner of exhaust pipe for cleaning ashes. Design on depth of chamber and opening diameter must not be taken lightly, as the stove must cater for daily routines and coal-saving.

Health Education

The major contents of health education include publicizing the harm of endemic fluorosis to human health and how to prevent and treat it; qualified stove standard and handling instructions; and correct way to dry and store corns and chili.

Damage of coal-burning type of endemic fluorosis and its prevention/treatment: publicizing and introducing how people are affected with this endemic, what damages it inflicts on people and how to prevent and treat the disease. Prevention plays an essential role as there is still no specific drug for this disease, which must be stressed.

Qualified stove standard and handling instructions requirement: it must be made clear to the residents that qualified stove must hold the following features, i.e., durable, smoke tight, and fit for construction safety code; grate, core, metal plate, lid, and exhaust system are well equipped; catering for cooking food, boiling animal feed, heating with higher thermal efficiency than open stove; specially designed stoves operate on methane, natural gas or electricity must fit for national quality technology standard; residents are suggested to clean the stove and exhaust pipe regularly, cover the lid while burning coals to avoid in-door smoke; if the stove will be lay idle, they need to be cleaned first, oil metal parts surface and then stored in dry place.

Correct way to dry and store corn and chili: drying corn and chili without contaminating them with coal-burning smoke, such as using sun-cure, flue-cured tobacco chamber, or on stoves with well-equipped exhaust system; then dried corn and chili must be sealed airtight in-door (such as using bag, box, or warehouse to avoid soot contamination); since corn and chili might absorb air particles with fluoride during storage, washing the fluoride dust on surface before eating is vital.

Clinical Treatments

Dentists and surgeons in Guizhou have carried out clinical treatments on the patients with dental and skeletal fluorosis from endemic areas. Medication studies on lowering oxidative stress level have made preliminary achievement (Guan et al. 2018).

Personalized Treatment of Dental Fluorosis

Dental fluorosis is the most obvious feature on endemic fluorosis patients with high incidence rate. Apart from damage to teeth and pathological changes, dental fluorosis puts tremendous psychological pressure on patients due to the change of color and shape on teeth. Hence, dental fluorosis treatment shoulders dual clinical and social medicine obligations. Concerning damage characteristics of dental fluorosis on dental morphology and color of patients, dental restoration plays an important role in treating dental fluorosis. Chinese dentists established effective dental-colorimetric method and upgraded the existing restoration technique to porcelain-fused-to-metal crown restoration and UV compound resin direct repair. These techniques specifically targeted dental fluorosis patients featuring with easy technological requirement and low cost, which were easier to be accepted by patients. In addition, since restored teeth shared similar color with adjacent teeth, and patients are satisfied with such aesthetical improvement. These effectives, targeting and practical dental restoration for the patients with dental fluorosis is well received.

Surgical Treatment of Skeletal Fluorosis

The main surgical treatment for skeletal fluorosis is joint replacement for severely damaged bone joints. Since skeletal fluorosis generates much osteophytosis, osteophyte loose soft tissues must be first erased, then positioned and installed with the equipment for osteotomy as compared with common joint replacement. Comprehensive treatment must be adopted before, during, and after surgery, as it prevents post-surgery joint adhesion or loss of joint function after being discharged. The combined application of internal and external use of traditional Chinese medicine enhanced the flexibility of joint replacement technique in the patients with skeletal fluorosis. Based on substantial studies, Chinese orthopedists propose that artificial joint replacement is an effective method to treat knee and hip joint with severe damage caused by endemic fluorosis and establish the corresponding systematic and effective technical method. Through popularization and application for the technique, more than 1000 patients with severe damages of knee and hip joints due to chronic fluorosis received artificial joint replacement. Through targeted evaluation, the results indicated that the joint function and quality of life were significantly improved.

Pharmaceutical Study on Targeting Endemic Fluorosis

So far, there is still no specific drug for the treatment of endemic fluorosis. At present, according to the mechanism of "free radical theory" of endemic fluorosis systemic injury, drugs to improve antioxidant capacity and reduce free radicals are mainly studied. Scientists from Guizhou Medical University worked with pharmaceutical companies are developing compound traditional Chinese medicine, based on integrative medicine theory, which could improve activity of antioxidant enzymes, defend the injury of free radicals and promote fluoride excretion. After laboratory experiments and partially clinical research, such medicine proved to be effective, showing a promising potential.

Other Measures

The areas with coal-burning type of endemic fluorosis in China were invariably found in remote mountainous regions at high altitude, featuring low residential density, underdeveloped transportation and economy, many of which are minority regions, such as Miao and Yi. It is far from enough to rely on just improving stoves and health education in such areas for the overall prevention and treatment of coal-burning type of endemic fluorosis. The decisive factor lies in the combined force of relevant government departments and social power. In this way, it is hopeful to establish leading committee and work mechanism targeting prevention and control of endemic fluorosis according to the reality of individual endemic areas. Therefore, financial and policy support from the local government is irreplaceable in this endeavor. While engaging on stoves improving projects, health education and changing traditional habits, it is of vital importance to work in the local economy, culture, education, and health cares. The change of living habit and improving quality of life are the answers to guarantee the goal of the projects, finally eradicating the harm of endemic fluorosis.

Conclusion

The endemic fluorosis is caused by excessive intake of fluoride from drinking water, food or air, drinking tea, etc. Its main manifestations are dental and skeletal fluorosis, together with comprehensive pathological injury on multiple systems and organs. Of all the three types of endemic fluorosis (drinking water, coal-burning and brick-tea types of endemic fluorosis), coal-burning and brick-tea types of endemic fluorosis are only found in southern or western China. Coal-burning type of endemic fluorosis is the result of human body taking in dozens or hundreds of times high of fluoride than normal over health standard from fluoride contaminated food and in-door air (cooking and drying food as well as heating on open stoves). China has stipulated the national standards for division of endemic areas and control of the disease based on its own reality. After decades of effort, comprehensive measures (improving stoves, health education, improving local economy, and prevention ability) have effectively detained the outbreak of coal-burning type of endemic fluorosis.

References

Dhar V, Bhatnagar M. Physiology and toxicity of fluoride. Indian J Dent Res. 2009;20:350–5.

Editorial. Food-induced skeletal fluorosis. Fluoride. 1981;14:49–50.

Guan ZZ, Wang LH, Sun DJ. Endemic fluorosis. In: Sun DJ, et al., editors. Endemiology. Beijing: People's Medical Publishing House of China; 2017.

Guan ZZ, Wang Y, Duan QH, Liu RG, Li FC, Xu SX, et al. Basic investigation and clinic treatment for the coal-burning type of endemic fluorosis in Guizhou, China. Fluoride. 2018;52:83.

Lyth O. Endemic fluorosis in Kweichow, China. Lancet. 1946;247:233–7.

Peckham S, Awofeso N. Water fluoridation: a critical review of the physiological effects of ingested fluoride as a public health intervention. Sci World J. 2014;2014:293019.

Srivastava S, Flora SJS. Fluoride in drinking water and skeletal fluorosis: a review of the global impact. Curr Environ Health Rep. 2020;7:140–6.

Susheela AK, Toteja GS. Prevention and control of fluorosis and linked disorders: developments in the 21(st) century - reaching out to patients in the community and hospital settings for recovery. Indian J Med Res. 2018;148:539–47.

Waldbott GL. The preskeletal phase of chronic fluoride intoxication. Fluoride. 1988;31:13–20.

Wei ZD, Zhou LY, Bao RC. Endemic food-borne fluorosis in Guizhou, China. Fluoride. 1981;14:91–3.

Whitford GM. The physiological and toxicological characteristics of fluoride. J Dent Res. 1990;69:539–49.

Zheng B, Wu D, Wang B, Liu X, Wang M, Wang A, et al. Fluorosis caused by indoor coal combustion in China: discovery and progress. Environ Geochem Health. 2007;29:103–8.

Part II
Pathological Changes of Endemic (Chronic) Fluorosis

Chapter 3
Skeletal Fluorosis

Jie Xiang, Xiao-Xiao Zeng, and Zhi-Zhong Guan

Abstract Skeletal fluorosis is the most serious pathological change of endemic fluorosis, in which its molecular pathogenesis is rather complicated. Obviously, the abnormal bone turnover plays an important role in the underlying mechanism of skeletal fluorosis, manifesting as the dynamic imbalance between osteoblastogenesis and osteoclastogenesis, as well as the abnormal changes of cytokines and multiple signal transduction pathways. In bone metabolism, abnormal distribution and regulation of calcium ions due to the deposition of large amounts of fluoride in bone have a significant impact on the occurrence of skeletal fluorosis. Interestingly, high level of oxidative stress and elevation of endoplasmic reticulum stress play an equal key role in the molecular mechanism of skeletal fluorosis. In addition, the abnormal modifications in hormone regulation, genetic susceptibility, and epigenetic change are also involved in the occurrence and development of skeletal fluorosis.

Keywords Skeletal fluorosis · Osteoblast · Osteoclast · Pathological changes

---

J. Xiang · X.-X. Zeng
The Key Laboratory of Endemic and Ethnic Diseases (Guizhou Medical University), Ministry of Education, Guiyang, Guizhou, People's Republic of China

Department of Pathology at the Affiliated Hospital of Guizhou Medical University, Guiyang, Guizhou, People's Republic of China

Z.-Z. Guan (✉)
The Key Laboratory of Endemic and Ethnic Diseases (Guizhou Medical University), Ministry of Education, Guiyang, Guizhou, People's Republic of China

Department of Pathology at the Affiliated Hospital of Guizhou Medical University, Guiyang, Guizhou, People's Republic of China

The Provincial Key Laboratory of Medical Molecular Biology, Guiyang, Guizhou, People's Republic of China
e-mail: zzg@gmc.edu.cn

Z.-Z. Guan (ed.), *Coal-burning Type of Endemic Fluorosis*,
https://doi.org/10.1007/978-981-16-1498-9_3

Introduction

Skeletal and dental fluorosis are the most typical clinical manifestations of endemic fluorosis. The endemic skeletal fluorosis refers to the condition that residents in the areas of endemic fluorosis suffer from chronic disorder in bone metabolism characterized by pain in the large joints of the neck, waist, and extremities, and limb motor dysfunction with abnormal X-ray signs of bones and joints due to excessive fluoride intake. In China and India, there are a large number of patients with endemic skeletal fluorosis. In China, 1.5 million patients with skeletal fluorosis live in the area of coal-burning type of endemic fluorosis. In Guizhou Province of China, where coal-burning type of endemic fluorosis is rampant, 15 million people in the area are affected, including 800,000 patients with skeletal fluorosis and 10 million with dental fluorosis (Guan et al. 2017). Skeletal fluorosis is the most detrimental form of endemic fluorosis to the human body, in which severe cases can lead to paralysis and contribute to substantial loss of labor force in the area. Therefore, it is of great clinical significance to understand the pathogenesis of skeletal fluorosis and provide a theoretical basis for the clinical therapy.

Clinical Manifestations and Pathological Features of Skeletal Fluorosis

Fluoride Metabolism and Its Correlation with Bone Tissue

Bone tissue is composed of a variety of cells and extracellular matrix (ECM). The cells include five types: osteoprogenitor cells, osteoblasts (OB), osteoclasts (OC), osteocytes, and bone overlying cells. The ECM of bone tissue is called bone matrix after mineral deposition. Bone matrix is composed of organic matter and inorganic matter (also known as bone salt). Organic matter is mainly secreted and formed by OB, accounting for about 35% of the weight of diaphysis, including a large number of collagen fibers and a small amount of amorphous matrix. Collagen fibers account for 90% of organic matter and are mainly composed of type I collagen (Col-I). The amorphous matrix is gelatinous and contains a variety of glycosaminoglycans (GAGs) and glycoproteins. GAGs include chondroitin sulfate, keratan sulfate, and hyaluronic acid. There are many kinds of glycoproteins and they all have special roles. For example, osteonectin is a glycoprotein attached to collagen, which is prone to bond with hydroxyapatite and can combine the inorganic components of bone with bone collagen. Such combination may act as the core of bone salt deposition. Bone gamma carboxyglutamic acid containing proteins (BGPs) is a calcium-bonding protein with high affinity to hydroxyapatite and Ca^{2+}, and its role is related to bone calcification and calcium transport. The organic component makes the bone tough. Osteocytes are distributed within the bone matrix, and other cells are located on the surface of bone tissue.

Bone salts in bone tissue account for 65 to 70% of diaphyseal weight, and their main components are calcium phosphate (accounting for 84%), followed by calcium carbonate (accounting for 10%), together with calcium citrate, magnesium phosphate, sodium monobasic phosphate, etc. About 60% of the bone salt is in the form of crystalline hydroxyapatite [$Ca_{10}(PO_4)_6(OH)_2$], and the remaining 40% is amorphous hydrated calcium hydrogen phosphate [$CaHPO_4$]. Hypothesis indicated that the latter can be transformed into the former. Hydroxyapatite is a fine crystal, also known as bone crystal. Under the electron microscope, the crystals are fine needle-shaped 15 nm in length, and tightly and regularly arranged along the long axis of collagen fibers. Once bone salts bond with organic components, the bone matrix is very hard to adapt to its supporting function. Each gram of bone salt contains about 1016 crystals, with a total surface area of up to 100 m^2, and other ions in body fluids such as Ca^{2+}, Mg^{2+}, Na^+, Cl^-, HCO_3^-, and F^- can be adsorbed through lattices of hydroxyapatite and perform rapid ion exchange in its surface layer. Bone tissue not only supports other soft tissues, but also stores a large number of trace elements such as calcium, phosphorus, sodium, and magnesium. Therefore, bone tissue plays an important role in maintaining fluid and electrolyte homeostasis.

Fluoride is a trace element with a very narrow safety threshold since it has strong affinity to bones. Therefore, when fluoride ions enter the body, more than 80% of the ions accumulate in bone tissue and play an important role in the growth and development of bones as well as the maintenance of physiological functions. However, when fluoride accumulates in the body for a long time, it can cause chronic fluorosis and afflict the bone tissue of the whole body. Then, bone metabolism disorders and systemic bone lesions are inevitable, which is termed skeletal fluorosis.

The effect of fluoride on bone tissue is complex, which can bidirectionally regulate the proliferation and differentiation of OB. Low-dose fluoride shows promoting effect, and high-dose fluoride shows inhibitory effect. However, different fluoride concentrations and treatment times, as well as the application of different experimental models make it difficult to determine the critical value of this bidirectional effect of fluoride. Similarly, the fluoride effect on OC regulation was also found to be promoting or inhibition. Fluoride can affect the content of trace elements including calcium, magnesium, etc., and inhibit the activity of enzymes dependence on calcium or magnesium ions which affect a variety of enzyme metabolic processes in bone cells. In addition, fluoride can replace the hydroxyl group of hydroxyapatite and affect the bone salt properties. The new fluorapatite crystal has bigger size, higher crystallinity, and lower solubility than the original crystal. All the factors affecting the metabolism of bone tissue cells and bone formation will promote the occurrence of osteoporosis or osteosclerosis in skeletal fluorosis. In the process of detecting bone metabolic state, it is necessary to use different experimental indicators. For example, alkaline phosphatase (ALP) is a marker of OB maturation and activity, which is used to determine the OB differentiation and activity. The secretion of BGP is used to judge the effect of various fluoride concentrations on bone matrix mineralization.

Clinical Manifestations of Skeletal Fluorosis

Skeletal fluorosis refers to a chronic, metabolic, and systemic bone disease caused by long-term excessive fluoride intake and then afflicting bone tissue. Fluoride is mainly absorbed through the gastrointestinal tract and respiratory tract. Once it enters the bone, its half-life is up to 7 years, which can affect bone strength through a variety of mechanisms, and the damaged bone returns to its original state. Skeletal fluorosis is usually mildly symptomatic, and patients experience vague pain in the joints of the hands, feet, and lower back. Muscle weakness, chronic fatigue, and joint stiffness with reduced range of motion gradually develop as the condition worsens. The bones and joints then become fragile, making movement difficult and painful, and many parts of the spine become fused. Limited spinal mobility is complicated with kyphotic deformity, lower extremity flexion contracture deformity, and limited chest wall expansion. In the final stage, patients suffer from disability, increased risk of fractures and complications such as radicular myelopathy (Sellami et al. 2020). In coal-burning endemic fluorosis areas, patients with severe skeletal fluorosis may find with gonyectyposis and gonycrotesis. Knee varus deformity is also known as the "O" type leg. In stand-up position, the lower limbs show arcuate protrusion deformity to the lateral side, the medial malleolus of both feet is close, the knees cannot be close together; when legs are spread, the space in between is similar to the shape of the character "O." Knee valgus deformity is also known as the "X" type leg. In stand-up position, the lower limbs show arcuate concave deformity to the medial side, the knees can be close together, the medial malleolus of both feet cannot be close but with large spacing, similar to the shape of the character "X." The most serious clinical manifestations of skeletal fluorosis are gonyectyposis and gonycrotesis.

Pathological Features of Skeletal Fluorosis

The pathological features of skeletal fluorosis are found to be different. Some scholars have performed histomorphological analysis on undecalcified sections from transiliac needle biopsies of patients with skeletal fluorosis. The results indicated that cortical thickening with narrowed bone marrow cavity, reduced bone resorption, and loss of osteoclasts. Others have described osteoid thickening of the bone surface and significant bone resorption. OB appears to be more common in fluorinated bone, but its activity and functional pathology are not clearly visualized. Bone condensation is also frequently reported, with increased osteoid circumference and width, defective bone formation, and prolonged OB circumference. In addition, pseudo-osteomalacia was found to be possible with increased osteoid thickness index, and this type of hyperostosis appeared to be caused by too slow juxtaposition of bone mineralization.

There are a variety of skeletal fluorosis manifestations in clinical pathology. Sclerosing skeletal fluorosis is a type of skeletal fluorosis mainly characterized by osteosclerosis (featuring extensive hyperosteogeny, sclerosis, and joint stiffness) and dyskinesia caused by periosseous soft tissue ossification, spinal fixation, thoracic fixation and limb ankyloses. Osteoporosis skeletal fluorosis is characterized by metabolic bone lesions with normal bone tissue calcification, normal calcium salt-to-matrix ratio, and reduced bone tissue per unit volume. Malacic skeletal fluorosis is characterized by osteomalacia, which refers to delayed bone calcification formation from new bone and metabolic bone lesions (in which the originally calcified bone is replaced by osteoid tissue). Mixed skeletal fluorosis is often found in clinical practice, which is manifested with osteosclerosis, osteoporosis, and softening at the same time. Skeletal fluorosis with osteomalacia in children is often found in coal-burning endemic fluorosis areas. The specific pathological changes of skeletal fluorosis are determined by multiple factors such as fluoride concentration, gender, age, and nutritional status.

In addition, periosteal soft tissue ossification, also known as periosteal soft tissue calcification, is a pathological manifestation of calcification or ossification in the peripheral soft tissues including joint capsule, inter-periosteum, ligament, tendon attachment, etc. Such change is also a characteristic change of skeletal fluorosis. Degenerative change of joint is joint function degeneration and movement disorder caused by degeneration, necrosis, exfoliation of articular cartilage, and hyperosteogeny of joint periphery. This change often occurs with skeletal fluorosis.

Imbalance of Bone Turnover in Chronic Fluorosis

Dynamic Imbalance of Bone Turnover

Bone metabolism is mainly involved in the process of OB forming new bone and OC resorbing old bone, and this renewed replacement of bone is often called bone turnover. Bone turnover occurs during growth and development, when bone formation and resorption are active to complete bone formation and remodeling. In the mature bone turnover phase, bone turnover is slow but persists in life. Bone remodeling occurs on the basis of basic multicellar units (BMU), and in each BMU, the continuous homeostasis of OB and OC activity is key to maintain normal bone mass and bone function. Specific processes include OC attachment to the old bone part, secretion of acidic substances to dissolve minerals, secretion of proteases to digest bone matrix, and formation of bone resorption pits. Subsequently, the OB migrates to the resorbed site and secreted bone matrix, and then the bone matrix is mineralized to form a new bone.

Excessive fluoride disrupts the homeostasis of OB and OC activity and affects their differentiation, which leads to the development of bone lesions. Skeletal fluorosis is bone lesion characterized by imbalance in bone turnover, which can manifest as osteosclerosis, osteomalacia, osteoporosis, periosseous soft tissue ossification, and degenerative changes in cartilage and joints. Active osteogenesis and accelerated bone turnover are important features in the progression of skeletal fluorosis and the pathological basis for the diversity of osteoblastic lesions. In recent years, research on pathogenesis of skeletal fluorosis focused on the regulatory mechanisms of fluoride on various cells involved in the process of bone turnover.

OB originates from pluripotent bone mesenchymal stem cells (BMSCs), which are the main functional cells of bone formation and are responsible for the synthesis, secretion, and mineralization of bone matrix. OB undergoes processes of OB proliferation, ECM maturation, ECM mineralization, and OB apoptosis during bone formation. The proliferation of OB is regulated by genes related to cell cycle and growth regulation. After that OB is slowed down, the differentiation phase is begun. The expression of genes related to cell proliferation decreases, while the expression of genes encoding proteins that promotes ECM maturation is begun. In the early stage of OB differentiation, ALP expression is increased. ALP is considered to be an early marker of ECM maturation and OB differentiation. The expression of messenger ribonucleic acid (mRNA) of ALP can be increased by more than tenfold at this time. During ECM maturation, OB secretes ALP and calcium crystals into the ECM and ALP increases the local phosphate content. ALP also promotes matrix mineralization accompanied by continued collagen synthesis, cross-linking, and maturation with each other. In the late stage of OB differentiation, the expression of ALP is decreased, while the expression of genes related to hydroxyapatite deposition in ECM, like osteopontin (OPN), BGP and bone sialoprotein (BSP) genes, peaks, and enters the mineralization stage. Non-collagenous proteins such as BGP are secreted into the ECM, bond with calcium and phosphorus along the long axis of collagen molecules. Then, they bond with collagen amino acid residues on the side chains of collagen molecules to form hydroxyapatite crystals. Multiple factors regulate these stages, which ultimately regulate bone formation.

Fluoride can regulate the proliferation and differentiation of OB in both directions. Collagen deposition and OB phenotypic markers such as ALP gene begin to increase during OB transit from proliferation to differentiation stage under the effect of fluoride. During the subsequent mineralization phase, BGP and OPN expressions are increased. Fluoride can directly stimulate the proliferation of OB cells, enhance ALP activity, and increase the synthesis of BGP and various collagens by promoting the increase of deoxyribonucleic acid (DNA) synthesis in OB cells. With the increase of fluoride concentration, the over-expressed osteogenic molecular markers (ALP, BSP, BGP, and OPN) may be further regulating the differentiation of OB precursor cells into OB, OB proliferation can be promoted, the activation of more OB will be accelerated, and OB activity will be increased. So that more OB participates in osteogenic activity, the aggravation of osteogenic activity will be promoted, and bone mass can be increased. The level of markers such as ALP and BGP of patients with skeletal fluorosis is higher than in controls. However, fluoride may also delay new bone mineralization while stimulating OB activity (Shen et al. 2019). In addition, fluoride does not indiscriminately make the proliferation of OB line active. When the fluoride dose exceeds the tolerance limit of cells, it can cause proliferation inhibition or even apoptosis. In general, OB activity will first increase and then decrease with the increase of fluoride exposure dose, showing an inverted U-shaped "dose-dependent" curve. However, this phenomenon was observed in OB isolated in vitro.

OC originates from the lineage monocyte-macrophage system. It is a special terminally differentiated cell, which can be fused by its mononuclear precursor cells to form giant multinucleated cells in a variety of ways. It is the main functional cell of bone resorption and plays an important role in bone development, growth, repair, and reconstruction. In the process of OC resorbing organic and mineral substances of bone matrix, it causes irregularities on the surface of the matrix and forms pits that resemble the shape of cells, known as Howship pits. On the side of the lacuna facing the bone, the cells are found with hair-shaped protrusions, much like the longitudinal and bristled edges of the epithelial cell surface. Under the electron microscope, irregular microvilli close to one side of the bone or cell processes are called ruffled edges. OC releases lactic acid and citric acid to the local area. Under acidic conditions, inorganic minerals in bone are swallowed by the ruffled edge, and some swallowed vesicles or phagocytic vesicles are formed in the matrix of the ruffled edge. Within the OC, inorganic matter is degraded and excreted in the bloodstream in the form of calcium ions. The loss of inorganic matter exposes collagen fibers within the bone matrix. OC secretes a variety of lysosomal enzymes, especially cathepsin K (Cath-K) and collagenolytic cathepsins. After the OC leaves the bone surface, its ruffled edges disappear, and the changed cells enter the static phase. It is known that the proliferation, differentiation, and maturation of OC are regulated by the state of OB. Osteoprotegerin (OPG) and OC differentiation factor are secreted by OB. The former mainly inhibits the production of OC, while the latter stimulates the differentiation and maturation of OC, activates mature OC and inhibits its apoptosis.

High expression of tartrate resistant acid phosphatase (TRACP) and Cath-K (a cysteine protease irreplaceable in the process of bone resorption) are the main markers of OC. It can degrade bone matrix proteins, including Col-I, OPN, and osteonectin, and is abundantly expressed when OC undergoes bone resorption. Cath-K is found in the OC crease at the site of active bone resorption, while its expression is found in the OB and osteocytes. It was confirmed that the serum level of BGP and Cath-K in the high fluoride group was significantly higher than that of control group. Such results indicated that skeletal metabolism was disturbed, OB and OC metabolism was active, and bone turnover was accelerated during fluorosis, which lead to skeletal fluorosis.

The activation or inhibition of OC by fluoride is still controversial. Some scholars believe that with the increase of fluoride exposure dose, OC cultured in vitro shows direct damage, shortened life cycle, increased cell death. In the meantime, its size shrinks, and the number and area of bone resorption pits decrease with the increase of fluoride concentration, thus inhibiting the bone resorption capacity of OC. In contrast, others believe that in the occurrence and development of skeletal fluorosis, OC function is active. The number and pit area of bone resorption increase with the increase of fluoride dose, and bone resorption is enhanced. The mechanism involves: enhancing the expression of TRACP and matrix metaloproteinase-9 (MMP-9) in OC or up-regulating expression of macrophage colony-stimulating factor (M-CSF) and receptor-activator of nuclear factor-kappa B ligand (RANKL). The fluorosis threshold theory suggests that low dose of fluoride can promote the proliferation of mouse OC cultured in vitro, and the promoting effect of fluoride is weakened with the increase of dose. Exceeding certain threshold might produce an inhibitory effect on OC proliferation and activity (Jiang et al. 2019). The reasons for the contradictory conclusions of the above studies may be related to multiple factors such as cell origin and different fluoride exposure doses.

It is worth noting that the occurrence of skeletal fluorosis is the long-term result of the dynamic changes of OB and OC, which interact and regulate each other. Their imbalance and the mechanism of imbalance is the key to determine how skeletal fluorosis develops. Osteosclerosis is mainly the result of increased OB activity accompanied by unchanged or weakened bone breaking activity, while osteoporosis in skeletal fluorosis is active osteoporosis with accelerated bone turnover, bone breaking and absorption are accompanied by active osteogenesis. In addition, pathological development of bone turnover is subject to the following features: different fluoride intake time, fluoride intake concentration, patient's age, gender, nutritional status, and many other factors. Thus, it is a complex regulatory network affecting the final state of the disease together.

Important Factors and Signal Pathways Associated with Imbalance in Bone Turnover

The mechanisms of fluoride-induced imbalance in bone turnover are complex and diverse, involving regulation of participating cells by different cytokines as well as related signals. OB and OC are important participating cells in the process of bone turnover. It has been confirmed that the abnormality of OB and OC plays an important role in the occurrence and development of skeletal fluorosis. The following content concerns important factors and signal pathways that affect osteogenesis and osteogenesis.

Transforming growth factor-β (TGF-β) is a general term of a class of superfamily members, which include TGF-βs, activin, bone morphogenetic protein (BMP), and growth differentiation 1 factors (regulating the growth and differentiation of various cells like OB). Of all the members, TGF-β1 and BMPs play the most critical regulatory role in the growth and differentiation of bone tissue include, while BMP-2 and BMP-7 in BMPs are alternatively considered to be the most osteogenic active BMPs family members. In the bone tissues of rats treated with sodium fluoride, the expressions of bone formation inducers (TGF-β1, BMP-2, and BMP-7) were significantly higher than those in the normal control group, and the expression sites in the long bone tissues were basically the same.

TGF-β1 is generated by OB and activated during bone resorption, which can increase the activity of ALP to varying degrees, regulate the formation of mineral nodules, and regulate the differentiation and function of OB and OC through the typical drosophila mothers against decapentaplegic protein (SMAD)-dependent pathway and mitogen-activated protein kinase (MAPK) pathway. TGF-β1 may play a key role in the regulation of fluoride on OB activity and differentiation. In addition, fluoride has been found to increase runt-related transcription factor 2 (RUNX2) mRNA expression significantly in MC3T3-E1 cells (Lee et al. 2017). Runx2 is the main transcription factor driving OB differentiation, which can activate and initiate BMSCs to differentiate into OB lineage, regulate OB differentiation and function, and plays an important role in the formation of bone tissue and bone remodeling. TGF-β1 can control BMSCs by converging on the Runx2 gene through the SMAD and p38 MAPK pathways to promote the proliferation and differentiation of OB.

BMP-2 not only aggregates undifferentiated BMSCs to the center of bone formation and promotes their differentiation into bone cells, but also regulates the secretion of certain specific proteins in cells, such as fibroblasts and myoblasts. These two differentiate into OB and hypertrophic chondrocytes, respectively. They can promote the reverse differentiation of a series of non-bone cells into bone cells and promote matrix calcification. For OB, BMP-2 also allows OB to maintain its unique cellular phenotype and induces an increase in OB markers, which can promote ECM mineralization. In the later stage of bone formation, BMP-2 also acts as an OC differentiation factor together with other OC differentiation factors to directly or indirectly stimulate OC differentiation and participate in bone remodeling. In addition, BMP-7 is involved in the development of tooth germs, heart, brain, and kidney during the embryonic period. Recombinant human BMP-7 has shown highly potent osteoinductive activity both in vivo and in vitro, allowing healing of bone defects in a variety of experimental animals. For example, it promoted OB proliferation and ALP expression, and promoted chondrocyte proteoglycan expression and repaired articular cartilage defects (Long et al. 1995). In conclusion, TGF-β family members regulate bone formation, and fluoride can induce the expression of TGF-β family in the pathogenesis of skeletal fluorosis, which is an important factor mediating the development of skeletal fluorosis.

MAPK, a serine/threonine kinase, is a type of protein kinase with dual phosphorylation ability of serine and tyrosine distributed in the cytoplasm. MAPK signal transduction pathway is one of the channels in the nucleus caused by extracellular signals, which can involve in various physiological processes such as cell formation, motility, apoptosis, differentiation, growth, and proliferation. MAPK is within each cascade, and the signal is propagated by continuous phosphorylation and continuous activation of kinases. In general, the classical MAPK cascade includes two processes: mitogen-activated protein kinase kinase kinase (MAPKKK) activating mitogen-activated protein kinase kinase (MAPKK), and then dual phosphorylation of threonine/tyrosine of MAPK activating MAPK. At present, four MAPK signal transduction pathways have been identified with eukaryotic cells: extracellular signal-regulated kinases (ERKs), c-Jun N-terminal kinase (JNK), p38, and ERK5 pathways. The ERKs, p38, and JNK pathways are known to be involved in the signal transduction of OB proliferation and differentiation, and play important roles in stress, apoptosis, bone metabolism, and inflammation. An increasing number of studies have shown that the molecular mechanism of fluoride bone injury involves multiple pathways in MAPK signal pathway.

Fluoride can inhibit unique fluoride-sensitive tyrosine phosphatase in OB, which leads to a sustained increase in tyrosine phosphorylation level of key signaling proteins in the MAPK mitotic transduction pathway and results in enhanced OB proliferation initiated by growth factors. In addition, fluoride and aluminum have a synergistic effect and can form fluoroaluminate. Fluoroaluminate activates pertussis toxin-sensitive Gi/o proteins on bone tissue cell membranes and then lead to the activation of cellular protein tyrosine kinases, which in turn increases protein tyrosine phosphorylation in the MAPK signal transduction pathway and ultimately promotes cell proliferation.

P38 MAPK is involved in the differentiation of BMSCs into OB, matrix collagen synthesis, and promotes OB differentiation and maturation. It also inhibits OC differentiation and bone resorption, and plays an important role in the regulation of the OB proliferation cycle. The expression level of p38 protein and mRNA in the OB of rats treated with fluoride was significantly increased, accompanied by a significant increase in the proportion of phosphorylated p38. In addition, it was more pronounced in the high fluoride group than in the low fluoride group, indicating that fluoride ions in the OB of rats could promote p38 protein expression, and p38 activation was dependent on the change of fluoride concentration. Activated p38 protein could further to activate downstream substrates, such as transcription factors, to promote the differentiation and activity of OB, which was involved in the pathogenesis of skeletal fluorosis (Yu et al. 2012).

Fluoride induced and increased phosphorylation of ERK1/2, JNK, p38, and ERK5 in MC3T3-E1 cells was consistent with experimental findings in fluoride-treated MDPC-23 rat odontoblast-like cells and A549 human lung epithelial cells. However, each MAPK was activated in a different manner, which ERK1/2 and ERK5 phosphorylation could be initiated in a shorter incubation time than JNK and p38. Notably, the above mentioned MAPKs could all respond to a variety of extracellular stimuli, increase cellular oncogene fos (c-Fos) transcription and post-translational phosphorylation of c-Fos protein to better stabilize c-Fos protein. In particular, ERK1/2 and ERK5 pathways were the major cause of c-Fos expression and/or c-Fos accumulation.

The c-Fos gene is originally the cellular homologue of the retroviral gene v-fos, isolated from mouse osteosarcoma, is a member of the Fos family. c-Fos gene and Jun family members (c-jun, JunB, and JunD) constitute activator protein-1 (AP-1). AP-1 is a dimeric transcription factor complex involved in cell proliferation, death, survival, differentiation, and stress responses. It has been found that increased c-Fos expression and increased AP-1 activity may inhibit the expression of OPG in some cells, including OB. Each member of the Fos family responds differently to cellular stress in response to fluoride exposure, which c-Fos expression is the most sensitive. Treatment of MC3T3-E1 cell line with fluoride induced c-Fos expression and increased DNA bond with c-Fos protein, accompanied by up-regulation of AP-1 transactivation activity are involved in the development of fluoride bone damage.

OB generated OPG can prevent RANKL from bonding to its receptor on the OC to inhibit OC activation and subsequent bone resorption. NaF treatment induced OPG expression in cultured MC3T3-E1 cells, OS732 human osteoblast-like cells, and rat bones in vitro; the up-regulation of NaF induced OPG mRNA was further enhanced in c-Fos knockout MC3T3-E1 cell lines. Increased c-Fos expression and AP-1 activity might inhibit OPG expression in bone cells, including OB, but the underlying mechanism remained unclear. Paradoxically, if c-Fos was inhibitory to OPG expression, then fluoride increased c-Fos expression, which reduced OPG expression, but this was not the case. Furthermore, in addition to modulating receptor activator of nuclear factor kappa B (RANK)-RANKL interaction in bone metabolism, OPG could also stimulate cell survival as a receptor for tumor necrosis factor-related apoptosis-inducing ligands. Silencing of OPG reduced the viability of MC3T3-E1 cells treated with 1 mM or 10 mM NaF for 24 h, but c-Fos knockdown was accompanied by a significant increase in OPG mRNA, which failed to prevent the decreased viability of MC3T3-E1 cells treated with 5 mM NaF for 24 and 48 h. Therefore, OPG was closely related to c-Fos, which together maintained normal bone metabolism. However, the complex regulation of OPG and c-Fos expression by fluoride, and c-Fos on OPG expression was not fully elucidated, which might be an important factor influencing fluoride induced bone injury (Iwatsuki and Matsuoka 2016).

Runx2, also known as core-binding factor a1, belongs to the transcription factor of Runt domain gene family. Many types of cells are known to be derived from common BMSCs, which differentiates in different directions under the induction of specific transcription factors, such as the muscle-specific transcription factor MyoD family, peroxisome proliferator-activated receptor γ (which determines the differentiation into adipocyte lines) while the transcription factor involved in the differentiation of BMSCs into OB lines is Runx2. Runx2 can not only promote the differentiation of undifferentiated BMSCs into OB, but also promote their transformation into osteogenic phenotype for already differentiated cells. Infection of BMSCs isolated from Japanese white rabbits in vitro with Runx2 recombinant adenovirus revealed that the expression of OB gene markers was up-regulated in Runx2 adenovirus infection, confirming that Runx2 could promote the differentiation of undifferentiated BMSCs into OB. Direct exogenous expression of transcription factor Runx2 demonstrated to induce myoblast trans-differentiation into mineralized OB lines (Gersbach et al. 2004).

In addition, Runx2 can not only regulate OB proliferation, growth, and differentiation, but also regulate the proteins gene expression of OB including ECM (controlling the physiological process of postnatal bone formation and development). Homozygous (Runx2 −/−) mice with Runx2 gene deletion died soon after birth due to inability to maintain respiration without rib formation. Such mice had no bone formation, their skeleton was composed of cartilage, lack of bone marrow, had neither blood vessels nor intercellular invasion of cartilage, and histological studies showed absence of OB in bone tissue (Otto et al. 1997). ECM proteins (*BGP*, *OP*, and *BSP* genes) all have Runx2 binding sites (two OB-specific cis-acting elements,

OSE1 or OSE2). Both α1 and α2 strand-encoding gene promoters of Col-I contain OSE2.

Previous studies showed that OB was actived at early phase and played a dominant role in skeletal fluorosis. Fluoride-treated animals were observed at the end of the experiment, at the end of 4 and 8 months. Runx2 expression appeared to be enhanced at mRNA and protein level after fluoride intake. Using OB cultured in vitro, it was found that different doses of fluoride (1, 2, and 4 ppm) could increase the expression of Runx2 mRNA in rat OB after 48 h and 72 h, and there were significant differences compared with the control group. Different indicators such as survival assay, electron microscopy, and ALP activity assay verified the activation of OB by fluoride.

Osterix was a newly discovered key transcription factor for OB differentiation, containing zinc finger structure following Runx2, which could act as downstream factor of Runx2 on bone formation. Runx2 could up-regulate Osterix promoter activity. No transcription product of Osterix gene was detected in the periosteum of Runx2 knockout mice. However, the expression of Runx2 gene was normal in Osterix inactivated embryos, which indicated that Osterix gene was not a transcription factor necessary for Runx2 gene expression. As a downstream factor of Runx2 in the OB differentiation pathway, Runx2 gene is required for Osterix gene expression.

Osterix has a zinc finger motif domain that can bind with eukaryotic promoters, regulate gene transcription, and regulate the expression of many important OB advanced phenotypic and functional proteins, such as Col-I, BGP, OPN, and BSP (which contribute to the synthesis of matrix and mineralization, and then promote bone formation). In addition, fluoride stimulated fibroblasts (FB) initiate the high expression of the osteogenic marker gene Osterix, which is dormant under normal physiological conditions, and further induced the conversion of FB to OB. Osterix plays an important role in the process of fluoride-induced heterotopic ossification, which may be the main cause of fluoride-induced bone injury such as bone and joint soft tissue calcification and ligament ossification.

For fluoride treated animals, it was found that fluoride could promote the transcription of Runx2 and Osterix genes in rat spine at transcriptional level compared with the control group. At protein level, fluoride treatment could promote the translation of Runx2 and Osterix genes in rat spine, but Runx2 may inhibit the translation of Osterix genes when Runx2 protein reaches a certain amount. The Osterix protein content was higher in the 45 mg/kg dose group than in the control group, but lower in the 110 mg/kg dose group than in the control group, which may be related to the bidirectional regulation of Osterix gene expression due to different concentrations of fluoride. In the study of MC3T3-E1 cells, it was found that fluoride significantly enhanced the proliferation, ALP activity, and mineralization of MC3T3-E1 cells, accompanied by the up-regulation of mRNAs such as Runx2, Osterix, and OPN. Even though Runx2 and Osterix protein action was inhibited by Runx2 silencing, it could be reactivated after fluoride treatment. So, the ion stimulated the osteogenic function of OB at least in part by increasing the expression level of Runx2 and Osterix (Lee et al. 2017).

Runx2 and Osterix are major transcription factors with crucial roles in OB gene expression and osteogenesis. They are usually combined to study various injuries of bone tissue. The involvement of both factors has been clearly found in the pathogenesis of skeletal fluorosis, and there is also a more complex regulatory network with other molecules, which is worth further exploration.

The Wnt/β-catenin canonical signal pathway belongs to the Wnt signal pathway. When there is no Wnt protein signal outside the cell, free β-catenin in the cytoplasm is phosphorylated by proteasome degraded glycogen synthase kinase-3β (GSK-3β), allowing free β-catenin in the cytoplasm to be maintained at a low level. When there is Wnt protein signal outside the cell, Wnt protein bonds with Wnt receptor complex, composed by low density lipoprotein receptor-related protein (LRP) and human frizzled protein on the cell membrane, and transmits the signal into the cell, the activity of GSK-3β is reduced, β-catenin degradation is inhibited. Wnt protein is stably accumulated in the cytoplasm and acts with the transcription factor to initiate and regulate the expression of target genes.

Studies have confirmed that Wnt/β-catenin canonical signal can regulate the proliferation and differentiation of OB and is closely involved in the process of bone formation. Firstly, Wnt/β-catenin canonical signal pathway plays an important role in the proliferation and differentiation of BMSCs, and Wnt3a in Wnt signal pathway can activate CCNl/Cyr61 gene and promote the differentiation of BMSCs into OB. Moreover, Wnt3a regulates the Wnt/β-catenin signal pathway to promote BMSCs to different stages in the process of OB differentiation. However, it may produce inhibitory effects in the OB terminal stage of differentiation. The β-catenin promotes the differentiation of BMSCs into OB while inhibiting their differentiation into adipocytes or chondrocytes (Rodda and McMahon 2006). In knocked out β-catenin gene of mice, the pre-OB no longer continued to differentiate and mature but turned to chondrocyte differentiation. The OB proliferation rate was reduced, the mineral deposition rate was decreased, and osteoporosis occurred while the proliferation rate of OC was increased. Similarly, mice with LRP5 knocked-out in the Wnt/β-catenin signal pathway also developed marked osteoporosis, decreased mineral density of spine bone, and tibial fractures at 2 months of age. Alternatively, activation of the canonical Wnt/β-catenin signal pathway could also increase target gene Runx2 expression, which in turn promoted OB differentiation and maintained the process of bone formation (Chen et al. 2017). Under normal conditions, stable β-catenin also has a synergistic effect with BMP-2, which can significantly up-regulate the gene expression of BGP in OB and promote matrix mineralization and new bone formation. Activation of Wnt/β-catenin signal pathway promotes bone formation and is even associated with its ability to inhibit programmed death in OB. LRP5 of rat demonstrated a 30% increase in cortical bone thickness, a 4-fold increase in total bone area, and a 40% increase in mineralized surface compared with wild-type mice, accompanied by a significant decrease in programmed cell and OB mortality.

Fluoride can regulate the metabolism of bone tissue through Wnt/β-catenin signal pathway. Rats exposed to fluoride for a long time are often prone to sclerosing skeletal fluorosis, characterized by increased bone mass. Scholars have confirmed that the expression of Wnt3a and β-catenin mRNA, and protein in the bone tissue of fluorosis rats is higher than that of the control group. There is a rising trend with the increase of fluoride dose, which demonstrates that fluoride can activate the Wnt/β-catenin signal pathway, up-regulate the production of ALP and OPG, promote bone formation and reduce bone resorption. It has been pointed out that fluoride induces the phosphorylation of serine 473 of protein kinase B (PKB/Akt) and serine 9 of GSK-3β, resulting in the inactivation of GSK-3β. Thus, such process stabilizes the expression and causes the nucleus aggregation of β-catenin protein, achieving the regulation of downstream genes. In addition, the positive effects of fluoride on ALP activity and mRNA expression of ALP, Col-I, BGP, and Runx2 are abolished by DKK-1 (a Wnt/β-catenin receptor blocker). It was certain that the Wnt/β-catenin signal pathway had a core regulatory role during fluoride-induced OB differentiation (Pan et al. 2014).

Notably, fluoride regulation of Wnt/β-catenin signal pathway in bone tissue was closely related to its concentration, and different doses of fluoride were found to bidirectionally regulate Wnt/β-catenin signal pathway in MC3T3-E1 cell experiments. Compared with the control group, β-catenin and LRP5 protein expressions were decreased in MC3T3-E1 cells in the high-dose experimental group, while increased in the low-dose experimental group. Such results indicated that fluoride exceeded certain cut-off value could downregulate β-catenin and LRP5 protein expression levels, which in turn inhibited the Wnt/β-catenin pathway signal transduction pathway and affected the differentiation of the cells (He et al. 2019). Given the modulatory effect of fluoride on Wnt/β-catenin signaling activity and the central role of Wnt/β-catenin signaling in bone tissue formation, this study found that Wnt/β-catenin signaling may be an important molecular mechanism in fluoride effects on bone formation.

Existing studies on the effects of fluoride on OC showed different conclusions. Some suggested that fluoride could promote OC formation and enhance bone resorption capacity, while others suggested that fluoride could reduce the number of OC and bone resorption. OC proliferation, differentiation, and maturation were known to be dependent on OB, and OB secreted OPG, RANKL, and M-CSF could significantly regulate OC status. The exact mechanism of OC differentiation was not well understood, but it was confirmed to be closely related to OPG which was secreted by OB. OPG was required for OC development, activation, and survival. Normally, OB expressed and released RANKL, bound to RANK on the surface of OC precursor cells, TNF recruited-receptor associated factor (TRAF) to bind with cytoplasmic region of RANK. It also initiated and transmitted differentiation signals of OC through the JNK pathway, nuclear factor-kappa B (NF-κB) pathway, and Akt pathway to promote OC proliferation and differentiation. Some scholars found that fluoride could promote the release of OPG from OB. OPG could competitively bond with RANKL. Its binding ability was better than RANKL, which could effectively block RANK/RANKL signal pathway, inhibit OC differentiation, and weaken OC bone resorption. RANKL/OPG ratio was related to the intensity of OC differentiation, and RANKL/OPG maintained as a certain ratio, which was important for maintaining OC differentiation and bone metabolism balance.

Fluoride can cause dysregulation of RANKL/OPG ratio. A recent study found that fluoride could significantly increase the proportion of RANKL/OPG in goldfish bone tissue, promote downstream target gene transcription, and increase tissue bone resorption (Sato et al. 2016). The RANKL mRNA/OPG mRNA ratio in the distal femoral joint tissue of rats was significantly increased after 6 months of fluoride exposure, and the increase of the ratio increased the likelihood of RANKL binding to RANK and OC differentiation, which was consistent with the findings in bone metabolism of fluoride-exposed rats. This could result in osteolysis and greater bone resorption activity, indicating fluoride OC-induced bone destruction. However, Liu et al. established an OB-OC co-culture model in vitro and pointed out that OB secreted OPG could be up-regulated by fluoride. Thus, OPG competitive inhibition hindered the binding of RANKL and RANK, excessively inhibited the differentiation and maturation of OC, and weakened bone resorption, which could also lead to fluoride bone damage (Liu et al. 2015). OC formation in bone marrow cells of C57BL/6 mice treated with M-CSF and RANKL verified that OC was sensitive to 0.5–8 ppm fluoride treatment. 0.5 ppm fluoride significantly reduced bone resorption activity of OC. Fluoride concentration was negatively correlated with OC bone resorption activity, but its effect on OC formation was not found (Pei et al. 2016). Irrespective of the final outcome, upregulation or downregulation of the RANK/RANKL/OPG axis by fluoride resulted in bone tissue damage. Increase or decrease in bone resorption activity ultimately leads to skeletal fluorosis manifestations such as osteoporosis or osteosclerosis.

RANKL acts synergistically with M-CSF, which has been found to be a key factor in OC formation. The lack of M-CSF synthesis and secretion of congenital osteopetrosis mice will result in changes of OC activity-weakening of bone resorption function manifests as osteopetrosis. In contrast to osteopetrosis, osteolysis and osteoporosis are caused by excessive OC formation in the body, resulting in stronger bone resorption than bone formation. Excessive fluoride can increase RANKL and M-CSF expressions to promote OC formation and OC resorptive activity, which then cause bone damage. Certain special drugs can also regulate bone remodeling by down-regulating RANKL and M-CSF expression and treat skeletal fluorosis.

NF-κB was discovered in mature B lymphocytes in 1986, and all NF-κB/Rel proteins have a homology domain composed of approximately 300 amino acids in the amino terminus. Normally, different inhibitors of NF-κB (IκB) bind to inhibit NF-κB/Rel dimers. When cells are stimulated by external factors, IκBα is autophosphorylated and rapidly degraded, NF-κB and IκBα dissociate, and NF-κB even enters the nucleus as a dimer to bind to specific gene promoters to initiate cytokine transcription.

It has been pointed out that RANKL can activate classical and non-classical NF-κB signal pathways in stem cells through TRAF6 and NF-κB-inducing kinases, which in turn regulate OC number as well as differentiate and affect normal bone metabolism. Since the role of NF-κB in bone was first recognized in the mid-1990s, the role of NF-κB signal pathway in bone development and related diseases has widely supplemented. It demonstrated that NF-κB could regulate the proliferation, differentiation, and apoptosis of OB, OC, osteocytes, and chondrocytes, as well as have an important role in bone development and common bone diseases. The results showed that the expression of Fas and FasL increased with the increase of fluoride concentration, while the expression of nuclear factor NF-κB decreased with the increase of fluoride concentration in the process of fluoride-induced OC apoptosis. Both showed a dose-dependent relationship. It was found that the expression of NF-κB was increased by the influence of low dose fluoride, which could reduce the apoptosis of OC, promote the generation of OB and accelerate the process of bone metabolic turnover. However, small doses were mainly used to stimulate OB production, so hyperosteogeny was likely to occur. In contrast, in high concentration of fluoride, NF-κB expression was inhibited, OC apoptosis increased, and then osteosclerosis aggravated. Therefore, bone injury in chronic fluorosis might result from abnormal OC apoptosis, in which changes in NF-κB could play a major regulatory role (Piva et al. 2006).

The formation of bones and development of the skeletal system require the coordinated activities of multiple signaling proteins. To investigate the molecular mechanism of bone tissue damages caused by long-term excessive intake of fluoride, a variety of advanced molecular techniques were employed. Wei et al. submitted bone samples as quantitative isobaric tags for relative and absolute quantification proteomics approach. They aimed to study the effects of excess fluoride on rat bones after determining the severity of skeletal fluorosis with hematoxylin and eosin staining. Increased bone mass was found in the medium and high dose groups, and decreased bone mass was found in the low dose group. 13, 35, and 34 differentially expressed proteins were identified in the low-, medium-, and high-dose groups, respectively. The protein expression pattern was more similar in the medium-dose and high-dose groups. These proteins were mainly related to collagen metabolism, proteoglycans, and MMPs. The effect of fluoride on SD rats was dose dependent (Wei et al. 2018).

A previous study used mRNA differential display polymerase chain reaction to compare the differential expression of genes between nomal OB and OB of fluoride-exposed rats. Among six cloned differentially expressed gene fragments, the rat's ribosomal protein L5 gene, Na^+-, K^+-ATPase β3 gene, nuclear localization signal protein receptor α2 gene, and Golgi P28 gene were expressed with lower levels in fluoride-exposed OB, and rat's ubiquitin-conjugating enzyme gene as well as a new gene fragment were up-regulated in fluoride-exposed OB. These genes were mostly related to protein synthesis, transport, and processing. In addition, the expression of mRNA in the peripheral blood of patients with fluorosis was detected by gene chip technology. The results showed that compared with the control group, 32 genes were up-regulated and 32 genes were down-regulated in the mild group, 53 genes were up-regulated and 20 genes were down-regulated in the severe group; compared with the mild group, only 4 genes were up-regulated and 11 genes were down-regulated in the severe group; compared with the control group, 2 genes were significantly up-regulated in the mild fluorosis group and the severe group, which were RAS gene and transmembrane protein 9 family 1 gene, and 14 genes were significantly down-regulated in the mild and the severe groups, such as transcription factor p38 protein, lactoferrin, keratin 4, adenosine receptor A2b, ATP lyase, and 3 genes related to protein synthesis, namely, mitochondrial translation initiation factor 2, single-stranded DNA-binding protein, and cytotoxic granule-associated protein.

In addition to the changes in the mRNA of fluorine-ingesting animals, recent studies found that micro-ribonucleic acids (miRNAs) played a crucial role in regulating mRNA transcription and protein expression in the cell growth and differentiation. In the study of fluoride, it was clear that some miRNAs were closely related to bone turnover by regulating the differentiation, growth and biological activity of bone tissue cells, and were involved in the development of skeletal fluorosis. Vitro studies of miRNA arrays showed that human osteosarcoma (HOS) cells treated with 8 ppm and 20 ppm NaF showed 82 and 62 differentially expressed miRNAs, respectively, compared to the control group. The miRNA profile of MC3T3-E1 cells treated with 8 ppm NaF was also found to be deregulated, with 45 miRNAs upregulated and 31 miRNAs downregulated. Most importantly, 31 miRNAs were differentially expressed between skeletal fluorosis cases and controls, and 85 miRNAs were differentially expressed between skeletal fluorosis cases and controls. In addition, there were three miRNAs (miR-200C-3p, miR-1231, and miR-3185) significantly upregulated in patients with skeletal fluorosis. These differential miRNAs played an important regulatory role in the development and differentiation of OB and OC, thus making bone turnover imbalanced (Pei et al. 2019).

Disorder of Calcium

Ca^{2+} is an important component constituting bones and teeth, and 99% of Ca^{2+} in human body is presented in bones. The homeostasis of bone tissue cells refers to the continuous flow of Ca^{2+} among extracellular fluid, intracellular solutes and intracellular calcium stores, which not only maintains their relatively stable Ca^{2+} concentration, but also maintains a quite different ion gradient difference. The basis of Ca^{2+} signal generation and transduction, and the completion of physiological functions of cells lies in this process. Ca^{2+} homeostasis of osteocyte plays an important role in various aspects of physiological functions of bone tissue. Fluoride can interfere with the normal metabolism of Ca^{2+}, resulting in extracellular Ca^{2+} concentration below the normal range and increased intracellular Ca^{2+}. These two aspects are involved in the etiological network and pathogenesis of endemic fluorosis. At the same time, the body's tolerance to fluoride will be lower under the condition of low calcium and low nutrition, which is the main precipitating and aggravating factor of skeletal fluorosis. Ca^{2+} supplementation was proven to alleviate fluoride damage to bone tissue cells.

In the occurrence and development of skeletal fluorosis, there are three basic links: systemic calcium deficiency; secondary hyperparathyroidism, increased parathyroid hormone (PTH) secretion; and increased cellular Ca^{2+} influx.

Low Ca^{2+} in External Environment

Normally, the extracellular Ca^{2+} concentration is 0.1–10 mmol/l and the cytosolic Ca^{2+} concentration is 0.1–1 μmol/l, with a 10,000–100,000-fold difference between the two. Excessive fluoride enters the body mainly through the following ways to cause low calcium in the extracellular fluid: fluoride can immediately bind to Ca^{2+} resulting in temporary decrease in serum Ca^{2+}; excessive fluoride activates osteoblastic cells, increases bone formation, and increases the need for Ca^{2+} at the front of the osteoblastic front; and fluoride is incorporated into bone salt crystals, increasing crystallinity, decreasing solubility, causing slow release of bone calcium, and reducing the Ca^{2+} concentration in body fluids. The imbalance of intracellular and extracellular Ca^{2+} concentration is an important mechanism for the occurrence and development of skeletal fluorosis. Even if the calcium intake from the diet is not reduced, excessive fluoride can increase the body's demand for calcium, which will also cause the body's relative calcium deficiency.

In addition, it is worth noting that Ca^{2+} nutrient level also has a decisive effect on the pathogenesis of skeletal fluorosis. From the epidemiology aspect of endemic fluorosis, severe and disabling skeletal fluorosis is always closely related to malnutrition, especially dietary calcium deficiency. Skeletal fluorosis cases with low Ca^{2+} intake often show osteopenia changes to osteoporosis and osteomalacia. Osteomalacic skeletal fluorosis (endemic genu valgum) is more common for children in some parts of India have significantly low calcium intake in their diet in addition to high fluoride intake. In areas of different degrees of fluorosis, the lower the Ca^{2+} concentration in drinking water, the higher the prevalence of skeletal fluorosis. The two were negatively correlated. If excessive fluoride intake is accompanied by low dietary calcium, the calcium deficiency will be significantly exacerbated, and the symptoms of skeletal fluorosis are more obvious and severe.

Low calcium in the external environment of the body can induce and aggravate skeletal fluorosis, which can increase the intestinal absorption of fluoride and bone fluoride deposition. More importantly, the role of the two is homotropic or consistent. Homotropy is mainly reflected in two aspects: both of them can cause increased PTH secretion, and both of them can activate osteogenesis, bone breaking function, and accelerate bone turnover. The specific manifestations are as follows: increased serum PTH; increased ALP activity reflecting osteogenic activity in serum; active OB on each bone surface, increased number of osteogenic and osteoclastic precursor cells in bone marrow; periosteal fibrosis indicating accelerated bone turnover; increased expression of MMP-9, a key enzyme reflecting bone resorption and bone remodeling in bone tissue; decreased Ca^{2+}-ATPase activity in the membrane system. The combined effects resulting from the homotropic nature of low calcium with high fluoride may not simply be superimposed but multiplied.

Increased Intracellular Ca^{2+}

Intracellular Ca^{2+} plays an important role in triggering mitosis, DNA synthesis and controlling cell cycle. The increased Ca^{2+} level of OB enhances the expression and activity of various growth factors. Ca^{2+} signal system is also utilized by many growth factors that stimulate cell proliferation. The metabolic effects of fluoride are associated with altered ion currents and membrane potentials. Whole-cell calcium channel currents in osteoblast-like cells were measured by patch clamp, and Ca^{2+} channel opening (by exposure to 25 ng/ml fluoride) was observed. With the increase of intracellular Ca^{2+} content, osteogenic activity increased in a dose-dependent manner. The increase of Ca^{2+} in OB might be an early mechanism by which fluoride activated OB function and promoted the development of skeletal fluorosis. Studies showed that fluoride could increase the amplitude and opening of selective K^+ ion channels, but the function could also be blocked by Ca^{2+} channel blockers. Such process indicated that the cascade of second messenger Ca^{2+} could also affect the activity of selective K^+ ion channels, which might also mediate the early effects of fluoride on bone tissue.

Increased intracellular Ca^{2+} retention in tissue cells during chronic fluorosis was caused by, at least in part, inhibition of intracellular Ca^{2+} excretion mechanisms. In vitro studies revealed that the intracellular Ca^{2+} concentration in MC3T3-E1 cell lines was significantly increased in each fluoride exposure group. Fluoride might lead to a decrease in the expression level of plasma membrane calcium ATPase 4B in the Ca^{2+} pump cell membrane by inhibiting calmodulin expression, which affected the transport of Ca^{2+} to the extracellular space and ultimately led to an increase in the Ca^{2+} concentration in the OB. After 6 weeks of chronic fluoride intake in drinking water, following changes were detected: the transport ability of renal cell membrane system to Ca^{2+} was impaired, the activities of Ca^{2+} pump in plasma membrane and Ca^{2+} pump in rough endoplasmic reticulum (ER) were decreased, and the contents of both pump proteins were decreased. The plasma membrane Ca^{2+} pump (Ca^{2+}-ATPase) degrades ATP to transport Ca^{2+} across the membrane to the extracellular space, while the role of the ER Ca^{2+} pump was to pump Ca^{2+} from the cytoplasm into the ER by hydrolyzing ATP against a concentration gradient. Impaired function of these two pumps results in increased intracellular Ca^{2+} (Xu et al. 2007). Fluorosis mediated the increase of intracellular Ca^{2+}, exceeded cellular tolerance, and finally led to cell apoptosis.

Another concern is that skeletal fluorosis is often accompanied by increased PTH secretion, while PHT is mainly responsible for the regulation of body bone mass and calcium homeostasis. Excessive PHT can mobilize Ca^{2+} from bone and transport increased Ca^{2+} from body fluids into soft tissue cells, which increases intracellular Ca^{2+} influx and causes Ca^{2+} overload. Imbalance in PHT skeletal fluorosis is another important cause of increased intracellular Ca^{2+} concentration, which plays a decisive role in driving bone lesions towards osteomalacia and osteoporosis. That meant that fluctuations in PTH, by regulating intracellular Ca^{2+} concentrations, were involved in the "calcium paradox" pathological process, meanwhile promoted the activation of the OB and the acceleration of bone turnover in skeletal fluorosis (Xu et al. 2010). PTH increased intracellular Ca^{2+} concentration in osteoblast-like cells and was considered to be the second signal PTH activated transduction pathway in OB, in addition to the cyclic adenosine monophosphate (cAMP) pathway.

In conclusion, regardless of the pathway fluoride increases the intracellular Ca^{2+} concentration, the mechanism depends on four aspects: increased influx of extracellular Ca^{2+}, blocked transport of intracellular Ca^{2+} to the outside of the plasma membrane, increased Ca^{2+} release from intracellular Ca^{2+} stores, and blocked intake of Ca^{2+} from the cytosol.

Oxidative Stress and Endoplasmic Reticulum Stress Correlated with Skeletal Fluorosis

Free radicals refer to atoms or groups with unpaired electrons formed by covalent bond cleavage of compound molecules under external conditions, and the free radicals mainly exist in biological systems are oxygen free radicals. Oxygen free radicals include superoxide anion radical ($O_2^{-\bullet}$), highly reactive hydroxyl radical (•OH), lipid oxygen radical and nitric oxide (NO), hydrogen peroxide (H_2O_2), singlet oxygen and ozone, commonly known as reactive oxygen species (ROS, brings potential damage to body surface). Oxidative stress (OS) refers to the excessive production of ROS when the body suffers various harmful stimuli or the weakened level of antioxidant system, which imbalances the two and causes a cascade reaction in the body.

Under normal physiological conditions, the body can produce ROS with potential cytotoxicity and instability, which play an indispensable role in host defense, gene transcription, and regulation of synaptic plasticity and apoptosis. Due to the coordination of antioxidant system in the body, a small amount of ROS maintains normal homeostasis without causing damage to the body, which maintains balance between oxidants and antioxidants. The antioxidant system in the body is mainly composed of antioxidant enzymes, including catalase, superoxide dismutase (SOD), glutathione peroxidase (GSH-Px), and glutathione S-transferase, as well as non-enzymatic antioxidant factors, such as melatonin, carotenoids, and some trace elements. These factors cooperate with each other to jointly protect against damage caused by ROS. To neutralize ROS in vivo, these antioxidant enzymes comprise a well-established barrier to regulate the balance of ROS in vivo. However, once ROS production excessively inhibits cellular antioxidant activity, OS damage leads to the accumulation of cytotoxic compounds and ultimately causes protein disintegration. Such changes disable the function of various enzyme actions, and cause the destruction of overall cell structure.

How does ROS cause body damage? It can react with a variety of biomolecules (tissue cell biofilm, protein, nucleic acid, etc.) and destroy the normal tissue structure and body function. First, biofilms are the basis of cellular life activities and are mainly composed of lipids, proteins, and sugars. Lipids are rich in polyunsaturated fatty acids. Since unsaturated fatty acids contain multiple weak bonds, they are vulnerable to ROS attack and undergo lipid peroxidation, which can then result in membrane degeneration. Therefore, the structure and function of the membrane are disrupted due to the changed membrane permeability and fluidity. Second, ROS will mutate certain amino acid residues, resulting in the subsequent breakage, polymerization or cross-linking of protein polypeptide chains. Such mutating process also changes protein structure and function. In addition, polynucleotides are the target molecules attacked by ROS. Purines and pyrimidines on the outside of the DNA double helix are most sensitive to ROS, which will lead to irreversible DNA damage: base modification or loss, double-stranded or single-stranded DNA breakage, cross-linking of DNA after being damaged. ROS interfere with many cellular signal pathways, such as cysteinyl aspartate specific proteinase (caspase) pathway of apoptosis, JNK/ Silent information regulator factor 2-related enzyme 1 (sirtuin 1, Sirt1)/autophagy pathway, etc., to affect many cellular functions. ROS is highly reactive with very short half-life, and it is very difficult to directly measure the levels of these ROS. Therefore, the research commonly uses the oxidation products after ROS reacting with other molecules as markers to reflect the levels of ROS. For example, ROS attacks polyunsaturated fatty acids in biofilms and triggers lipid peroxides (LPO), including malondialdehyde (MDA), which can reflect the degree of lipid peroxidation in the body and indirectly reflect the degree of cell damage. In addition, the detection of antioxidant systems in the body can also indirectly detect whether the body is in an OS state. For example, antioxidants SOD, glutathione (GSH), GSH-Px, and glutathione reductase (GR) were used as detection indicators of whether fluoride induced excessive ROS production.

To date, the OS theory is one of the relatively complete theories that explain the pathogenesis of chronic fluorosis. In our early study, skeletal changes with the elevated level of lipid peroxidation were observed in the rats with chronic fluorosis (Guan et al. 1989). Therefore, we first proposed that increased levels of free radicals may be the central link of the different types of damage associated with chronic fluorosis. After that, we found the elevated levels of lipid peroxide and reduced activities of antioxidant enzymes both in the serum and a variety of organs of animals with chronic fluorosis and in the serum of patients suffering from coal-burning type of endemic fluorosis (Guan and Yang 1991; Guan et al. 1991, 1992).

Fluoride is an extremely negative element with active chemical properties. However, it does not form ROS with oxygen in the body itself. On the contrary, it interferes with the normal oxygen metabolic process, induces the respiratory burst of cells, and strengthens a series of free radical-related reactions after a large amount of fluoride intake in the body, resulting in increased ROS, consumption of a large number of antioxidant enzymes and non-enzymatic antioxidant substances to lead cells into OS state. This may be a central link in the pathogenesis of systemic damage in fluorosis. Regarding fluoride-induced ROS generation, the earliest reports back in 1975 when Curnutte and Babior studied anaerobic conditions and the effects of some inhibitors on the production of O^{2-} by human polymorphonuclear leukocytes. They accidentally found that cultured in 20 mM NaF solution could produce O^{2-} at a rate of 20-fold higher than that of the control group (Curnutte and Babior 1975). Over the years, fluoride research experts in China scrutinized endemic fluorosis areas and collected blood and urine samples from patients. They found that patients with fluorosis had increased serum levels of lipid peroxidation, including oxidation products with reactive lipid peroxidation levels (e.g., 8-iso-prostaglandin F2α (8-iso-PGF2α) and MDA), and products of reactive protein damage such as advanced oxidized protein products. Besides, 8-hydroxy-2′-deoxyguanosine (8-OHdG) was increased, which would react to DNA attack (Varol et al. 2013). In addition, the activities of antioxidant enzymes (SOD, GSH-Px, and total antioxidant capacity) decreased indicating the population in fluorosis area was generally in OS state, and there were various damages of lipid biofilm, protein, and DNA. At the same time, it was found in experimental fluorosis animals and cells that the OS injury of fluoride to the body involved a variety of organs and cells. For example, the liver, serum, and red blood cells of fluoride-treated rats showed decreased GSH-Px activity, decreased GSH content, and increased LPO content. In the study of renal tissues of fluorosis rats, it was found that in addition to the changes of traditional OS indicators such as increased MDA, there were some newly discovered antioxidant proteins such as transferrin expression levels. The OS levels of brain tissues of fluorosis rats and cultured nerve cells treated with excessive fluoride were increased, the $OH^•$, H_2O_2, $O_2^{-•}$, and MDA levels were significantly increased, while the SOD and GSH-Px activities were decreased (Guan et al. 1998; Gao et al. 2008). In addition, the phospholipids and coenzyme Q of cell membranes were decreased, and the lipid composition of biofilms was changed (Guan et al. 2000). Combined with various studies, scholars verified that fluoride would lead to the imbalance of oxidative and antioxidant systems in the body, cause OS, and finally lead to changes in the biological structure of cells and dysfunction. OS is one of the major mechanisms, by which fluoride leads to various chronic pathological changes in the body.

The most obvious pathological changes in patients with fluorosis were changes in bone tissue and teeth. In a study of the interrelationship between OS and skeletal fluorosis, it was found that bone and teeth were more significantly sensitive to OS damage in response to fluoride. At the same concentration of fluoride, bone and teeth were the earliest to develop OS in all organs of the body, and the dose of fluoride required to develop OS was also low. OS injury was considered to be one of the major causes of skeletal fluorosis and dental fluorosis. Over the years, it was found in the bone tissue of fluoride-exposed rats that ROS was increased and the activities of antioxidant enzymes such as SOD was decreased, and biological structures and functional molecules were damaged. Treatment of primary mouse OB with fluoride could cause them to generate excessive ROS, experience OS damage, and even trigger apoptosis, which were consistent with studies in vitro (Wang et al. 2011). Alternatively, mitochondria were considered to be a major source of intracellular ROS. First, fluoride ions could be protonated extracellularly to form hydrofluoric acid, which was able to freely diffuse through the membrane. After entering the mitochondria, fluoride ions mainly inhibited the activity of mitochondrial complex IV and partially inhibited the activity of complexes I-II and complexes I-III. Since mitochondrial complexes I and III were the main sites of O^{2-} synthesis, fluoride ions could enhance their production by inhibiting mitochondrial activity at the level of the respiratory chain. In addition, fluoride ion inhibition at the level of mitochondrial complex IV could promote the formation of ubiquinone radicals, which in turn could react with molecular oxygen and increase the production of O^{2-}. Ultimately, fluoride ions led to increased O^{2-} production by inhibiting the mitochondrial electron transport chain and subsequent formation of H_2O_2, which could be converted to reactive HO^- to cause oxidative damage to cells (Fina et al. 2014).

OS damage was known to manifest in ROS damage to biofilms, proteins, and DNA structure and function, while ROS also interfered with organelle functions including the ER. Fluoride could cause endoplasmic reticulum stress (ERS) in OB, which mediated the early abnormal activation of OB during the pathogenesis of skeletal fluorosis. ERS refers to the folding and assembly of newly synthesized secretory and transmembrane proteins in the ER in eukaryotic cells, and ER function disturbance leads to the accumulation of ER-folded and misfolded proteins. OS leads to ERS, while excessive ROS is also produced during ERS. The downstream signal pathways of the two partially overlapped, and OS and ERS may jointly mediated cellular stress damage in bone tissue. For example, both could induce nuclear factor erythroid 2-related factor 2 (NRF2) nuclear translocation, and activate its downstream antioxidant target genes, as well as to induce SIRT1/autophagy pathway up-regulation to counteract the toxic effect of fluoride on cells through ROS, but fluoride would eventually lead to body "decompensation" (Zhou et al. 2019; Suzuki et al. 2015). At present, it is confirmed the significant correlation between OS and ERS in the pathogenesis of skeletal fluorosis, and the specific mechanism requires further study.

NO is the only nitrogen-containing ROS in the body, involving in the occurrence and development of a variety of diseases including skeletal fluorosis. It is also the smallest, lightest, and simplest gaseous messenger molecule in the body, involving in a variety of physiological processes: the transmission of signals between nerve cells, the maintenance of constant blood pressure, and the body's defense response. Nitric oxide synthase (NOS) is a specific enzyme that catalyzes the production of NO using L-arginine as a substrate. NO generated by inducible nitric oxide synthase (iNOS) is the most important source of NO in vivo under pathological conditions. NO can regulate both OB and OC. With the absence of NO, the ALP activity of OB is significantly decreased. After adding appropriate amount of exogenous NO, the ALP activity is significantly increased, but NO excess can trigger OB oxidative damage. Similarly, low concentrations of NO promote cytokine-induced bone resorption and may be necessary for OC to exercise its function, while high concentrations of NO inhibit bone resorption by inhibiting OC formation and the resorption of mature OC. INOS can provide a large amount of NO to participate in the regulation of OB and OC during bone turnover during fluorosis, and play its role as a messenger molecule in bone turnover and can participate in the mechanism of bone injury. Notably, NO excess can also activate the pro-apoptotic protein Bax, which can cause mitochondria to release cytochrome C. Cytochrome C activates initiator caspase-9 by binding to apoptotic protein activator-1, and then the caspase-9 reactivates executive caspase-3. After caspase-3 cleaving the proteasome protein, this protein (an endonuclease) enters the nucleus and begins to degrade the DNA chain, ultimately leading to apoptosis. In addition, NO excess caused by fluoride could also inhibit glucose 6 phosphate dehydrogenase activity and pentose phosphate oxidation pathway, resulting in impaired cellular energy metabolism, which together led to the initiation of apoptosis (Bergandi et al. 2011).

Hormonal Regulatory Mechanism in the Procedure of Skeletal Fluorosis

Endemic fluorosis is a systemic disease. In addition to the direct effects of fluoride on cells such as OB and OC, fluoride can also indirectly regulate various cells through hormones secreted by the endocrine system (PTH, glucocorticoid (GC), calcitonin (CT), etc.) together with other active substances including cytokines. This is also one of the important mechanisms in the development of skeletal fluorosis.

PTH is an important hormone, secreted by the parathyroid gland. PTH can regulate calcium and phosphorus metabolism and has a bidirectional regulatory effect on bone formation and resorption depending on the time and concentration of its action. Fluoride in the blood circulation can cause the increase of intracellular Ca^{2+} level in parathyroid cells through the corresponding signal pathway and promote PTH secretion. At the same time, fluoride can also cause the decrease of cAMP level or direct damage to parathyroid glands, thereby inhibiting the secretion of PTH. In general, excessive fluoride enters the body will result in low extracellular Ca^{2+}; while the parathyroid cell membrane contains Ca^{2+}-sensitive receptors, so the increased secretion of PTH is mainly regulated by the blood Ca^{2+} concentration. Transient decrease of serum Ca^{2+} concentration is considered to be the most important cause of increased PTH secretion. Secondary increased PTH secretion can further affect the metabolism of Ca^{2+}, so the fluctuation of blood Ca^{2+} level is involved in the occurrence of skeletal fluorosis together with the change of PTH level. Chronic fluorosis is often accompanied by secondary hyperparathyroidism, in which parathyroid cell proliferation is obvious, serum PTH level is significantly higher than normal level, OB and OC are in an activated state. Hence, bone turnover is enhanced. Alternatively, during the pathogenesis of skeletal fluorosis, PTH regulates BMSCs differentiation into OB and increases their activity, promotes the conversion of bone lining cells into OB, and inhibits mature OB apoptosis. Therefore, bone formation is promoted.

CT is a single-chain polypeptide hormone synthesized and secreted by thyroid parafollicular cells. Its main biological effect is to reduce blood calcium, and the main target organ is bone tissue. CT can act directly on the CT receptor (CTR), rapidly inhibit OC activity, and inhibit OC precursor cell proliferation, differentiation, and aggregation. CT can reverse the OC formation process, so that the decreased number of OC leads to the decreased release of bone Ca^{2+}. PTH and prostaglandin E2 have the effect of increasing OC acidity, but this effect can be inhibited by CT. In the development of skeletal fluorosis, it can inhibit CT secretion and decrease CT level regardless the decrease of serum Ca^{2+} caused by fluoride or Ca^{2+} malnutrition, or the secondary increase of PTH secretion. There was also a reduction in CT level in patients with endemic genu valgum caused by fluorosis in Indian region. However, there were also cases in which no obvious changes in CT level were observed in bone turnover studies in subchronic fluorosis rats.

1,25-(OH) $_2D_3$ is an active metabolite formed after vitamin D is metabolized by the liver and kidney. Its main biological role is to regulate calcium and phosphorus metabolism and maintain calcium homeostasis in the body. Its synthesis and secretion are regulated by PTH, Ca^{2+}, and phosphate levels. 1,25-(OH) $_2D_3$ also has a bidirectional effect on bone turnover and can regulate OB and OC activities as well as the synthesis of bone matrix proteins. Long-term treatment of low doses of 1,25-(OH) $_2D_3$ can activate OB and its precursor cells to promote bone formation and inhibit bone resorption. Large doses of 1,25-$(OH)_2D_3$ can promote the activation and recruitment of OC and its precursor cells by indirect action in the short term, while inhibiting OC apoptosis and promoting bone resorption, but it turns into promoting bone formation after a few days. 1,25-$(OH)_2D_3$ enhanced fluoride-stimulated bone formation, and the combination of 1,25-$(OH)_2D_3$ with fluoride has been found to be useful in the treatment of osteoporosis-related diseases. Studies in a rat model of osteomalacia skeletal fluorosis showed under low Ca^{2+} feeding conditions that decreased blood Ca^{2+} levels caused increased serum 1,25-$(OH)_2D_3$ levels. The increased level of 1,25-$(OH)_2D_3$ induced by severe low Ca^{2+} has a synergistic effect with PTH, promoting the formation of bone turnover and aggravating the pathological damage caused by fluoride.

GC is a sterol hormone synthesized and secreted by adrenocortical cells. It is involved in a variety of metabolic processes in the body, including bone metabolism. GC acts throughout normal bone turnover and abnormal bone turnover processes in disease states. The study found that serum corticosterone levels were significantly lower in fluorosis rats compared with the control group. The decrease of GC level weakened the bone resorption mediated by GC, and also eliminated the inhibitory effect on bone formation. Thus, it helped bone metabolism to progress towards osteogenic activity. In addition, relatively low dose of endogenous GC is necessary to promote the differentiation of BMSCs into OB, so the reduced GC level contributes to the differentiation of OB precursor cells into OB, which leads to the enhanced osteogenic activity.

Sex hormones in the body are mainly estrogens and androgens, which play an important role in the development and relative stability of bone. Both hormones have receptors distributed in bone tissue cells, including BMSCs, OB, osteocytes, and OC. Both of them regulate bone metabolic activity by the corresponding receptors, and the bone metabolic activity of androgens is partially accomplished by estrogens. Fluoride was reported to reduce serum androgen level, and serum testosterone level decreased in fluorosis regardless of the occurrence of skeletal fluorosis. The decrease in serum testosterone level was more pronounced in patients with skeletal fluorosis changes. Skeletal fluorosis itself is in a state of high bone turnover, and decreased androgen level can cause further enhancement of bone resorption, aggravate the pathological damage of skeletal fluorosis, and develop the lesions towards osteoporosis and osteomalacia. In addition, clinical and animal experiments found that estrogen could reduce bone turnover and maintain bone mass. Fluoride decreases estrogen secretion at the gonadal level and then affects bone metabolism through changes in estrogen, which may also be one of the damaging factors of skeletal fluorosis.

Insulin is an anabolic hormone that plays an important role in bone growth and metabolism. Its receptor is expressed in both OB and OC, and the receptor has tyrosine kinase activity. Normally, insulin can promote the proliferation and differentiation of OB, enhance the function of OB, promote bone formation and maintain normal bone mineralization through its receptor mediation. At the same time, it can reasonably inhibit the bone resorption of OC. Low concentration of fluoride has the same direction as insulin on bone metabolism. Insulin can not only stimulate OB activity itself, but also enhance the stimulating effect of fluoride on OB activity. Insulin combined with fluoride can significantly stimulate OB proliferation, but some scholars found that the two have no significant effect on inducing early differentiation of OB. In addition, treatment with high doses of fluoride had the potential to inhibit insulin secretion as well as the expression of insulin receptors in bone tissue. The mechanism of fluoride promoting osteogenesis activity is multifactorial linked, and insulin is one of the main factors affecting its action.

Thyroid hormones (THs) are systemic sex hormones that maintain calcium balance and regulate bone turnover. TH has effects on both bone formation and resorption. Although OC also has TH receptor expression, the bone metabolic activity of TH is mainly mediated by the OB. TH itself can stimulate OB proliferation and differentiation and activate OB, and then affect OC function through OB secretion of local factors (interleukin 6 and interleukin 8). Fluoride is an effective stimulator of thyroid follicular epithelial cells and affects TH secretion mainly through phosphoinositide and cAMP signal pathways. However, it was found that low fluoride (0.01–1 mm) stimulated cAMP activity in thyroid cells, while high fluoride (1–100 mm) inhibited cAMP activity, and high fluoride also generated direct toxic effects on thyroid cells. Some scholars found that high fluoride could promote thyroid function and TH secretion in the early stage, but the effect of long-term high fluoride on thyroid function might make the thyroid gland lose its responsiveness to fluoride or have decreased TH secretion.

Disruption of hormone balance by fluoride led to dysregulation of bone metabolism and was widely involved in the pathology of skeletal fluorosis (Perumal et al. 2013). This hormonal dysregulation was fluoride treatment concentration- and time-dependent. It deserved further research since fluoride had different effects on bone metabolism under different conditions.

Genetic Susceptibility and Epigenetics

In recent years, genetic epidemiological studies showed evidence of an association between genetic polymorphisms in candidate genes and differences in patterns of patients with different types of skeletal fluorosis living in the same community and exposed to the same environment. In specific environmental exposures, including dietary patterns of fluoride intake and the same nutrients condition, an individual's genetic background may play an important role in influencing the risk of fluorosis, and genetic factors may contribute to inter-individual differences in susceptibility or resistance to fluoride exposure.

Animal studies found that different strains of mice might experience varying degrees of dental fluorosis when they had the same fluoride content in their mineralized tissues. The effects of increasing fluoride doses on susceptibility to enamel fluorosis in three inbred mouse strains (susceptibility, resistance, and intermediate) were evaluated. It was found that although there was no significant change in bone mineral density in the three strains, there was a significant change in bone mass in the susceptive strains, a moderate change in the middle strains, and no change in the resistant strains. These results indicated that genetic factors might contribute to the differences in bone quality alterations in different strains of inbred mice with different susceptibility to fluorosis under the same fluoride exposure.

Some studies investigated the relationship between polymorphisms in candidate's genes and fluorosis risk in populations living in endemic areas of countries (China and India). Researchers conducted a lot of research about the genes that were involved in bone formation and development such as *BGP*, estrogen receptor 1 (*ESR1*), and collagen type I α 2 (*COL1A2*), genes that encode receptors involved in bone formation and metabolism (vitamin D receptor (*VDR*), *CTR*, genes that encode hormones (*PTH*, prolactin (*PRL*)), genes that encode detoxification enzymes (glutathione S-transferase pi 1 (*GSTP1*)), genes that encode extracellular matrix proteins (e.g., *MMP*), genes that are involved in cognitive functions such as catechol-O-methyltransferase (*COMT*), and genes that are involved in immune functions (e.g., MPO). Some researchers found that *COL1A2*, *CTR*, and *ESR* gene polymorphisms were positively associated with the risk of dental fluorosis, while ESR1 and COMT gene polymorphisms were found to be significantly associated with the risk of non-skeletal fluorosis in the population of endemic area. Some other studies showed that *GSTP1*, *MMP-2*, *PRL*, *VDR*, and *MPO* gene polymorphisms had a significant association with an increased risk of endemic skeletal fluorosis. A research team also found evidence that genetic factors were associated with skeletal fluorosis in aluminum industry workers. However, larger sample studies of candidate genes in different populations living in endemic area were required before the association between candidate genes and fluorosis risk could be confirmed (Pramanik and Saha 2017).

Epigenetic changes in endemic fluorosis remained as an underexplored area, but limited studies found that methylation is involved in the development of skeletal fluorosis. Methylation refers to the catalytic transfer of methyl groups from active methyl compounds (S-adenosylmethionine) to other compounds. It can form various methyl compounds or chemically modify certain proteins or DNA to form methylation products. Within biological systems, methylation is enzymatically catalyzed, and this methylation involves the regulation of gene expression, the regulation of protein function, as well as RNA processing. DNA methylation is one of the main epigenetic mechanisms that regulate many cellular processes and is catalyzed by DNA methyltransferase 1 (DNMT1) to covalently modify cytosines at the 5′ end. CpG methylation of promoter regions regulate the expression of various genes. Currently, there were limited studies showing exposure to fluoride would alter DNA methylation status at the genetic level. Changes in the methylation of DNA repair genes O-6-methylguanine-DNA methyltransferase and mutL homolog 1 were found in the blood of adult fluorosis patients and chronic fluorosis rats, which may play a role in fluorosis-induced liver injury. Fluoride can disrupt DNA methylation of H19 and Peg3 imprinted genes during early development of mouse embryos.

In a study related to skeletal fluorosis, whole genome bisulfite sequencing assay was used to investigate the changes in DNA methylation levels in HOS cells exposed to fluoride. Fluoride exposure was found to cause hypermethylation of BMP-1, methionyl aminopeptidase 2 (*METAP2*), *MMP-11*, and BTB domain and CNC homolog 1 (*BACH1*) gene promoter regions, resulting in decreased expression of *BMP-1*, *METAP2*, *MMP-11,* and *BACH1*. This epigenetic inhibition might delay fracture repair and bone formation by reducing cartilage reabsorption, new blood vessel formation, vascular system development, ECM secretion, and bone porosity (necessary to provide elastic and mechanical strength of bone). Therefore, newly formed bone would deform and had an increased risk of fracture—the main hallmark and condition of skeletal fluorosis. This study suspected that fluoride-induced DNA methylation of *BMP-1*, *METAP2*, *MMP-11*, and *BACH1* promoter regions might be associated with the development of skeletal fluorosis (Daiwile et al. 2019). In addition, analysis of DNA methylation in peripheral blood leukocytes of children (coal-burning type skeletal fluorosis wards at the genome-wide level) revealed that a total of 15 differentially CpG hypermethylated sites and 22 hypomethylated CpG sites were detected in children with skeletal fluorosis compared with controls, involving 29 known genes. Among them, DNMT1 was considered necessary to be responsible for maintaining the methylation pattern of the genome. High expression of *DNMT1* transcription and translation enhanced the expression of human OB-related genes, which might be an early molecular event in skeletal fluorosis. Meanwhile, gene hypomethylation of transcriptional coactivator 2/3 (*MAML2/3*) might be a potential biomarker for skeletal fluorosis in children. *MAML2/3* was a transcriptional coactivator of Notch receptors, and Notch signaling had an important role in skeletogenesis and bone homeostasis. In addition, differential methylation of the coactivator BMP-8B and activating transcription factor 6 (ATF6) were also demonstrated in peripheral blood leukocytes of children with skeletal fluorosis (Deng et al. 2019).

Fluoride-induced histone trimethylation also plays an important role in the pathogenesis of skeletal fluorosis. Histones have long undergone covalent modifications including methylation, acetylation, phosphorylation, ubiquitination, threomethylation, citrullination, ribosylation, altering the interaction between DNA and nuclear proteins. These modifications can regulate different biological processes, such as gene regulation, DNA repair, chromosome condensation, spermatogenesis, and so on. Modification of histones by fluoride exposure was reported earlier and administration of fluoride resulted in reduced levels of H3K9 and H3K18 acetylation in mouse embryos. Alternatively, exposure to fluoride resulted in increased levels of H3K9 and H3K4 dimethylation in mouse early embryos. These changes in histones resulted in impaired oocyte maturation and adversely affected the development of mouse embryos. Some scholars linked fluoride-induced histone modifications to the pathogenesis of skeletal fluorosis, and fluoride-induced histone methylation was gradually appreciated.

By investigating the role of fluoride-induced histone modifications in the development of skeletal fluorosis using human HOS cell lines, it was found that the expression of histone methyltransferases (panchromatic histone lysine methyltransferase 1, enhancer of zeste homolog 2 and global histone trimethylation levels, including H3K9 and H3K27) were significantly increased after fluoride exposure (8 ppm). These histone methyltransferases catalyzed the transfer of methyl groups at positions 9 and 27 of H3 histone lysine (K). H3K9 chip polymerase chain reaction (PCR) array analysis determined that H3K9 trimethylation levels in the promoter regions of two key genes, transforming growth factor beta receptor 2 (TGFBR2) and SMAD family member 3 (*SMAD3*), increased with fluoride exposure. The epigenetic inhibitory effects of H3K9 trimethylation on the expression patterns and functional associations of the identified genes were determined by quantitative real-time PCR and STRING analysis. Gene expression profile also showed a significant decrease in the expression profiles of *TGFBR2* and *SMAD3*, which was consistent with histone trimethylation-mediated gene silencing and confirmed the epigenetic repression by promoter H3K9 trimethylation. Similar decreases in *TGFBR* and *SMAD3* expression profiles were observed in rats exposed to fluoride. In addition, the expressions of two essential genes involved in *TGFBR2* and *SMAD3* downstream signal pathways such as collagen type Iα1 (*COL1A1*) and *MMP-13* were also down-regulated. While *TGFBR2*, *SMAD3*, *COL1A1*, and *MMP-13* genes were functionally associated and involved in collagen and cartilage development/morphogenesis, connective tissue formation, biomineral tissue development, endochondral bone formation, and bone morphogenesis (Daiwile et al. 2018).

Conclusions

This chapter reviewed the recent literature addressing fluoride damage to bone tissue and discussed the molecular mechanisms that might be involved in the development of skeletal fluorosis. This part analyzed the imbalance process of osteogenesis and osteoclast in fluoride exposure, the related factors and signal pathways that might be involved, abnormal intracellular and extracellular Ca^{2+} levels, and changes in various cellular stress indicators. This chapter also introduced the indirect regulation of hormones on systemic bone tissues and the influence of oxidative stress, and explained the genetic susceptibility and epigenetic changes which described the pathological process of skeletal fluorosis comprehensively. However, the pathogenesis of skeletal fluorosis is complex, and many factors such as fluoride exposure concentration, time, age, gender, and nutritional status of fluoride consumers can be involved in regulating the occurrence and development of skeletal fluorosis. Thus, the final state of the disease could be affected together with a variety of mechanisms. Further experimental elucidation is required on the mechanism of skeletal fluorosis.

References

Bergandi L, Aina V, Malavasi G, Morterra C, Ghigo D. The toxic effect of fluoride on MG-63 osteoblast cells is also dependent on the production of nitric oxide. Chem Biol Interact. 2011;190:179–86.

Chen Y, Hu Y, Yang L, Zhou J, Tang Y, Zheng L, et al. Runx2 alleviates high glucose-suppressed osteogenic differentiation via PI3K/AKT/GSK3β/β-catenin pathway. Cell Biol Int. 2017;41:822–32.

Curnutte JT, Babior BM. Effects of anaerobiosis and inhibitors on O_2-production by human granulocytes. Blood. 1975;45:851–61.

Daiwile AP, Sivanesan S, Tarale P, Naoghare PK, Bafana A, Parmar D, et al. Role of fluoride induced histone trimethylation in development of skeletal fluorosis. Environ Toxicol Pharmacol. 2018;57:159–65.

Daiwile AP, Tarale P, Sivanesan S, Naoghare PK, Bafana A, Parmar D, et al. Role of fluoride induced epigenetic alterations in the development of skeletal fluorosis. Ecotoxicol Environ Saf. 2019;169:410–7.

Deng J, Zeng XX, Liu XH, Chen D, Dong YT, Song H. Screening and verification of differential DNA methylation genes in peripheral blood cells of children with skeletal fluorosis in Guizhou Province of China. Fluoride. 2019;52:135–48.

Fina BL, Lombarte M, Rigalli JP, Rigalli A. Fluoride increases superoxide production and impairs the respiratory chain in ROS 17/2.8 osteoblastic cells. PLoS ONE. 2014;9:e100768.

Gao Q, Liu YJ, Guan ZZ. Oxidative stress might be a mechanism connected with the decreased alpha 7 nicotinic receptor influenced by high-concentration of fluoride in SH-SY5Y neuroblastoma cells. Toxicol In Vitro. 2008;22:837.

Gersbach CA, Byers BA, Pavlath GK, García AJ. Runx2/Cbfa1 stimulates transdifferentiation of primary skeletal myoblasts into a mineralizing osteoblastic phenotype. Exp Cell Res. 2004;300:406–17.

Guan ZZ, Yang PS. Changes of lipid peroxidation and antioxidant substances in serum and erythrocytes of rats with chronic fluorosis. Chin J Endem. 1991;9:4–6.

Guan ZZ, Yang PS, Yu ND, Zhuang ZJ. An experimental study of blood biochemical diagnostic indices for chronic fluorosis. Fluoride. 1989;22:112–8.

Guan ZZ, Yang PS, Wang YN, Wei ZD, Zhou LY, Li FC. Study on antiaging index in plasma and erythrocytes of adult patients with fluorosis and aluminum poisoning. Guizhou Med. 1991;15:265–7.

Guan ZZ, Yang PS, Su YS, Wang YN. Changes of lipid peroxide content and anti-oxidant level in blood of the children living in the F-AI combined toxicosis area of Shuicheng of Guizhou. J Guiyang Med Coll. 1992;16:198–200.

Guan ZZ, Wang YN, Xiao KQ, Dai DY, Chen YH, Liu JL, et al. Influence of chronic fluorosis on membrane lipids in rat brain. Neurotoxicol Teratol. 1998;20:537–42.

Guan ZZ, Xiao KQ, Zeng XY, Long YG, Cheng YH, Jiang SF, et al. Changed cellular membrane lipid composition and lipid peroxidation of kidney in rats with chronic fluorosis. Arch Toxicol. 2000;74:602–8.

Guan ZZ, Wang LH, Sun DJ. Endemic fluorosis. In: Sun DJ, editor. Endemiology. Beijing: People's Medical Publishing House of China; 2017.

He H, Ping J, Jiang CH. Effects of sodium fluoride on osteoblast MC3T3-E1 differentiation and Wnt signal transduction pathway. Chin J Clin Pharmacol. 2019;35:45–7.

Iwatsuki M, Matsuoka M. Fluoride-induced c-Fos expression in MC3T3-E1 osteoblastic cell. Toxicol Mech Methods. 2016;26:132–8.

Jiang NN, Guo FY, Shi LQ, Xu H. The effect of sodium fluoride on the main bone cell types. Chin J Osteoporos. 2019;25:1389–93.

Lee M, Arikawa K, Nagahama F. Micromolar levels of sodium fluoride promote osteoblast differentiation through Runx2 signaling. Biol Trace Elem Res. 2017;178:283–91.

Liu XL, Song J, Liu KJ, Wang WP, Xu C, Zhang YZ, et al. Role of inhibition of osteogenesis function by Sema4D/Plexin-B1 signaling pathway in skeletal fluorosis *in vitro*. J Huazhong Univ Sci Technolog Med Sci. 2015;35:712–5.

Long MW, Robinson JA, Ashcraft EA, Mann KG. Regulation of human bone marrow-derived osteoprogenitor cells by osteogenic growth factors. J Clin Invest. 1995;95:881–7.

Otto F, Thornell AP, Crompton T, Denzel A, Gilmour KC, Rosewell IR, et al. Cbfa1, a candidate gene for cleidocranial dysplasia syndrome, is essential for osteoblast differentiation and bone development. Cell. 1997;89:765–71.

Pan L, Shi X, Liu S, Guo X, Zhao M, Cai R, et al. Fluoride promotes osteoblastic differentiation through canonical Wnt/β-catenin signaling pathway. Toxicol Lett. 2014;225:34–42.

Pei JR, Li BY, Gao YH, Xu JX, Darko GM, Sun DJ. Relationship between fluoride exposure and osteoclast markers during RANKL-induced osteoclast differentiation. Environ Toxicol Pharmacol. 2016;46:241–5.

Pei JR, Xu JX, Gao YH, Sun HN, Lv CP, Jiang YT. Identification of pathogenesis-related microRNA profiles in skeletal fluosis. Fluoride. 2019;52:29–41.

Perumal E, Paul V, Govindarajan V, Panneerselvam L. A brief review on experimental fluorosis. Toxicol Lett. 2013;223:236–51.

Piva R, Belardo G, Santoro MG. NF-kappaB: a stress-regulated switch for cell survival. Antioxid Redox Signal. 2006;8:478–86.

Pramanik S, Saha D. The genetic influence in fluorosis. Environ Toxicol Pharmacol. 2017;56:157–62.

Rodda SJ, McMahon AP. Distinct roles for Hedgehog and canonical Wnt signaling in specification, differentiation and maintenance of osteoblast progenitors. Development. 2006;133:3231–44.

Sato M, Hanmoto T, Yachiguchi K, Tabuchi Y, Kondo T, Endo M, et al. Sodium fluoride induces hypercalcemia resulting from the upregulation of both osteoblastic and osteoclastic activities in goldfish, Carassius auratus. Comp Biochem Physiol C Toxicol Pharmacol. 2016;189:54–60.

Sellami M, Riahi H, Maatallah K, Ferjani H, Bouaziz MC, Ladeb MF. Skeletal fluorosis: don't miss the diagnosis! Skelet Radiol. 2020;49:345–57.

Shen LM, Feng CY, Xia SJ, Wei YW, Zhang H, Zhao DQ, et al. Progressive research in the molecular mechanisms of chronic fluorosis. In: Fluoride pollution-facts and fundamental findings of health and environmental remediation. London: IntechOpen; 2019.

Suzuki M, Bandoski C, Bartlett JD. Fluoride induces oxidative damage and SIRT1/autophagy through ROS-mediated JNK signaling. Free Radic Biol Med. 2015;89:369–78.

Varol E, Icli A, Aksoy F, Bas HA, Sutcu R, Ersoy IH, et al. Evaluation of total oxidative status and total antioxidant capacity in patients with endemic fluorosis. Toxicol Ind Health. 2013;29:175–80.

Wang Z, Yang X, Yang S, Ren G, Ferreri M, Su Y, et al. Sodium fluoride suppress proliferation and induce apoptosis through decreased insulin-like growth factor-I expression and oxidative stress in primary cultured mouse osteoblasts. Arch Toxicol. 2011;85:1407–17.

Wei Y, Zeng B, Zhang H, Chen C, Wu Y, Wang N, et al. Comparative proteomic analysis of fluoride treated rat bone provides new insights into the molecular mechanisms of fluoride toxicity. Toxicol Lett. 2018;291:39–50.

Xu H, Zhou YL, Zhang JM, Liu H, Jing L, Li GS. Effects of fluoride on the intracellular free Ca^{2+} and Ca^{2+}-ATPase of kidney. Biol Trace Elem Res. 2007;116:279–88.

Xu H, Liu QY, Zhang JM, Zhang H, Li GS. Elevation of PTH and PTHrp induced by excessive fluoride in rats on a calcium-deficient diet. Biol Trace Elem Res. 2010;137:79–87.

Yu YN, Yang D, Zhu HZ, Deng CN, Guan ZZ. Expression of mRNA and protein of p38, Osx, PI3K and Akt1 in rat bone with chronic fluorosis. Chin J Pathol. 2012;41:622–6.

Zhou X, Chen Z, Zhong W, Yu R, He L. Effect of fluoride on PERK-Nrf2 signaling pathway in mouse ameloblasts. Hum Exp Toxicol. 2019;38:833–45.

Chapter 4
Medical Imaging Manifestation of Skeletal Fluorosis

Qing-Hong Duan, Xu-Guang Chen, and Zhi-Zhong Guan

Abstract There are three types of endemic fluorosis, namely drinking water type, coal-burning type, and brick-tea type of endemic fluorosis. With skeletal fluorosis and dental fluorosis as the major manifestations, skeletal fluorosis is the key symptom of the disease. Medical imaging results of skeletal fluorosis caused by these three types show similar manifestations: changes in sclerotin density, peripheral osseous ligament, tendon, periosteum, and joints change. Medical imaging manifestations objectively represent the pathological conditions of bone damage. Thus, it serves as a key measure for clinical diagnosis of skeletal fluorosis and prevention of endemic fluorosis.

Keywords Endemic fluorosis · Radiography · Skeletal fluorosis · Osteoporosis · Osteosclerosis

Introduction

Skeletal fluorosis is the major manifestation of endemic fluorosis. Its clinical manifestation includes arthralgia in lower back and leg, ankylosis, skeletal deformity, nerve root, and spinal cord compression. Although theses clinical manifestations

Q.-H. Duan (✉)
Department of Radiology at the Affiliated Cancer Hospital of Guizhou Medical University, Guiyang, Guizhou, People's Republic of China
e-mail: duanqinghong@gmc.edu.cn

X.-G. Chen
Department of Radiology at the Affiliated Hospital of Guizhou Medical University, Guiyang, Guizhou, People's Republic of China

Z.-Z. Guan
The Key Laboratory of Endemic and Ethnic Diseases (Guizhou Medical University), Ministry of Education, Guiyang, Guizhou, People's Republic of China

Department of Pathology at the Affiliated Hospital of Guizhou Medical University, Guiyang, Guizhou, People's Republic of China

Z.-Z. Guan (ed.), *Coal-burning Type of Endemic Fluorosis*,
https://doi.org/10.1007/978-981-16-1498-9_4

share certain clinical features, they show no specificity. Epidemiology data and medical imaging results are key parts in the diagnosis of skeletal fluorosis. Medical imaging plays an irreplaceable role in the staging and subtyping of skeletal fluorosis. Therefore, research on imaging manifestations of skeletal fluorosis is vital for the prevention of endemic fluorosis.

Manifestation and Epidemic Characteristic of Skeletal Fluorosis

Prevalence of Skeletal Fluorosis

Endemic fluorosis (fluorosis hereafter) is a chronic fluorosis caused by long-term fluoride in-take from drinking water and food. It is a symbiotic endemic with the geographic environment with the symptoms of dental fluorosis and skeletal fluorosis. Skeletal fluorosis refers to chronic erosion and systemic bone disease caused by fluorosis and afflicts osseous tissues due to long-term fluoride in-take. Fluorosis afflicted teeth is termed as dental fluorosis. In 1901, Eaqer was the first one to report dental fluorosis. In 1932, Motlev and Gudijonsse conjured the term "fluorosis" and proposed that the disease was often accompanied with osteosclerosis.

Skeletal fluorosis in China was first reported in areas of Beijing and Taiyuan in 1930. In 1932, Anderson reported 54 cases of dental fluorosis in Langfang area. At the same time, Kilbomrn G found numerous cases of patients with dental fluorosis and rachitis in Dusai area in the east of Guizhou province. After 1960, a large number of epidemiology surveys on skeletal fluorosis were conducted and confirmed that morbidity in northern China was closely tied with fluoride content in drinking water. These surveys covered areas in Shandong, Shanxi, Hebei, Inner Mongolia, northeastern provinces, Ningxia, Shaanxi, Tianjin, and Beijing, and results indicated that high fluoride content areas were widely distributed in China.

Etiology, Pathology, and Pathogenesis of Skeletal Fluorosis

The human body requires 1.0–1.5 mg intake of fluoride. The causes of chronic fluorosis include drinking high fluoride content water, digesting vegetables contaminated by fluoride-containing pesticide or high fluoride contaminated food. Plus, drinking tea with extremely high fluoride content and long-term digestion of high fluoride-containing sea products also contribute to trigger chronic fluorosis. The widespread use of fluoride in industry has also led to the increase of industrial caused fluorosis patients.

Fluorosis can be categorized into acute and chronic type as it is both endemic and industrial disease. Patients with acute fluorosis are often found in production workers of hydrofluoric acid since they inhale hydrogen fluoride gas. Chronic fluorosis is caused by long-term digestion of high fluoride-containing food and drinks in small amount or inhaling fluoride-containing mineral dust. Coal-burning type of endemic fluorosis is caused by high fluoride content in coal and coal products in endemic areas: indoor burning of high fluoride-containing coal produced fluoride-containing soot, then such soot contaminates farm products (corns, chili, and so on) hung in the room. It is these foods consumed by local people to cause chronic fluorosis. After fluoride entering human body through digestion or respiratory tract, it circulates around the whole body. Fluoride lowers the concentration of magnesium and calcium ion in blood by bonding with them, which is the cause of convulsion, muscle cramps, and pain in patients with endemic fluorosis. Meanwhile, enzyme activities rely on magnesium or calcium ions are inhibited, thus multiple enzyme metabolism processes are affected.

High fluoride ions show toxicity on cytoplasm. Such ion can affect cellular function, replace hydroxyl group of hydroxyapatites in bone. It can also affect the bone metabolism in the form of osteoporosis or osteosclerosis (or both). Manifestations of the effect are osteopenia, bone fracture or deformity and then lead to skeletal fluorosis. Teeth are susceptible to fluoride during development, and dental fluorosis would be triggered by high fluoride content in blood condition.

Clinical Manifestation of Skeletal Fluorosis

Coal-burning type of endemic fluorosis shares the same clinical manifestations with other types of endemic fluorosis: damage on dental and bone tissues. Progression of skeletal fluorosis can last for decades and is often found in young adults. Its major clinical manifestations and clinical signs include arthralgia in lower back and legs, ankylosis, skeletal deformity, nerve root, and spinal cord compression.

The following symptoms are common for patients with skeletal fluorosis: full body muscle pain, dizzy, palpitation, tiredness, reduced appetite, nausea, vomiting, abdominal distension, diarrhea, constipation, muscle atrophy, and changed electromyography (EMG). There would be corresponding symptoms and clinical signs if thyroid, adrenal gland, gonadal gland, crystalline lens, and central nervous system are injured. A study conducted research on 37 patients with skeletal fluorosis by testing their serum fluoride ion, thyroid hormone, and thyroid-stimulating hormone (TSH) concentration. It was found that serum T_4 and T_3 of fluorosis patients were significantly lower than that of control group, TSH was obviously higher than T_3 concentration and T_3/T_4 ratio of control group. It was also found that TSH was negatively correlated with serum fluoride ion concentration, indicating fluorosis could lower thyroid function and such result was related to fluoride in serum.

The following bone and arthralgia symptoms are found in skeletal fluorosis: consistent rest pain arthralgia in neck, lower back, and extremities' large joints regardless of the season or weather. In addition, convulsion, numbness, and morning stiffness in joints often come along. An investigation in Yina-jia (Zhijin county) on fluorosis confirmed 76 patients with coal-burning type of fluorosis out of 428 subjects in this area. Clinical symptoms statistics revealed that impaired mobility rate of extremities joints and positive rate of lower back pain reached 85.7% to 100%, pressing pain in lumbo-sacral joint reached 64.2%, sciatic pressing pain of both sides reached 17.1%, positive rate of straight leg raise accounted for 11.4% (both knees accounted for the majority and patellar reflex disappeared). Dizziness (77.1%) was the major subjective symptom for patients, followed by tinnitus, abdominal pain, numbness, and headache.

In general, female patients with skeletal fluorosis showed more severe symptoms than male patients. Studies proposed that 50% of female patients would suffer from spine stiffness, while 7% for that of male under the condition of 10 mg/l fluoride-containing drinking water. In this condition, 22.2% of female patients would suffer from scoliosis, hunchback, or paralysis, while 7% for males. Such differences might be related to pregnancy and breast-feeding. Despite clinical manifestations of patients with skeletal fluorosis showed certain features, diagnosis required combined epidemiology data since such manifestations showed no specificity. In addition, medical imaging test was also important for the diagnosis.

Medical Imaging Check

X-ray Results

In osteoporosis type, rough and sparse bone texture is the sole manifestation of early stage of skeletal fluorosis.

In osteomalacia type, spine and pelvic show the most obvious reduced bone density and ambiguous texture. Scoliosis and hunchback are also observed. Double concaves of vertebral body are deformed, pelvic is constricted and deformed, pseudo-fracture line is formed. All these results lead to osteomalacia. Osteomalacia, osteoporosis, osteosclerosis, and soft tissue calcification are presented in hybrid type.

In osteosclerosis type, osteosclerosis is often found in spine, pelvic, costa, and cranium. Bone texture of this type resembles sackcloth or granule. Severe case is found with fused bone texture with ambiguous structure. X-ray transparency is low and looks like ivory. Osteosclerosis is often found with epiphysis osteoporosis in extremities, secondary hyperparathyroidism is found in deficient calcium intake. Extremities' bone manifests in osteitis fibrosa cystica.

Computerized Tomography (CT) Manifestation

CT results of patients with skeletal fluorosis show bamboo joint-shaped spine, thoracic, and upper spine segments in particular. Attachment of vertebral body show increased density, together with osteophyte, and ligament ossification. Bone stenosis is also observed.

Magnetic Resonance Imaging (MRI) Manifestation

Significant changes in morphology and signal of vertebral body in patients with skeletal fluorosis often come with abnormal manifestation of MRI. Vertebral body shows even or uneven hypointensity changes. Uneven hypointensity is different from healthy signal in the form of plaque-shape or needle-shape in vertebral body. Hypointensity area of peripheral part of the vertebral body is thickened. Paracentral sagittal view is found with facet joint osteophyte, hypotrophy ossification, and cascade-shape ligamentum flava. The following features are observed: spinal stenosis, compressed, and deformed dural sac, compressed spinal cord with edema, weighted image of T2 showed hyperintensity.

Single Photon Emission CT (SPECT) Manifestation

Radioactive signal of osteophyte is strong with wide scope and shows no specificity. Bone biopsies show disarranged bone lamella after decalcification, content of bone fluorine, calcium and magnesium are increased, bone phosphate and serum phosphorus remain normal. Ultrastructure shows specific changes.

Diagnosis Criteria of Endemic Skeletal Fluorosis in China

Scope

The criteria standardized principle of diagnosis and division for clinical and X-ray check in endemic skeletal fluorosis. Such criteria can be used for clinical diagnosis and X-ray check division for endemic fluorosis, as well as prevalence general investigation and monitoring, prevention and treatment efficacy evaluation.

Terminology and Definition

The following terminology and definition can be applied to the criteria. (1) Endemic skeletal fluorosis: residents of the endemic area are found with chronic metabolic bone disease due to excessive fluoride intake. Manifestations of this disease include: arthralgia in neck, lower back and extremities, motor dysfunction, and abnormal X-ray results in bone and joints. (2) Big joints of four limbs: joints of shoulder, elbow, wrist, pelvis, knee, and ankle. (3) Rest pain: joint pain in non-working, non-weight-loading or non-motion condition.

Evidence of Diagnosis

Epidemiology history is very important, including being born and raised in endemic fluorosis area or migrating to endemic area for more than 1 year after birth.

Clinical manifestation: (1) bone and joint pain symptom includes continuous rest pain in neck, lower back and extremities regardless season and weather; the pain is often accompanied with convulsion, numbness, and morning stiffness. (2) deformed extremities and motion dysfunction include (a) limited motion in neck: limited flexion, backward extension, and swivel; (b) limited motion in upper limb: middle finger fails to touch acromion of the same side when elbow is bend, middle finger fails to touch the pinna of opposite side from occiput posterior, middle finger fails to touch contralateral inferior angle from behind, arms cannot raise 180 °C. limited motion in lower back: flexion, backward extension, and swivel are all limited, spine is deformed; (d) limited motion in lower limbs exhibits that legs cannot be straightened, gonyectyposis or gonycrotesis and the patients walk slowly and even paralyzed.

Manifestation of bone and joints X-ray results: osteosclerosis, osteoporosis, osteomalacia, bone turnover, ossification of osseous peripheral soft tissues, and degenerative changes in joints, which refer to Table 4.1.

Principle of Diagnosis

Diagnosis is based on epidemiology history, clinical symptoms, signs and the results of skeletal and articular X-ray. If clinical diagnosis disagrees with X-ray results, X-ray results shall be taken as final.

Division Criteria of Diagnosis

Clinical Diagnosis and Division

Clinical diagnosis and division include follow items.

1. Mild symptoms: continuous rest pain is only found in neck, lower back, and big joints of extremities (more than three parts) regardless of season and weather. Symptom is often accompanied with convulsion, numbness, morning stiffness, and lower back stiffness.
2. Moderate symptoms: in addition to the bone and articular pain mentioned previously, it is often accompanied with motor dysfunction (in neck, lower back, joints of upper and lower extremities). Thus, living and working ability of the patients is decreased.

Table 4.1 X-ray manifestations of endemic skeletal fluorosis and division

Types	X-ray manifestations
Osteosclerosis	
Mild	Gravel-like bone structure; granule-like bone structure; bone spot
Moderate	Coarse-grained trabecular; fine-grained trabecular; sackcloth-like trabecular; partially fused fine-grained trabecular; coarse bone manifestation
Severe	Fused general coarse-grained trabecular; fused general fine-grained trabecular; fused extra-large and scarce trabecular; iliac bones turtle scale-like trabecular; coarse net-like trabecular; ivory-like osteosclerosis
Osteoporosis and osteomalacia	
Mild	Attenuated trabecular become sparse or ambiguous
Moderate	General osteoporosis, coupled with mild ossification of interosseous membrane of forearm or shank
Severe	Loose belt below epiphysis; brush-shaped manifestation of metaphysis; vertebra body double frame manifestation; pseudo-fracture line; double concave of vertebral body together with osteosclerosis
Hybrid changes (bone turnover)	
Mild	Diaphysis metaphysis sclerosis band
Moderate	Obvious disorganized and ambiguous trabecular in extremities' metaphysis; bone cortex of attachment of the pronator teres muscles become osteoporosis
Severe	Cortex become osteoporosis; cancellous bone is evenly sclerosis; cotton wadding-like bone structure; worn out carpet-like trabecular; pelvic density is obviously increased and also deformed by osteomalaica
Osseous peripheral and articular changes	
Mild	Radius crest is enlarged, peripheral part is ossified, surface becomes rough; calcification of shank interosseous membrane and presented as plumule breaking the ground
Moderate	Ossification of forearm, shank and pelvic osseous peripheral soft tissues; ossification of elbow flexion tendon
Severe	Obvious ossification of forearm, shank and pelvic osseous peripheral soft tissues; obvious ossification of elbow flexion tendon; big joints of extremities go through obvious degenerative changes and deformed

3. Severe symptoms: bone and articular pain is accompanied with severe motor dysfunction (in neck, lower back, joints of upper and lower extremities). Patients' extremities are deformed, and their living and working ability is significantly decreased, lost or patients are paralyzed.

X-ray Results of Diagnosis and Division

1. Mild symptoms: any one of the following manifestations can confirm patients with mild symptoms. (a) Abnormal structure of trabecular manifested in gravel-like or granule-like bone structure and bone spot; (b) attenuated trabecular with sparse, ambiguous, and disorganized structure, or diaphysis metaphysis become ossified together with mildly ossified osseous peripheral soft tissues of forearm

and shank; c, radius crest is enlarged, peripheral part is ossified, surface becomes rough; e, calcification of forearm or shank interosseous membrane and presented as plumule breaking the ground.

2. Moderate symptoms: any one of the following manifestations can confirm patients with moderate symptoms. (a) Abnormal structure of trabecular manifested in coarse-grained, fine-grained, sackcloth-shaped trabecular or partially fused trabecular; (b) general osteoporosis together with calcification of forearm or shank interosseous membrane; (c) trabecular structure of extremities' metaphysis is obviously disorganized and ambiguous, bone cortex of attachment of the pronator teres muscles become osteoporosis; (d) obvious calcification in shank, forearm interosseous membrane or pelvic tendon and attachment of ligament are observed.
3. Severe symptoms: any one of the following manifestations can confirm patients with moderate symptoms. (a) Most of trabecular fusion is presented in ivory-like osteosclerosis; (b) obvious osteoporosis or osteomalacia is coupled with calcification of shank or forearm interosseous membrane; (c) worn out carpet-like trabecular or cotton wadding-like bone structure, increased density coupled with deformed bone, and cortex becomes osteoporosis are observed; (d) multiple big joints go through sever degenerative changes or are deformed, osseous peripheral soft tissues are obviously ossified.

X-ray manifestations of endemic skeletal fluorosis and division please refer to Table 4.1.

Prerequisites of Bone and Articular X-ray Check

1. General check can resort to anteroposterior X-ray check (including elbow joint).
2. Endemic monitoring or epidemiology check can check anteroposterior X-ray results of forearm (including elbow joint) and shank (including knee joint).
3. Treatment and prevention efficacy evaluation should include X-ray results of anteroposterior X-ray results of forearm and shank, together with that of pelvic.

Differential Diagnosis

Clinical and X-ray manifestations of osteoarthritis, rheumatic arthritis, ankylosing spondylitis, and rheumatoid arthritis are similar to that of endemic skeletal fluorosis; therefore, it is worth differential diagnosis.

1. Osteoarthritis: this disease, also called osteoarthropathy, is a degenerative disease on articular cartilage. It is often found in patients over 50 years old afflicting distal interphalangeal joint and weight-bearing joints (knee and pelvic). The key symptom is local pain in interphalangeal joint, which can be aggravated by motion or

weight-bearing, and alleviated by rest. Common manifestations are swollen joint, tenderness, clicking sound or friction sound in motion. X-ray test results show narrowed joint space, sclerosis and deformed articular surface, osteophyte in articular peripheral parts, free bodies in articular cavity.

Patients with skeletal fluorosis often have endemic area living history. They have continuous rest pain in multiple big joints of the body, together with convulsion, numbness, and morning stiffness. Multiple articular dysfunctions can be found in neck, shoulder, elbow, lower back, pelvic, and knee. X-ray test results show skeletal fluorosis manifestation in sclerotin and osseous peripheral parts.

2. Rheumatic arthritis: such disease is often found in adolescents. Infection of the upper respiratory tract is found before onset of the disease. The disease afflicts multiple big joints, with the manifestations of being symmetrical, migrating, multiple joints red, swelling, broiling, pain or pressing pain, with limited motion. The disease is closely related with weather changes. Acute stage will not cause deformed joints. The disease is often accompanied with myocarditis and increased anti-streptolysin "O." X-ray test shows no abnormal results in bone and joints.

   Endemic skeletal fluorosis shows slow onset progress and no acute stage. Pain in the bone and joints is not accompanied with sings of red, swelling, broiling, and pressing pain. Location of pain is fixed and it shows no obvious relationship with weather changes. X-ray results of sclerotin and joints show skeletal fluorosis manifestations.

3. Ankylosing spondylitis: this disease is a chronic no-specific inflammatory disease featuring poker spine with unknown causes. Age of onset is 15 to 30, patient over 40 with such disease is rare. The disease afflicts sacroiliac joints and creeps to spine and then causes articular bony ankylosis. Early stage of onset is found with blunt pain which is hard to locate in the lower back. The pain is sharp and accompanied with lower back stiffness. The pain in the morning is even worse and could be aggravated in cold and damp environment. In advanced stage, hip joint is found with flexion contracture and characteristic fixed gait. X-ray results found that sacroiliac joints are the parts with earliest attack; in early stage, peripheral subchondral bone becomes ambiguous with insect bite-like damage. It also shows local erosive sclerosis, narrowed joint space, and fused sacroiliac joints (bony ankylosis). When the disease afflicts spine, the manifestations are general osteoporosis in spine, vertebral facets space becomes ambiguous and narrowed, vertebral body becomes square-shaped; in advanced stage, intervertebral disc and paravertebral ligament becomes calcified (ossified), spine is bamboo joint-shaped.

   Endemic skeletal fluorosis is often found in patients over 30 with no previously mentioned progress. Their clinical manifestations are multiple big joints pain and motor dysfunction. Articular fibrous ankylosis is its characteristic. X-ray test finds that bone texture and bone density are abnormal, osseous peripheral soft tissues of forearm, shank interosseous membrane are ossified.

4. Rheumatoid arthritis: this disease is multiple systems autoimmune disease. It mainly and symmetrically afflicts phalangeal and palm facet joints. Its clinical manifestations are articular pain, numbness, peripheral skin is hot, red, and swelling, enlarged joints, and limited motor function. Morning stiffness is obvious that lasts over 1 h. Joints are swelling in spindle-shape, the rest joints are deformed and morning stiffness is most characteristic. X-ray test finds that peripheral articular soft tissues are swelling in early stage, articular epiphysis is osteoporosis. It is possible to find articular cartilage bursa change or articular peripheral osseous erosion. As it develops, deformities like obvious cartilage burse damage, narrowed joints space, articular facet erosive damage, muscle atrophy, and subluxation can be found. In advanced stage, fibrous or bony ankylosis is confirmed.

   Endemic skeletal fluorosis shows manifestations of multiple big joints pain and motor dysfunction. Patients show no red, swelling, and hot joints. They show short-time morning stiffness, accompanied convulsion, and numbness. X-ray test finds osteosclerosis, osteoporosis, and osteomalacia in pelvic parts. Osseous peripheral soft tissues of extremities are ossified.

Imaging Changes of Skeletal Fluorosis

Correlation Among All Types of Endemic Fluorosis

Endemic fluorosis is generally categorized into drinking water type, coal-burning type, and brick-tea type according to the geographical conditions. Drinking water type endemic fluorosis is a chronic fluorosis with the widest distribution and largest number of patients both at home and the world. Coal-burning type and brick-tea type endemic fluorosis are unique to China (Sun et al. 2001).

Pathogenesis of all three types of endemic fluorosis is excessive fluoride intake. The only difference is the content of fluoride in food or water. Many scientists believed that X-ray manifestations of coal-burning type or drinking water type endemic fluorosis had three syndromes: change of sclerotin density, changes of osseous peripheral ligament, tendon and periosteum, and articular change. X-ray results showed no fundamental differences between these two types of endemic fluorosis. In addition, there was no significant difference of X-ray between drinking-tea type of endemic fluorosis and those of the two endemic fluorosis mentioned above. In 2017, differences between coal-burning type of skeletal fluorosis and industrial skeletal fluorosis were compared and found that their X-ray manifestations were similar with slight differences (Du et al. 2017).

Mechanism of Endemic Skeletal Fluorosis Imaging Manifestation

For decades, scientists from both home and abroad conducted extensive research on mechanism of endemic skeletal fluorosis imaging manifestation (Lian and Wu 1986; Gruber and Baylink 1991; Wang and Wu 1992; Kleerekoper 1996; Li et al. 2020). A study concerning X-ray manifestation of skeletal fluorosis from basic pathological perspective and its pathological basis concluded that gravel-live changes resulted from increased sclerotin in intersections of trabecular. It was discovered that osteoporosis of extremities was early changes of skeletal fluorosis and while sclerotic spine was the basis of osteoporosis (Lian and Wu 1986). Another research conducted comparison between X-ray manifestation and pathology of animals with skeletal fluorosis (Wang and Wu 1992) found that basic pathology of cartilage X-ray changes was formed by transverse plate below cartilage and chondrocyte hypertrophy, atrophy and necrosis. X-ray changes of bone are caused by comprehensive yet drastic osteoclastic and osteoblast bone turnover. The osteoporosis was simultaneous with ossification of periosteum and endosteum. A research proposed that medical imaging manifestations of endemic skeletal fluorosis were diversified, including calcification, or ossification, osteosclerosis, reduced bone mass, incremental line, and metaphysis osteomalacia in soft tissue attachment parts (Wang et al. 1994).

The following study and research were conducted by Chinese scientists. Abnormally increased expression of BMP-2 and BMP-7 in synovium of rats with fluorosis was discovered. It suggested that these two factors play vital roles in occurrence and development of fluorosis arthrosis. Another research proved fluoride effect on activity of osseous BMP in rats and its role to prevent reconstruction of osseous tissues. There was also a research proposed that severity of medical imaging manifestations of skeletal fluorosis might be subject to the age of fluorosis onset, based on the study of infant rats exposed to excessive fluoride and showed more severe symptoms than adult rats. It was also suggested that fluoride affected estrogen secretion at gonadal level and affected bone metabolism via changes of estrogen. Such results provided a new perspective to explain imaging manifestations of skeletal fluorosis.

A study reviewed 24 skeletal fluorosis cases in northern India and found that all patients suffered from hypocalcemia and increased SAP in serum inorganic phosphorus. Such results revealed the reason of osteomalaica in skeletal fluorosis. Biomechanical properties of fluorosis rats were proved to be significantly lower than that of health ones (degree of elongation, impact toughness, shear and bending strength). It revealed that biomechanical cause of manifestations, like skeletal deformity, pseudo fracture-line. There was also a study on endemic areas with varied nutrition conditions. It was discovered that osseous peripheral sclerosis and osteosclerosis (no osteomalacia) were major symptoms in areas with good nutrition condition; while osteoporosis and osteomalacia were major symptoms in areas with poor nutrition. This study concluded that X-ray manifestations types of fluorosis were closely related with nutrition condition. Stress relaxation amount of anterior cruciate ligament 7200s of fluorosis rats was studied and found to be significantly lower than control group, while another study on fluorosis rabbit discovered that stretching biomechanics properties index of its elbow ulnar nerve was significantly reduced. These results might serve to demonstrate the pathogenesis of fluorosis arthrosis. In another study on rabbits with excessive fluoride intake, it was found that ligamentum flavum of thoracic vertebra showed reduced and fractured elastic fibers, disorganized and substantial collagenous fibrogenesis, together with degenerative changes and calcification of ligamentum flavum. It might explain spinal stenosis was often found in skeletal fluorosis.

One research specifically pointed out that fluoride promoted osteogenic differentiation by affecting expression of ossification related genes. A study explained the mechanism of bone injury caused by chronic fluorosis from imbalanced antioxidant system due to high fluoride content, and increased free radical content in bone caused such injury. It was suggested that excessive fluoride caused osteoblastic and osteoclastic activities all contributed to the high bone turnover state. Fluoride could affect bone resorption degree by changing relative expression ratio of osteoprotegerin and osteoprotegerin ligand, which provided pathway for enhanced bone turnover effect. Another research on skeletal fluorosis proved that fluorinated and calcium deficiency would all lead to bone injury. The change in chemical elements in bone was involved in pathogenesis of fluorosis, which explained that skeletal fluorosis was the results of multiple factors.

As for the research by foreign scientists, it was discovered that excessive fluoride would stimulate osteoblastic and osteoclastic activity, which manifested in simultaneous accelerated new bone formation and old bone resorption. When the fluoride deposited in vivo reached certain degree, osteoporosis and new sclerotin without normal structure were both formed. There would also be deposition of calcium fluoride and fluorapatite, which would deform and sclerosis the bone and trigger changes of bone density (Gruber and Baylink 1991; Kleerekoper 1996). In a study conducted in vitro, fluoride was proved to promote mitosis of osteocyte and bone formation (Farley et al. 1983). Another study proposed that mitosis-promoting amount of fluoride could also specifically inhibit the phosphokinase in osteoblast but increase mitogen, increase activity of protein kinase aminoacyl phosphorylation to promote proliferation of osteocyte (Lau et al. 1995). At the same year, a study proposed that fluoride could increase specific tyrosine kinase to promote osteocyte proliferation (Burgener et al. 1995).

In human or animal tests, studies reported that bone mass increase in fluorosis was triggered by increased number of osteoblasts, enhanced activity, and prolonged life cycle. An early study suggested that bone mass increase in fluorosis was triggered by more than just enhanced activity of osteoblast, but also reduced number of osteoclast count and shortened life cycle (Bely 1991). A recent study reviewed the past research and designed cohort study and experiment in vitro. It aimed to understand the effect of miR-122-5p by targeted cellular cycle cyclin-dependent kinase (CDK4) on fluorosis osteoblast activation (Li et al. 2020).

It was found that miR-122-5p expression was downregulated, CDK4 expression was upregulated, and miR-122-5p was negatively correlated with the expression of CDK4 in coal-burning type of fluorosis population. In human osteoblast treated by sodium fluoride, it was proven that miR-122-5p mediated osteoblast activation over fluorosis by upregulating CDK4 protein. To further prove such results, the study conducted Dual-Luciferase experiment and indicated that miR-122-5p indeed regulated protein level of CDK4. They proposed that miR-122-5p involved in occurrence and development of skeletal fluorosis by targeting CDK4.

A recent research experimentally studied mechanism of parathyroid hormone (PTH) on skeletal fluorosis (Jiang et al. 2020). The study concluded that parathyroid hormone could enhance the fluoride effect on osteoclastic formation of relevant molecular expression, but it could not enhance fluoride effect on stimulation of osteocyte cocultured and driven osteoclast formation. When the fluoride affected precursor of osteoclast was cocultured with osteocyte-like cells, genetic expression level of TRAP, RANK, JNK, and NFAtc1 was obviously increased. Fluoride effect on differentiating osteoclast was more powerful than parathyroid hormone. To sum, osteocyte played a vital role in the mechanism of RANK-JNK-NFATc1 signal pathway in fluoride affected osteoclast formation, and PTH also played a key role in this process. Due to the inconsistent conclusions, its mechanism remained unclear.

Medical Imaging Characteristics of Endemic Skeletal Fluorosis

Chronic fluorosis in adult patients manifests different degrees of osteosclerosis, in which in the late stage the symptoms are more obvious. Osteosclerosis affects the whole body, and the more easily affected parts are spine, pelvic, and costa. Extremities' diaphysis could also be affected. Trabecular is thickened by osteosclerosis of cancellous bone, and medullary cavity is shrunk and even disappeared. Periosteal osteophyte makes the osseous edge uneven, which is found to be enlarged or thickened.

Calcification of tendon is a specific manifestation of this disease. It is often found in less deployable or fixed joints. The calcification is first found in paravertebral parts. Severe changes can transform the spine into bamboo joints-shape, which is similar to ankylosing spondylitis manifestations but specific changes in sacroiliac joints are absent. Sacrosciatic sacrotuberous ligament, sacrosciatic sacrospinous ligament, and other ligaments would all become calcified. It was also proposed that bone changes in endemic fluorosis were mainly found in spine, pelvic, and costa (Gupta et al. 2016). Radiographic features of osteosclerosis include increased bone density (osteosclerosis), reduced bone mass and osteoporosis, ossification of tendon and ligament, calcification of interosseous membrane, and ossification of posterior longitudinal ligament.

Skeletal fluorosis bone shows multiple pathological changes and X-ray manifestations: osteosclerosis, osteoporosis, osteomalacia, incremental disorder line, and degenerative changes in joints.

Research on X-ray results of endemic skeletal fluorosis was most concerned in 1980s to 1990s. Chinese scientists conducted substantial research to this regard and made outstanding contribution to X-ray diagnosis of endemic fluorosis. It was proposed that X-ray manifestation of skeletal fluorosis were as follows: osteosclerosis, osteoporosis, osteomalacia, periosteal osteophyte, ligament ossification, articular cartilage necrosis, and bone growth disorder. The study concluded that the diagnosis was based on bone density change, periosteal osteophyte, and osteophytosis. Another study tried to pin-point the 6 X-ray manifestations of skeletal fluorosis: osteosclerosis, osteoporosis, osteomalacia, discontinuous growth trace, articular degenerative changes, calcification, and ossification of interosseous tendon attachment. Such study proposed that systemic osteosclerosis and comprehensive ligament ossification shared great value for diagnosis. Another study reviewed 24 skeletal fluorosis cases in northern India and concluded that previously mentioned multiple symptoms were found in individual patients. Osteosclerosis and ligament calcification were the most commonly found ones in 90% and 50% of the patients. There was a study proposed that coal-burning type of endemic skeletal fluorosis showed the worst articular damage, and fluorosis olecranarthropathy had higher diagnosis value for this type of endemic.

There were reports suggesting that different morphological ossification of interosseous membrane as key manifestation of skeletal fluorosis. For decades, Chinese scientists' discovery and findings on X-ray manifestation and diagnosis of skeletal fluorosis have long been the guideline of diagnosis and treatment of endemic fluorosis in China. *Endemic Skeletal Fluorosis Diagnosis Criteria of China (WS192-2008)* issued by Ministry of Health categorized X-ray manifestation into: osteosclerosis, osteoporosis, osteomalacia, bone turnover, articular degenerative changes, degenerative changes in osseous peripheral soft tissues. These manifestations were further divided into mild, moderate, and severe degrees with 37 types of X-ray manifestations, which standardized medical imaging diagnosis of skeletal fluorosis. Yet, past research was conducted by traditional X-ray imaging technology. The fast-developing imaging technology and digital X-ray imaging equipment were widely used for clinical purposes nowadays. Adopting modern digital X-ray imaging equipment for endemic skeletal fluorosis research remains blank. Whether or not modern digital CR and DR equipment could discover more valuable X-ray manifestations required further research.

Three types of skeletal fluorosis could be categorized according to the changes in bone density and structure.

1. Osteosclerosis: (a) increased bone density, thickened trabecular, fused bone cortex becomes thickened, medullary cavity is narrowed or disappeared; (b) ossified interosseous membrane and peripheral ligament.

2. Osteoporosis: (a) reduced bone density, sparse trabecular. Sclerotin shows resorption and decalcification or deformed bone in different degrees; (b) ossification of interosseous membrane or osseous peripheral ligament.
3. Mixed type: with both features of osteosclerosis and osteoporosis, and different degrees of bone hyperplasia and bone resorption, as well as web-shaped cancellous bone or cyst-shaped structure, sparse cortical structure and obviously reduced trabecular count in unit area.

All of the above types have varying degrees of change.

1. Early-stage change: any one of the following manifestations can be diagnosed as early stage skeletal fluorosis: (a) flaky spot-shaped texture or thickened and disorganized bone texture are observed obviously in diaphysis epiphysis and pelvic bones; (b) two and more plumule break-ground-shaped bone spur are observed in cortical margin of extremities' diaphysis. Wave-shaped osteophytes in radius spine are observed.
2. Mild degree: (a) bone density is slightly higher than normal, trabecular is coarse-grained. Bone spot can be observed, trabecular is attenuated and its density is decreased; (b) cuspidated ossification is observed in muscle ligament attachment, obvious osteophyte is observed in interosseous membrane.
3. Moderate degree: (a) bone density is obviously increased, cortex and trabecular are thickened, and they are partially fused; or cortex is attenuated, trabecular is thin and sparse, density is lowered, or trabecular is sparse, with thick and large texture and increased density; (b) interosseous membrane and osseous peripheral ligament are ossified in relatively wide scope.
4. Severe degree: (a) trabecular is thickened and the majority of it is fused into flaky, medullary cavity cortex boundary is ambiguous; osteoporosis is observed in patchy resorption, partial cortical bone is disrupted and disappeared, deformed bone is obvious. (b) Ossification of interosseous membrane and peripheral ligament is more obvious and even in bridge-shaped. Calcification of other soft tissues (such as vessel) can be observed.

CT and MRI research fell short in skeletal fluorosis. Past research analyzed that CT features of spinal stenosis of skeletal fluorosis patients as thickened ligamentum flavum, ossification caused spinal stenosis in multiple intervertebral space, and accompanied with osteosclerosis in intervertebral facet joints. Such analysis was in accordance with a report on CT and MRI manifestations (Kumar et al. 2009). Another study reviewed CT manifestations of 37 skeletal fluorosis patients and discovered that 33 cases were accompanied with osteosclerosis in intervertebral facet joints in different degrees. A research on spinal cord compression symptoms of fluorosis discovered different degrees of spinal stenosis in skeletal fluorosis, bone density of spine was obviously higher than that of healthy one. In addition, multiple intervertebral anterior and posterior longitudinal ligaments were found to be thickened and calcified. It was also found that intervertebral facet joints osteosclerosis, narrowed intervertebral foremen, together with transverse diameter, anteroposterior diameter and spinal canal size were significantly smaller than normal value. A past research concluded that fluoride showed extraordinary damage on axis skeletal in spine and peripheral soft tissues, the calcification or ossification of soft tissues in spinal cord caused spinal stenosis and compressed spinal cord and nerve (Fisher et al. 1989).

X-ray and MRI manifestations of 18 patients were analyzed. The results showed that major imaging manifestations of fluorosis arthrosis were thickened synovium, swollen joint capsule, joint fluid, broadened joint space, and calcified articular cartilage. Hence, joint space was narrowed, osseous articular surface was eroded and became ambiguous or sclerotic joints; free bodies were observed in articular cavity, osteosclerosis was observed in joint edge, peripheral ligament and joint capsule became calcified. Even joint space was ambiguous. A report reviewed MRI manifestations of spinal stenosis in skeletal fluorosis (Li et al. 2009) and indicated that skeletal fluorosis spinal stenosis was mainly caused by thickened and ossified ligamentum flavum and posterior longitudinal ligament. The manifestation was signal changes in long T1 short T2. Herniated disc served as another reason. In addition, other researches showed that vertebral body of skeletal fluorosis patients appeared abnormal signal, showing all vertebral bodies with reduced signal in T1WI and T2WI and suggesting that MRI should have a great value on diagnosis of skeletal fluorosis stenosis and such disease was multifocal or multicentric (Zhu et al. 2017; Ahsan et al. 2016). Some studies disagreed with MRI results on ligament calcification. MRI mainly served to confirm the severity and scope of spinal stenosis, find out the location of spinal cord compression and spinal cord condition. Such results were significant to differentiating diagnosis and surgery instruction. In recent years, with the popularization of CT and MR equipment, the advantageous CT and MR might have pivotal clinical significance in further study of skeletal fluorosis imaging manifestations, and early stage manifestations in particular.

Diagnostic Value of Olecranarthropathy for Skeletal Fluorosis

Articular changes in skeletal fluorosis are called fluorosis arthrosis. Medial epicondyle and lateral epicondyle are the attachment of extensor tendon and flexor muscle of forearm and wrist. Ossification of these two parts is obvious and typical in olecranarthropathy induced by skeletal fluorosis, but rare in other types of olecranarthropathy. Therefore, ossification in this part could be considered with relatively higher diagnostic value for skeletal fluorosis. X-ray results of forearm, shank and pelvic of the patients with coal-burning type of fluorosis suggested that fluorosis olecranarthropathy was most representative in X-ray diagnosis of this type of fluorosis. It also proposed that calcification of tendon attachment in elbow was the specific manifestation of coal-burning type of endemic fluorosis and had relatively higher differentiating diagnostic value. The study also considered that severity of fluorosis olecranarthropathy could serve as one of the division criteria of this type of fluorosis. It supported the diagnostic value of ossification of medial epicondyle and lateral Epicondyle on skeletal fluorosis.

Ossification of radius annular ligament is formed on specific osteophyte on lateral edge of radius. Radius annular ligament attachment is located in leading-edge and trail-edge of ulna radial notch. Together with ulna, it forms a bone fiber ring with big upper opening and small lower opening, containing radius rotating in the ring without prolapse. Previously mentioned ossification shadow happens to be the upward section extension of lateral radius annular ligament ossification. Radius annular ligament ossification had key value in fluorosis olecranarthropathy diagnosis (Duan et al. 2012). A large amount of clinical and research data indicated that fluorosis shared relevance with degenerative ligament and ossification. In patients with ossification of ligamentum flavum, fluoride content in the ligamentum flavum tissues is significantly increased, even without skeletal fluorosis.

It is pivotal to differentiate fluorosis olecranarthropathy from degenerative arthropathy. Sometimes, both diseases can be found in the same joint of the patient. Modern histopathology and electron microscopy observation have already proved that: fluorosis olecranarthropathy is due to excessive fluoride directly caused articular cartilage degeneration, necrosis, and particle deposition of matrix calcium salts. Articular osteosclerosis and hypotrophy, fibrous cartilage degeneration, osteophyte, ossification, articular deformation and function disorder occur in multiple parts (joint capsule, articular peripheral tendon, and ligament attachments). Degenerative arthropathy is also called arthrosis. It is the resorption, osteophyte reaction (to degeneration, atrophy, and necrosis of articular cartilage due to aging) or secondary to the articular cartilage degeneration and post-degeneration osseous reaction due to physical injury or inflammation. These two shared different pathogenesis and different imaging manifestations. The latter showed little calcification in peripheral ligament of vertebral body and facet joints when it was found in spine. This paper believed that degenerative arthropathy in elbow joints would observe little ossification of tendon.

Fluorosis arthropathy in elbow joint show irregular, disrupted, and sclerotic osseous facet joints. Narrowed joints space, cyst below osseous facet joints, and formation of multicentric osteophyte in epiphysis are the manifestations of fluorosis arthropathy as well as the common manifestation of degenerative arthropathy. The ossification of peripheral tendon attachment in elbow joints is the characteristic manifestation of fluorosis olecranarthropathy, especially ossification of extensor tendon and flexor muscle of radius annular ligament, medial epicondyle, and lateral epicondyle. These manifestations can be adopted as diagnosis basis and features for skeletal fluorosis (Fig. 4.1).

Differentiating Diagnosis

Differentiating diagnosis from skeletal fluorosis should identify osteopetrosis, osteoblast metastatic cancer, and renal bone disease. Osteopetrosis can be observed with increased bone density, transverse stripe-shaped shadow in tubular bone, multi-layer wave-shaped dense shadow in the ilium and tarsal bones, which are much clear than

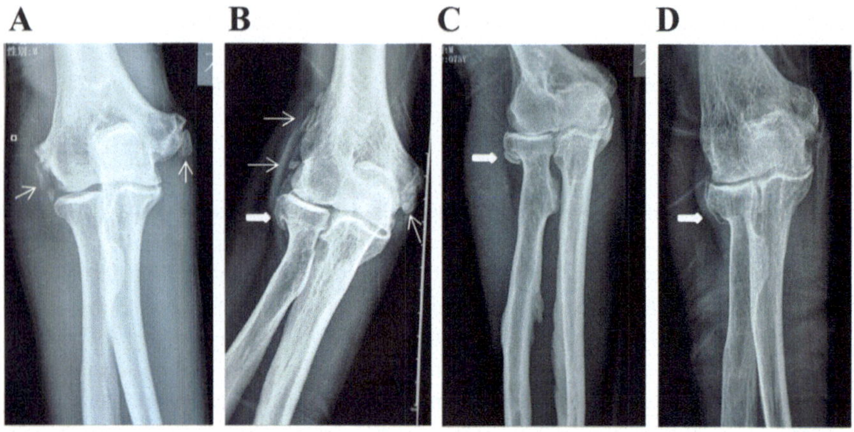

Fig. 4.1 X-ray examination of skeletal fluorosis. (**a**) The formation of ossified free bone blocks at the attachment of the tendon of the medial and lateral epicondylar ligament of the humerus (thin arrow). (**b**) Distortion of elbow joint with visible ossification of radial annulus ligament (thick arrow). (**c**) The interosseous membrane of the ulnar and radius bone is obviously ossified (thick arrow). (**d**) Ossification of the annulus radial ligament (thick arrow)

those of skeletal fluorosis. Sclerotic changes in osteoblast metastatic cancer are not evenly distributed and the modification of osseous structure is often observed. Renal bone disease is very similar to some skeletal fluorosis, with generally dense and/or loose bone, coarse, and fuzzy trabecular bone, etc. It is often difficult to distinguish from X-ray signs only, and it needs to be differentiated by combining epidemiology, clinical manifestations and renal function examination. No calcification of ligaments is found in any of the above diseases.

Value of Digital Radiography in Skeletal Fluorosis by X-ray Check and Screening

Digital radiography (DR) was one of the modern digital imaging technologies. With the fast-development of medical imaging technology, DR test was widely used in clinical practice and widely promoted. It had become the mainstream technology for clinical X-ray check. Over the past decades, imaging screening of endemic skeletal fluorosis in endemic areas relied on traditional X-ray simulated imaging technology, which was film photosensitive image forming. In recent years, DR digital imaging technology was adopted by this study on screening endemic skeletal fluorosis in endemic areas. In addition, past endemic skeletal fluorosis imaging data was combined to investigate the value of DR in endemic fluorosis screening.

The study took photo quality and evaluation criteria into consideration to conduct research on skeletal DR photos and traditional X-ray simulated imaging photos. It was concluded that the former was better than the latter.

Since Roentgen, the German scientist, discovered X-ray in 1895, traditional X-ray simulated imaging technology (film photosensitive imaging technology) was adopted in clinical practice for more than 100 years. *Diagnosis Criteria of Endemic Skeletal Fluorosis* (WS 192-2008) of China specifically requires X-ray photos of patient's forearm, shank, and pelvic to diagnose skeletal fluorosis. Since the 70s to 80s, screening and diagnosis of skeletal fluorosis heavily relied on film photosensitive imaging assay. As a simulated imaging technology, the quality of the photo was subject to multiple factors (such as performance of X-ray machine, quality of X-ray, skills of technician, quality of film, quality of intensifying scree, development, fixing solution). Even the cartridge handling way would affect the quality and definition of the photo. Mishandled film was common. Because of these factors, photosensitive imaging technology posed tremendous difficulty on the diagnosis of skeletal fluorosis due to its direct effect on accuracy and positive rate of the disease.

In 1997, Philips Electronics Co. (Holland) launched the first digital radiography (DR) device in the world. Such device laid a foundation for the comprehensive image digitization of medical imaging. With the fast-development of medical technology, numerous hospitals in China shifted from traditional film photosensitive imaging to digital radiography. Such shift availed these hospitals with cutting-edge technology in imaging screening and diagnosis in endemic skeletal fluorosis. Unlike the traditional simulated imaging, DR was better at effective detective quantum efficiency (eDQE) (Bertolini et al. 2012). DR directly transformed X-ray photon into digital images via electronic cassette, and the computer organized the data into a complete X-ray diagnosed digital image. In such process, data loss was avoided and man-made interference could be reduced. Thus, image quality was obviously improved, details of symptoms were more obvious, and it was better for locating the disease and qualitative diagnosis. The application of DR technology in screening and diagnosis in endemic skeletal fluorosis would better presenting imaging manifestations of skeletal fluorosis with its reliable technology. In addition, the image could clearly present affected histological structure of skeletal fluorosis with choices of different sizes, location of the image, and zoom-in options. These functions were preferrable to distinguish healthy ones from diseased ones. Post-processing platform of DR image could further improve the quality of image by obviously lowering or eliminating grid artifact. In this way, it could significantly improve accuracy and positive rate of skeletal fluorosis diagnosis. For now, portable DR technology was mature and user-friendly. It was incomparably excellent in work intensity, process, quality of image, radiation dose, data management, and online transmission.

Compared with traditional X-ray simulated imaging, DR imaging was more advantageous and valuable in screening and diagnosis of endemic skeletal fluorosis. It was suggested that such technology could be widely used in general check, endemic monitoring and treatment efficacy evaluation.

Scholars of medical imaging made some achievements in imaging application on endemic skeletal fluorosis. However, with the advancement of science and technology, medical imaging devices were also evolving. It was clinically possible to study occurrence and development of disease from molecular level with the assistance of modern imaging device. Thus, such technology would lend support to the research of endemic skeletal fluorosis and enable medical imaging to contribute more to the prevention and control of endemic fluorosis.

Conclusion

Skeletal fluorosis triggered by endemic fluorosis (regardless coal-burning type or drinking water type) showed X-ray manifestations of changes in sclerotin density, peripheral osseous ligament, tendon, periosteum, and joints. Imaging manifestations of these three syndromes were vital for diagnosis and severity determination since they showed certain specificity in X-ray, CT, or MRI imaging. Imaging manifestations of CT and MRI imaging were in accordance with these three syndromes. CT results showed more detailed results, and MRI imaging was capable to show changes ahead of X-ray and CT detection competence. MRI test produced no radiation and would play a more important role in early diagnosis of skeletal fluorosis. It might serve as a major imaging measure for early diagnosis, endemic determination, and prevention works of endemic skeletal fluorosis.

References

Ahsan T, Jabeen R, Hashim S, Bano Z, Ghafoor S. Fluorosis... causing paraplegia... mutilating life. J Pak Med Assoc. 2016;66:213–6.

Bely M. The inhibited enchondral ossification in experimental osteofluorosis in rats. Fluoride. 1991;24:17.

Bertolini M, Nitrosi A, Rivetti S, Lanconelli N, Pattacini P, Ginocchi V, et al. A comparison of digital radiography systems in terms of effective detective quantum efficiency. Med Phys. 2012;39:2617–27.

Burgener D, Bonjour JP, Caverzasio J. Fluoride increases tyrosine kinase activity in osteoblast-like cells: regulatory role for the stimulation of cell proliferation and Pi transport across the plasma membrane. J Bone Miner Res. 1995;10:164–71.

Du FM, Duan QH, Chen XG, Li Y, Zuo YH, Jjao J, et al. The differences of X-ray findings of skeletal fluorosis between coal-burning type endemic fluorosis and industrial fluorosis. J Pract Radiol. 2017;33:83–6.

Duan QH, Chen XG, Wang XL, Jiao J, Hu J, Guan ZZ, et al. Ossification of ligament and tendon attached around the elbow joint in diagnosis of skeletal fluorosis. J Pract Radiol. 2012;28:1738–40.

Farley JR, Wergedal JE, Baylink DJ. Fluoride directly stimulates proliferation and alkaline phosphatase activity of bone-forming cells. Science. 1983;222:330–2.

Fisher RL, Medcalf TW, Henderson MC. Endemic fluorosis with spinal cord compression: a case report and review. Arch Intern Med. 1989;149(3):697–700.

Gruber HE, Baylink DJ. The effects of fluoride on bone. Clin Orthop Relat Res. 1991;267:264–77.

Gupta N, Gupta N, Chhabra P. Image diagnosis: dental and skeletal fluorosis. Perm J. 2016;20: e105–6.

Jiang N, Guo F, Xu W, Zhang Z, Jin H, Shi L, et al. Effect of fluoride on osteocyte-driven osteoclastic differentiation. Toxicology. 2020;436:152429.

Kleerekoper M. Fluoride and the skeleton. Crit Rev Clin Lab Sci. 1996;33:139–61.

Kumar H, Boban M, Tiwari M. Skeletal fluorosis causing high cervical myelopathy. J Clin Neurosci. 2009;16:828–30.
Lau KH, Wu LW, Yoon HK. Inhabition of phosphotyrosine dephosphorylation (p-tyr) leads to increase p-tyr phosphorylation of MAP Kinase (MAPK) activity, and human bone cell proliferation. Bone. 1995;16:96.
Li SF, Liu BG, Kang LQ, Sun LY, Ran FD, Zhang L. MRI appearances of endemic skeletal fluorosis of spine. J Pract Radiol. 2009;25:1290–3.
Li C, Qin Y, Ouyang T, Yao M, Zhang A, Luo P, et al. miR-122-5p mediates fluoride-induced osteoblast activation by targeting CDK4. Biol Trace Elem Res. 2020;199(4):1215–27.
Lian ZC, Wu EH. Osteoporosis - an early radiographic sign of endemic fluorosis. Skelet Radiol. 1986;15:350–3.
Sun DJ, Shen YF, Zhao XH, Wei HL, Zhang XF, Sun YF, et al. Analysis on the disease trend and present state of endemic fluorosis in China mainland. Chin J Endem. 2001;20:429–33.
Wang Q, Wu SX. Bone X-ray findings and pathological basis of experimental fluorosis. Chin J Radiol. 1992;26:552–4.
Wang Y, Yin Y, Gilula LA, Wilson AJ. Endemic fluorosis of the skeleton: radiographic features in 127 patients. AJR Am J Roentgenol. 1994;162(1):93–8.
Zhu C, Zhou S, Liu L, Song Y. Multiple levels of spinal canal stenosis in endemic skeletal fluorosis. Joint Bone Spine. 2017;84:737.

Chapter 5
Epidemic Characteristic and Pathogenesis of Dental Fluorosis

Jian Liao

Abstract Dental fluorosis, as an endemic disease, is an enamel development disorder caused by excessive exposure to fluoride during development stage of teeth. Such endemic shows differences in patients' age, geographic location, and dwelling places. Dental fluorosis is often found in child patients and poses severe damage on the patients. Prevalence rate in children with dental fluorosis aged 8–12 is precise to mirror the overall prevalence rate and damage of this endemic. The pathogenesis of dental fluorosis is yet to be determined. Its major effects are enhancing the stability of crystal structures, affecting teeth development, and promoting the re-mineralization of teeth.

Keywords Dean's fluorosis index · Dental fluorosis · Caol-burning type of endemic fluorosis · Epidemiologic feature · Pathogenesis

Introduction

Dental fluorosis is a special symptom as well as the most typical symptom (Gu et al. 2020) of endemic fluorosis manifests on teeth. This special type of enamel development disorder (Sun et al. 2018) is caused by excessive intake of fluoride in teeth development stage. Clinical manifestation of dental fluorosis is the absence of health shade of teeth in white foggy appearance or yellow/brown coloring. Severe cases come with irregular enamel defect or even deformed tooth (Fig. 5.1).

Premolars are the worst affected teeth in patients with chronic endemic fluorosis (Moller et al. 1970). It was found that exposure to fluoride before children was 3 years old could increase the risk of dental fluorosis in upper central incisor (Do et al. 2014), and 2–8 years old are susceptible age for the later erupted permanent teeth (except third molar) (Bhagavatula et al. 2018).

J. Liao (✉)
The Key Laboratory of Head and Neck Cancer Research, Hospital and School of Stomatology, Guizhou Medical University, Guiyang, Guizhou, People's Republic of China
e-mail: liaojian@gmc.edu.cn

Z.-Z. Guan (ed.), *Coal-burning Type of Endemic Fluorosis*,
https://doi.org/10.1007/978-981-16-1498-9_5

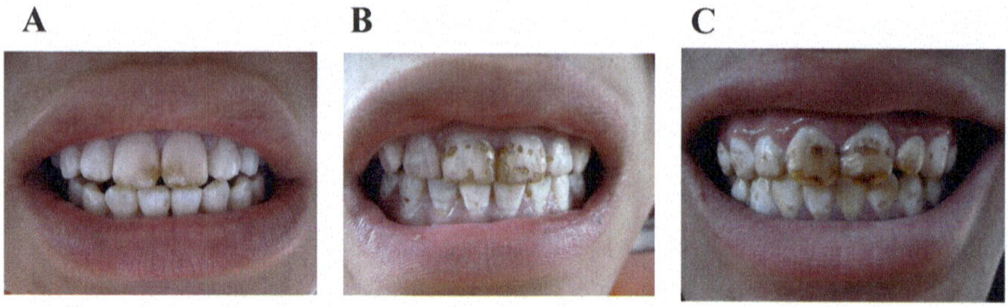

Fig. 5.1 Irregular enamel defect or deformed tooth. Patients with dental fluorosis in the area of coal-burning type of endemic fluorosis, showing irregular enamel defect or even deformed tooth

Endemic fluorosis is categorized into three types: drinking water type, brick-tea type, and coal-burning type. Dental fluorosis cases are confirmed and reported in all over the world. In China, coal-burning type of endemic fluorosis areas are found in 15 provinces (cities) affecting 16.5 million patients with dental fluorosis, such as Guizhou, Hubei, Sichuan, Chongqing, Yunnan, Hunan, and Shaan Xi.

In numerous essays, dental fluorosis and mottled teeth were considered as the same thing. However, there are certain differences concerning the two: mottled teeth include dental fluorosis triggered by fluoride, mottled teeth triggered by non-fluoride factors, and specific enamel opaque of unknown cause. Therefore, mottled tooth is an umbrella yet ambiguous concept.

General Epidemiological Characteristics of Dental Fluorosis

Age Distribution

Age is an independent risk factor for dental fluorosis occurrence (Brahman et al. 2014; Guissouma et al. 2017). Since the difference in development of ameloblast and enamel in deciduous teeth, permanent teeth, and other teeth, dental fluorosis onset is featured with obvious age characteristics.

Deciduous Teeth

The deciduous or permanent teeth have their own characteristics due to their different susceptibility to fluoride. Placenta serves as barrier against fluoride to a certain degree, hence, only limited amount of fluoride penetrates the placenta barrier and cannot cause dental fluorosis in deciduous teeth. However, excessive fluoride would still penetrate placenta barrier and cause dental fluorosis in deciduous teeth. Permanent teeth are mainly affected by endemic fluorosis. The permanent teeth erupt after the age of 6, which add to the risk of morbidity rate. Around 12 years old, when all the permanent teeth are erupted and dental fluorosis have irreversible damage on teeth, and morbidity rate of this endemic maintains at a relatively stable level. Dental caries or periodontal diseases are more likely to cause the permanent teeth loss at middle age, and morbidity rate might decrease after such age.

For people migrate from low fluoride-containing area to high fluoride one, their dental fluorosis morbidity rate and lesion degree are closely related to the age of migration. For those migrate younger than 2 years old, their dental fluorosis morbidity rate and lesion degree of permanent teeth after years of migration show no significant difference with that of locally born and raised people in high fluoride-containing area. For those migrate over 2 years old, dental fluorosis morbidity rate would be lowered and lesion degree would be reduced with the increase of age. For those migrate over 7 years old, since the enamel is basically mineralized, they show no significant differences with those born and raised in low fluoride-containing area. As the third molar mineralize at later stage of life, for those migrate into endemic area after their teens might still be affected by dental fluorosis.

Gender Distribution

There are no significant differences in gender distribution of dental fluorosis. The oral epidemiology survey indicated that dental fluorosis morbidity rate and dental fluorosis index of male and female were both 11.7% and 0.25.

Geographical Distribution

Coal-burning type of endemic fluorosis areas are widely found in in over 15 provinces/cities/autonomous regions (Yunnan, Guizhou, Sichuan, Hubei, Hunan, Guangxi, Jiangxi, Henan, Shaanxi, Shanxi, Ningxia, Hebei, Beijing, and Liaoning) of China. Over 30 million people reside in 197 inflicted counties, including 16.5 million patients with dental fluorosis and 1.08 million patients with skeletal fluorosis. Severe endemic areas are often found in southwest China (such as northeast of Yunnan, west and northwest of Guizhou, east of Sichuan and the Three Gorges area), which feature with dire contaminated food and air. Northeast of China (Benxi county in Liaoning) features with contaminated air, thus manifests with relatively mild symptoms.

As the major manifestation of endemic fluorosis, dental fluorosis features with geographic aggregation, a close relationship with local high fluoride nature and access to high fluoride content. According to a research conducted by *Institute of Geochemistry Chinese Academy of Sciences*, contributory factor of severity differences in coal-burning type of endemic fluorosis area in southwest China lies in altitude. Regions on 1000–1500 m above sea level are categorized as alpine regions, featuring cold climate and 6 months of long heating period (by fire). What's more, raining season in summer time would also require heating by fire. Since the heating heavily relies on anthracite coal and open flame, fluoride has a long window to contaminate air. To make it worse, the short frost-free period in such areas is only suitable to grow corn. Hence, the short harvest time collides with unbroken spell of wet weather leaves open flame as the only option for corn dehydration. Corns processed in this way are inevitably contaminated by the fluoride, thus endemic areas with such features often turn out as severe endemic areas. Regions on 500–1000 m above sea level are often characterized with warm climate and short heating period. Oryza sativa and corns are both grown in such area. Solarization is often chosen for corns and oryza sativa dehydration, which could avoid fluoride contamination. Anthracite coal baking is only alternative, and its fluoride contamination time is shorter than previously mentioned regions. Thus, degree of fluoride contamination is reduced. In addition, corn only composes part of the staple food, so fluoride contaminated food intake in such region is less than the previously mentioned region and region with such altitude is not evaluated as severe endemic area. Regions on less than 500 m above sea level often feature with oryza sativa growing and rice as staple food. Hence, fluoride intake from food is limited. Source of fluoride is limited to cooking and short-term heating by anthracite coal. Thus, fluoride intake is less and such endemic area is evaluated as mild one. The storage of corn is also related with fluoride contamination. For example, certain areas have the habit of storing dehydrated corn or rice in jars, which is an ideal way to prevent the food from fluoride contamination. Some areas are accustomed to hang the corn from the ceiling or on the scaffolding, which facilitates the time and amount of fluoride contamination. Pepper is a major component of diet for southwestern residents. No matter for solarization approach area or baking (coal-burning) approach area, residents all hang the pepper on outdoor scaffoldings. Such practice is prone to fluoride-containing soot contamination but with differences in its content. Therefore, fluoride content in pepper is also a contributory factor in endemic fluorosis areas distribution.

Fluoride content of in-door air in coal-burning type of endemic fluorosis area varies drastically due to multiple reasons (living condition, ventilation, coal-burning habit and quantity, heating habit and time, as well as diet habit). As a result, there could be significant differences in fluoride intake time in the same family. It is impossible to determine the fluoride intake of individuals on daily basis but requires more assays to do so. It is advisable to compare the urinary fluoride levels of residents in northern drinking water type of endemic fluorosis area with that of residents in coal-burning type area to measure the fluoride intake level of residents.

Urban and Rural Distribution

Dental fluorosis is possible to affect both urban and rural residents. The *3rd National Oral Health Epidemiology Survey* indicated that morbidity rate of such disease in countryside was higher than that of urban areas, for example, morbidity rate in

12-year-old age group was 9.0% in city and 14.3% in countryside. Such differences could be resulted from economic disadvantages of the countryside. Since local residents used underground water with super fluoride content, morbidity rate of dental fluorosis in countryside was relatively high (Liu et al. 2016). Plus, such difference could be caused by asymmetric fluoride content in air and food from coal-burning contamination.

In the areas of coal-burning type of endemic fluorosis, the air is severely contaminated by fluoride. Peak concentration and average concentration of fluoride content in air of severe endemic area could reach as much as 40 times of single maximum allowable concentration and daily average concentration according to national standard. Hence, air contamination is a major source of fluoride in endemic area. Fluoride intake in respiratory tract and dental fluorosis morbidity rate presents a fine linear dose–effect relationship. In endemic areas with less than 3.5 mg/day of total fluoride intake, the fluoride intake via respiratory tract reaches 0.3–0.4 mg/day, which is 3.8–5 times more than that of allowance. Such result causes obvious prevalence of dental fluorosis (morbidity rate of such disease reached 50–70%) due to the excessive fluoride content in air. Fluoride content in air of mild/moderate/severe endemic areas present linear dosage–effect relationship with morbidity rate of dental fluorosis, correlation coefficient reaches 0.93 ($P < 0.01$).

In coal-burning type of endemic fluorosis areas, the fluoride content in food is increased significantly because of the fluoride contamination. For example, fluoride content in corns reaches 87 mg/kg and people's fluoride intake reaches 4–20 mg/day through food. Since the low fluoride content in water, fluoride intake via food is the major component of total fluoride intake. Dental fluorosis morbidity rate reaches over 98%. The fluoride intake via staple food, food and digestive tract shows obvious linear dosage effect relationship with dental fluorosis morbidity rate and lesion severity.

Epidemiological Significance and Medical Application of Dental Fluorosis Morbidity Rate

Morbidity Rate of 8- to 12-Year-Old Children

If dental fluorosis morbidity rate of individual has any significance for individual diagnosis, the morbidity rate of certain population/location can be applied for the diagnosis of population fluorosis, i.e., reflecting, evaluating, and determining the prevalence and harmfulness of fluorosis. Therefore, morbidity rate of dental fluorosis is pivotal for endemic fluorosis prevention and control. For example, it can be used to evaluate and determine the endemic areas of fluorosis. At the meantime, the difficulty in investigation, low credibility, and low comparability in data of morbidity rate in certain location (migrants, mobile population, age, and living habits) pose tremendous challenges on the prevention and control of the endemic. Hence, a relatively concentrated and stratified group should be chosen to represent the whole population with the features of low mobility, stable age structure, minimum affecting factors, and strong comparability. Population aged 8–12 is the perfect choice to represent the morbidity rate and harmfulness of disease on the whole population. Mineralization of permanent tooth basically complete between 7 and 8 years old. The eruption of permanent teeth marks little influence of age on dental fluorosis morbidity rate. Thus, the total dental fluorosis morbidity rate of children aged 8–12 should be in line with the whole population. Such group boasts with the following advantages: stable age structure, stay at school with little mobility and are easy to be organized for medical check-up; such group also shows fewer affecting factors and migrants are easy to rule out for data credibility; year by year data and data from other sources can be compared directly. These are the reasons that dental fluorosis morbidity rate of 8–12 years old children is preferred in practical research. In 1981, the 1st version of *Endemic Fluorosis Prevention and Control Standard* (*Experimental*) in China stated: 30% and above of dental fluorosis morbidity rate of locally born and raised 8–15 years old children indicating such area as mild endemic area. The current criteria revised according to actual situations: 30% and above of dental fluorosis morbidity rate of 8–12 years old children indicating such area as mild endemic area.

Dean's Index for Dental Fluorosis and Epidemic Grading

Dean proposed epidemic grading criteria for dental fluorosis based on Dean's calculation concerning dental fluorosis index of the population (Dean 1990). Such criteria categorized dental fluorosis endemic into the following levels: negative, questionable, mild, moderate, moderately significant, severe, and most severe. Such criteria are very detailed in endemic grading. Such criteria has been used in some areas of endemic fluorosis, which is considered to be desirable in regional division of the disease. Content of the *Dean's Index* as follows.

1. Dean's Index for dental fluorosis: such index presents endemic intensity of dental fluorosis quantitatively. It is calculated according to the following equation: Dental fluorosis index = (suspected cases × 0.5 + extremely mild cases × 1 + mild cases × 2 + moderate cases × 3 + severe cases × 4)/tested cases, where 0.5, 1, 2, 3, and 4 were the weighting coefficients.
2. Dean's dental fluorosis epidemic grading: endemic intensity in endemic area is categorized quantitatively according to the morbidity rates. Specific criteria are shown in Table 5.1.

Table 5.1 Standard for prevalence and grading of dental fluorosis determined by Dean's index

Epidemic process in the area of endemic fluorosis	Morbidity rate of dental fluorosis				Dental fluorosis index
	Extremely mild (%)	Mild (%)	Moderate (%)	Severe (%)	
Negative	<10				<0.4
Questionable	10–35				0.4~
Mild	>35	<50			0.6~
Moderate	>50	<35			1~
Moderately severe			35–50	<35	2~
Severe			>50	<35	3~
Most severe				>35	>4

Dental Fluorosis Morbidity Rate Served as Major Index for Determining and Classifying Endemic Areas

Endemic area confirmation is a major step for the prevention and control, which makes it possible to invest substantial amount of personnel and monetary support for the prevention. The so-called endemic area refers to endemic fluorosis rampant areas. Dental fluorosis morbidity rate of local residents is a key condition to determine whether or not an area was endemic fluorosis rampant. Fluorosis area grading criteria states that: dental fluorosis morbidity rate of locally born and raised between 8 and 12 years old population exceeds 30% would be identified as endemic fluorosis area. It must be noted that confirmation of coal-burning type of endemic fluorosis is determined by the condition of patients. Hence, dental fluorosis morbidity rate shoulders great significance in determining endemic area as well as prevention and control.

Dental Fluorosis Morbidity Rate Served as Major Index to Evaluate Prevention and Control Efficacy

The confirmation and classification of endemic area is based on the condition of endemic. Prevention and control efficacy evaluation is also based on change of endemic conditions. Previous study (Mohd Nor et al. 2020) indicated that fluorosis morbidity rate significantly decreased after fluoride concentration in water was reduced or ceased to use such water source. Theoretically, dental fluorosis morbidity rate of 8–12 years old children should be decreased and fluorosis severity should be alleviated years after improved water projects or stoves projects. Such changes shared a close relationship with the duration of water/stove improvement projects and age of tested children. The reasons were simple: development and mineralization of permanent teeth occurred between 7 and 8 years old. If such water/stove improvement projects lasted over 7–8 years, and target areas would be up to hygiene standard, there should be a drastic difference concerning dental fluorosis morbidity rate between children born before and those born 7–8 years after the projects. Dental fluorosis morbidity rate of the former was supposed to maintain at a relatively high level, while that of latter should be decreased to within 30% with alleviated lesion degree and increased chalk structure. Children born between such period (between 1 and 7 years of improved projects) would show a gradually decreased morbidity rate and endemic degree. Such results lead to the distinction of dental fluorosis morbidity rate and endemic degree among elder and young children, indicating age characteristics unique to the endemic areas received improved water/stove projects.

Acceptance and utilization of the general public on prevention and control measures were the prerequisites of prevention efficacy. However, such efficacy must be tested. If the 7–8 years or longer water/stove improvement projects failed to significantly decrease dental fluorosis morbidity rate in 8–12 years old children, and children born 7–8 years after the water/stove improvement initiative showed morbidity rate over 30%, following aspects must be reviewed for better efficacy: implementation of prevention and control measures, hygiene standard of prevention facilities quality and habitual use of such facilities in general public. In addition, possible flaws of prevention and control measures must be pondered to sever major sources of fluoride. Hence, the condition changes of fluorosis, represented by dental fluorosis morbidity rate of 8–12 years old children, served to evaluate prevention and control efficacy, and lent support to analyze relationship between endemic conditions and prevention efficacy together with utilization conditions. In this way, locating the cause would lead to right measures. All in all, dental fluorosis morbidity rate of 8–12 years old children is direct and authoritative in evaluating prevention and control efficacy.

Dental Fluorosis Morbidity Rate Mirrored the Severe Damage of Fluorosis on Children

Chronic fluorosis in children manifested in dental fluorosis, and damage on their growth and development. Although children with fluorosis showed skeletal changes and non-bone damage, such manifestations were typically one time and self-limited or reversible in later development. To this regard, dental fluorosis was the manifestation of fluorosis in children and its morbidity rate of children represented harmfulness of fluorosis on children (part of the harm). Such harmfulness would pose further damage on their growth and development, as well as their physical and mental health, such as deformed physique, mental retardation or mental disorder (Lou et al. 2021; Molina-Frechero et al. 2017; Saeed et al. 2020). Such damage would worsen with the increase of fluoride intake and prolonged duration. The damage could also progress to adulthood and lead to adult skeletal fluorosis. With effective measures to reduce the overall fluoride intake, such damage in children could be alleviated or even eradicated.

Dental fluorosis morbidity rate of 8–12 years old children was an accessible and reliable index to evaluate fluorosis rate in population. Investigation on dental fluorosis morbidity rate of 8–12 years old children was quite practical, and collecting urine (once) sample randomly is easy to practice. Therefore, such morbidity rate had pivotal significance in prevention and control of endemic fluorosis.

Relationship Between Prevalence of Dental Fluorosis and Urinary Fluoride Levels

Studies have shown that there is a significant dose–effect relationship between the degree and prevalence of dental fluorosis and the amount of fluorine taken in the body (Ding et al. 2011). The majority (around 85%) of fluoride absorbed in the body, especially those in blood, would be discharged by urine. There could be individual differences in urinary fluoride value, and such difference would also be found in different time of the day on the same individual. However, urinary fluoride level of population maintained at a stable one in a relatively longer period. Such level showed direct relationship with amount of fluoride intake. It increased with the augment of fluoride intake and mirrored the fluoride intake level in the population. It could be concluded that the higher fluorosis morbidity rate (such as dental fluorosis morbidity rate) was, the higher the urinary fluoride level in population was. These two shared a parallel relationship. Despite urinary fluoride level could not be adopted as diagnosis standard for individual fluorosis, it could serve as reference for endemic fluorosis diagnosis in population. It would be of great convenience if urinary fluoride level and dental fluorosis morbidity rate could be chosen as the key index to confirm and classify endemic areas. Compared with the endemic condition changes, urinary fluoride level showed advantages in quick response and obvious recent development towards fluoride intake. For example, in water/stove improvement endemic areas, dental fluorosis morbidity rate of 8–12 years old children might not show obvious changes in the short term, but urinary fluoride level in children could have been decreased significantly. Hence, urinary fluoride level in children could serve as key index to evaluate efficacy of prevention and control measures, as well as the facilities. Due to the slow response and longer response time (7–8 years), dental fluorosis morbidity rate could be chosen as reference for medium- and long term-efficacy. These two indexes could comprehensively describe the changes of endemic condition and evaluate prevention efficacy objectively.

Physiological Effect of Fluoride on Teeth

Biological Effects of Fluoride on Teeth

Fluoride could enhance the stability of crystal structure. If there was low concentration of fluoride in tissue fluid during teeth development, fluoride ion could enter apatite which was mineralizing and replace the hydroxyl radical (•OH) in hydroxyapatite in enamel to form fluorapatite ($Ca_{10}(PO_4)6F_2$). Such process enhanced the stability of apatite crystal and hence would prevent dental caries. Mineralizer in dental hard tissues was an impure hydroxyapatite. Such hydroxyapatite presented in the form of hexagonal prisms. Its crystal structure showed Ca^{2+} circled C axis to form triangle shape and distributed on two surfaces. Radius of oxygen atom of •OH was big, three Ca^{2+} could not circle •OH on the same surface, but could coexist with H^+ on the same surface. O^{2+} was 0.003 nm away from the surface, either above or below the surface. After F replaced the •OH, since the radius of F was relatively small, it could coexist with three Ca^{2+} on the same surface. The strong negative charge of F^- formed strong electrostatic attraction with Ca^{2+}. Such change enhanced the stability of crystal structure, promoted the order and density of crystal alignment, lowered the solubility of enamel, and enhanced the resistance of enamel against organic acid from oral microorganism. Meanwhile, the alteration of crystal stability also improved the morphology of teeth in aspects of reducing the diameter of teeth and height of cusp, and making the fissure caries become shallow. Such morphological changes were beneficial to the dental self-cleaning and resistance against dental caries. The formulated apatite crystal then involved in the composition of teeth in enamel development phase before eruption of teeth. Fluoride enters apatite crystal by replacing •OH, bicarbonate (CO_3^{2-}), and $C6H5O7^-$, or directly filled the vacancy of hydroxyapatite.

Effect of Fluoride on Tooth Development

Fluoride could affect bio-mineralization process of mineralized tissues (including mesenchymal origin and ectodermal origin) with various effects on teeth. Effects of fluoride include affecting balance of calcium, the bio-synthesis, secretion of enamel matrix or composition of minerals; working on matrix protein directly or indirectly to prevent the removal of protein; having special effects on cellular metabolism and its functions. Previous study indicated that fluoride affected ameloblast function by directly working on ameloblast in development period or indirectly working on extracellular matrix. Excessive intake of fluoride can inhibit the synthesis of protein and reduce the amount of enamel in secretory phase (Lyaruu et al. 2014). Fluoride could bond with ATP enzymatic activity to affect the transfer of amino acid. In this way, cellular intake of amino acid would decrease and then inhibit synthesis of protein. Long-term fluoride intake can enable the bond between fluoride and enamel matrix protein, thus affecting the synthesis of secretory enamel. Such bond could also serve as fluoride reserve to release fluoride and bond with minerals by protein hydrolysis in enamel maturation process. In enamel maturation process, the protein removal was the key step of final enamel mineralization. Fluoride interferes with the protein removal and mineral deposition process in this enamel maturation process and makes the protein in matured enamel slip and then prohibits the formation of hydroxyapatite crystals.

Dentin development could only be affected in condition of fluoride concentration far exceeding the concentration which might trigger enamel fluorosis. Overall manifestation of such effect is dentin hypocalcification. Hard tissue of teeth is the target organ of fluorosis effect. Enamel is the tissue with the highest degree of mineralization and is most sensitive to fluoride. During the development process of dental germ, excessive fluoride intake triggers enamel dysplasia and then forms dental fluorosis. Fluoride can affect dentin with low degree of mineralization and then affect the formation of hybrid layer and rensin tag. Such result may lower adhesion of dentin.

Promoting Re-mineralization of Teeth by Fluoride

Appropriate amount of fluoride can prevent disease due to its positive effect on metabolism of calcium and phosphorus. Small amount of fluoride can accelerate the formation of apatite (mineralization element) in bones and dental hard tissue, and fluoride can enhance their stability. In addition, fluoride deficiency can trigger the calcium and phosphorus metabolic disorder. A substantial number of studies proved that appropriate amount of fluoride sustains the healthy teeth, while fluoride deficiency can increase the susceptibility of dental caries.

Fluoride promotes re-mineralization in ways of: (1) fluoride affects the nature, amount and deposition location of minerals in artificial dental caries; (2) low concentration fluoride (<l mg/l) can phase invert minerals in caries into fluorapatite and reduce the susceptibility of dissolving on acid; (3) fluoride generates relatively big crystals on the surface of caries and reduce the susceptibility of crystal on acid erosion; (4) calcium and phosphate in re-mineralization liquid can deposit in various parts below surface layer, such as surface band and dark band; (5) fluoride processes demineralized caries and re-mineralization on 100–150 μm outmost layer of enamel surface.

Pathological Changes of Dental Fluorosis

Currently, pathogenesis of dental fluorosis was yet to be determined (Li et al. 2020). Existing studies showed the following results.

Changes in Tooth Enamel with Chronic Fluorosis

Enamel change of fluorosis equals to dental fluorosis, which is the chronic fluorosis caused dysplasia on enamel during teeth development. Manifestations of such disease are chalk, discoloration, defects, and so forth on enamel surface.

Dental fluorosis is often found in permanent teeth in multiple symmetric pairs. Such disease is often discovered in labial side of incisor, or labial and buccal side of teeth, sometimes all the teeth are affected. Lingual side can be affected only mildly. Affecting rate of teeth in decreasing order is from incisor to molar. Fluoride seems to have toxic effect on teeth selectively with unknown mechanism. Deciduous teeth can be affected but with relatively low morbidity rate. Changes of dental fluorosis visible to naked eyes are as follows. (1) Chalk change: enamel on labial side at early stage of changes from healthy translucent shade to dull shade. Such results present in opaque chalk changes, such as mottle-, stripe-, or flake-shaped chalk. Such chalk change can even cover the whole surface of teeth. (2) Enamel surface coloring: color changes of different degrees are found on enamel surface, ranging from light yellow, yellow, brownish yellow, dark brown to black. Such coloring presents in mottle- or flake-shape covering the whole tooth surface. (3) Enamel defect: pitting of different sizes or depths is found on the glaze. These features present in mottle-, flake-, or fused flake-shape. There is possibility that surface enamel fall off from tooth in relatively large area, resulting the loss of healthy, smooth, continuous, and complete glaze structure. Enamel defect is limited to the enamel surface in 1/4 to 1/3 of the total depth of enamel, which rarely erodes deep layer.

The study on microstructure of dental fluorosis enamel suggested that its pathological changes resulted from the dysplasia of enamel rod and its interstitial. As early as 1916, scientists observed the glaze of different degrees dental fluorosis teeth under microscope. They discovered that organic matter in glaze of fluorosis teeth was relatively abundant, but there was absence of mesenchyme between enamel rod and its interstitial (either empty or filled with air). Microscope radiology test and teeth grinding test found that hypocalcification and misalignment in enamel rod, and it lost continuity because of intermittent spherical amorphous structure. The enamel interstitials were expanded, mesenchyme was thickened, lines of Retzius disappeared, enamel rod presented in ripple shape, and lateral edge of lesion reached 3/4 of enamel thickness. Enamel turned into opaque chalk due to the misalignment of enamel rod and existence of interstitial air. Ultraviolet irradiation proved that pigment within enamel rod was exogenous, while absence of enamel mesenchyme availed pathway for pigment penetration. Later, it was discovered that pigment deposited at 1/5 of enamel thickness. With the presence of enamel lamella, pigment could penetrate into the depth of enamel through the lamella, even to the dentin. Enamel hypocalcification and pitting could worsen with the development of endemic fluorosis. X ray was adopted to check dental fluorosis slice and revealed that perviousness of X ray was increased in outer layer of enamel. Such increase was directly related with severity of dental fluorosis. Transverse section of enamel hypocalcification part was observed under high resolution transmission electron microscopy (HRTEM), and large size pothole was found in crystal core along crystal C axis. In addition, multiple defects were observed on crystal surface of a and b axes. If there were exogenous pigmentation in enamel rods interstitials, dental surface presents with various degrees of coloring. Dye and silver nitrate were relatively easier to penetrate enamel of fluorosis teeth. Pigmentation within stained enamel could be removed by alcohol and gasoline. These results indicated enamel dysplasia and hypocalcification. Fluoride mainly works on enamel and ameloblast in development period. It could trigger disorder in ameloblast secretion and matrix deposition, disturbance of matrix mineralization in later stage. All these results contributed to enamel dysplasia. The dysplasia degree is related to the amount of fluoride intake.

Dentin Changes

Dentin is susceptible to the damage of fluorosis. Multiple layers of sphere interstitials are found in dentin of fluorosis teeth. The sphere interstitials are found in large amount and huge size and featured with pigmentation. Some sphere interstitials align in line along dentin outside; some align in scattered manner; obvious dotted lines are also found within dentin. These findings suggested that calcification process of dentin is disordered or hypocalcified. Past study observed fluorosis teeth by scanning electron microscope (Foreman 1988) found that incremental line of dentin showed aggravated tendency, hypercalcification could be found in intertubular dentin. At the meantime, hypocalcification and hypercalcification were also observed. Such results presented in strip-shaped uncalcified area and abnormally aligned collagen fiber contributed to interstitials in different sizes. Certain dentinal tubules were poorly formed or even blocked by minerals.

Cementum Changes

Cementum fluorosis shares similar changes with skeletal fluorosis. Tooth root of patients with fluorosis become tiny due to decreased thickness or become rough and irregular due to increased thickness. Root tip contains calcified block formed by skeletal cementum with excessive fluoride. Previous study proved that hardness of cementum of fluorosis teeth was lower than that of normal teeth (Lakhani and Laxman 2017). Cemento-enamel junction was observed under microscope and found to be diffused and irregular. With the development of dental fluorosis, thickness of Tomes layer was also increased (Ramesh et al. 2017).

Pathogenesis of Dental Fluorosis

Damage of Fluoride to Enamel

Currently, pathogenesis of dental fluorosis was yet to be determined (Li et al. 2020). Existing studies showed that enamel, with 96% inorganic substance and no cellular component, is the hardest substance in human body. Basic component of this inorganic substance is apatite. Enamel presents in translucent form, which looks creamy white because of the yellow dentin inside.

Enamel is calcified from ameloblast secretion. In the development period of permanent teeth, ameloblast secrets organic substance continuously, and it deposits and calcifies along ameloblast synapse faced toward dentin to form enamel rod. Ameloblast also generates enamel rod mesenchyme continuously to fill in the interstitials. It is the enamel rods and interstitial mesenchyme that forms enamel. Mineralization components of enamel rod and interstitial mesenchyme are all crystals with definite forms. Enamel presents with certain optical characteristics because of the mineral crystals in enamel aligned in certain order. That is why the surface of enamel is translucent with features of being smooth and glossy.

Damage of fluoride on ameloblast can concern various development stages. The effect of fluoride on hypoplasia and differentiation of ameloblast is yet to be determined. More evidence is required to prove the normal physiological level of fluoride will affect morphology of tooth in development or hyperplasia and differentiation of ameloblast. Pre-ameloblast in hyperplasia and differentiation has significant resistance against fluoride.

Fluoride (3–7 mg/kg) can affect cellular structure of secreting ameloblast at early stage in rats and reduce dose-dependent temporary protein synthesis. High dose of fluoride (9 mg/kg) was injected into hamster once to induce molar neck ring epithelial parts secreted ameloblast at early stage to form cyst. Fluoride can affect the ameloblast structure and lead to high degree of mineralization of ameloblast in lower strata of secreted ameloblast at early stage which already formed cyst. The enamel matrix (the fluorosis plaque matrix) which is secreted during fluoride exposure cannot be mineralized. If the calcium ion concentration in the medium is relatively low when exposed to fluoride, those effects will be aggravated. While in high calcium ion concentration condition, fluoride effect on matrix and cellular structure will be significantly reduced or erased. These results indicated that lowering plasma calcium ion level and injecting high dose of fluoride can enhance the damage of fluorosis plaque.

Compared with ameloblast in early and later secretory phase, ameloblast in full secretory phase has stronger resistance against acute fluoride exposure. The secretory ameloblast can only be changed when given high dose of fluoride.

If the teeth in developmental phase are exposed to excessive fluoride in the long term, thickness of enamel will be reduced by 10%. Even though more evidence is required, this suggested that long-term exposure to fluoride can reduce the secretory ameloblast synthesis with matrix. Long-term fluoride exposure can also affect mineralization in secretory phase, while such effect is less obvious than acute exposure to high fluoride content. Long-term fluoride exposure can induce multiple under-mineralized or over-mineralized lines on teeth, which are the contrast between highly-mineralized parts in inner layer and non-mineralized fluorosis matrix in outer layer. It suggests that severity of mineralization defects at this stage is highly dependent on the dosage and exposure time of fluoride.

In later stage of secretory phase, cells are more sensitive to fluoride peak value compared with early stage and full secretory phase of ameloblast. If the hamster was given a certain dose of fluoride (4.6 mg/kg), the later stage and transitional ameloblast of secretory phase in its molars could be occasionally induced to isolate from enamel and then formed ameloblast cyst. In contrast, ameloblast of early stage of secretory phase showed no such manifestation.

If fluoride is introduced at this stage, it will lead to molar germs cyst. According to the observation of scanning electron microscope, such cyst could aggravate damage from under-mineralization of enamel on erupted molars. These damages are mainly manifested by the appearance of pitting and depressions on the enamel surface of erupted teeth.

Previous study indicated that enamel structure showed no obvious changes when exposed to fluoride, despite the fact that mature enamel was most sensitive to fluorosis. Structure of mature enamel of rat could only be changed when the fluoride was injected repeatedly for days. Another research proved that long-term fluoride exposure could achieve the following results: mildly shortened structure of mature enamel, reduced or erase typical orange pigmentation on incisor, mildly reduced the number of lysosome and phagosome, decreased the activity of lysosome, and induced protein expression of endoplasmic reticulum stress.

Fluoride can damage the regulation of mature ameloblast. There are two different forms of ameloblast in mature period enamel, which are smooth fundus ameloblast and rough fundus one. These two types of ameloblast arrange in layers, respectively, to form ameloblast layers with smooth fundus and rough fundus. The former can be stained by glyoxal-bis-(2-hydroxyanil) (GBHA) while the latter cannot. Currently, their functions are yet to be determined. Calcein or GBHA staining revealed that after the contact between fluoride and ameloblast in mature period, smooth fundus ameloblast showed no regulatory band, while rough fundus ameloblast formed dyeing band gradually was extended. After being exposed to fluoride, the most incisal edge of smooth fundus ameloblast regulatory band disappeared the first. With the exposure time being prolonged, other smooth fundus ameloblast band gradually disappeared along incisal edge till the neck direction. Apart from altering the regulation, fluoride was found to reduce the circulatory intake of biomarker 45 Ca in similar manner. When fluoride exposure ended, ameloblast band was recovered to smooth fundus from newest to old ones and neck part to incisal direction. Such results indicated that fluoride effect on regulation of ameloblast in mature period was reversible, and new cells recovered faster than old cells.

Damage of Fluoride on Amelogenin

The damage of fluoride on amelogenin occurs the secretory period of enamel development since amelogenin is secreted by enamel. When mass fraction of fluoride was bigger than 50 ppm, enamel matrix synthesis was reduced. There was also assumption that fluoride showed no influence on amelogenin in secretory phase. Den Besten fed rats with drinking water (mass fraction as 0, 10, 25, 50, and 100 ppm) for five consecutively weeks. The results revealed that there was no difference in amelogenin content of enamel in secretory phase of all groups. However, if the fluoride concentration in plasma was high, it could block ameloblast from secreting amelogenin.

The maturation of enamel includes reduce of organic matter, increase of inorganic matter, and the process of enamel crystal replacing enamel matrix. Fluorotic enamel distinguishes itself from healthy enamel in content increase of organic substance (such as protein) and decrease of inorganic substance. Fluoride can hold retention of amelogenin in mature enamel and thus block the formation of hydroxyapatite. The dominant mechanism is the fluoride effect on the activity of enzyme. Hydrolase plays a vital role in decomposition of amelogenin (such as protein metabolic enzyme and serine proteinase). The latter only shows activity in mature enamel but fluoride can decrease its activity. Previous study on enzyme proved that high concentration of fluoride could decrease the activity of enzyme, but it was non-exclusive that low concentration of fluoride might have the same effect.

In addition, certain amount of amelogenin functions via endocytosis of ameloblast. For example, high concentration of fluoride could decrease the activity of dipeptidyl peptidase. After lysosome enzymatic activity is lowered, cells cannot play its role in capturing granules and large amount of amelogenin aggravated outside the cell. Meanwhile, as high concentration fluoride enters mature enamel, enamel matrix forms fluorapatite instead of hydroxyapatite, and it increases the local pH value. Since amelogenin is not prone to dissolve and aggregate in neutral environment, it cannot be effectively distributed in enamel matrix.

As the major component of tooth, hydroxyapatite is the most important one in $M_{10}(XO_4)_6Z_2$ compounds. Of which, M belongs to various metallic ions or H_3O^+, X belongs to P, Si, Ge, S, Cr, and Z can be divided into •OH, Cl^-, Br^-, CO_3^{2-}. All these compounds share the same hexagon crystal structure. Molecular formula of hydroxyapatite is Ca_{10} $(PO_4)_6(OH)_2$. This crystal is formed by hexagonal prisms. Growth dynamics of enamel crystal, size, and form of crystal are determined by matrix protein generated in the process of enamel formation. Over the years, it had been proven that fluoride could trigger under-mineralization of enamel and lower its hardness. Fluoride can alter multiple features of crystal (size, number, form, and quality) through interfering enamel formation. However, such hypothesis is yet to be proven.

A study discovered that diameter of fluorosis teeth crystal was enlarged by observing healthy enamel crystal and fluorosis enamel crystal via High Resolution Transmission Electron Microscopy, X ray diffraction, and scanning microscope.

There were also studies revealed that fluorosis enamel showed no difference with healthy enamel. The comparison between the two showed typical differences or not still required further study.

Nano-meter level observation showed that surface of fluorosis crystal was rougher than that of healthy one. The degree of roughness on crystal surface increased with the increase of fluoride content in drinking water. If being exposed to higher level of fluoride, fluoride can alter the form and size of crystal on microscopic level.

Fluoride has different effects on enamel crystal from different development periods. Fluoride affects the prism-shaped enamel formed in early phase of secretion to be thicker and shorter, and the edge will be completely fused without interstitials. Transitional enamel crystal in later phase of secretion shows similar changes under fluoride effect. Such structure is too fragile to be sliced. Fluorosis crystal in secretory enamel shows features as thicker crystal and is unlikely to be affected by fluorosis, but it has bigger crystal interstitials and retains individual crystal structure compared with early secretory phase.

Damage of Fluoride on Dentin

Damage to dental fluorosis occurs in both enamel and dentin. The manifestation of enamel hypocalcification and hypercalcification can be observed in both tissues. There are two simultaneous steps in the formation of dentin, which are formation and mineralization of early phase dentin. Mass fraction of collagen fiber accounts for 90%, non-collagen fiber, and proteoglycan account for 10% of macromolecular substance synthesized by odontoblast. Odontoblast secrets collagen fiber and proteoglycan to its outside and then forms uncalcified early dentin matrix. Other non-collagenous proteins, such as phosphoprotein, are transferred to dentin synapse and then secreted to pre-mineralization front, which is the part about to occur local mineralization and then full mineralization. Fluoride effect on dentin focuses on alteration of its structure, matrix, dentin protein, and odontoblast.

Acute or chronic exposure to excessive fluoride will all lead to dentin structure changes. Manifestation of these change are obvious incremental line, increased dentin in sphere interstitials, simultaneous hypocalcification, and over-calcification. Abnormally giant tubules are found in dentin. Giant tubules function to enhance the penetration, form secondary mineralization as well as stimulate endodontium surface to form third phase dentin. Mineralization of dentin surface and giant tubules is found in severe fluorosis teeth, which is beneficial to endodontium protection and exercise masticatory function. Such results turn out to be a defensive response.

Dentin matrix is formed by collagen and non-collagen. Fluoride can affect both of them. Collagen is the target component of fluorosis, since fluorosis leads to the formation of abnormal collage. Such change is the pathological basis of abnormal calcification of the development of bone, peripheral osseous tissue, and teeth. Biochemistry and X-ray investigation also found that collagen content in condyloid cartilage of domestic rabbits with fluorosis obviously decreased. X-ray diffraction found abnormal collagen alignment in skeletal tissues of patients with fluorosis. Proteoglycans (PGs) is another group of components of dentin matrix. PGs is bond by covalent bond of glycosaminoglycans (GAGs) with specific core protein, which then forms mineralized skeleton with type I collage. Chronic exposure to high dose of fluoride will trigger bio-chemical changes of PGs structure of bones and dentin matrix in the form of obvious diminished relative molecular mass (Mr). Hass et al. isolated PGs extract from fluorosis dentin. They investigated core protein and GAGs. The results revealed that no change was found in corn protein Mr, and GAGs Mr obviously diminished. GAGs molecular chain was shortened which explained that the affection of fluorosis increased short chain synthesis or decomposed long chains into more short chains. Short chain came with more negative electric charge, which can firmly bond on the surface of hydroxyapatite to affect mineralization. By the end of 1950s, autoradiographic technique was adopted to discover that sodium fluoride could inhibit the bond between phosphorus and dentin PGs. Rats with fluorosis were found with incomplete dentin matrix structure of incisor.

Non-collagenous protein of dentin includes a group of dentin specific protein, which plays a vital role in mineralization. Till now, study on fluoride effect on dentin specific protein was limited to dentin phosphoprotein (DPP). Fluoride can induce alteration of bio-chemical structure in DPP and then affect its phosphorylation. Such change is the key part in fluoride interference in dentin mineralization. The effect of fluoride on GAGs and DPP explained that fluoride affect the post-protein translation process with unclear mechanism. Effect from other enzymes was not reported.

The influence of fluoride on odontoblast had been wildly proven. Dental germ of rat cultivated in high fluoride cultivation medium revealed that cellular morphological form of odontoblast was shortened, some synapse disappeared, inter- or intracellular mesenchyme was full of black particles. Male Waister rats received subcutaneous injection of 84 mg/g sodium fluoride. One to twelve hours later, it was found that cytoplasmic granule in odontoblast was swollen, vacuole was formed, the number of Golgi apparatus and cytosome-4 increased. Dentin in early phase was also affected with the deposition of amorphous substance containing GAGs. This amorphous substance was found with two specific collagenous fibers: aggregated collagenous fiber segment and collagenous fiber with relatively longer fibrilla. With the progress of time, these two fibers migrated from cavity end of odontoblast to dentin direction, and then were wrapped in dentin. Calcification of collagenous fiber began to deposit for needle- or panel-shaped edge of crystal, later on, edge of crystal would gradually cover collagenous fiber.

Conclusions

Dental fluorosis is a special and typical manifestation of endemic fluorosis on teeth. The epidemiological features indicated that age was an independent risk factor affecting occurrence of dental fluorosis. Dental fluorosis morbidity rate of 8–12 years old children could serve as a convenient and reliable index to evaluate dental fluorosis morbidity rate in the population. Such endemic showed no difference in gender. In addition, such endemic was featured with regional aggregation, and morbidity rate in countryside was obviously higher than that of city. The prevalence of dental fluorosis could be classified according to *Dean's Dental Fluorosis Index*. Morbidity rate of this endemic was also a major index to evaluate the prevention and control efficacy for the disease. Currently, pathogenesis of dental fluorosis was yet to be proved. Biologically, appropriate amount of fluoride has positive effect on human body. However, excessive fluoride will affect formation of enamel and mineralization of dentin in developmental period. It will then cause chalk, coloring, and defects on the teeth from pathological aspects, such as hypocalcification of dentin and overcalcification. It would also lower the hardness of bone mass.

References

Bhagavatula P, Curtis A, Broffitt B, Weber-Gasparoni K, Warren J, Levy SM. The relationships between fluoride intake levels and fluorosis of late-erupting permanent teeth. J Public Health Dent. 2018;78:165–74.

Brahman KD, Kazi TG, Baig JA, Afridi HI, Khan A, Arain SS, et al. Fluoride and arsenic exposure through water and grain crops in Nagarparkar, Pakistan. Chemosphere. 2014;100:182–9.

Dean HT. Endemic fluorosis and its relation to dental caries. Nutrition. 1990;6:435–45.

Ding Y, Gao YH, Sun H, Han H, Wang W, Ji X, et al. The relationships between low levels of urine fluoride on children's intelligence, dental fluorosis in endemic fluorosis areas in Hulunbuir, Inner Mongolia, China. J Hazard Mater. 2011;186:1942–6.

Do LG, Miller J, Phelan C, Sivaneswaran S, Spencer AJ, Wright C. Dental caries and fluorosis experience of 8-12-year-old children by early-life exposure to fluoride. Community Dent Oral Epidemiol. 2014;42:553–62.

Foreman PC. Fluorescent microstructure of mineralized dental tissues. Int Endod J. 1988;21:251–6.

Gu LS, Wei X, Ling JQ. Etiology, diagnosis, prevention and treatment of dental fluorosis. Chin J Stomatol. 2020;55:296–301.

Guissouma W, Hakami O, Al-Rajab AJ, Tarhouni J. Risk assessment of fluoride exposure in drinking water of Tunisia. Chemosphere. 2017;177:102–8.

Lakhani N, Laxman KV. Microhardness of nonfluorosed and fluorosed dental cementum: an *in vitro* study. SRM J Res Dent. 2017;8:74–7.

Li C, Qin Y, Ouyang T, Yao M, Zhang A, Luo P, et al. miR-122-5p Mediates fluoride-induced osteoblast activation by targeting CDK4. Biol Trace Elem Res. 2020;199(4):1215–27.

Liu Y, Wang Q, Lai JF, Chen JG, Wang CL, Ye YL. Logistic regression analysis on risk factors of dental fluorosis. Chin J Dis Control. 2016;20:1144–6.

Lou D, Luo Y, Liu J, Zheng D, Ma R, Chen F, et al. Refinement impairments of verbal-performance intelligent quotient in children exposed to fluoride produced by coal burning. Biol Trace Elem Res. 2021;199:482–9.

Lyaruu DM, Medina JF, Sarvide S, Bervoets TJ, Everts V, Denbesten P, et al. Barrier formation: potential molecular mechanism of enamel fluorosis. J Dent Res. 2014;93:96–102.

Mohd Nor NA, Chadwick BL, Farnell DJ, Chestnutt IG. The impact of stopping or reducing the level of fluoride in public water supplies on dental fluorosis: a systematic review. Rev Environ Health. 2020;35:419–26.

Molina-Frechero N, Nevarez-Rascón M, Nevarez-Rascón A, González-González R, Irigoyen-Camacho ME, Sánchez-Pérez L, et al. Impact of dental fluorosis, socioeconomic status and

self-perception in adolescents exposed to a high level of fluoride in water. Int J Environ Res Public Health. 2017;14:73.
Moller IJ, Pindborg JJ, Gedalia I, Roed-Petersen B. The prevalence of dental fluorosis in the people of Uganda. Arch Oral Biol. 1970;15:213–25.
Ramesh M, Narasimhan M, Krishnan R, Aruna RM, Kuruvilla S. The Effect of fluorosis on human teeth under light microscopy: a cross-sectional study. J Oral Maxillofac Pathol. 2017;21:345–50.
Saeed M, Malik RN, Kamal A. Fluorosis and cognitive development among children (6-14 years of age) in the endemic areas of the world: a review and critical analysis. Environ Sci Pollut Res Int. 2020;27:2566–79.
Sun WJ, Feng SY, Zhang H, Zhang HZ, Deng J. Effect of 2 bleaching therapies on decoloring of stained dental fluorosis. Shanghai J Stomatol. 2018;27:195–9.

Chapter 6
Damage in Nervous System

Na Wei and Zhi-Zhong Guan

Abstract Excessive intake of fluoride in the body for a long time will cause toxic effects on the brain and other nerves, such as increased fluorine content in the central nervous system, changes in the morphology and ultrastructure of nerve cells, abnormal synthesis of neurotransmitters, receptor function, and signal transduction pathways. These changes will damage the function of nerve cells and show corresponding neurological symptoms and the decline of cognitive ability. The main manifestations of spinal nerve damage are pain and dysfunction. In addition, it may also damage the auditory nerve and cause hearing damage. Excessive fluoride intake during pregnancy can also penetrate placental barrier and have adverse effects on the offspring. In particular, if excessive fluoride is taken during the growth and development period of the offspring, it will not only affect the body development, but also damage the nervous system development, showing growth retardation and mental retardation. In order to understand the mechanism of nerve damage caused by fluoride, oxidative stress, apoptosis, neurotransmitter receptor function, and epigenetics have been under scrutiny, which also provided experimental basis for the intervention and treatment of nerve damage caused by fluoride.

Keywords Damage mechanism · Growth and development · Morphological function · Neurotoxicity

---

N. Wei (✉)
Department of Pathology at the Affiliated Hospital of Guizhou Medical University, Guiyang, Guizhou, People's Republic of China
e-mail: weina@gmc.edu.cn

Z.-Z. Guan
Department of Pathology at the Affiliated Hospital of Guizhou Medical University, Guiyang, Guizhou, People's Republic of China

The Key Laboratory of Endemic and Ethnic Diseases (Guizhou Medical University), Ministry of Education, Guiyang, Guizhou, People's Republic of China

Z.-Z. Guan (ed.), *Coal-burning Type of Endemic Fluorosis*,
https://doi.org/10.1007/978-981-16-1498-9_6

Introduction

For a long time, it has been noticed that chronic fluorosis can affect the central nervous system (CNS) and cause brain and spinal cord injury. Fluoride can penetrate the blood–brain barrier with the increase of content and time of fluoride exposure, and induce an increase of fluoride concentration in brain tissue, which could damage the nervous system. It is confirmed that patients with endemic fluorosis have symptoms of CNS dysfunction, such as headache, dizziness, memory loss, tremor, and so on. The changes of children's mental ability in endemic fluorosis areas have been confirmed in several countries and the negative effects of high fluoride content on learning and memory capacity by animal experiments. Neurotoxic manifestations, such as tremor, tension, paralysis and so on also occur in the animal models with chronic fluorosis. Many changes concerned in metabolism, morphology, and mobility in nervous system induced by chronic fluorosis include the abnormal histopathology, the decreased intelligence, the increased level of oxidative stress, the decreased function of cholinergic neurotransmitter receptors, the damaged cell biofilm structure, and the abnormal nerve signal transduction pathway. In August 2008, the *28th International Society for Fluoride Research* was held in Toronto, Canada. The theme of the conference was: the effect of fluoride on bone and brain (Fluoride: Bone and Brain Effect) indicated that nervous system research was the most important non-osseous target of fluorosis for the international fluorosis research community besides skeletal and dental fluorosis.

Clinical Changes of Nervous System in Patients with Endemic Fluorosis

Fluoride can affect many physiological functions of the nervous system. Earlier studies had found that 84% of the 68 cryolite workers exposed to fluoride for a long time suffered from skeletal fluorosis, while 22% of patients with skeletal fluorosis had neurological symptoms, including excessive fatigue, drowsiness, headache, and vertigo (Roholm 1937). Patients with endemic fluorosis might show symptoms of CNS dysfunction such as memory loss, dizziness, headache, and ataxia, which suggested that fluoride may have a direct toxic effect on the CNS. In addition, there were significant differences in reaction behavior, language fluency, comprehension, and learning ability (between the patients and the control group in the endemic fluorosis areas of coal-burning pollution). It suggested that patients with endemic fluorosis suffered from cognitive impairment (Shao et al. 2003). Patients with endemic fluorosis can also suffer from different degrees of hearing impairment, such as tinnitus, hearing loss, etc., and major clinical manifestation as high-frequency hearing loss, which were categorized under nervous deafness. It was speculated that the fluoride caused hearing damage was the result of direct damage to the auditory nerve (Induswe et al. 2018). Fluorosis caused spinal cord nerve damage was one of the important causes of paralysis and disability in patients with chronic fluorosis. Clinically, the spinal cord nerve injury is mostly manifested as spinal canal and nerve root injury. Osteosclerosis is a common symptom in patients with skeletal fluorosis, which is caused by the deposition of fluoride on the bone. Such fluoride accumulation causes bone sclerosis and osteophyte formation, resulting in narrowing of intervertebral foramen, stimulating or compressing the spinal nerve roots, which are common symptoms of skeletal fluorosis patients. Patients with skeletal fluorosis may show weakness and stiffness of both lower limbs and flexion disorders in the early stage or gradually show varying degrees of paralysis of both lower limbs, increased muscle tension of paralyzed limbs, positive Babinski sign and numbness of both lower limbs (Xia et al. 2005).

Histopathological Changes of Nervous System in Patients with Chronic Fluorosis

A large number of studies had proven that excessive fluoride in the body could accumulate in specific important areas of the brain through the blood–brain barrier, thus affecting the normal morphology and function of brain tissue. Animal experiments showed that with the increased content of fluoride in cerebral cortex, midbrain, cerebellum, and hippocampus of chronic fluorosis rats, and there was an obvious dose–effect relationship. The results showed that the demyelinating changes were observed in the cerebral cortex and subcortical area of rats exposed to fluoride, and there were absence of local nuclear membrane and gathering of chromatin edge; the number of Purkinje cells in cerebellum decreased, Nissl body staining was irregular, dendrites were lengthened and disappeared, and cell pyknosis occurred (Shashi and Patiala 2003). The ultrastructure of neurons in each region also showed significant changes, including: severe mitochondrial degeneration, increased presynaptic active area, edema, increased electron density of synaptic space, and integration of synaptic space and postsynaptic membrane (Mullenix et al. 1995). In addition, the ultrastructural observation of neurons in dorsal raphe nucleus of chronic fluorosis rats showed that the ratio of nucleus to cytoplasm increased, heterochromatin increased and gathered, mitochondria dilated, cristae ruptured or disappeared, endoplasmic reticulum became rough and Golgi complex flat sac dilated. There was a significant increase of lipofuscin in the neuronal cell bodies, especially in the processes, and more autophagosomes appeared. All these suggested that high fluoride intake could directly damage the nerve cells of the dorsal raphe nucleus. Electron microscopic observation on the cochlear spiral of chronic fluorosis rats showed that the column cells of the cochlear spiral were slightly damaged, the static-cilia of the inner hair cells were not arranged normally (seriously tilted and disordered, showing fused mass and partially missed). The static-cilia of outer hair cells showed disordered arrangement (adhesion, and even breaking and falling off in varying degrees) which further confirmed that the hearing impairment was caused by chronic fluorosis rather than bone changes. There was also possibility that it was caused by excessive fluoride acting on the cochlear spiral of the inner ear and lead to hair cell damage (Seng et al. 2008).

In the autopsy of patients with skeletal fluorosis, researchers found a decrease in the number of nerve cells in the anterior horn of the spinal cord, the formation of nuclear vacuoles and the decrease of Nissl body in the cell body, which may be the result of direct nervous system damage caused by fluoride. The experiment also confirmed that there were a series of pathological manifestations of spinal cord nerve damage in animals with fluorosis: uneven distribution of lateral motor neurons in spinal cord, the lateral anterior neurons have shrunk significantly, pyknosis of axons, light staining of Nissl body, fragmentation or agglutination of Nissl corpuscles, breaking loss of terminal ganglion, expansion or medullary degeneration of terminal nerve axons, retrograde injury of proximal axons, and so on. Mice fed with high fluoride content water also showed vacuolar degeneration, pyknosis, nerve phagocytosis, and interstitial edema of nerve cells in spinal cord, which were aggravated with the increase of fluoride concentration and feeding time (Shen et al. 2015). In addition, there were pathological changes such as focal hemorrhage, a small amount of exudative hemorrhage and subarachnoid hemorrhage in the parenchyma of the spinal cord. According to the pathological observation of ulnar nerve in rats, it was found that the myelinated nerve fiber bundles of rats (fed with high fluoride water for 180 days) were broken, disordered, and irregular. The ulnar nerve of fluorosis rats was composed of myelinated nerve fibers and unmyelinated nerve fibers. Under electron microscope, the sheath of myelinated nerve fibers thickened, rich microfilaments were seen, mitochondria decreased, part of myelin sheath thickened and showed anesthesia-like changes. Combined with clinical investigation, it was considered that the motor dysfunction and deformity of upper limb in patients with fluorosis was closely related to the damage of ulnar nerve caused by fluorosis and the combined effects of calcification and ossification of joint and its surrounding soft tissue.

Changes of Learning and Memory Capacity in Experimental Fluorosis Cases

Learning and memory is an advanced neural behavior of the brain. Animal experiments have shown that learning and memory ability impairment is one of the main manifestations of brain damage caused by sub-chronic or chronic fluorosis. Recently, this study conducted a Morris water maze test on SD rats and their 28-day-old offspring fed with fluoride in drinking water. The results showed that the escape latency and stay-quadrant time of fluorosis rats and their offspring were significantly different from those of the control group, suggesting that the accumulation of fluoride in brain tissue leads to learning and memory impairment of experimental rats and their offspring. At the same time, it was also found that oxidative stress increased and the expression of neuronal nicotinic acetylcholine receptors (nAChRs) $\alpha7$ and $\alpha4$ subunits decreased in rat's brain and cultured neurons, which were closely related to each other (Zeng et al. 2020).

Transgenic mice with high expression of amyloid precursor protein (APP) (animal model of Alzheimer's disease) were studied. After being fed with different concentrations of fluoride, the number of senile plaques in brain tissue increased significantly, the ability of learning and memory decreased, the activity of enzymes related to the production of amyloid peptide increased significantly, the level of synaptic protein in hippocampus decreased, and the level of oxidative stress increased (Cao et al. 2019).

Effects of Chronic Fluorosis on the Development of Nervous System

Changes of Fetal Development in Endemic Fluorosis Area

If the pregnant mother is exposed to high content fluoride for a long time, fluoride is absorbed into her blood through the respiratory system and digestive system. The fluoride accumulated in the mother will not only affect her own nervous system, but also accumulate in the offspring's nervous system through the placental barrier, resulting in a certain degree of adverse effects on the development and differentiation of the CNS of offspring (Wei and Guan 2013).

The researchers studied the possible effects of fluoride on fetuses in pregnant women with coal-burning type of endemic fluorosis who were 4 months or more into pregnancy. The results showed that with the increase of urinary fluoride level of pregnant women, the content of fluoride in amniotic fluid also increased. In the first trimester, amniotic fluid mainly comes from maternal transudate. Since the second trimester, fetal urine has become the main source of amniotic fluid, so the components in amniotic fluid can reflect the metabolic level of the fetus. On the one hand, the content of fluoride in amniotic fluid can reflect the level of fluoride that is transmitted from mother to fetus and excreted by metabolism in the fetus. On the other hand, the amniotic fluid can exchange with the mother through fetal membrane, thus the fluoride in the amniotic fluid also includes those enter the amniotic fluid from the mother. Therefore, the content of fluoride in amniotic fluid is considered to be a good index to study the vertical transmission of fluoride, and the increase of fluoride content directly confirms that fluoride can enter the fetus through the placental barrier. In this study, the content of fluoride in amniotic fluid of pregnant women from endemic fluorosis area was positively correlated with the level of fluoride in urine; the pathological changes such as active osteogenesis and disorder of fetal femur were observed under light microscope and scanning electron microscope, indicating that the damaging effect of fluoride on human body occurs in fetal period. In addition, the embryonic stage and early childhood are the most critical periods for the development and differentiation of brain cells. During this period, the development of the blood–brain barrier is not perfect and is easily contaminated by high fluoride. Therefore, excessive fluoride content in the brain will lead to relative brain growth retardation; on the other hand, the fluoride will cause varying degrees of toxic damage to nerve cells.

Experimental Study on the Effect of Chronic Fluorosis on Fetal Development

In animal experiments, rat embryos at 8.5 days of pregnancy were treated with different doses of sodium fluoride in a rotating culture system in vitro. The results showed that with the increase of fluoride dose, the development and differentiation of yolk sac and vascular network (which reflect the growth and development of embryos and organ differentiation) showed a downward trend and certain dose–effect relationship. Using micromass to culture mesencephalon cells or limb bud cells of rat and mouse embryos, it was found that certain concentrations of fluoride significantly inhibited the proliferation and differentiation of mesencephalon cells and limb bud cells of both kinds of animal embryos. When the concentration of fluoride was equal to or greater than 15 mg/l, the occurrence of embryo malformation increased significantly, mainly manifested by neural tube insufficiency and body inversion. It is confirmed that fluoride has potential teratogenicity and developmental toxicity to the development of nervous system (Zhang and Gao 2004).

Degeneration of nerve cells was found in hippocampus, amygdala, cortex and cerebellum of fetal rats, neonatal rats, and 10-week-old rats fed with high fluoride drinking water, which showed poor development of nerve cells in CNS. Cells become smaller and densely distributed; the edge of cells is irregular, chromatin condenses into clumps, cells are pyknotic, and apoptotic bodies are formed. Long-term excessive fluoride intake can cause demyelination in cerebral cortex and subcortex, decreases in the number of cerebellar Purkinje cells, scarcity of dendrites, and the number of synapses, and so on. The brain weight of neonatal mice exposed to fluoride and 28-day-old offspring exposed to fluoride was significantly lower than that of the normal control group. The density of nerve cells in the cerebral cortex of the fluoride-exposed group was higher than that of the control group and the percentage of undifferentiated neuroblasts in the total number of cells was higher than that of normal mice of the same age, both were the basis for poor development of nerve cells. Electron microscopy showed that most of the nerve cells in the cerebral cortex were immature, with less cytoplasm, less organelles, smaller mitochondria, reduced cristae, less rough endoplasmic reticulum and free ribosomes, and demyelination in severe cases. At the same time, the study on the offspring of coal-burning type of endemic fluorosis found that the ultrastructure of the offspring of the high fluoride group showed the accumulation of chromatin in the nucleus of nerve cells in the cerebral cortex compared with the control group. There were changes such as local absence of nuclear membrane, fusion of synaptic space, blurred synaptic structure, and so on. These histological changes indicate that high fluoride intake can have toxic effects on the growth and differentiation of brain cells and sub-cells in the early stage of body development. Fluorosis can reduce the ability of nerve cell proliferation and induce nerve cell apoptosis. The primary cultured rat brain cells were treated with fluoride for 24 h, the S-phase cell proliferation stopped, the transcription factor, nuclear factor-kappa B (NF-κB) was up-regulated and there was DNA damage. In brain development, excessive fluoride can reduce the number of cells entering S phase, affect DNA synthesis, lead to a decrease in the number of nerve cells, and promote the development of nervous system. The contents of fluoride, copper, iron, and other metal ions in the brain tissue of 21-day-old offspring increased neonates born with chronic fluorosis; while the contents of manganese and zinc ions decreased, the activities of antioxidant enzymes including catalase (CAT), superoxide dismutase (SOD) and glutathione peroxidase (GSH-Px) decreased, and the content of lipid peroxide increased.

Behavioral ability and active muscle tone are closely related to long-term psychomotor development. Therefore, some data show that the intelligence level of primary and secondary school students in fluorosis areas decreases. In the study of animals with coal-burning type of fluorosis, it was also found that compared with the control group, the learning and memory capacity as well as brain cholinesterase activity of 30-day-old offspring born with chronic fluorosis were significantly decreased; the expression levels of nAChR protein and mRNA were also significantly decreased, which was significantly correlated with the learning and memory capacity of offspring. Recently, this study discovered that the expression of muscarinic acetylcholine receptors (mAChRs) in the brain tissue of the offspring of rats with chronic fluorosis was low, and the low level was closely related to the learning and memory capacity of the offspring. After giving male rats 50 ppm and 100 ppm fluoride water for 30 days, the habitual learning of rats was impaired, and the times of active avoidance behavior of rats fed with 100 ppm fluoride water decreased, indicating that fluorosis had the potential to weaken the capacity of learning and memory. The open-field behavior and water maze test were used to observe the effects of different concentrations of sodium fluoride solution on spontaneous behavior, learning and memory of offspring of pregnant rats. The results showed that compared with the control group, the learning capacity of rats in medium fluoride group and high fluoride group decreased. It suggested that fluoride exposure on pregnant rats could pass through the placental barrier, affect the nervous system function of offspring and damage their learning and memory capacity. The learning and memory capacity of the offspring of the rats fed with coal-smoked food in the disease area was lower than that of the control group in water maze test. Recent study showed that the body weight of offspring of rats with chronic fluorosis at different developmental stages showed a downward trend compared with the control group of the same age, and the body weight of rats exposed to fluoride increased slowly. In addition, the time for the completion of nerve reflexes were prolonged, and the spatial learning and memory capacity decreased, suggesting that the mother and its offspring were exposed to high fluoride for a long time. The development of neural behavior and learning and memory capacity are retarded.

Changes of IQ of Children in Endemic Fluorosis Areas

As an accumulative toxic, fluoride can be poisoning to the nervous system when it accumulates through the "blood–brain barrier" in the brain to a certain amount. It not only changes the morphological structure of brain tissue, but also affects multiple physiological functions of the brain, including intellectual development, learning and memory capacity. As in the stage of growth and development, children are susceptible to high content harmful substances from the external environment, and their bodies are extremely vulnerable to damage. The international community has paid total attention to the significant impact of high fluoride environment on the IQ of children in endemic fluorosis areas (Liu et al. 2014).

The study found that children's intelligence significantly decreased in drinking water contaminated fluorosis areas such as Shanxi, Inner Mongolia, Xinjiang and Tianjin. There was a negative correlation between high fluoride intake and children's intellectual development. At the same time, the height of children in this area was significantly lower than that in the control group. In the investigation of coal-burning type of fluorosis areas, it was also found that the morphological indexes such as height, body mass, and chest circumference of the children in the endemic area were at a low level, especially height development bottomed level, while the intelligence level of the children in the endemic area was significantly lower than that in the non-endemic area. It was proven that long-term exposure to high content fluoride in coal-burning fluorosis areas will have adverse effects on children's intellectual development, mainly the damage on logical thinking ability.

In endemic fluorosis areas (fluoride content in drinking water was 5.55 ppm), Eighty-nine school-age children were selected as the experimental groups for the study, while in low fluoride areas ((fluoride content in drinking water was 2.01 ppm)) with the same educational levels and economic conditions, 101 children were selected as control group. The results showed that the IQ of children in high fluoride content areas was significantly lower than that in low fluoride content areas, and the decrease was negatively correlated with urinary fluoride content; IQ level was irrelevant with education level and economic conditions. We also investigated the fluorosis condition and intelligence level of 8–12 years old children in a typical coal-burning endemic fluorosis area of Qixingguan Area in Bijie City, Guizhou Province. The results showed that the detection rate of dental fluorosis in children in the endemic area was significantly higher than that of non-endemic areas. The urine fluoride content of children in the endemic area increased significantly. Through the "Raven Standard Reasoning Test," the IQ of children without dental fluorosis in the control group was significantly higher than that of children from endemic areas, and the decrease of children's IQ was negatively correlated with the increase of urinary fluoride content. It was confirmed that no matter drinking water type of fluorosis or coal-burning type of fluorosis, the intelligence level of children in endemic areas decreased due to high fluoride exposure.

The brain topographic map is related to the physiological function changes of intelligence-related factors, and the temporal lobe is related to memory function, and its pathological changes are often accompanied by mental retardation. Frontal lobe is the main functional area of mental intelligence, developmental disorders of frontal lobe and temporal lobe will directly affect the development of intelligence. From the findings of routine electroencephalograph, the brain electrical activity of mental retardation children in high fluoride area was assisted by topographic map technique. It was found that the lesion area of mental retardation in high fluoride area reached central frontal lobe and left temporal lobe. The percentage of θ energy in central area, parietal lobe, occipital lobe, and left temporal lobe of experiment group was significantly higher than that of normal control group, while the percentage of α energy in the central and occipital lobe of experiment group was significantly lower than that of normal group.

In the drinking water type of endemic fluorosis area in India, the researchers investigated 1145 12-year-old children and 18-year-old teenagers from different areas with varied fluoride levels in drinking water. The results showed that 9.48% of the children in the areas with high fluoride in drinking water had headaches, 1.21% had insomnia, and 3.23% had neurological symptoms such as tiredness. Long-term consumption of drinking water with excessive fluoride levels in children can cause brain damage, decreased IQ, as well as long-lasting learning, memory, and behavioral disorders (Eswar et al. 2011). In Iran, Pakistan, and other regions (Sadia et al. 2020), if children in the growth and development stage drank too much high fluoride water for a long time, and the fluoride content in children's urine increased, the intelligence level decreased significantly compared with subjects from low fluoride drinking water areas. Such results served to prove that high fluoride levels could damage children's intellectual development. In Mexico, the intelligence level of children in three villages with different levels of fluoride and arsenic in drinking water decreased compared with those in areas with low levels of fluoride and arsenic, suggesting that children's exposure to high levels of fluoride or arsenic will increase the risk of decreased intelligence.

Molecular Mechanism of Nerve Injury in Chronic Fluorosis

Mechanism of Free Radical Damage

Under normal circumstances, the generation and scavenging of oxygen free radicals in the body are balanced. If oxygen radicals produce too much or the antioxidant capacity decreased, reactive oxygen (ROS) attacks polyunsaturated fatty acids in the biofilms, causing a range of oxidative damage. Fluorine is a relatively active trace element, which can cause respiratory bursts and reactions related to the production of free radicals. Long-term intake of excessive fluoride can interfere with oxygen metabolism and lead to the increase of oxygen free radicals. Oxygen radical is a group with high chemical reaction activity, which can react with a variety of molecules in the body and destroy the structure of normal cells or tissues, then affecting its metabolism and function. At the same time, fluoride may be an inhibitor of a large number of antioxidant enzymes, which can affect other trace elements in antioxidant enzymes and decrease the activity of antioxidant enzymes (Rzeuski et al. 1998). Excessive free radicals can attack macromolecular substances such as lipids, proteins, and nucleic acids, resulting in abnormal cells and tissues, and causing damage to the body.

In the development of chronic fluorosis, the content of malondialdehyde (MDA), which is the intermediate product of lipid peroxidation, is increased in blood and brain tissue of rats and their offspring. The activities of GSH-Px, SOD, and CAT in brain are decreased, which can destroy the scavenging ability of brain tissue on free radicals and cause a series of free radical reactions of polyunsaturated fatty acids in nerve cell membrane lipids, resulting in increased membrane brittleness. Then, it ultimately causes damage to nerve cells and affects the normal physiological function of the brain (Vani and Reddy 2000). When SOD inducer is added, the excretion rate of fluoride in urine will be increased and the symptoms of toxication can be alleviated. The neurotoxicity of fluoride in animals with chronic fluorosis or nerve cells treated with high fluoride in vitro can be reduced by antioxidant treatment, such as supplementation of vitamin E, reduced glutathione, melatonin, selenium, and other antioxidant measures (Oner et al. 2020).

Mechanism of Structural Changes of Cellular Biofilm Lipids

Brain cell biofilm structure lipids can be divided into three categories: phospholipids, neutral lipids, and glycolipids. There are more phospholipid subclasses, mainly phosphatidylcholine, phosphatidylethanolamine, phosphatidylinositol, phosphatidylserine, cardiolipin, sphingomyelin, and so on, which are the main body of the bilayer structure of cell membrane lipid. Neutral lipids include coenzyme Q and polyterpene alcohols. Coenzyme Q is one of the main substances in the cellular oxidative phosphorylation respiratory chain and it has strong endogenous antioxidation. Glycolipids are divided into glycosyl glycerol and glycosphingolipids, and the latter is further divided into neutral and acidic glycosphingolipids. The normal composition of biofilm structure lipids is the basic guarantee for maintaining the normal function of cells. The functional states such as signal transduction, ion channel opening, neurotransmitter binding with corresponding receptors and intracellular enzyme activity are closely related to the normal composition of biofilm lipid structure.

Brain tissue is not only the most phospholipid-rich organ in the body, but also the organ with the highest oxygen consumption, which is easy to produce ROS and cause lipid peroxidation. Therefore, the brain tissue is the most vulnerable organ prone to the damage of free radicals in chronic fluorosis. It has been found that chronic fluorosis can cause abnormal changes in the biofilm lipid structure of brain cells and nerve cells cultured in vitro, which is characterized by a decrease in the content of phospholipids (such as phosphatidylethanolamine). The composition of phospholipid fatty acids is disordered, the content of polyunsaturated fatty acids is decreased, and the content of neutral lipid (coenzyme Q) with endogenous antioxidant function on biofilm is decreased (Guan et al. 1998). Therefore, the changes in the composition of membranous lipids caused by chronic fluorosis are bound to trigger disorders of cell function and metabolism together with damage to the nervous system. As mentioned above, one of the mechanisms of nerve injury in chronic fluorosis is that fluoride causes lipid peroxidation through free radicals, and the decrease in the composition of polyunsaturated fatty acids in phospholipid structure further confirms the attacking effect of free radicals on these fatty acids; the decrease in the content of coenzyme Q confirms the weakening of antioxidant capacity in brain tissue.

Mechanism of Neurotransmitter Receptor Alteration

At present, it is believed that nAChRs are closely related to brain learning and memory capacity, and is the molecular basis of brain dysfunction and intellectual impairment caused by chronic fluorosis (Chen et al. 2003). The nAChRs are a group of non-secondary messenger-mediated neurotransmitter receptors in the nervous system, which can not only bind with the corresponding neurotransmitters, but also forms ion channels on the cell membrane. Nicotine cholinergic receptors are composed of two subunits: α and β subunits. Up to now, nine subunits (α2-α10) and three subunits (β2-β4) have been found, and each nicotine cholinergic receptor subclass is composed of five subunits. The most common subtype of nicotine cholinergic receptor in brain tissue is the subtype (α4)2(β2)3 (composed of two α4 subunits and three β2 subunits) and subtype (α7)5 (composed of five α7 subunits). The function of neural nicotine cholinergic receptors is not only related to learning, memory, and cognitive ability, but also has obvious neuroprotective effects and regulatory effects on other neurotransmitter receptors. A large number of studies have shown that the expression levels of neuronal nicotine cholinergic receptors α3, α4, and α7 subunits proteins decreased in the brain tissue of chronic fluorosis rats and cultured nerve cells treated with high fluoride in vitro, and the expression level of mRNA changed to some extent. The bonding rate of nicotine cholinergic receptor with radioactive ligand [^{3}H]epibatidine or [^{125}I]-bungarotoxin was significantly lower than that of the control group (Shan et al. 2004). Epibatidine is a ligand that bonds specifically with nicotine cholinergic receptor subclass containing α4 or α3 subunits, while -bungarotoxin is a ligand that specifically bonds with nicotine cholinergic receptor α7. The decreased binding density of epibatidine and α-bungarotoxin with nAChRs indicates that the function of the main types of the receptors in brain tissue is seriously affected by fluorosis. More importantly, it has been found that the expression of neuro-nicotine cholinergic receptor in the brain tissue of the offspring of rats with chronic fluorosis is significantly decreased, which may be an important mechanism for the intellectual changes of children in endemic fluorosis areas. The mechanism of nicotine cholinergic receptor reduction involves the gene expression of the receptor, the change of the receptor after transcription, the assembly of the receptor on the cell membrane, and the destruction of the receptor protein itself.

The mAChRs belong to the super-large receptor family coupled with G protein. After activation, it transduces subsequent signals through G protein coupled with G protein, which causes a variety of physiological and biochemical changes in the cell, and participates in the modulation of learning, memory, and neuronal plasticity in CNS. It also participates in important physiological functions such as respiration, circulation, movement, and thermoregulation. There are five subtypes of muscarinic receptors. Common types of mAChRs in the brain are M1, M2, and M3 subtypes, each of them can cause different physiological and biochemical effects. The experiments of the research showed that the protein and gene expression levels of M1 and M3 receptors in the brain tissue of rats with chronic fluorosis were decreased in varying degrees, and the degree of decrease was related to the decrease of learning and memory capacity of animals (Dong et al. 2015; 2017).

In addition to the changes of cholinergic receptors, it was also found that chronic fluorosis caused abnormal changes in cholinesterase activity, which was characterized by the decrease of acetylcholinesterase (AChE) and butyrylcholinesterase (BuChE) activities in the brain tissue of fluorosis rats. Studies had also shown that other neurotransmitter receptor systems were also affected by fluorosis. Changes in the contents of neurotransmitters such as norepinephrine, epinephrine, and 5-hydroxytryptamine were found in brain tissue, and so did changes in glutamate receptors (especially N-methyl-D-aspartate receptors and adenosine α1-adrenoceptors) (Wei et al. 2018). The changes of these transmitters and receptors may also be the material basis of nervous system dysfunction in chronic fluorosis.

Mechanism of Neuronal Apoptosis

With the in-depth study of chronic fluorosis, people showed greater interest in apoptosis caused by fluorosis. Apoptosis is an autonomous and orderly cell death controlled by genes, which involves the activation, expression, and regulation of a series of genes, for example Fas, Bax, p53 , Bcl-2, and so on. The Caspases family is a group of cysteine proteases that exists in the cytoplasm. An important common feature of Caspases is the specific breaking of the peptide bond after interacting with aspartic acid residue, resulting in the activation or inactivation of a protein. It has been confirmed that there are at least 11 types of caspases family (caspases-2, -8, -9, and -10 are the initiators of apoptosis, while caspases-3, -6, and -7 are the executors of apoptosis). It had been reported that the apoptosis level of brain tissue and nerve cells cultured with fluoride in chronic fluorosis animals increased significantly (Zhang et al. 2007a, b), caspase-3 overexpression was observed, and the apoptosis rate of nerve cells increased significantly in the brain tissue of rats fed with high fluoride water for three consecutive generations, especially the brain damage of offspring rats was particularly serious. The expression of IL-1β converting enzyme in the brain of rats with chronic fluorosis significantly increased. Its mechanism might be related to the activation of G protein or oxidative stress, and Bcl-x might play a regulatory role in the damage of nerve cells caused by chronic fluorosis. Chronic fluorosis can increase the production of nitric oxide and H_2O_2 in nerve cells, which can induce much apoptosis. Caspase-3 is activated by fluoride and then further cleaves poly-ADP-ribose polymerase, which leads to DNA fragmentation and cell apoptosis. ROS can damage mitochondria, decrease the mitochondrial transmembrane potential, and release cytochrome C into the cytoplasm. In the presence of ATP, cytochrome C assists apoptotic protease activating factor to activate caspase-9 to cause cascade reaction, resulting in Caspase-3 activation and chromosome damage. Lipid peroxidation enhances the brittleness of the membrane structure due to the amino groups connecting proteins and phospholipids. If free radicals are not removed in time, it will lead to a chain reaction and apoptosis, in other words, apoptosis caused by fluorosis can be reversed by supplementation of antioxidants. Recent studies have found that protein kinase C β (PKC13)/p66shc oxidative stress pathway can mediate oxidative damage and apoptosis through the mitochondrial pathway. Apoptosis-inducing factor (AIF)-mediated caspase-independent apoptosis is also involved in the process of neuronal apoptosis (Deng et al. 2016).

Mechanism of Nitric Oxide

Nitric oxide (NO) is a small molecular substance with various biological activities found in recent years, which is catalyzed by nitric oxide synthase (NOS) in vivo. NOS are a group of isozymes, and there are differences in its enzyme activity, characteristics, and genes generated in different tissues and cells. According to the tissue or cell origin, it can be divided into neuronal nitric oxide synthase (nNOS), endothelial NOS (eNOS), and leukocyte/macrophage NOS (iNOS). The nNOS exists widely in neurons such as cerebral cortex, hippocampus, and amygdala. The eNOS mainly exists in vascular endothelial cells, while iNOS exists widely in various kinds of cells (mainly leukocytes and macrophages). NO has a variety of functions in the CNS. Normal physiological amount of NO plays a role in intercellular information transmission, mainly affecting learning and early memory. NO can enhance the long-term synaptic effect of hippocampus (an important structural area of learning and memory in mammals). However, when the concentration of NO in the brain generated by nNOS and iNOS increases significantly, it can react with ROS to form nitro peroxide, which is a powerful oxidant. Excessive NO can mediate the toxicity of glutamate and lead to DNA damage. Recent studies have shown that fluoride can increase the activity of NOS in rat's brain, acute fluorosis can also enhance the expression of NOS positive neurons in rat's cortex, and damage

DNA and RNA in cortical neurons. High dose fluoride can significantly increase the content of NO in rat's brain compared with the control group, and its mechanism might be related to the abnormal expression of iNOS stimulated by fluoride. It has been reported that long-term fluoride exposure can inhibit the expression of nNOS positive neurons in rat's cortex. Short-term exposure to low dose fluoride increased the number of nNOS positive neurons in cerebral cortex of mice, while exposure to high dose fluoride decreased the number of nNOS positive neurons. It was suggested that low dose fluoride could enhance the expression of nNOS positive neurons in cerebral cortex of mice with acute fluorosis, while high dose fluoride could inhibit it. The changes of NO in the development of fluorosis indicate that it plays certain role in the mechanism of brain injury triggered by chronic fluorosis.

Mechanism of Signal Transduction Pathway Alteration in Nervous System

There are many reports on the effect of fluoride on signal transduction of nerve cells in recent years. Fluorosis can induce changes in the expression of mitogen-activated protein kinase (MAPK), protein kinase C, NF-κB, and nuclear factor erythroid 2-related factor 2, which is also an important aspect of the molecular mechanism of nerve injury caused by fluorosis.

At present, most studies on the relationship between neural signal transduction pathway and chronic fluorosis are MAPK series, which are one of the most important pathways in eukaryotic signal transduction network and play a key role in gene expression regulation and cell function. MAPK chain signal pathways include MAPK, MAPK kinase (MEK or MKK or MAP2K), and MEK kinase (MEKK, MKKK or MAP3K), MAPK chain protein kinase MAP3K-MAP2K-MAPK), which transmit upstream signals to downstream response molecules by phosphorylation. There are three main subgroups of MAPK: extracellular regulated protein kinases (ERK), c-Jun N-terminal kinase (JNK), and p38MAPK. ERK signal transduction pathway regulates cell growth and differentiation. JNK and p38MAPK signal transduction pathway play important roles in stress responses such as inflammation and apoptosis. The induction and maintenance of long-term potentiation in CNS depend on the activation of ERK and play important roles in regulating the function of learning and memory. The results showed that the expression levels of ERK and Phospho-ERK in the brain tissue of experimental fluorosis rats were higher than those in the control group, while the activation rate of ERK showed the opposite trend—decreased in brain tissue and the learning and memory capacity of animals decreased in the fluoride exposure group (Liu et al. 2011a, b). This phenomenon is possibly related to the stress or compensatory state caused by high dose fluoride intake, which leads to the increase of ERK and phospho-ERK protein expression; however, the increase of its activation is not enough to make up for the long-term potentiation of fluorosis, which still shows the decline of learning and memory capacity in fluorosis rats. On the other hand, the binding of external stress signal to cell surface receptor (tyrosine kinase) and activation of ERK signal pathway can cause a series of cellular effects, controlled by substrate phosphorylation in downstream cytoplasm and nucleus. Such binding will result in high phosphorylation of many microtubule-related proteins, cytoskeleton destruction and apoptosis. Recent studies have also shown that chronic fluorosis can increase the activation of phospho-JNK in many brain regions such as cerebral cortex and hippocampus of experimental animals, and it is related to the accumulation of fluoride in the body. Activated JNK can regulate the activity of members of the Bcl-2 family and cause apoptosis directly (Liu et al. 2011a, b). Chronic fluorosis may lead to abnormal development and differentiation of nerve cells and increase apoptosis through this series of stimulation pathways, thus affecting the physiological function of cells and reducing the learning and memory capacity of the brain.

In addition, chronic fluorosis may upregulate NF-κB signal pathway in rat hippocampal neurons, resulting in nerve cell DNA damage (Zhang et al. 2007a, b). Some studies have detected an increase in the expression of NF-κB protein in the brain of rats which was exposed to fluoride. The expression of the receptor for advanced glycation end products (RAGE) and NF-κB protein/mRNA increased in the hippocampus of rats exposed to fluoride. It is speculated that excessive fluoride can enhance the oxidative stress caused by RAGE activation, which then leads to the activation of NF-κB. It causes the expression and release of a large number of pro-inflammatory cytokines, growth factors, and adhesion molecules, then resulting in lasting inflammatory response, neuronal toxicity, and cell death, and finally leading to nerve injury.

Mechanism of Developing Damage of Nervous System Caused by Chronic Fluorosis

Mechanism of synaptic damage: in the process of nervous system development, neurons first complete tissue morphogenesis and migrate to their designated location. Then, in order to correctly perform the function of transmitting and processing information, neurons begin to develop and precisely construct their axons and dendrites, and select their target tissue in the extremely complex surrounding tissue environment. In this way, neurons can form precise synaptic connections with appropriate cells in the target tissue. This highly specific connection requires an extremely precise controlling mechanism in such development. The ability of synaptic connections to change and adjust transmission efficiency is called the plasticity of neural development. Plasticity plays a key role in the development of nervous system and considered as the basis of the development of neural circuits (which functions for learning and memory). Stimuli, such as changes in external environmental factors during neural development, can cause spontaneous neural activity within the network to trigger a variety of molecular signal mechanisms.

Thus, it results in changes in synaptic plasticity, which in turn has adverse effects on the normal physiological function of nerve cells. It was found that long-term fluoride accumulation could significantly change the synaptic microstructure parameters of hippocampus: the curvature of synaptic interface, the width of synaptic gap, the thickness of postsynaptic density (PSD), the length of synaptic active zone; meanwhile, the thickness of postsynaptic dense area became smaller, the width of synaptic gap became larger, the mitochondria inside of presynaptic membrane enlarged, cristae ruptured, dissolved, disappeared, and showed vacuolization. In the study of neonatal SD rats exposed to different concentrations of fluoride, it was found that in rats with chronic fluorosis: the spontaneous activity and inquiry behavior were significantly inhibited, the ability of spatial learning and memory was significantly decreased, the fluidity of synaptic membrane and the expression of postsynaptic density PSD-95 were significantly decreased. The mechanism is reasoned that: fluorine is the most electronegative element, and it is very easy to obtain an electron to become F^-, which is more active and can generate free radicals. Unsaturated fatty acids on the membrane are easy to be attacked by free radicals to form lipid peroxidation product--malondialdehyde, while it can reduce membrane fluidity by binding with proteins, lipids, and sugars on the membrane. PSD-95 is one of the major proteins in the membrane-associated guanosine kinase family, which is distributed in cortical neuronal cell bodies and nerve dendrites. PSD-95 and GKAP (proteins associated with PSD-95) first form a postsynaptic skeleton, and then glutamate receptor subunits attach to the skeleton and jointly regulate synaptic activity. In addition, PSD-95 has a specific regulatory effect on the clustering and localization of K^+ channels and signal transduction. Functional synaptic plasticity requires accurate clustering and correct activation of postsynaptic sites of neurotransmitter receptors. The decrease of PSD-95 expression can reduce the number of N-methyl-D-aspartate receptor, thus affect the transmission of neural information, and ultimately damage the learning and memory of rats. It is suggested that the changes of synaptic membrane fluidity and PSD-95 expression in hippocampus caused by chronic fluorosis may be one of the synaptic mechanisms of learning and memory impairment caused by chronic fluorosis.

Mechanism of oxidative stress is as follows: animal studies showed that the oxidative stress was enhanced; the activities of SOD, CAT, and GSH-Px were significant decreased, and the content of MDA and the final product of lipid peroxidation were significantly increased in chronic fluorosis rats and their offspring. When investigating the peroxidation and antioxidant system of fluoride in piglets, the same results were obtained: MDA and NO in thyroid, liver, kidney, and other tissues of piglets increased, while the activities of antioxidant enzymes decreased. The mechanism might be that fluoride directly attacked oxygen, interfered with intracellular oxygen metabolism. Thus, such process generated a large amount of ROS, reduced the activities of antioxidant enzymes such as SOD and GSH-Px, disrupt the scavenging ability of brain and other soft tissues on free radicals, and lead to the imbalance of redox balance. Therefore, oxidative stress damage is an important way of fluoride-induced soft tissue toxicity. In the study on 21-day-old offspring of pregnant rats with fluorosis, it was found that manganese and zinc ions decreased, antioxidant enzymes such as CAT, SOD, GSH-Px decreased, while fluoride, copper, and iron ions increased, lipid peroxides increased. Oxidative stress markers were positively or negatively correlated with metal ions to some extent. Researchers investigated and analyzed the changes of blood zinc content in children and adults with coal-burning fluorosis. The results showed that the urine fluoride content in the fluorosis group was higher, while the blood zinc content was lower than that of control group. Zinc is a type of activator and the component/activator of SOD and other enzymes in the body. These enzymes play very important roles in antioxidation, which indirectly confirms that oxidative stress is one of the important factors lead to fluorosis.

Mechanism of neurotransmitter receptor alteration is as follows: the nervous system is a complex and sophisticated network, and the accumulation of fluoride will cause abnormal changes in a variety of neurotransmitters, receptors, and signal transduction pathways in the brain. The study determined the receptors for neurotransmitters in the brains of fetuses in endemic fluorosis areas. The results showed that the contents of norepinephrine, 5-hydroxytryptamine, and 21-receptor in the brain of fetuses in endemic fluorosis areas were significantly lower than those in non-endemic areas, while the content of epinephrine increased significantly. It was found that the AChE activity in synaptic plasma membrane of fluorosis rats and their offspring decreased significantly with the increase of dose and exposure time, thus the content of acetylcholine in brain tissue increased. The accumulation of acetylcholine may also enhance the excitotoxicity of glutamate, block neuromuscular conduction, cause muscle tension and stiffness, and slow down the learning speed at the same time. In previous studies, we found that the spatial learning and memory ability of rats fed with high fluoride in drinking water was decreased, the expressions of $\alpha4$ and $\alpha7$ nAChRs at protein level and total antioxidant capacity were declined, and MDA content was increased. In the study concerning the brain nicotinic receptors and cholinesterase activity of offspring of rats with coal-burning fluorosis, it was found that the learning and memory capacity of the offspring from coal-burning fluorosis rats was significantly decreased, which might be related to the decrease of nicotine receptor protein content and brain cholinesterase activity (Gui et al. 2010). The expression of mAChRs in the brain tissue of mice in developmental period was low, which was closely related to the learning and memory capacity of the offspring (Dong et al. 2015). It was speculated that the dysfunction of cholinergic nervous system might be one of the reasons for the decline of learning and memory capacity. High levels of fluoride also significantly inhibited the synthesis of RNA and DNA in the brain of rats' offspring, thus inhibited protein synthesis, and affected the development and function of the brain.

The mechanism of epigenetic change theory is as follows: at present, researchers have made an in-depth exploration on the molecular pathogenesis of endemic fluorosis from perspectives of basic medicine, preventive medicine, environmental science, etc. However, due to the complexity of environmental–genetic interaction and regulatory mechanism, the molecular pathogenesis remains unclear. In recent years, based on the research findings of genomics and proteomics, epigenetics has become the focus of scholars at home and abroad. Epigenetics is defined as a heritable change in gene expression without involving DNA sequence changes. Epigenetic mechanisms mainly include DNA methylation, histone modification, nuclear chromatin remodeling, and micro-RNA (miRNA). DNA methylation is an important epigenetic reprogramming mechanism, which is very important for normal embryonic development. The researchers observed the effect of sodium fluoride on DNA methylation of rat's early embryos, acid sulfite sequencing and ELISA detection of mouse spermatozoa and liver. The results showed that the methylation level of *H19*, an imprinted gene derived from father in fluoride treatment group, was lower than that of normal control embryos. There was no difference in maternal imprinted gene and Line-1 repeat sequence between fluoride treatment group and control group, and there was no significant change in genomic DNA methylation of liver and sperm. The methylation level of *H19* in female mice mated with male mice exposed to fluoride for 35 days decreased, but *PEG3* methylation increased. The expression of H19 was significantly increased in blastocyst stage of male mice treated with sodium fluoride before mating, but no *H19* expression was detected in embryonic stage. It was suggested that fluoride could disrupt the gene imprinting of normal embryos through direct interaction during pregnancy. Although long-term exposure to sodium fluoride does not directly affect DNA methylation levels of sperm and liver, it may lead to abnormal genetic imprint in early embryos. Other studies reported that sodium fluoride could affect methylation modification and induce apoptosis of mice's early embryo cells. Kunming mice were treated with different doses of fluoride. The results showed that all fluoride treatment groups could inhibit oocyte maturation, fertilization, and blastocyst formation. At the same time, the results of TUNEL analysis showed that in comparison with the control group, low and middle dose fluoride could significantly induce apoptosis in blastocyst stage, while high dose fluoride could induce apoptosis in 8-cell stage with a certain dose–effect relationship. In addition, immunostaining showed that DNA methylation increased, and the gene expression of *H3K9m2* and *H3K4m2* increased with the increase of fluoride dose. At the transcriptional and protein levels, DNMT1, DNMT3a, G9a, LSD1, and MLL1 generally increased, while JHDM2a levels decreased. The imprinted gene H19 from the paternal side showed hypomethylation, while the imprinted gene *IGF2* from the maternal side showed hypermethylation, which was consistent with the results of high expression of *H19* and low expression of *IGF2* confirmed by qPCR. Such results confirmed that fluoride's damage on mouse's oocyte maturation and embryonic development through epigenetic.

Conclusion

Excessive fluoride can penetrate the blood–brain barrier and accumulate in the brain tissue, which can damage the nervous system and cause pathological and functional changes of the nervous system. The molecular mechanism of nerve injury in chronic fluorosis is complicated, which is mainly related to the increased level of oxidative stress, the change of neurotransmitter receptor function, the damage of cell biofilm structure, and the abnormality of nerve signal transduction pathway. These changes affect the morphology, metabolism, and function of the nervous system, which may be the key mechanism of nerve injury triggered by chronic fluorosis. Among the mechanisms listed above, the theory of free radical damage can better explain the mechanism of CNS injury, that is, excessive intake of fluoride can accumulate in brain tissue through the blood–brain barrier, inducing excessive production of free radicals and causing lipid peroxidation or protein oxidation. The results can destroy the lipid structure of cell biofilm, affect the binding of neurotransmitters and receptors, affect the function of neural signal transduction system, affect the activity of cellular protease, and cause nervous system damage and brain dysfunction. The differentiation and development of nervous system is a complex process regulated by many factors and steps. The critical period of human brain development is from the beginning of pregnancy to the age of 3 years old after birth, during which the development and differentiation of the brain is rapid, and are very sensitive to the changes of internal and external environment and stimulation. The excessive accumulation of fluoride in the body can enter the fetal brain through the placental barrier and blood–brain barrier, which can have adverse effects on its morphological structure, substance metabolism, binding of transmitters and receptors, signal transduction, and epigenetics. All these could lead to fetal nervous system differentiation and development disorder, resulting in toxic damage to children's intellectual development. It will also lead to abnormal changes of CNS growth, such as nervous system symptoms, decreased capacity of learning and memory, mental retardation, and so on.

References

Cao K, Xiang J, Dong YT, Xu Y, Li Y, Hong W, et al. Exposure to fluoride aggravates the impairment in learning and memory and neuropathological lesions in mice carrying the APP/PS1 double-transgenic mutation. Alzheimers Res Ther. 2019;11:35.

Chen J, Shan KR, Long YG, Wang YN, Nordberg N, Guan ZZ. Selective decreases of nicotinic acetylcholine receptors in PC12 cells exposed to Fluoride. Toxicology. 2003;183:235–42.

Deng CM, Tan LC, Deng J, Liu YJ, Qiu ZW, Guan ZZ. Influence of chronic fluorosis on protein kinase Cβ/p66shc signal pathway in the brain of rats. Chin J Endemiol. 2016;35:333–7.

Dong YT, Wang Y, Wei N, Zhang QF, Guan ZZ. Deficit in learning and memory of rats with chronic fluorosis correlates with the decreased expressions of M1 and M3 muscarinic acetylcholine receptors. Arch Toxicol. 2015;89:1981–91.

Dong YT, Wei N, Qi XL, Liu XH, Chen D, Zeng XX, et al. Attenuating effect of vitamin E on the deficit of learning and memory of rat with chronic fluorosis: the mechanism may involve muscarinic acetylcholine receptors. Fluoride. 2017;50:354–64.

Eswar P, Nagesh L, Devaraj CG. Intelligence quotients of 12-14 year old school children in a high and a low fluoride village in India. Fluoride. 2011;44:168–72.

Guan ZZ, Wang YN, Xiao KQ, Dai DY, Chen YH, Liu JL, et al. Influence of chronic fluorosis on membrane lipids in rat brain. Neurotoxicol Teratol. 1998;20:537–42.

Gui CZ, Ran LY, Li JP, Guan ZZ. Changes of learning memory ability and brain nicotinic receptors of rat offspring with coal burning fluorosis. Neurotoxicol Teratol. 2010;32:536–41.

Induswe B, Opinya G, Khasakhala LI, Owino R. The auditory working memory of 13-15-year-old adolescents using water with varying fluoride concentrations from selected public primary schools in north Kajiado Sub county. Am J Med Sci. 2018;8:274–90.

Liu YJ, Gao Q, Long YG, Yu YN, Guan ZZ. Effect of chronic fluorosis on phospho-Elk-1 expression in rat brain. Chin J Endem. 2011a;30:251–5.

Liu YJ, Guan ZZ, Gao Q, Pei JJ. Increased level of apoptosis in rat brains and SH-SY5Y cells exposed to excessive fluoride--a mechanism connected with activating JNK phosphorylation. Toxicol Lett. 2011b;204:183–9.

Liu F, Ma J, Zhang H, Liu P, Liu YP, Xing B, et al. Fluoride exposure during development affects both cognition and emotion in mice. Physiol Behav. 2014;124:1–7.

Mullenix PJ, Denbesten PK, Schunior A, et al. Neurotoxicity of sodium fluoride in rats. Neurotoxicol Teratol. 1995;17:169–77.

Oner AC, Dede S, Yur F, Oner A. The effect of vitamin C and vitamin E on DNA damage, oxidative status, and some biochemical parameters in rats with experimental fluorosis. Fluoride. 2020;53:154–63.

Roholm K. Fluoride intoxication: a clinical-hygiene study with a review of the literature and some experimental investigations, vol. 138. London: HK Lewis; 1937.

Rzeuski R, Chlubek D, Machoy Z. Interactions between fluoride and biological free radical reactions. Fluoride. 1998;31:43–5.

Sadia Z, Humayun A, Shafiq R, Shan E, Amer S, Farhat Y, et al. Effect of excess fluoride consumption on urine-serum fluorides, dental state and thyroid hormones among children in "Talab Sarai" Punjab Pakistan. Open Chem. 2020;18:119–28.

Seng DJ, Chen QM, Liu XL, Gong ZP, Liu M, Ma YX. The experimental observations on hearing damage caused by chronic fluorosis in rats. Chin J Otol. 2008;6:162–5.

Shan KR, Qi XL, Long YG, Nordberg N, Guan ZZ. Decreased nicotine receptor in PC12 cells and rat brains influenced by fluoride toxicity--a mechanism relating to a damage at the level in post-transcription of the receptor genes. Toxicology. 2004;200:169–77.

Shao QL, Wang YL, Li LW, Li J. Study of cognitive function impairment caused by fluorosis. Chin J Endem. 2003;22:336–8.

Shashi A, Patiala G. Histopathological investigation of fluoride-induced neurotoxicity in rabbits. Fluoride. 2003;36:95–105.

Shen QF, Xia YP, Xu TT, Tian R, Li HN. Spinal neurocyte apoptosis in rats with chronic fluorosis and improvement after defluorination. Chin J Endem. 2015;34:790–4.

Vani ML, Reddy KP. Effects of fluoride accumulation on some enzymes of brain and gastrocnemius muscle of mice. Fluoride. 2000;33:17–26.

Wei N, Guan ZZ. Effect of chronic fluorosis on nervous system development. Chin J Endem. 2013;32:347–50.

Wei N, Dong YT, Deng J, Wang Y, Qi XL, Yu WF, et al. Changed expressions of N-methyl-D-aspartate receptors in the brains of rats and primary neurons exposed to high level of fluoride. J Trace Elem Med Biol. 2018;45:31–40.

Xia YP, Xu TT, Zhang LL, Du WJ, Tian CR. Surgical treatment of fluorosis cervical stenosis. Chin J Endem. 2005;24:85–6.

Zeng XX, Deng J, Xiang J, Dong YT, Cao K, Liu XH, et al. Protections against toxicity in the brains of rat with chronic fluorosis and primary neurons exposed to fluoride by resveratrol involves nicotinic acetylcholine receptors. J Trace Elem Med Biol. 2020;60:126475.

Zhang BZ, Gao XL. Experimental study of the effect of fluoride on embryonic development of mouse *in vitro*. J Lanzhou Med Coll. 2004;30:1–3.

Zhang M, He P, Wang AG, Chen XM, Yang KD. Effects of fluoride on cell cycle, apoptosis and NF-κB expression in cultured rat hippocampal cells. J Environ Health. 2007a;24:669–71.

Zhang M, Wang A, He W, He P, Xu BY, Xia T, et al. Effects of fluoride on the expression of NCAM, oxidative stress, and apoptosis in primary cultured hippocampal neurons. Toxicology. 2007b;236:208–16.

Chapter 7
Modification of Digestive System

Jian Zou and Zhi-Zhong Guan

Abstract In the digestive system influenced by endemic (chronic) fluorosis with non-bone injury, the liver appears the mainly typical alteration and manifests pathological changes of microstructure, modifications of enzyme activity, and abnormal expressions of the related protein molecules. The effects of chronic fluorosis not only show direct damages on gastrointestinal tract, but also indirectly impair its metabolism and function by affecting the normal balance of bacterial flora distribution. Up to now, there are few studies on the effects of chronic fluorosis on pancreas, but mainly focusing on the influence on pancreatic endocrinology. Pathogenesis of digestive system damage caused by chronic fluorosis is relatively complex. At present, it is mainly based on the theory of oxidative stress. The increased level of free radicals leads to pathological changes in the histology, metabolism, and function of digestive system.

Keywords Digestive system · Fluorosis · Gastrointestinal tract · Liver · Pancreas

J. Zou
The Key Laboratory of Endemic and Ethnic Diseases (Guizhou Medical University) of the Ministry of Education, Guiyang, Guizhou, People's Republic of China

Department of Pathology at the Affiliated Hospital of Guizhou Medical University, Guiyang, Guizhou, People's Republic of China

Z.-Z. Guan (✉)
The Key Laboratory of Endemic and Ethnic Diseases (Guizhou Medical University), Ministry of Education, Guiyang, Guizhou, People's Republic of China

Department of Pathology at the Affiliated Hospital of Guizhou Medical University, Guiyang, Guizhou, People's Republic of China

The Provincial Key Laboratory of Medical Molecular Biology, Guiyang, Guizhou, People's Republic of China
e-mail: zzg@gmc.edu.cn

Z.-Z. Guan (ed.), *Coal-burning Type of Endemic Fluorosis*,
https://doi.org/10.1007/978-981-16-1498-9_7

Introduction

Chronic fluorosis causes extensive pathological changes in multiple systems and organs of the body, and digestive system is a common target of non-bone injury. Concerning effects of chronic fluorosis on the digestive system, liver damage is particularly prominent. Liver is the main detoxifying organ of the body, and excessive fluoride entering into the body inevitably produces toxic effects on the liver, then on the gastrointestinal tract and pancreas. Recently, it has been found that chronic fluorosis can also indirectly affect the relative abundance of intestinal flora, thus affecting gastrointestinal function. In this paper, the pathomorphological changes of digestive organs caused by chronic fluorosis, abnormal changes of functional indexes, and the damage mechanism of chronic fluorosis are discussed to explore the influence of chronic fluorosis on the digestive system.

Influence of Chronic Fluorosis on Liver

Normal Function and Metabolism of Liver

Liver is the largest digestive gland and the central station of metabolism in the body. The main function of liver is to metabolize three major nutrients (decomposition and synthesis of sugar, protein, and fat) as well as the metabolism of vitamins and hormones. Metabolic activities in liver are very active since there is the abundance of enzymes in the organ. The clinical discriminant indexes of liver function concern serum glutamic pyruvic transaminase (GPT) or alanine transaminase (ALT), glutamate oxaloacetic transaminase (GOP) or aspartate aminotransferase (AST), lactate dehydrogenase (LDH), alkaline phosphatase (AKP), acid phosphatase (ACP), γ-glutamyl transpeptidase, total bilirubin (T-BIL), cholinesterase (ChE), albumin (A), globulin (G), ratio of A/G, etc. These indexes can help to judge whether liver appears injured including functional and metabolic changes.

Effects of Chronic Fluorosis on Liver Morphology, Function, and Metabolism

Epidemiological investigation and animal experiments showed that excessive intake of fluoride could induce abnormal changes in hepatic tissue structure and function. The histopathological examination of liver from animals with chronic fluorosis indicated that the hepatocytes were fused into fragments with unclear boundaries; the nuclei in the cells slightly stained or condensed; nuclear membranes broken down; and mitochondria vacuolated or disrupted (Zhou et al. 2020).

It was found that the levels of AKP, ACP, and GPT in serum were obviously raised in patients with endemic fluorosis (Wang 2016; Liang et al. 2000). The activities of cyclic adenosine monophosphate (AMP), AKP, ACP, GOT, and GPT were significantly increased in patients with chronic fluorosis. A significant variance in the activities of cAMP, AKP, ACP, GOT, and GPT in fluorosis patients was determined with elevation of fluoride level in drinking water (Shashi and Bhardwaj 2011). It presents a positive correlation between fluoride contents in drinking water or serum and the levels of GOT, GPT, ALP or ACP in serum of fluorosis patients in such endemic area of the disease.

The content of transaminase is abundant in liver. When liver is damaged, the transaminase is released into the blood from the damaged hepatocytes, resulting significant increase of transaminase content in blood. It was reported that the levels of GPT, GOT, APK, LDH, and T-BIL in serum of mice with chronic fluorosis were significantly increased (Zhou et al. 2015; Lu et al. 2017). In addition, activities of GPT and GOT in cultured medium of incubated hepatocyte exposed to fluoride were significantly increased with a fluoride dose-dependance.

The results from rats exposed to fluoride in vivo at the prenatal and postnatal stages showed a raise of fluoride concentration in liver and the alterations in antioxidant enzyme activity, and cyclooxygenase-2 and prostaglandin E2 levels in liver (Dec et al. 2018). Also, hepatocytes were diffusely altered in some parts of rat liver exposed to fluoride, with changes resembling microvesicular steatosis. The results indicated that long-term exposure to fluoride during the system development induces fluoride storage in liver, and the changed activity of antioxidant enzymes and cyclooxygenase. Chronic fluorosis is toxic to the liver, which results in disturbances in the homeostasis in liver. The exposure to fluoride might contribute to the changes of the related parameters with liver in American adolescents (Malin et al. 2019). However, a reverse causality may not be ruled out, in which altered liver function might also influence the absorption of fluoride and metabolism.

Mechanism of Liver Changes Caused by Chronic Fluorosis

The mechanism of various damages caused by chronic fluorosis is complicated. Many scholars tried to explain the pathogenesis of chronic fluorosis from abnormal changes of nucleic acid, protein, metabolic enzyme, collagen, and imbalance of major and trace elements caused by fluorosis, but none of them could fully explain the manifestation of systemic damage. The main mechanism of fluoride-induced liver injury may be related to oxidative stress, inflammatory reaction, changes in signal transduction pathway, abnormal expression of receptors, increased apoptosis, changes in cell membrane structure and function, disturbance of calcium metabolism and changes in mitochondria. Among them, the theory of oxidative stress can fully explain the mechanism of systemic injury of chronic fluorosis and has been confirmed by most scholars.

The Increased Oxidative Damages in Liver

Oxidative damage usually indicates the excessive production of free radicals (especially reactive oxygen radicals) when the body is subjected to various harmful stimuli or the weakening of antioxidant system level (including antioxidant enzyme activity and the quality of antioxidant substances) occurs. The damage is caused by the imbalance between these two systems. In the process of oxidative stress, various substances such as lipids, proteins, DNA, and other macromolecular substances that make up tissues and cells undergo various degrees of oxidation due to the oxidation of free radicals. Such process results in the damage of lipid peroxidation, protein oxidation, and DNA oxidation (denaturation, cross-linking, cleavage, etc.) and producing corresponding metabolites. Oxidation indexes are as follows: reactive oxygen species (ROS), malondialdehyde (MDA), 4-hydroxynonenoic acid (HNE), carbonyl, 8-hydroxydeoxyguanosine (8-OHdG), advanced oxidation protein products (AOPPs), 8-isoprostaglandin F2α (8-iso-PGF2α); while antioxidant indexes include antioxidant enzymes, such as superoxide dismutase (SOD), glutathione peroxidase (GSH-Px), thioredoxin peroxidase (Prx), and catalase (CAT), as well as the antioxidant substances, such as vitamin E, vitamin C, coenzyme Q (or ubiquinone), bioflavonoids, selenium (Se), and carotene.

It has been found that various oxidative injuries on lipids, proteins, and DNA are appeared in residents from endemic fluorosis areas. By observing oxidative stress products in serum and urine from the residents in endemic fluorosis areas, the level of urinary fluoride and serum MDA, 8-OHdG, 8-iso-PGF2α, and AOPPs were higher than those of control group. Since lipid damage occurred earlier, serum 8-iso-PGF2α might reflect the earlier state of oxidative stress, which could be used as one of the indexes of oxidative damages for screening early fluorosis cases (Chen et al. 2017). The content of ROS was increased significantly together with the elevated levels of nitric oxide (NO) and MDA in liver and serum of mice exposed to fluoride, whereas SOD and GSH-Px activities as well as total antioxidant capability decreased significantly (Zhou et al. 2015). The effect of sodium fluoride on oxidative stress and apoptosis as well as their relationship in mouse liver were investigated (Lu et al. 2017). The data indicated that oxidative stress induced by fluorosis is characterized by increased levels of ROS and MDA, decreased activities and mRNA expressions of SOD, CAT, GSH, GSH-Px, and glutathione S transferase, which are involved in the process of apoptosis.

The relationship of eating normo- or hyper-caloric diets and oxidative stress concerning lipid metabolism induced by fluorosis in rat liver was evaluated (Pereira et al. 2016). In the results, expression of target proteins related to oxidative stress in liver was enhanced in the groups exposed to fluoride, especially the animals treated with hyper-caloric diet. Most lipid disorders and pathological changes were observed at the earlier period in the damaged liver influenced by fluorosis. In addition, the reduced plasma triglyceride was found in the group exposed to fluoride. The declined level of apolipoprotein E with a high fat diet-dependent pattern might be considered as a particular decrease in steatosis of the rats treated with hyper-caloric diet and fluoride. These data suggest a changed lipid homeostasis in liver due to the stimulated oxidative stress induced by fluorosis, especially the animals treated with both hyper-caloric diet and fluoride.

Excessive fluoride intake causes significant changes by stimulating oxidative stress and apoptosis. Interestingly, some useful microelements, such as Se, might enhance the activities of antioxidant enzymes and attenuate cell apoptosis (Li et al. 2020). In animal model, when Se was added to the diet of fluorosis rats, the fluoride levels of liver and kidney decreased, lipid peroxidation declined and antioxidant enzyme activities recovered. These data indicated that Se, an antioxidant, could not only antagonize the high fluoride in rats, but also correct the metabolic disorder of free radicals caused by high fluoride. Se also alleviates fluoride-induced apoptosis by increasing the expression of heat shock protein 70 (HSP70) and reduces oxidative stress by regulating the levels of sirtuin 1 and antioxidant enzymes.

Aggravated Inflammation

The occurrence and development of chronic fluorosis leads to tissue and cell damages, in which an important manifestation is inflammatory reaction. Treatment with antioxidants can effectively inhibit DNA oxidative damage induced by chronic inflammation. The increase of oxidative stress can increase the generation of inflammatory cytokines. Similarly, the increase of inflammatory cytokines can also stimulate the generation of free radicals, thus increasing the level of oxidative stress. Under light microscope, it was observed that there were degenerative changes in the liver of rats exposed to fluorosis: dilatation and hyperemia of central vein of hepatic lobule, balloon-like changes or punctate necrosis of some hepatocytes, eosinophilic degeneration and eosinophilic necrosis, nuclear pyknosis, and deep staining. Under electron microscope, the following were observed: swelling mitochondria, vacuolar changes, rupture or disappearance of cristae, unclear structure of cell membrane, and inflammatory cell infiltration in the hepatic portal vein area. These changes indicated an existence of inflammatory reaction. The mechanism is that excessive fluoride interferes with the metabolism of oxygen and leads to the elevated level of free radicals. The liver contains a large amount of unsaturated fatty acids, which is vulnerable to free radicals and is an organ prone to lipid peroxidation damage caused by fluorosis. The damage of chronic fluorosis to hepatocytes is mainly the destruction of ultrastructure (mitochondria and endoplasmic reticulum), which resulted from the oxidative stress response of hepatocytes. Therefore, oxidative stress is involved in the inflammatory process and serves as a prerequisite for the generation of inflammatory factors. The increase of inflammatory cytokines in turn can stimulate the production of free radicals, thus increasing the level of oxidative stress.

The Elevating Level of Apoptosis

As an important factor in tissue damage, chronic fluorosis often leads to apoptosis. Apoptosis refers to spontaneous programmed cell death, which is strictly controlled by multiple genes, including Bcl-2, caspases, oncogenes, tumor suppressor gene p53, and so on. Among them Bcl-2 family is an important inhibitor of apoptosis, locating in mitochondrial outer membrane, perinuclear membrane, and smooth endoplasmic reticulum. Bcl-2 is the core molecule of apoptosis suppression. As an important member of the Bcl gene, Bcl2-associated X protein (Bax) can form dimers with Bcl-2 to accelerate cell apoptosis and is an apoptosis promoting gene. The relative expression level of Bcl-2/Bax is a very important link in the regulation of apoptosis.

In chronic fluorosis, apoptosis aggravates with the increase of fluoride dose. Such aggravation results in the gradually declined Bcl-2 and increased Bax, indicating that fluoride can affect cell apoptosis. Excessive fluoride accumulation can lead to DNA damage, while DNA strand breakage can activate DNA-dependent protein kinase, resulting in phosphorylation and activation of p53 protein. The activation of p53 can enhanced the downstream expression of pro-apoptotic genes such as Bax or P53-induced gene and initiate apoptosis. The increased level of apoptosis in liver of rats with chronic fluorosis was confirmed (Fan et al. 2015). Death receptor (Fas)/ death receptor ligand (FasL) was an important member of the apoptotic induction family. The expression of Fas at protein and mRNA levels in liver of animals exposed to fluoride were significantly increased, indicating that the excessive exposure of fluoride to liver can induce apoptosis and thereby result in the damage of liver (Miao et al. 2013).

The mechanism concerning apoptosis induced by chronic fluorosis may be involved in in the increased protein and mRNA levels of tumor necrosis factor recptor-1 (TNF-R1), Fas- or TNFR-related death domain and cysteine aspartate specific proteases (caspase-8 and -3) with dose- and time-dependences. Importantly, oxidative stress, triggered by fluorosis, stimulates apoptotic occurrence, which significantly impairs hepatic function (Lu et al. 2017). In the liver of the mice exposed to fluoride, the obvious increases in mRNA expressions of cytochrome C oxidase, and caspase-9 and -3, as well as TUNEL-positive nuclei were found. In addition, excessive fluoride can increase mitochondrial fission of hepatocytes through the GTPase dynamic associated protein 1/mitochondrial fission factor signaling pathway, which brings serious damage to the structure of mitochondria and then leading to apoptosis (Zhou et al. 2020). Fluoride could also induce liver apoptosis and autophagy by activating interleukin 17 signal pathway, which eventually lead to liver function damage (Zhao et al. 2018). In the liver of animals with chronic fluorosis, apoptosis-related genes including *caspase-8* and *-9*, *Bim*, *CYGb*, *GSTP1*, and *NCF1* were increased, while *caspase-3*, *Bcl-2*, *Bak*, *GPX1*, *SOD1*, and *SOD2* decreased. It was indicated that the mechanisms relating apoptotic and oxidative in liver of the animals with fluorosis were activated, in which these parameters play an important role in understanding the molecular basis of damage by fluoride (Efe et al. 2021). Nevertheless, the mechanism of hepatocyte apoptosis induced by chronic fluorosis should be further explored.

Proteins and Genes Related to Changes in Liver Metabolism and Function

With the increasing maturity of protein and genomics technology, a variety of protein and gene changes related to liver injury induced by chronic fluorosis have been found.

Hedgehog (Hh) is a covalent cholesterol-binding secretory protein with an important effect in development. At present, researchers believe that hepatic stem cells can proliferate and differentiate into immature cells such as hepatocytes and bile duct epithelial cells. Such results are responsible for renewing senescent and dead liver cells under physiological conditions, maintaining the integrity of liver structure and function, and repairing liver injury by proliferation and differentiation under pathological conditions. Hh signal pathway is activated to repair the liver by stimulating the differentiation and proliferation of hepatic stem cells and myofibroblasts. The expressions of Indian hedgehog protein (Ihh), sonic hedgehog protein (Shh), transmembrane protein (Smo), nuclear transcription factor protein (Gli1 and Gli3) in the signal pathway were increased in hepatocytes of rats with fluorosis. Such pathway was inhibited by cyclopamine, suggesting that Hh signal pathway is involved in the repair process of liver injury caused by chronic fluorosis (Zhao et al. 2014).

Phosphatidylinositol 3-kinase-protein kinase B α3 (PI3K) is a kind of intracellular phosphatidylinositol kinase with regulatory subunit P85 and catalytic subunit P110. Scrinc/thrconinckinasc and phosphatidylinositol kinase participate to regulate many cellular functions, e.g., cell proliferation and differentiation, apoptosis, and glucose transport. The activity of protein kinase B (Akt), a serine/threonine specific protein kinase, also concerns the variety of cell growth processes, including glucose metabolism, and cell proliferation, transcription, and cell migration, as well as apoptosis. The increased expressions of PI3K and Akt1 at mRNA and protein level in hepatocytes of rats exposed to fluoride were obviously observed, which was related to the increase of apoptosis and intracellular calcium concentration. The increased expressions of these factors might correlate to the pathogenesis of liver injury due to fluorosis (Fan et al. 2015).

Osteopontin (OPN), a negatively charged non-collagen bone matrix glycoprotein, is widely distributed in a variety of tissues and cells, and can participate in tissue repair, self-metabolism, cell adhesion, cellular immunity, signal transduction, and apoptosis. In normal liver tissue, OPN is rarely expressed but highly expressed in the liver with injury. OPN is found to be mainly expressed in hepatic macrophages, stellate cells, and activated Kupffer cells. Fluoride exposure can promote expression of OPN in rat liver (Zhou et al. 2009). Excessive fluoride, low calcium, and low protein had a synergistic effect on promoting the expression of OPN in rat liver. Interestingly, it was observed that OPN was overexpressed in hepatocytes of the fluorosis rats by feeding corn containing high amount of fluoride because of the burning of coal, which enhanced calcium deficiency and less protein intake (Zhou et al. 2012). There are many mechanisms regarding the damage of fluorosis influenced by OPN. Excessive amount of fluoride can cause an interference with calcium mobilization in the bone, results in a large number of free Ca^{2+} accumulated in body fluids and soft tissues. It then induces OPN to activate Ca^{2+} channels and followed by an increase in intracellular Ca^{2+} ions, which results in hepatocyte apoptosis and injury. The structure of OPN determines its high sensitivity to the change of free Ca^{2+} concentration, which is possible to influence the interaction of OPN with integrin αVβ3. Thus, OPN affects dependence of integrin αVβ3 on its ligand structure with characteristics of signals produced by the cells, then increasing intracellular Ca^{2+} concentration and eventually leading to apoptosis of hepatocytes.

Effect of Fluorosis on Gastrointestinal Tract

Function and Metabolism of Gastrointestinal Tract

It is the smooth muscle contraction activity in gastrointestinal tract to complete food mechanical digestion and push food forward digestive tract. Gastrointestinal tract contains many digestive glands, which secrete digestive fluid to promote digestion and absorption of food.

Effect of Chronic Fluorosis on Gastrointestinal Function

Fluorosis can not only directly damage gastrointestinal tract but also indirectly affect the balance of bacterial flora in the tract. It eventually leads to damage of gastrointestinal function and structure. Excessive intake of fluoride can cause damage to the digestive tract. The gastric mucosa of patients with chronic fluorosis in endemic fluorosis area was observed by endoscope, in which the gastric mucosa of these patients had pathological changes including erosion, necrosis, punctate hemorrhage, and epithelial cell exfoliation (Xie et al. 1995).

Mechanism of Gastrointestinal Dysfunction Caused by Fluoride Exposure

In the homeostasis of the immune system and the health of the host, intestinal microflora is important. A study concerning changes in the structure and microflora of the rectum due to fluorosis showed that fluoride seriously damaged the structure of the rectum and significantly inhibited the proliferation of rectal epithelial cells (Wang et al. 2020). After fluoride exposure, the distributions of goblet cells and mast cells, and the content of glycoproteins were significantly decreased; the relative richness of microflora was changed, presenting increased Bacteroidetes and decreased Firmicutes. In addition, the effect of fluoride on intestinal flora of silkworm was studied. It was found that after fluorosis, the flora of Enterococci and Bacillus were decreased. Also, the increased microflora of Staphylococci indicated that fluoride could destroy the balance of intestinal flora of silkworm by changing the diversity and proportion of bacteria (Li et al. 2015). In order to understand the effect of fluoride on the intestinal flora of mice, the transcriptional spacers in colonic feces of fluoride-exposed mice were studied by sequencing (Cao et al. 2020). The results indicated that the smut fungi changed significantly in the mice exposed to fluoride. At the genera level, fluoride significantly influenced Epicoccuum, Penicillium, Microdochium, Plectosphaerella, and Pluteus, among which, a strong positive correlation was observed between Penicillium and Pluteus. Therefore, it can be concluded that fluoride mainly affects the relative species abundance of the intestinal microflora of mice at the genus level.

Ingested fluoride is absorbed mainly in the small intestine, which is controlled by the enteric nervous system. After excessive fluoride exposure, there was not only important intestinal symptoms, but also certain effects on the enteric nervous system. The effects of chronic fluoride exposure were evaluated on the duodenums of rats through proteomic and morphological analyses. Immunofluorescence assays were performed in the myenteric plexus of the duodenum to detect HuC/D, neuronal nitric oxide, vasoactive intestinal peptide, calcitonin gene related peptide, and substance P. Proteomic analysis of the duodenum demonstrated the alterations in the expression of several proteins, especially those related to important biological processes (Melo et al. 2017). It suggested that excessive fluoride also had certain effects on the autonomic nerves in the digestive tract.

Effect of Fluorosis on Pancreas

Function and Metabolism of Pancreas

Pancreas is a gland with both exocrine and endocrine functions. The endocrine function of pancreas is mainly related to the regulation of glucose metabolism and exocrine, one of which is pancreatic juice with strong digestive ability, and secreted by acinar cells and ductal wall cells of the pancreas.

Effects of Excessive Fluoride on the Morphology and Function of Pancreas

There are few reports on the effect of fluoride on pancreatic exocrine, but focusing on the effect of fluoride on pancreatic endocrine. Excessive intake of fluoride causes damage to the endocrine and exocrine functions of the pancreas to a certain extent. Morphological effects of fluoride on pancreatic exocrine cells were observed under electron microscopy in rats exposed to fluoride (Matsuo et al. 2000). The results showed that in the exocrine pancreatic cells of rats exposed to fluoride, the expanded or globular-shape rough endoplasmic reticulum (rER), and the broken lamellar arrangement were determined. In addition, intracisternal granules were formed in the globular-shape rER. Many autophagosomes were also found in exocrine cells. The autophagosomes were bi-membranous or multi-membranous and contained cytoplasmic organelles and/or intracellular particles. The inner and outer lobules of the autophagosome bilayer are usually separated by a distinct electron transmittance region. During a long period of osmium fixation, the two membranes of autophagosomes were filled with osmium-reactive deposits. Many autophagosomes are surrounded by a single or multiple osmiophilic layers. In some cases, osmium-positive small balloons surround the free surface of the spherical rER containing intracerebral particles. These results suggest that fluoride blocks the proenzyme output of the rER, leading to the formation of granules and autophagosomes. Osmiophilic cysts may participate in the isolation of cytoplasmic organelles to form autophagosomes.

To observe the effects of increasing fluoride dose on bone formation markers, changes in serum insulin level and pancreatic secretion, and mRNA expression of insulin receptor (INSR) and osteocalcin (OCN) in vitro were investigated (Hu et al. 2012). The results showed that the activity of ALP and the level of OCN were significantly increased with the increase of bone fluoride concentration, indicating that fluoride had a significant stimulating effect on the osteogenesis of rats. The serum insulin level increased significantly and glucagon decreased in the fluoride exposure group. The insulin-positive region of pancreatic tissue and the size of islets were obviously increased. In addition, fluoride significantly promoted the expression of InsR mRNA in vitro. The results point out that insulin secretion is closely related to fluoride exposure, and insulin signaling pathway may be one of the important factors in the developing procedure of skeletal fluorosis.

The serum insulin level in rats exposed to fluoride was significantly raised, while glucagon generally decreased (Hu et al. 2012). Histomorphological analysis also exhibited that the insulin-positive region and the islet volume were increased. In addition, it was found that AKP/ALP activity and osteocalcin level was increased significantly in rats with chronic fluorosis, providing evidence to support the correlation between insulin signaling pathway and developing procedure of skeletal fluorosis. Excessive exposure of fluoride can interference carbohydrate metabolism, which may result in promoting hyperglycemia, insulin resistance, and changed insulin signaling (Pereira et al. 2017). Furthermore, chronic exposure to fluoride increased concentrations of insulin, fluoride, OCN, and TNF-α in plasma and decreased trabecular bone area of the tibia, causing changes of bone metabolism markers in ovariectomized rats.

Conclusion

Chronic fluorosis can cause digestive system damage, especially liver injury. Excessive fluoride exposure leads to pathological changes in liver tissue and abnormally functional indexes of liver. Such mechanism may be involved in the increase level of hepatocyte oxidative stress and apoptosis, and modified level of liver-related protein and gene due to the excessive exposure of fluoride. The effect of chronic fluorosis on gastrointestinal tract not only has direct or indirect damage, but also destroys the balance of intestinal flora by regulating the diversity and proportion of bacteria in gastrointestinal tract, which then affects gastrointestinal function. Excessive fluoride also has different effects on the morphology and function of pancreas, resulting in changes in insulin signal pathway, which is related to abnormal bone metabolism.

References

Cao QQ, Li R, Fu R, Zhang XH, Yue BJ, Sun ZL, et al. Intestinal fungal dysbiosis in mice induced by fluoride. Chemosphere. 2020;245:125617.

Chen XW, Wan CW, Zhang SY, Luo Y, Qi XL, Li Y. Serum oxidative stress products among residents living in endemic area of coal-burning fluorosis. Chin J Public Health. 2017;33:824–7.

Dec K, Łukomska A, Baranowska-Bosiacka I, Pilutin A, Maciejewska D, Skonieczna-Żydecka K, et al. Pre-and postnatal exposition to fluorides induce changes in rats liver morphology by impairment of antioxidant defense mechanisms and COX induction. Chemosphere. 2018;211:112–9.

Efe U, Dede S, Yüksek V, Çetin S. Apoptotic and oxidative mechanisms in liver and kidney tissues of sheep with fluorosis. Biol Trace Elem Res. 2021;199:136–41.

Fan B, Yu Y, Zhang Y. PI3K-Akt1 expression and its significance in liver tissues with chronic fluorosis. Int J Clin Exp Pathol. 2015;8:1226–36.

Hu CY, Ren LQ, Li XN, Wu N, Li GS, Liu Y, et al. Effect of fluoride on insulin level of rats and insulin receptor expression in the MC3T3-E1 cells. Biol Trace Elem Res. 2012;150:297–305.

Li G, Xia X, Sendegeya P, Zhao H, Long Y, Zhu Y. Effect of fluoride on gut microflora of silkworm (Bombyx mori). Acta Microbiol Sin. 2015;55:926–34.

Li W, Dong S, Chen Q, Chen C, Dong Z. Selenium may suppress peripheral blood mononuclear cell apoptosis by modulating HSP70 and regulate levels of SIRT1 through reproductive hormone secretion and oxidant stress in women suffering fluorosis. Eur J Pharmacol. 2020;878:173098.

Liang CK, Zhang SL, Li WH, Ma F, Wu YP, Li Y, et al. Blood chemistry and electrolyte effects of exposure to fluoride from burning coal on population. Chin J Endemiol. 2000;19:47–50.

Lu Y, Luo Q, Cui H, Deng H, Kuang P, Liu H, et al. Sodium fluoride causes oxidative stress and apoptosis in the mouse liver. Aging. 2017;9:1623–39.

Malin AJ, Lesseur C, Busgang SA, Curtin P, Wright RO, Sanders AP. Fluoride exposure and kidney and liver function among adolescents in the United States: NHANES, 2013-2016. Environ Int. 2019;132:105012.

Matsuo S, Nakagawa H, Kiyomiya K, Kurebe M. Fluoride-induced ultrastructural changes in exocrine pancreas cells of rats: fluoride disrupts the export of zymogens from the rough endoplasmic reticulum (rER). Arch Toxicol. 2000;73:611–7.

Melo CGS, Perles J, Zanoni JN, Souza SRG, Santos EX, Leite AL, et al. Enteric innervation combined with proteomics for the evaluation of the effects of chronic fluoride exposure on the duodenum of rats. Sci Rep. 2017;7:1070.

Miao K, Zhang L, Yang S, Qian W, Zhang Z. Intervention of selenium on apoptosis and Fas/FasL expressions in the liver of fluoride-exposed rats. Environ Toxicol Pharmacol. 2013;36:913–20.

Pereira HA, Dionizio AS, Fernandes MS, Araujo TT, Cestari TM, Buzalaf CP, et al. Fluoride intensifies hypercaloric diet-induced ER oxidative stress and alters lipid metabolism. PLoS One. 2016;11:e0158121.

Pereira AG, Chiba FY, de Lima Coutinho Mattera MS, Pereira RF, de Cássia Alves Nunes R, Tsosura TVS, et al. Effects of fluoride on insulin signaling and bone metabolism in ovariectomized rats. J Trace Elem Med Biol. 2017;39:140–6.

Shashi A, Bhardwaj M. Study on blood biochemical diagnostic indices for hepatic function biomarkers in endemic skeletal fluorosis. Biol Trace Elem Res. 2011;143:803–14.

Wang T. Study on serum chemical elements and serum enzyme content in patients with endemic fluorosis. Chin J Control Endemic Dis. 2016;31:185–6.

Wang HW, Miao CY, Liu J, Zhang Y, Zhu SQ, Zhou BH. Fluride-induced rectal barrier damage and microflora disorder in mice. Environ Sci Pollut Res Int. 2020;27:7596–607.

Xie HZ, Gui YF, Wang L. Effect on gastric mucosa in fluorosis and arsenic-fluorosis. J Xinjiang Med Univ. 1995;18:18–20.

Zhao LN, Yu YN, Deng CN. Protein and mRNA expression of Shh, Smo and Gli1 and inhibition by cyclopamine in hepatocytes of rats with chronic fluorosis. Toxicol Lett. 2014;225:318–24.

Zhao Y, Li Y, Wang J, Manthari RK, Wang J. Fluoride induces apoptosis and autophagy through the IL-17 signaling pathway in mice hepatocytes. Arch Toxicol. 2018;92:3277–89.

Zhou YX, Yu YN, Wang JQ, He J, Zhang H, Guan ZZ. The study of osteopontin expressions in hepatocytes of fluorosis rats with different levels in protein and calcium. Chin J Control Endemic Dis. 2009;4:245–9.

Zhou YX, Mo SL, Wang RH, Liu JC, Liu WF, Wang JQ, et al. Corn baked by burning coal triggered overexpression of osteopontin in hepatocytes of rats following fluorosis. Toxicol Ind Health. 2012;28:195–202.

Zhou BH, Zhao J, Liu J, Zhang JL, Li J, Wang HW, et al. Fluoride-induced oxidative stress is involved in the morphological damage and dysfunction of liver in female mice. Chemosphere. 2015;139:504–11.

Zhou BH, Wei SS, Jia LS, Zhang Y, Miao CY, Wang HW. Drp1/Mff signaling pathway is involved in fluoride-induced abnormal fission of hepatocyte mitochondria in mice. Sci Total Environ. 2020;725:138192.

Chapter 8
Modification of Urinary System

Xiao Xiao and Zhi-Zhong Guan

Abstract Kidney is a vital organ of body and its basic function is to generate the urine, so as to remove the body metabolic products and some toxic wastes. At the same time the organ retains the moisture absorption function and other useful substances to regulate water and electrolyte balance for maintaining the acid-base balance. In addition, kidney still has endocrine function to produce renin, erythropoietin, active vitamin D3, and prostaglandin. These functions of the kidney ensure the stability of the body's internal environment, so that metabolism can be normally carried out. About 50% of fluoride ingested in the body is removed by kidney every day, and its removal rate is related to the pH of urine. Excessive accumulation of fluoride can lead to damages of renal morphology, function, and metabolism, including the morphological abnormality of renal tubular tissues, the function inhibition of renal tubular reabsorption, the dysfunction of renal excretion, etc. Severe chronic fluorosis can cause a significant change of the epithelial cell cycle of renal tubules, the elevated level of oxidative stress, the excessive alteration of autophagy, and even apoptosis.

Keywords Apoptosis · Chronic fluorosis · Excretion · Kidney · Renal tubules

X. Xiao
The Key Laboratory of Endemic and Ethnic Diseases (Guizhou Medical University), Ministry of Education, Guiyang, Guizhou, People's Republic of China

Department of Pathology at the Affiliated Hospital of Guizhou Medical University, Guiyang, Guizhou, People's Republic of China

Z.-Z. Guan (✉)
The Key Laboratory of Endemic and Ethnic Diseases (Guizhou Medical University), Ministry of Education, Guiyang, Guizhou, People's Republic of China

Department of Pathology at the Affiliated Hospital of Guizhou Medical University, Guiyang, Guizhou, People's Republic of China

The Provincial Key Laboratory of Medical Molecular Biology, Guiyang, Guizhou, People's Republic of China
e-mail: zzg@gmc.edu.cn

Z.-Z. Guan (ed.), *Coal-burning Type of Endemic Fluorosis*,
https://doi.org/10.1007/978-981-16-1498-9_8

Introduction

Excessive intake of fluoride can have adverse effects on many organs of human body, of which in the non-bone organs kidney is one of the most vulnerable to chronic fluorosis. Kidney is the main excretory organ of fluoride, in which 75% of the total fluoride removed by the body is excreted through the organ. Fluoride may accumulate in the body when the intake of fluoride exceeds the normal excretion capacity of the kidneys. Long-term exposure to high level of fluoride in environment will cause pathological changes and dysfunction of the kidney, which results in changes in the level of related chemicals in blood and urine.

The rate of fluoride excretion in the kidney is quite fast. The fluoride ingested soon appears in the urine, and 20–30% of the fluoride ingested can be excreted in the urine within 3–4 h. In the first 4 h after fluoride intake, the rate of urinary fluoride excretion is the fastest, and then decreases gradually. The concentration of fluoride in blood and urine often increases in patients with chronic fluorosis, especially in urine, which is an important reference for the diagnosis of endemic fluorosis. However, there are many factors that can increase urinary fluoride, especially some foods with high fluoride content, so it is not possible to take an occasional increase of fluoride in urinary as a basis for the diagnosis of chronic fluorosis.

In severe cases, pathological changes such as apoptosis and necrosis of renal tubular epithelial cells have been observed. In addition, excessive amount of fluoride can induce an increase in the level of oxidative stress, causing mitochondrial damage, endoplasmic reticulum stress, and abnormal expression of related genes.

Effect of Chronic Fluorosis on Kidney

Renal Pathohistological Changes by Chronic Fluorosis

In the cases of chronic fluorosis, obvious pathological changes in morphology and structure of renal tissue have been found. The sensitive part of renal injury caused by chronic fluorosis mainly lies in renal tubules. Pathological changes of renal tubular epithelial cells keep an important link in causing renal function injury. The edema to varying degrees and lumen stenosis were observed in the renal tubular epithelial cells of rats with chronic fluorosis. The obviously hyperemia of renal tubulointerstitium, the swollen, dissolved or necrotic changes of the epithelial cells in the proximal convoluted tubule of the renal cortex, and the enlarged lumen as well as the exfoliated epithelial cells in the lumen were found in the rats exposed to high level of fluoride (Li et al. 2019). Under electron microscope, many spherical particles were formed or decomposed in renal tubule basement membrane of rats with chronic fluorosis. Renal tubular epithelial cells showed a series of apoptosis signals such as nuclear pyknosis and chromatin division. Occasionally, cell proliferation and endoplasmic reticulum vesiculation were observed. The swollen mitochondria, the reduced or even disappeared mitochondrial cristae, the ruptured mitochondrial membranes, the gathered mitochondrial chromatin, and the folded or broken nuclear membranes frequently occurred in the animals with chronic fluorosis. Some studies have confirmed that a number of lysosomes in proximal convoluted tubule epithelial cells are increased during chronic fluorosis, and irregular dense bodies formed by lysosomes are also observed in cytoplasm. The damages by fluorosis to kidney mainly happen in the renal tubules located at the junction of cortex and medulla. The dilation of renal tubules is likely due to the obstruction of the tubule at the lower part and the increased pressure in lumen, exhibiting the compensatory dilation of renal tubules (Wang and Fan 2014). Furthermore, the glomerular sacs also are expanded when the lumen pressure of renal tubules increases to a certain extent. The exposure to high concentration of fluoride in vitro also induces obvious pathological reactions of renal cells, such as chromosome aberration, cell cycle arrest, release of cytokines and arachidonic acid, apoptosis, etc.

Effect of Chronic Fluorosis on Renal Function and Metabolism

Many studies have clarified the role of the kidney in metabolism and fluoride removal. Kidney damage caused by chronic fluorosis has a negative effect on kidney metabolism and fluoride excretion, resulting in more fluoride retention, which in turn further aggravates kidney damage and forms a vicious circle. When chronic fluorosis causes severe impairment of renal function, urine fluoride excreted by the kidneys decreases and residual fluoride concentrations in the serum are further increased. Some studies have pointed out that daily intake of 12 ppm of fluoride would increase the risk of adverse renal reactions in people, while the drinking water with 2 ppm fluoride can lead to kidney damage in children, and this damage would worsen with the increase of fluoride content in water (Dharmaratne 2019). Long-term anesthesia with fluorine-containing compound methoxyflurane may lead to renal failure and death. Excessive fluoride can destroy the synthesis of collagen, resulting in accelerated decomposition of renal collagen, which may inevitably damage the kidney. In addition, fluoride also has certain inhibitory effects on kidney-related enzyme activities, which can aggravate kidney injury (Dharmaratne 2019).

Apart from the pineal gland, the kidney is the organ most exposed to fluoride in soft tissues, so the kidney has a higher risk of fluorosis. Some studies had found that after 40 days of exposure to 50 ppm fluoride (prepared with sodium fluoride), the brush edge of the proximal convoluted tubules disappeared, the renal tubular epithelium fell off, and the renal tubules flattened. The fact of hypertrophy and hyperplasia of renal proximal convoluted tubules in rats exposed to 5 ppm fluoride for 500 days indicates that when fluoride intakes exceeds renal excretion, high concentrations of fluoride could damage renal tubule-related areas and destroy renal tubule reabsorption function, and eventually lead to chronic kidney diseases

(Guo et al. 2020). The concentrations of fluoride in blood and urine of patients with chronic fluorosis often increase (Vale et al. 2019). Since most of the fluoride in the human body is excreted through the kidney, and its concentration in the form of urine can directly reflect the fluoride load in the body (Wu et al. 2015). Generally, the amount of urine fluoride is the highest in the first half of the night. Therefore, the increase of urine fluoride concentration in morning urine can be used as a more reliable reference basis for diagnosing endemic fluorosis groups. However, there are many factors that can temporarily increase urine fluoride, especially eating some foods with high fluoride content, more tests of urine fluoride concentration are necessary for diagnosis of chronic fluorosis.

It has been reported that the main manifestations of renal injury caused by endemic fluorosis are renal concentrated dysfunction (hypotonic urine) and acute renal tubular acidosis. Animal experiments also prove that high-dose fluoride intake can cause renal functional failure, such as the frequent urination, the increases of urinary fluoride and non-protein nitrogen contents in urine, etc. The mechanism may be involved in the activities of ATPase and alkaline phosphatase in renal tubular epithelial cells, which are inhibited and therefore induce the renal tissue structure changes due to the inhibition of renal tubular epithelial cells by fluorosis. A large amount of electrolyte is lost in the process of fluoride removal, resulting in imbalance of physiological indexes such as water, osmotic pressure, pH value, etc. The early life stage is the key period of renal development, and the separated epithelial cells are assembled into renal tubules during this period. Therefore, chronic fluorosis has the renal damage effect on early embryos. Exposure to fluoride during embryonic and early lactation in mice can also cause renal interstitial hemorrhage (Guo et al. 2020).

Mechanism of Renal Damage Caused by Chronic Fluorosis

Change of Renal Cell Cycle

High dose of fluoride intake has certain effects on the cell cycle of renal tissue cells. It has been found that the level of -glutamine transferase in the urine of rabbits with fluorosis was significantly increased. In addition, K^+, Ca^{2+}, and P^{3+} metabolic disorders were particularly obvious. The number of renal cells in G2 phase was significantly increased in the high-dose fluoride group and the middle-dose fluorine group compared with the control group, while the number of renal cells in G1 phase was significantly decreased in the high-dose fluorine group (Li et al. 2007). The study also showed that the G0/G1, S, G2/M phases, and cell proliferation index of the rats with chronic fluorosis were lower, especially in the G2/M phase. The main damage site of fluoride to renal tissue is renal tubules, that is, it may mainly affect the absorption function of kidney. Excessive fluoride affects the function of the kidney by affecting the synthesis of DNA, impeding the entry of cells into the division stage and promoting cell death. This inhibition to DNA synthesis is likely to be secondary to protein synthesis disorders or induced by affecting the normal functioning of DNA polymerases and other enzymes involved in DNA synthesis, in which we may thus suppose that effect of fluoride on the cell cycle is closely related to its inhibition of DNA synthesis, resulting in blocking cell entry into the division phase, thus leading to cell death and thereby impairs renal function.

Other evidence indicated fluoride induces G2/M cell cycle arrest accompanied by upregulation of expression levels of various related proteins, suggesting that fluoride can induce G2/M cell cycle arrest by activating signal pathways, thus inhibiting renal cell proliferation and renal development (Luo et al. 2018).

Disorder of Renal Calcium Balance

Calcium (Ca^{2+}) is an important substance for cells to maintain various physiological functions and participate in cell physiological and pathological activities, such as regulating intracellular signals and membrane channel functions, participating in cell differentiation and apoptosis, etc. Calcium has long been reported in the prevention and treatment of skeletal fluorosis as an antagonist and greatly applied in clinic. Studies have found that the damage of cells and tissues is accompanied by the influx and accumulation of calcium, and the growth process of various cells is directly and indirectly related to calcium.

There are L-type calcium channels (LTCC) in renal cell membrane. Related studies have shown that fluoride exposure can reduce the activation voltage of LTCC and prolong the opening time of calcium channel by increasing the expression of calcium channel inhibitor protein, resulting in calcium deficiency in the whole body and calcium overload in kidney cells, resulting in calcium contradiction. Extracellular calcium influx affects the downstream of relevant signal pathways, causing abnormal expression of calcium ion-regulated transcription factor nuclear factor-κB (NF-κB). NF-κB induces expression changes of apoptosis proteins such as B-cell lymphoma-2 (Bcl-2) and Bcl-2 interacting mediator (Bim), and promotes apoptosis leading to damage (Shao et al. 2020).

Calcium malnutrition has been reported to exacerbate fluoride-induced nephrotoxicity. Both protein folding and chaperone function maintenance require the endoplasmic reticulum cavity to maintain a high level of Ca^{2+}, while the change of intracellular Ca^{2+} concentration in renal cells after fluorosis will directly stimulate endoplasmic reticulum stress. Endoplasmic reticulum stress can cause Ca^{2+} to release from endoplasmic reticulum cavity into cytoplasm, promote Ca^{2+} to enter nucleus and mitochondria, further promote generation of reactive oxygen species (ROS), induce apoptosis and lead to injury. Adding appropriate amount of calcium after fluorosis can not only inhibit the apoptosis pathway of endoplasmic reticulum stress to a certain extent, but also reduce the toxic effect of fluoride on kidney in histological and hematological indexes. However, excessive calcium will aggravate the damage of kidney structure and functional changes.

High Level of Oxidative Stress

A large number of studies have shown that the mechanism of renal damage induced by chronic fluorosis also involves raised levels of oxidative stress, i.e. the increase of free radicals and the weakening of antioxidant defense system. Fluoride, as a kind of ionized element with strong reactivity and small molecular weight, can easily induce ROS production through simple diffusion through cell membrane. Therefore, chronic fluorosis can induce the body to be in a high level of oxidative stress. In addition, high level of fluoride can inhibit the activity of oxidases. With changing the content of reduced glutathione (GSH) by fluorosis, ROS at mitochondrial level is produced excessively, resulting in cell damage, oxidations of protein, lipid and DNA.

It has been pointed that lipid peroxides in serum of patients in endemic fluorosis area and in multiple organs, serum and red blood cells from rats with chronic fluorosis rats were significantly increased. Renal damage and apoptosis of carp in a dose-dependent manner indicated by the histopathological examination and ultrastructural observation were found (Cao et al. 2015). Furthermore, excessive amount of fluoride also induced remarkable enhancement of ROS production and oxidative stress, such as the increase of lipid peroxidation level and the depletion of intracellular GSH level in kidney of the fish. The contents of ROS, malondialdehyde (MDA), 8-hydroxy-2′-deoxyguanosine (8-OHdG), protein carbonyl, and cysteinyl aspartate specific proteinases 3 in serum, liver, and kidney were higher in broilers with high exposure of fluoride (Wang et al. 2018a, b). With the deepening of fluorosis, the decreased activities of antioxidant enzymes such as catalase (CAT), glutathione peroxidase (GSH-Px), and nitric oxide synthase (NOS), the declined content of nitric oxide and the increased level of lipid peroxidation were observed in the tissues of rat kidney.

Mitochondria are the centers of aerobic energy metabolism in cells, which play a vital role in cell survival. On the other hand, mitochondria might be the sensitive primary target of various environmental toxicants. Since mitochondrial respiratory chain is an important source of oxygen free radicals, if excessive peroxide cannot be neutralized by endogenous antioxidants or enzymes, the large amonts of excessive accumulation would result in oxidative damage and mitochondrial dysfunction, leading to cell death and organ damage. Importantly, the function of mitochondria is related to its dynamic behavior, i.e., the balance between fusion and fission. When mitochondrial fusion is attenuated by deficiencies in key components involved, such as mitochondrial GTPase-mitofusions (mitofusion 1 (Mfn1) and mitofusion 2 (Mfn2)) and optic atrophy type-1, and/or by elevations in the levels of key elements of mitochondrial fission (Fis1 and Drp1), these organelles become fragmented. Conversely, when the balance is shifted toward fusion, the mitochondria will be long and excessively interconnected. Mitochondrial fusion–fission balance has a valuable role in maintain mitochondrial function. When the dynamic balance of the mitochondria is broken, their morphology could change dramatically. In the renal tissues of rats with chronic fluorosis, expression of both Mfn1 protein and mRNA was clearly reduced, whereas that of Fis1 was elevated (Qing et al. 2015). Meanwhile, the level of MDA was increased and the total antioxidant capacity lowered. In addition, swollen or fragmented mitochondria in renal cells were observed under the electronic microscope. These findings indicate that chronic fluorosis can lead to the abnormal mitochondrial dynamics and changed morphology in the rat kidney, which in mechanism might be induced by a high level of oxidative stress in the disease.

The evidence above suggests that a large amount of fluoride accumulation could cause imbalance between renal oxidation and antioxidant defense mechanisms, make the body in an oxidative stress state, and thus induce renal peroxidative damage.

Excessive Autophagy of Renal Cells

The damage process of tissue or organ function not only involves excessive apoptosis of cells, but also is closely related to excessive autophagy of cells. Autophagy under physiological conditions is also called type II programmed cell death, which is an important mechanism for cells to maintain their homeostasis. Autophagy is an important decomposition process, which participates in metabolic activities of degradation and circulation. It can remove damaged organelles, pathogens in phagocytes, and misfolded proteins. However, excessive autophagy can lead to pathological disorders and even cell death. In the factors concerning the regulation of autophagy, Mammalian/mechanistic target of rapamycin (mTOR) is the most important upstream signal molecule. External stimuli such as hypoxia, oxidative stress, and poisoning can activate autophagy, which is mediated by mTOR signaling pathway. In addition, beclin-1 expression is up-regulated and interacts with various molecules to participate in autophagy formation. Class III phosphatidylinositol 3-kinase (PI3KC3) complex is another important autophagy signaling complex necessary for autophagy body formation and acts by recruiting downstream proteins. Bcl-2, as a class of anti-apoptosis proteins, can bind to Bcl 2 homology 3 binding domain of beclin1 to inhibit the formation of beclin1-PI3KC3 complexes downstream of autophagy. Furthermore, autophagy marker light chain 3 (LC3) is an important autophagy marker protein. The transformation of free LC3-I to lipid-bound membrane-bound LC3-II is considered as a key to autophagy formation.

Fluoride exposure can induce autophagy in different organs and cells by increasing ROS production and reducing oxidative stress induced by antioxidant system. Recent reports showed that high amount of fluoride increased mRNA expression levels of LC3, LC3I, LC3II, and Beclin-1 with a concomitant decrease in mTOR and Bcl-2 in mouse kidney tissue (Guo et al. 2020). Meanwhile, the protein expression levels of LC3-II/LC3-I and Beclin-1 were also significantly increased with a concomitant decrease in mTOR and Bcl-2 in the mice subjected to individual exposure of fluoride. It was observed that renal tissue of mice was highly sensitive to fluoride. Since the content of ROS in renal tissue was positively correlated with the ratio of LC3-II/LC3-I as well as mTOR and Beclin-1 expression, the excessive generation of

ROS in renal tissue of chronic fluorosis may induce excessive activation of autophagy process. In an established rat model of prenatal and postnatal exposure to fluoride, the significant increases of fluoride levels in blood and bone were found (Tian et al. 2020). In this study, fluoride disrupted histopathology and ultrastructure in the kidney, and also altered creatinine (Cr), blood urea nitrogen (BUN), and uric acid (UA) levels. Intriguingly, fluoride uptake induced the formation of autophagosomes in kidney tissue and resulted in the upregulation of genes encoding autophagy-related proteins. Collectively, these results suggest that nephrotoxicity of fluoride for offspring exposed to the pollutants from in utero to puberty is associated with deregulation of autophagy.

Modification of the Related Cytokines

Fibroblast growth factors (FGFs) are multifunctional growth factors and the proteins related to cell proliferation activities, and regulate cell proliferation, migration, and differentiation. Under normal circumstances, the expression of basic fibroblast growth factor (bFGF) in kidney is very low, which is only found in renal interstitial fibroblasts. During renal development, renal parenchymal cells transiently overexpress bFGF, which induces renal tubular differentiation. When renal tubular epithelial cells are damaged and regenerated, the body mediates the repair of renal tubular cells by enhancing the synthesis and secretion of bFGF. Interestingly, excessive fluoride can damage renal tubular epithelial cells, thus promoting the expression of bFGF in cells, which therefore induces the increase of fluoride dosage (Zhang et al. 2007).

Osteopontin (OPN) is a secretory phosphorylated glycoprotein. The expression of OPN is normal in the undamaged renal tissue, but high in the damaged renal tissue. The increase of OPN expression at a certain dose of fluoride can play a protective role in kidney, improve its tolerance to ischemia, inhibit the production of inducible NOS, reduce the production of nitric oxide in renal tubular epithelial cells and the level of cell peroxidation, inhibit cell apoptosis, and promote cell regeneration. However, OPN can promote the increase of intracellular free calcium level (Zhou et al. 2012). The high expression of OPN is increased with the increase of fluoride dosage and induces the aggravation of renal injury, which suggests that OPN is closely related to the degree of renal injury caused by fluorosis and can be used as a predictor of renal damage resulted from fluorosis.

Apoptosis of Renal Cells

Fluoride can have obvious negative effects on renal cells in a dose-dependent manner, whereas apoptosis is one of the important causes of renal injury caused by fluoride. Apoptosis is a process that involves many complicated steps and eventually leads to cell death. It directly destroys the normal biological function of cells and further affects the functions of corresponding tissues and organs. This process is induced by increasing the transformation of cytochrome C (Cyt C) into cytoplasm. The released Cyt C enters cytoplasm and is activated by forming a complex with caspase 9 precursor. Activated caspase 9 acts on downstream caspase 3. Furthermore, activated caspase 3 can degrade a variety of cytoplasm and nucleoproteins and stimulate nuclease to cause DNA degradation. In addition, the expression of caspase 8 has also increased. Caspase 8 collects adapter protein in the form of trimer and activates the initiating apoptosis protease. Activated caspase 8 can activate caspase 3 through proteolysis to induce apoptosis. Therefore, caspase 3 can be activated simultaneously by exogenous caspase 8 and endogenous caspase 9 to induce a series of characteristic changes of apoptosis, such as cytoplasmic concentration, nuclear agglutination, and DNA fragmentation (Song et al. 2014).

Apoptosis can also be mediated by apoptosis molecules such as Bcl family and calpain. Bcl family is a polygenic family. Bcl-2 can inhibit cell apoptosis. Bax promotes cell apoptosis by forming heterodimer with Bcl-2. By detecting the expression of two genes in fluorosis rats, it is found that Bax and Bax/Bcl-2 are both increased, suggesting that apoptosis regulated by Bcl family is also involved in renal damage caused by fluorosis (Yang et al. 2013). Calpain-1 is a kind of calcium-regulated apoptosis-promoting molecule. Excessive accumulation of fluoride can cause calcium overload and activate calpain-1. Activated calpain-1 can break X-linked inhibitor of apoptosis protein (XIAP) and make XIAP ineffective in inhibiting various caspases downstream, thus activating caspase-3 through cascade amplification reaction and initiating apoptosis process. Notch-1 signaling pathway is also closely related to apoptosis. The activated Notch-1 is translocated into the nucleus and induces the expression of target gene *Hes1*, thus affecting the growth process of cells. After fluoride exposure, the activity of antioxidant enzymes in the kidney tissues of experimental animals was significantly reduced, leading to excessive production of ROS, while the mRNA expressions of calpain-1, caspase-3, Notch1, and Hes1 in the kidney tissues were positively correlated with the ROS content. A recent study reported that fluorosis induced the decreased apoptotic levels of caspase-3 and Bax and caspase-8, Bcl-2, Bcl-2-like Protein 1, Bim, and Bcl-2 antagonist/killer, and the increased values of the oxidative stress genes, such as *Cygb, Gstp1,* and *Ncf1* genes, as well as the decreased *Gpx1, sod1*, and *sod2* genes in the kidney tissues of sheep (Efe et al. 2021). As a result, it was shown that apoptosis and oxidative mechanisms are activated in the kidney tissues with fluorosis and these parameters have an important role in understanding the molecular basis of tissue damage in fluorosis, which suggests that over-production of ROS in renal tissue of animals with chronic fluorosis can also cause over-activation of apoptosis process.

Endoplasmic reticulum stress (ERS) is a protective response to external unfavorable factors. Some studies have shown that ERS is closely related to renal injury mechanism caused by various renal toxicants. Under long-term or severe fluoride stimulation, ERS induces changes in related signal pathway responses with unfolded protein response as the core and induces apoptosis by regulating JNK and caspase12 signal pathways (Wang et al. 2018a, b).

The susceptibility of the kidneys to fluoride toxicity can largely be attributed to its anatomy and function. Kidney biopsies taken from patients with childhood nephrotic syndrome were subjected to ultrastructural and apoptotic analysis (Quadri et al. 2018). Various subcellular ultrastructural changes including nuclear disintegration, chromosome condensation, cytoplasmic ground substance lysis, and endoplasmic reticulum blebbing were observed. Increased levels of apoptosis were observed in the patients with high level of fluoride in urine than those without high urine fluoride. Various degrees of fluoride-associated damages to the architecture of tubular epithelia, such as cell swelling and lysis, cytoplasmic vacuolation, nuclear condensation, apoptosis, and necrosis, were observed.

In general, the process of apoptosis in kidney caused by chronic fluorosis is very complex and is regulated by multiple pathways and genes. The interaction of various factors in the process needs further research and clarification.

Correlative Detection Indexes for Judging Renal Function Changes in Chronic Fluorosis

Serological Detection Index

Urea nitrogen (UN) is the main component of non-protein nitrogen in blood, accounting for about 50% of the total non-protein nitrogen under normal circumstances. When non-protein nitrogen increases due to kidney diseases, UN increases most obviously. Therefore, UN test is clinically taken as a routine item of renal function test. Some scholars have found that the serum UN content of the residents who have been exposed to fluoride pollution for a long time is significantly higher than that of the control group by detecting their blood biochemical indexes and the serum UN content of experimental animals with fluorosis is also obviously increased.

Cr is mainly discharged out of the body by glomerular filtration. Under the condition that exogenous Cr is stable, Cr concentration in blood can be used as one of the indexes to detect glomerular filtration function. Since Cr is eliminated with urine through the kidney, when renal function is not complete due to chronic fluorosis the ability to remove them decreases, thus increasing their content in blood. Their content in blood is an objective indicator reflecting renal function and can reflect glomerular filtration function to a certain extent.

Cystatin C (CysC) can only be removed in the patient's kidney. Its concentration mainly depends on glomerular filtration rate, which is not affected by muscle mass, sex, and age. Moreover, the detection process is relatively simple and can be detected by collecting serum samples. Clinical studies have proved that because the kidney has a certain reserve capacity, the indexes of UN and Cr will only increase when the glomerular filtration rate decreases by more than 50%. CysC has higher specificity and sensitivity than UN and Cr, so it is more suitable to reflect the changes of glomerular filtration rate during the decompensation period of renal insufficiency. The use of CysC alone or in combination with Cr strengthens the association between the estimated glomerular filtration rate and the risks of death and end-stage renal disease across diverse populations (Shlipak et al. 2013), which may be considered as one sensitive parameter of kidney injury by chronic fluorosis.

Urine Test Index

According to the investigations, renal damage mainly depends on the level of circulating fluoride (related to urine fluoride) in the body, and the symptoms of renal tubular insufficiency in fluoride-induced renal damage appear earlier than glomerular damage.

Urine homocysteine, a very important factor causing vascular diseases, has a certain positive correlation with vascular injury induced by fluorosis.

Retinol binding protein (RBP) is a common protein with relatively low molecular weight, which can form complexes with related hormone proteins in blood. When renal function is abnormal, its content will greatly increase. Homocysteine and RBP indexes can directly reflect the degree of renal function of patients. Combining the two together can better indicate the degree of renal function and damage of patients (Xue et al. 2017).

Acid phosphatase (ACP) is a hydrolytic enzyme that catalyzes the hydrolysis of monoester phosphate to inorganic phosphate under acidic conditions. Fluorosis can cause kidney ACP activity significantly decreased, which in mechanism may be associated with the following factors. Metal ions have a special affinity for fluoride and make the bridge role of metal ions in the enzymatic reaction or activator function abate, which may result in the decrease of enzyme activity of ACP; fluoride may inhibit oxygen metabolism and interfere ACP metabolism.

Oxidative Stress Related Indicators

Fluorine is the most active nonmetallic element among nonmetals, which keeps highly oxidizing and can react with most hydrogen-containing compounds such as water, ammonia, and all liquid, solid and gaseous chemicals except helium, neon, and argon. In recent years, a large number of studies have shown that the renal damage caused by excessive fluoride intake may be due to the resulting enhanced oxidative stress. Because of their strong reactivity and short life, the half-life of free radicals is very short, so it is very difficult to directly measure the level of these free radicals. However, the oxidation products after the reaction of free radicals with other molecules can be used as a marker to reflect the kidney damages.

Oxygen free radicals can attack polyunsaturated fatty acids in biofilms, induce lipid peroxidation, and generate a metabolite of the reaction, MDA. High level of

MDA was found in kidney of rat with chronic fluorosis (Qing et al. 2015), which can cross-link with sugars, proteins, and lipids, denature cell membranes, and cause damage to cells (Zuo et al. 2018). In an experiment in vivo, the decreased proportions of polyunsaturated fatty acids in phospholipids and high levels of lipid peroxidation expressed by MDA were observed in kidney of the rats with fluorosis (Guan et al. 2000).

Advanced oxidation protein products (AOPPs) are newly discovered markers of oxidative stress in protein macromolecules. AOPPs are mainly protein crosslinked products formed by oxidation of tyrosine residues of serum albumin. Studies have shown that the level of AOPPs in many patients with chronic diseases is much higher than that of normal level (Witko-Sarsat et al. 1996; Yuan et al. 2018). Clinical studies have also observed that AOPPs levels in patients with chronic kidney diseases increase with the progress of renal insufficiency. Therefore, AOPPs may be used as a protein oxidative stress marker that indirectly reflects renal function damage caused by chronic fluorosis (Lu et al. 2008).

Under normal conditions, oxidative stress can cause DNA damage, inactivate tumor suppressor genes, and enhance the activity of proto-oncogenes, thus promoting cell carcinogenesis. DNA oxidative damage can produce a large number of purine and pyrimidine hydroxylation products. Among them, 8-OHdG is an oxidative adduct produced by reactive oxygen radicals attacking the 8th carbon atom of guanine base in DNA molecules. Similar to AOPPs, 8-OHdG also has obvious content changes in some chronic renal diseases, which has become the most commonly used biomarker in DNA oxidative damage (Wang et al. 2019).

SOD plays a vital role in the oxidation and anti-oxidation balance of the body. This enzyme can remove superoxide anion free radicals and protect cells from damage. Therefore, SOD activity can indirectly reflect the body's ability to remove oxygen free radicals. In many studies, the activities of antioxidative enzymes, including SOD, GSH-Px, and CAT, are significantly decreased in kidney of rats with fluorosis (Błaszczyk et al. 2010; Zhang et al. 2013). Fluoride exposure also induces a decrease in serum level of non-enzymatic antioxidants such as uric acid and of some oligoelements: zinc and copper, known to be cofactors of SOD (Bouaziz et al. 2006).

Conclusion

The chapter mainly discusses the effect of chronic fluorosis on kidney and its pathogenesis. Accumulation of fluoride can influence renal function, structure, and metabolism. Several mechanisms are involved in the damage of kidney induced by chronic fluorosis, i.e., the change of cell cycle, unbalance of calcium, effect of oxidative stress, excessive autophagy, alteration of related cytokines, and apoptosis of renal cells. In addition, for the diagnosis and judgment of clinic manifestation of renal damages, certain parameters in serum and urine are selected as detecting factors.

References

Błaszczyk I, Grucka-Mamczar E, Kasperczyk S, Birkner E. Influence of methionine upon the activity of antioxidative enzymes in the kidney of rats exposed to sodium fluoride. Biol Trace Elem Res. 2010;133:60–70.

Bouaziz H, Croute F, Boudawara T, Soleilhavoup JP, Zeghal N. Oxidative stress induced by fluoride in adult mice and their suckling pups. Exp Toxicol Pathol. 2006;58:339–49.

Cao J, Chen J, Xie L, Wang J, Feng C, Song J. Protective properties of sesamin against fluoride-induced oxidative stress and apoptosis in kidney of carp (Cyprinus carpio) via JNK signaling pathway. Aquat Toxicol. 2015;167:180–90.

Dharmaratne RW. Exploring the role of excess fluoride in chronic kidney disease: a review. Hum Exp Toxicol. 2019;38:269–79.

Efe U, Dede S, Yüksek V, Çetin S. Apoptotic and oxidative mechanisms in liver and kidney tissues of sheep with fluorosis. Biol Trace Elem Res. 2021;199:136–41.

Guan ZZ, Xiao KQ, Zeng XY, Long YG, Cheng YH, Jiang SF, et al. Changed cellular membrane lipid composition and lipid peroxidation of kidney in rats with chronic fluorosis. Arch Toxicol. 2000;74:602–8.

Guo Q, Sun ZL, Niu RY, Manthari RK, Yuan MK, Yang KD, et al. Effect of arsenic and/or fluoride gestational exposure on renal autophagy in offspring mice. Chemosphere. 2020;241:124861.

Li GC, Zhang WB, Shen YM, Xu H. Changes of kidney function and cell cycle in fluorosis rabbits. Chin J Lab Diagn. 2007;11:1591–2.

Li BY, Zhang Z, Zhou B, Guo LY, Wang XH, Yu Y, et al. Pathological changes of femur, liver, kidney and brain tissues in rats exposed to fluoride. Chin J Comp Med. 2019;29:99–104.

Lu YW, Zhu QY, Ding F, Gu Y, Lin SY. *In vitro* preparation and detection of a novel oxidative stress marker AOPPs. Fudan Univ J Med Sci. 2008;35:21–5.

Luo Q, Guo HR, Kuang P, Cui HM, Deng HD, Liu H, et al. Sodium fluoride arrests renal G2/M phase cell-cycle progression by activating ATM-Chk2-P53/Cdc25C signaling pathway in mice. Biochemistry. 2018;51:2421–33.

Qing SL, Deng J, Lou DD, Yu WF, Pei JJ, Guan ZZ. The decreased expression of mitofusin-1 and increased fission-1 together with alterations in mitochondrial morphology in the kidney of rats with chronic fluorosis may involve elevated oxidative stress. Biology. 2015;29:263–8.

Quadri JA, Sarwar S, Sinha A, Kalaivani M, Dinda AK, Bagga A, et al. Fluoride-associated ultrastructural changes and apoptosis in human renal tubule: a pilot study. Hum Exp Toxicol. 2018;37:1199–206.

Shao DD, Zhang JY, Tang L, Yu QL, Hu XX, Ruan Q, et al. Effects and molecular mechanism of l-type calcium channel on fluoride-induced kidney injury. Biol Trace Elem Res. 2020;197:213–23.

Shlipak MG, Matsushita K, Ärnlöv J, Inker LA, Katz R, Polkinghorne KR, et al. Cystatin C versus creatinine in determining risk based on kidney function. N Engl J Med. 2013;369:932–43.

Song GH, Gao JP, Wang CF, Chen CY, Yan XY, Guo M, et al. Sodium fluoride induces apoptosis in the kidney of rats through caspase-mediated pathways and DNA damage. J Physiol Biochem. 2014;70:857–68.

Tian XL, Xie JX, Chen XS, Dong NS, Feng J, Gao Y, et al. Deregulation of autophagy is involved in nephrotoxicity of arsenite and fluoride exposure during gestation to puberty in rat offspring. Arch Toxicol. 2020;94:749–60.

Vale G, Simões N, Santana G, Mota B, Moura M. Gastrointestinal absorption and renal excretion of fluoride after ingestion of a high-fluoride dentifrice. Biol Trace Elem Res. 2019;190:24–9.

Wang L, Fan HJ. Effects of excessive fluoride on animals. Anim Husb Vet Med. 2014;41:119–22.

Wang YX, Xiao X, Zhan XA. Antagonistic effects of different selenium sources on growth inhibition, oxidative damage, and apoptosis induced by fluorine in broilers. Poult Sci. 2018a;97:3207–17.

Wang JM, Xu HM, Zhang J, Gao YF, Zhao YF, Li YY. Effects of different calcium levels on the apoptosis pathways of renal tissue cells in rats with fluorosis. Acta Vet Zootech Sin. 2018b;49:2723–32.

Wang HY, Wei HF, Wang WF, Zhou XJ, Guo YX, Mi XG, et al. Expression of ICAM-1 and 8-OHdG in renal cell carcinoma and its relationship with clinicopathological parameters. Chin J Immunol. 2019;35:953–61.

Witko-Sarsat V, Friedlander M, Capeillere-Blandin C, Nguyen-Khoa T, Nguyen AT, Zingraff J, et al. Advanced oxidation protein products as a novel marker of oxidative stress in uremia. Kidney Int. 1996;49:1304–13.

Wu JH, Li DD, Yang D, Qing M, Li BY, Liu XN, et al. Correlation between urinary fluoride level and intaking of fluoride per day in Tibetan and Kazakh population in brick-tea-borne fluorosis areas. Chin J Endem. 2015;34:549–52.

Xue XY, Ye JJ, Li LL. Effect of Cys-c, Hcy and RBP in early diagnosis of diabetic nephropathy. Diabetes New World. 2017;14:171–2.

Yang SY, Zhang L, Miao KK, Qian W, Zhang ZG. Effects of selenium intervention on chronic fluorosis-induced renal cell apoptosis in rats. Biol Trace Elem Res. 2013;153:237–42.

Yuan D, Wang HP, Zhang R, Li DH, Jiang CM, Chen X, et al. Analysis of the difference between the cognitive function and blood oxidative stress between the healthy and the patients with high fluoride poisoning. Chin J Control Endem Dis. 2018;33:335–6.

Zhang JM, Xu H, Zhang XY. Effect of excessive fluoride on the expression of basic fibroblast growth factor in rat kidney. Chin J Control Endem Dis. 2007;22:414–6.

Zhang J, Song J, Zhang J, Chen X, Zhou M, Cheng G, et al. Combined effects of fluoride and cadmium on liver and kidney function in male rats. Biol Trace Elem Res. 2013;155:396–402.

Zhou YX, Mo SL, Wang RH, Liu WF, Liu JC, Wang JQ. Corn baked by burning coal triggered overexpression of osteopontin in hepatocytes of rats following fluorosis. Toxicol Ind Health. 2012;28:195–202.

Zuo H, Chen L, Kong M, Qiu LP, Lü P, Wu P, et al. Toxic effects of fluoride on organisms. Life Sci. 2018;198:18–24.

Chapter 9
Modification of Cardiovascular System

Jian-Chao Bian, Man Zhang, and Ji Bian

Abstract The endemic fluorosis is a systemic chronic disease. The excessive fluoride intake inflicts poisoning on practically all organs and tissues. Cardiovascular system injury triggered by fluorosis is the major manifestation of fluoride-induced non-bone damage. The investigation concerning the injury mechanism of fluoride on cardio-vascular system would further identify the non-bone damage induced by fluoride and its role in fluorosis pathogenesis. This chapter is devoted to the overall introduction of latest research on fluoride damage on cardiovascular system, and preliminary analysis on possible pathogenesis of its effect on cardiovascular system.

Keywords Apoptosis · Circulation system · Heart · Oxidative stress

Introduction

The damage of chronic fluorosis to cardiovascular system is a very important part of the non-bone tissues influenced by excessive intake of fluoride. The early reports concerning cardiovascular injury induced by fluoride were more involved in population in the areas of endemic fluorosis by epidemiological studies. From 1980 to 1990s, many literatures reported that the morbidity and mortality of cardiovascular system were higher in areas of endemic fluorosis than those without endemic fluorosis. Subsequent clinical examinations of patients with chronic fluorosis revealed palpitations, arrhythmias, reduced cardiac output, and other clinical symptoms, as well as abnormal electrocardiogram tests. Recently, many investigations by employing molecular epidemiological methods confirm that high concentration of

J.-C. Bian (✉) · M. Zhang
Shandong Institute of Prevention and Control for Endemic Disease, Jinan, People's Republic of China
e-mail: bianjc1008@163.com

J. Bian
Shandong Normal University, Jinan, People's Republic of China

Z.-Z. Guan (ed.), *Coal-burning Type of Endemic Fluorosis*,
https://doi.org/10.1007/978-981-16-1498-9_9

fluoride is closely related to the occurrence of essential hypertension in adults and that high fluoride exposure is significantly correlated with the detection rate of carotid atherosclerosis.

From the end of the twentieth century to the present, scientists have carried out a large number of relevant experimental studies in cardiovascular injuries caused by fluoride while carrying out field epidemiological investigations. Many studies in vivo and in vitro have confirmed that high fluoride has multiple effects on the cardiovascular system. Fluorosis results in the abnormal electrocardiogram and myocardial electrophysiological parameters in animal studies; affects hemodynamic parameters and metabolism of lipid and lipoprotein; induces the changed signaling pathways and the in-balance of antioxidant system in cardiovascular system; and causes the modifications in myocardial enzyme profile, which leads to heart injury and arteriosclerosis.

In recent years, scholars at the research area have put forward many theories on the mechanism of cardiovascular injury caused by fluorosis, such as oxidative stress, the changed myocardial ion channels, the abnormal biochemical metabolism of myocardial cells, the destruction of cytoskeleton, the damage of biofilm and the induction of apoptosis in the system. However, based on the comprehensive analysis of the current studies, the effects and mechanisms of chronic fluorosis on the injury of heart and blood vessel in detail may be the result of the interaction of multiple factors, which needs to be further studied and fully discussed.

Population-Based Study

Fluorosis affects the cardiac function. It is shown that the overall cardiac function and aortic elasticity are damaged in fluorosis patients compared with control group (Varol et al. 2010). Heart rate recovery index (HRRI) refers to the peak value of max exercise test heart rate minus the recovery heart rate. HRRI1, HRRI2, and HRRI3 are results of heart rate peak value minus 1st min, 2nd min, and 3rd min heart rate value in recovery stage. Comparing endemic fluorosis patients with healthy people, the abnormal HRRI1 incidence rate of fluorosis patients increased, and that of HRRI2 decreased significantly (Adali et al. 2013). It is also shown in another study that the left atrial diameter, left ventricular end-diastolic diameter and aortic dimension obviously increased, and these three values would increase accordingly with the severity of fluorosis (Liu et al. 2018).

Fluorosis would injure the cardiac muscle. Investigation was carried out on cardiac muscle of fetus with five to seven months gestational age in coal-burning type of endemic fluorosis area. It was found that myocardial mitochondrion in fetus was swollen, size was enlarged, crista was reduced or disappeared. Mitochondrial size and density of fetus in this endemic area obviously differed from that of non-endemic area group. It indicated that excessive fluoride had obvious effect on fetal myocardial cells (Wen and Yang 1997). Fluoride is potent for long-term effect on cardiac function. It was found in a study that 18 years after water improvement project in certain high fluorosis area, there were more cardiovascular disease symptoms found in skeletal fluorosis patients than in non-skeletal fluorosis people in this area. In addition, with the disease progressing, a rise could be found in the proportion of patients with such symptoms. This finding indicated that water improvement (lower the fluoride content) could not clear the influence of fluoride on cardiovascular system in skeletal fluorosis patients. In mild and moderate patients with skeletal fluorosis, abnormal changes in electrocardiogram (ECG) improved obviously, while this change improved slowly in patients with severe symptoms (Ji et al. 2004).

Fluorosis can affect blood lipid level. Past research discovered that fluorosis patients showed different indexes compared with control group: higher total cholesterol (TC), triglyceride (TG), low-density lipoprotein cholesterol (LDL-C) content and risen arteriosclerosis index, lowered high-density lipoprotein cholesterol (HDL-C). All these indicated that lipid metabolism pathways in fluorosis patients were impaired. In addition, lowered content in glutathione peroxidase (GSH-Px) and superoxide dismutase (SOD), and increased malondialdehyde (MDA) concentration indicated enhanced oxidative stress response (Zhang et al. 2018).

Fluorosis can trigger morphological change in red cells. Scanning electron microscope discovered abnormal protuberant in red cells in the majority of drinking water type of endemic fluorosis patients, echinocyte increased with shards, severe cell aggregation, red cell deformability rate increased in increasing order of control area, mild disease area and serious disease area. Red cell deformability rate of patients from water improvement project area was lower than that of serious disease area, yet below the level of control area (Chen et al. 2004). Erythrocyte membrane protein fraction in fluorosis patients also went through changes: patients from 5-year water improvement project area showed that their protein fraction has yet to be recovered to normal level. This study indicated that morphological features of red cell of patients from endemic area went through significant changes and deformability rate improved as the fluoride burden increased. Of which, the mercapto might play an essential role (Sun et al. 2003).

Experiment

Fluoride can affect cardiac electrophysiological index. The cardiomyocyte intro-microelectrode assay was used to detect the cardiac electrophysiological parameter of rats. Test results showed that following indexes changed in fluorosis rats: lowered resting potential (RP), action potential (AP), and action potential maximal shortening velocity (Vmax), shortened action potential duration (APD) of APD50 and APD90; the test also indicated that membrane potential of cardiomyocyte action potential of fluorosis rat decreased and period shortened together with lowered amplitude (Bian et al. 1996). Hemodynamics parameters of fluorosis rats revealed that high dose fluorosis rats group showed obviously lowered heart rate, reduced coronary flow, increased left ventricular pressure, and left ventricular end-diastolic pressure. These all proved that fluorosis obviously damaged the cardiac function of rats (Shen et al. 2003).

Fluoride can cause myocardial injury. Researchers fed sodium fluoride deionized solution to Wistar rats in three concentrations (125 mg/l, 250 mg/l, 500 mg/l) for two months. Later, they discovered that activity of lactate dehydrogenase (LDH) and α-hydroxybutyric acid dehydrogenase (α-HBDH) obviously increased, and LDH/HBDH increased with the rise of fluoride concentration. These were changes of cardiac enzymes when myocardial injury happened (Ma et al. 2009). Chronic exposure to fluoride also inflicts injury on cardiac muscle of Wistar rats with the following manifestation: cardiac muscle fibers attenuated, then dissolved and deformed; inflammatory cell infiltration increased, and calcium increased. Mitochondrion lesion, stromatolysis, edema, and vacuole were also found. Atomic Force Microscope captured deformation and bending. Those observed changes shared some similarity with changes in myocardial hypertrophy and myocardiopathy (Quadri et al. 2018).

Fluoride affects the blood lipid. Past study tested blood lipid of rats exposed to fluoride for four weeks and 12 weeks. The study discovered that fluoride group showed reduced HDL and no changes in TC and TG. Such results indicated the abnormal HDL was triggered by sub-chronic fluoride exposure (Yang and Xi 2018). Another study fed rats with 50 mg/l high dose fluoride solution for 10 months and divided the subjects into four groups: fluoride solution group, fluoride solution with selenium, switched to tap water group, switched to tap water with selenium. Ten months later, it was discovered that following parameters changed in fluorosis rats serum: TC and LDL-C increased, HDL-C and HDL-C/LDL-C lowered, antioxidant enzyme, and lipid metabolism improved in fluoride solution with selenium treatment; lipid metabolism index returned to normal after switched to tap water with selenium treatment (Bian et al. 1999). These studies proved that lipid metabolism in chronic fluorosis rats was impaired and promoted arteriosclerosis, while selenium antagonized fluoride caused lipid metabolism disorder.

Fluoride causes the changes of arteriosclerosis. Recent study conducted systemic research on fluoride caused vascular injury and arteriosclerosis (Bian et al. 2010a, b, c; Yang et al. 2010). New Zealand rabbits were used as research subjects and divided into four groups: control group, lipid-fed group, fluoride-fed group, lipid and fluoride-fed group. Six months later, research showed that high dose fluoride treatment could trigger aortic endothelium injury and a series of changes in blood indices and oxidation resistance; such injury and changes had synergism to the formation of atherosclerosis. These research findings showed that high dose fluoride treatment obviously impaired aortic endothelium, caused abnormality in cellular structure and trend of cell, lipid droplet was found, red blood cell and fibrin deposition increased, arteriosclerosis tendency was detected. High dose fluoride treatment could rise the level of serum TC, TG and LDL-C, increase blood viscosity and erythrocyte aggregation, decrease deformability of red blood cell, and cause abnormality of blood flow. Such treatment could also lower the activity of serum and antioxidant enzyme, increase lipid peroxide content, trigger overall lipid peroxidation impairment; in addition, this treatment could lower plasma 6-ketone-prostaglandin F1α content, increase endothelin-1 (ET-1) content in plasma, trigger vasoconstriction and platelet aggregation, lower the activity of nitric oxide synthase (NOS), increase the activity of liver NOS, cardiac muscle, liver inducible nitric oxide synthase (iNOS); the abnormal expression of NOS-mRNA could trigger content change in nitric oxide (NO).

The Cultivation of Cardiomyocyte

The cultivation of cardiomyocyte is a widely used in myocardial morphological, functional, pathological, and pharmacological research on cellular and molecular level. Scientists observed the effect of fluoride ion on cardiomyocyte cultured in vitro by using Wister newborn rats cardiomyocyte cultured in vitro. The findings showed that low concentration fluoride accelerated beat frequency of cardiomyocyte cultured in vitro, while high concentration fluoride slowed down the frequency; cellular fibrillation and arrest were also noticed (Bian et al. 1993). There was also research conducted to investigate effect of different doses of fluoride on human umbilical vein endothelial cells and antagonizing effect of selenium on fluoride triggered cellular injury. Research findings indicated that high concentration fluoride changed morphological features of endothelium, triggered apoptosis, and activated the expression of adherence factor of endothelium; while moderate concentration of selenium could inhibit the previous changes triggered by fluoride (Bian et al. 2010a, b, c).

Potential Pathogenesis of Fluoride Triggered Cardiovascular System Impairment

Oxidative Stress

Fluoride can integrate with the antioxidants and free radicals SOD once it enters the body. Such integration could disrupt the balance between oxidants and antioxidants to a certain degree: the body is forced into a state of oxidative stress, which would result in impairment of structure and function of macromolecules; in addition, such state would alter relevant signal pathways and impair tissues or cells (Cheng et al. 2015; Alper et al. 2015). A substantial number of researches proved that content of lipid peroxide and oxidation protein products increased in endemic fluorosis patients and experiment fluorosis animals; antioxidant enzyme or antioxidants level was lowered or compensatory changed with oxidative stress level risen (Wang 2015). Increased serum oxidative stress products, including lipid peroxidation, protein, and DNA damages, were observed in patients with fluorosis (Chen et al. 2017). Past research discovered that excessive fluoride intake could result in ET-1 expression increase in human or animals, as oxidative stress could regulate fluoride triggered ET-1 increase. Despite the multiple physiological functions of ET-1 in vascular biology, its main function is to improve vascular tone and blood pressure. ET-1 is considered to play a major role in various cardiovascular diseases like hypertension, congestive heart failure, and atherosclerosis (Sun et al. 2016). Another research found that the iNOS mRNA expression of white blood cell and NO content of fluorosis patients were obviously higher than control group (Liu et al. 2005). NO refers to the informational molecule with comprehensive physiological functions in living creatures. NO could either clean the free radicals or produce free radicals with stronger toxicity. Abnormal changes in NO would disrupt the balance of free radicals in the body, as it would result in producing too many reactive oxygen species or not cleaning enough and ultimately lead to body injury.

Apoptosis

High dose fluoride features its impairment on tissues and cells on changes in intracellular chromatin margination and cellular atrophy. Such characteristic is in line with the description on apoptosis by numerous scientists, who categorized apoptosis as the major factor of fluoride triggered tissue impairment in body (Bian et al. 2003). Another research finding further proved that fluoride acted on mitochondria and oxidative stress involves apoptosis (Anuradha et al. 2001). Several researches proved that fluoride had evident inducing effect on cardio- and vascular apoptosis (Bian et al. 2010a, b, c).

Elicitors of cardiomyocyte apoptosis activate survival or death related gene to regulate apoptosis through certain signal transmission pathways. Of all the pathways, inhibition or activation of each link has regulation effect on cardiomyocyte apoptosis. Cysteine aspartic acid protease (Casp) is the core regulating factor of apoptosis which belongs to the same family of cysteine proteinase. Fas Ligand activates promoter Casp-8 and 10 and then split to activate Casp-3, 6, and 7 to cause apoptosis. Exposure to fluoride activates these signal pathways and induces apoptosis. Furthermore, excessive fluoride would trigger oxidative stress or endoplasmic reticulum stress and promote apoptosis. Bcl-2 is the proto-oncogene isolated from rat's B lymphocyte, which could regulate the apoptosis directly, inhibit or block out apoptosis triggered by different factors and ensure cellular survival. Research has proved that Bcl-2 played a major role in the process of fluorosis triggered apoptosis (Barbier et al. 2010).

Signal pathways induce the expression of anti-apoptotic Bcl-2, including extracellular signal-regulated kinase 1/2 and phosphatidylinositol-3-kinases or protein-serine-threonine kinase. These Bcl-2 family members could protect the integrity of mitochondrion, block out the release of Cytochrome C, and activate Casp-9. Tumor necrosis factor α (TNFα) could activate apoptosis-promoting and apoptosis-inhibiting pathways; TNFα could activate Casp-8 and -10 to induce apoptosis or inhibit apoptosis by nuclear factor kappa-B. Exposure to fluoride could block out all these signal pathways (Wei et al. 2019).

Energy Metabolism

Fluoride inhibits the enzyme activity whose are related to energy metabolism, like succinate dehydrogenase and isocitrate dehydrogenase. Due to the energy metabolic disorder, calcium overload in cardiomyocyte triggered compensatory beats and then decompensation. Insufficient energy supply from cell would lead to actin microfilament cytoskeleton dissolution, and then trigger a series of ultra-morphological structure lesion. Research reported findings on cardiomyocyte changes after toxic effect of fluoride: microvillus reduced and later disappeared, protuberance dwindled, cell shriveled or become smaller and round. "Volcano shaped" dent and "cauliflower shaped" bulla were found on cell body. These morphological changes damaged the cellular skeletal system. Excessive fluoride could also inhibit activity of Na^{+}-K^{+} ATP and Ca^{2+}-Mg^{2+} ATP in cardiomyocyte, affect intracellular ion concentration of potassium, sodium, calcium, and magnesium, and ultimately impair physiological function of the cell (Yang et al. 2014). These findings were the physiopathology base of fluorosis rat ECG and cardiac electrophysiological disorder.

Conclusion

This chapter carried out a comprehensive review on cardiovascular injury by fluoride from fluorosis patient study and experiment study, as well as a review on the pathogenesis of fluoride triggered cardiovascular system injury. Existing researches have proved that excessive fluoride could result in cardiomyocyte, vascular injury, and even arteriosclerosis. Fluoride affects multiple aspects of cardiovascular system, including cardiac electrophysiology, cardiac-cellular structure, apoptosis, relevant enzyme activity, free radical metabolism, blood cytokine, and blood rheology. These results indicate that fluoride-induced cardiovascular damage is a complex process of multiple pathological reactions and mutual promotion, which is difficult to explain its pathogenic mechanism with a single reason.

References

Adali MK, Varol E, Aksoy F, Icli A, Ersoy IH, Ozaydin M, et al. Impaired heart rate recovery in patients with endemic fluorosis. Biol Trace Elem Res. 2013;152:310–5.

Alper M, Aydemir AT, Köçkar F. Induction of human ADAMTS-2 gene expression by IL-1α is mediated by a multiple crosstalk of MEK/JNK and PI3K pathways in osteoblast like cells. Gene. 2015;573:321–7.

Anuradha CD, Kanno S, Hirano S. Oxidative damage to mitochondria is a preliminary step to caspase-3 activation in fluoride-induced apoptosis in HL-60 cells. Free Radio Biol Med. 2001;31:367–73.

Barbier O, Arreola-Mendoza L, Del Razo LM. Molecular mechanisms of fluoride toxicity. Chem Biol Interact. 2010;188:319–33.

Bian JC, Guan SF, Liu YP. The effects of various concentrations of fluoride on cardiomyocytes *in vitro*. Chin J Ctrl Endem Dis. 1993;8:8–10.

Bian JC, Song SL, Yang XX, Xiang YZ, Ye P, Xian SM, et al. Effect of selenium on the electrophysiological changes of myocardial cells of the rats with fluorosis. Chin J Endem. 1996;15:386–49.

Bian JC, Liu YP, Yang XX, Ye P, Xian SM, Liu Y. Experimental study on effects of selenium in rats with fluorosis. Chin J Endem. 1999;18:93–6.

Bian JC, Wang HM, Liu CJ. Advances in the relationship between apoptosis and fluorosis. Endem Dis Bull. 2003;18:100–2.

Bian JC, Yang XX, Lin XY, Zhu QL, Fan T. Pathological morphology of vascular endothelial injury and arteriosclerosis caused by high fluoride and the effects of selenium. Chin J Endem. 2010a;29:64–7.

Bian JC, Lin XY, Yang XX, Hou XD, Fan T, Zhu QL. Study on the effect of fluoride on cultured endothelial cells and the role of selenium intervention. J Nutr. 2010b;32:474–9.

Bian JC, Lin XY, Yang XX, Fan T, Zhu QL. Changes of certain oxidative, anti-oxidative and vascular function indexes of New Zealand rabbit exposed by high-fluoride. J Hyg Res. 2010c;39:110–4.

Chen BL, Sun GX, Li ED, Dai GJ. Erythrocyte shape of patients with endemic fluorosis was observed by electron microscope. Chin J Endem. 2004;23:430–1.

Chen XW, Wan CW, Zhang SY, Luo Y, Qi XL, Li Y. Serum oxidative stress products among residents living in endemic area of coal-burning fluorosis. Chin J Public Health. 2017;33:824–7.

Cheng L, Zeng G, Liu Z, Zhang B, Cui X, Zhao H, et al. Protein kinase B and extracellular signal-regulated kinase contribute to the chondroprotective effect of morroniside on osteoarthritis chondrocytes. J Cell Mol Med. 2015;19:1877–86.

Ji F, Xu H, Zhang YX, Zhang J, Liu WY, Shao H, et al. Study on the cardiovascular damage of skeletal fluorosis patients. Chin J Ctrl Endem Dis. 2004;19:321–3.

Liu YH, Guo CH, Yu CY. The role of expression iNOS-mRNA of leukocyte and plasm NO in patients of fluorosis bone. Chin J Ctrl Endem Dis. 2005;20:86–8.

Liu JF, Zheng YJ, Wang J, Jia Q, Guo WJ. Echocardiographic analysis of 61 cases of aortic sclerosis due to endemic fluorosis. Chin J Ctrl Endem Dis. 2018;3:286–9.

Ma BH, Gao YS, Pan GL, Chen XD, Zhang YJ. Effects of different concentrations of fluoride on the contents of Ca, Mg and P in rat serum. J BaoTou Med Coll. 2009;25:4–5.

Quadri JA, Sarwar S, Kar P, Singh S, Mallick SR, et al. Fluoride induced tissue hypercalcemia, IL-17 mediated inflammation and apoptosis lead to cardiomyopathy: ultrastructural and biochemical findings. Toxicology. 2018;406-407:44–57.

Shen XY, Ruan Q, Zhang ZG, Xu XL. Influence of fluorosis on hemodynamics of rat. Chin J Ctrl Endem Dis. 2003;18:208–9.

Sun GX, Dai GJ, Zhang ZY, Liu T, Ji YB, Chen BL, et al. The analysis of the erythrocyte membrane protein ingredients of patients with endemic fluorosis. Chin J Endem. 2003;22:67–9.

Sun L, Gao Y, Zhang W, Liu X, Sun D. Mechanisms underlying endothelin-1 level elevations caused by excessive fluoride exposure. Cell Physiol Biochem. 2016;40:861–73.

Varol E, Akcay S, Ersoy IH, Koroglu BK, Varol S. Impact of chronic fluorosis on left ventricular diastolic and global functions. Sci Total Environ. 2010;408:2295–8.

Wang Y. The effect of chronic fluorosis on Nrf2/ARE signaling pathway. Chin J Endemiol. 2015;34:151–3.

Wei W, Pang S, Sun D. The pathogenesis of endemic fluorosis: research progress in the last 5 years. J Cell Mol Med. 2019;23:2333–42.
Wen PY, Yang WX. Effect of chronic fluorosis on myocardial cells of human fetuses. Chin J Endemiol. 1997;16:7–8.
Yang L, Xi SH. An analysis of serum biochemical indexes of rats chronically exposed to fluoride. Chin J Endemiol. 2018;37:112–5.
Yang XX, Bian JC, Lin XY, Zhu QL, Fan T. Experimental pathomorphological study of high fluorine on vascular injury and arteriosclerosis. Chin J Ctrl Endem Dis. 2010;25:179–83.
Yang X, Hao XH, Gao JP, Wang Y, Liu ML, Huang FB, et al. The study of damage mechanism of fluorosis on heart. Chin J Endem. 2014;33:583–5.
Zhang SM, Shen JX, Du ZW, Ge XZ. The damage of fluorosis to cardiovascular system and the relationship between fluorosis and atherosclerosis. Chin J Ctrl Endem Dis. 2018;3:280–2.

Chapter 10
Damage of Endocrine System

Xiao-Xiao Zeng, Jie Xiang, and Zhi-Zhong Guan

Abstract Fluoride is abundance in nature and can spread to the human body through the water, air, or food. Long-term intake of excessive amount of fluoride may damage skeleton and teeth, and also soft tissues, such as the endocrinological system. Here, we briefly focus on the influence of fluoride on thyroid gland, parathyroid gland, insulin sensitivity, and gonads, as well as the possible mechanism of fluoride toxicity. A large number of studies have shown that children in endemic fluorosis areas have significant thyroid dysfunction and changes in intelligence compared with children in normal areas, which may be influenced by minerals associated with inhibiting the expression of sodium iodide symporter and the activities of Na^+-K^+-ATPase. The majority of studies confirm that high accumulation of fluoride reduces the level of free calcium in the body and increases parathyroid hormone (PTH) and PTH-related peptide. In turn, more PTH may lead to abnormal calcium homeostasis and bone metabolism. Moreover, a correlation was observed between long term exposure with fluoride and the increased secretion of insulin resistance, which may cause the changes in bone metabolism. Probably, fluoride may exert a negative effect on the level of sex hormone and impair hypothalamic-pituitary-testicular axis and hypothalamic-pituitary-ovarian axis,

X.-X. Zeng · J. Xiang
The Key Laboratory of Endemic and Ethnic Diseases (Guizhou Medical University), Ministry of Education, Guiyang, Guizhou, People's Republic of China

Department of Pathology at the Affiliated Hospital of Guizhou Medical University, Guiyang, Guizhou, People's Republic of China

Z.-Z. Guan (✉)
The Key Laboratory of Endemic and Ethnic Diseases (Guizhou Medical University), Ministry of Education, Guiyang, Guizhou, People's Republic of China

Department of Pathology at the Affiliated Hospital of Guizhou Medical University, Guiyang, Guizhou, People's Republic of China

The Provincial Key Laboratory of Medical Molecular Biology, Guiyang, Guizhou, People's Republic of China
e-mail: zzg@gmc.edu.cn

Z.-Z. Guan (ed.), *Coal-burning Type of Endemic Fluorosis*,
https://doi.org/10.1007/978-981-16-1498-9_10

which may be resulted from the abnormal occurrences of genes, miRNA and estrogen receptor and gene polymorphism.

Keywords Endocrine system · Fluorosis · Gene · Thyroid · Toxicity

Introduction

Fluorine has a low stability and can form complex with hydrogen and various metals or exist in the form of fluoride ions. Fluoride readily associates with calcium, aluminum, and magnesium in nature. In the environment, fluoride can be released from mineral mining, volcanic eruption, and industrial manufacturing processes, and dispersed in the air, water, and soil. Calcium and magnesium are the most fluorine-bound complexes in the body. For years, water fluoridation is employed as a preventive method to decayed tooth in many countries in the world, including USA and some European countries. On the other hand, exposure to excessive fluoride for a long period through drinking water, food, air, tea, and other means in a given natural environment can cause endemic fluorosis.

Long-term exposure to high level of fluoride by above sources can give rise to chronic fluorosis. Chronic fluorosis covers a range of typical dental and skeletal fluorosis, characterized by incomplete mineralization of enamel and constant pains in the spine and limb joints and rigid joints, bone malformation. In addition, except dental and skeletal impairment, endemic fluorosis causes widespread systemic damages, being a problem with an increased disorder in public health and socioeconomic concern. It has been suggested that excessive intake of fluoride may trigger oxidative stress and numerous metabolic changes of endocrine system (Skórka-Majewicz et al. 2020). High concentrations of fluoride inhibit the function and metabolic procedure of thyroid, as an endocrine disruptor. Exposure to high amount of fluoride can lead to directly and/or indirectly inhibitory effects on the adrenal glands and the parathyroid, in which, the mechanism, however, is still not entirely clear. The review aims to provide evidence concerning toxicity of high level of fluoride to endocrinological system.

Influences of Fluoride Exposure on the Thyroid Gland

Iodine Metabolism and Thyroid Function

The thyroid gland is the largest endocrine gland in the human body, and its follicular epithelial cells are the sites for the synthesis of thyroid hormones (THs). There are two main forms of THs secreted into the blood circulation, including thyroxin (T_4) and tri-iodothyronine (T_3). All T_4 in the human blood is secreted by the thyroid gland, while 20% of T_3 by the thyroid gland, but 80% is produced by deiodination of

T_4 in peripheral tissues. Furthermore, the biological activity of T_3 is about 5 times that of T_4.

Water-soluble iodide is a key substrate for the synthesis of TH_S. Sodium/iodide symporter (NIS) concentrates iodide after transporting iodide from blood to the thyroid through the substrate surface. NIS also regulates the availability of iodine in the thyroid gland and mediates the transport of iodine across cell membranes. Iodide is absorbed in the digestive tract, respiratory tract, placenta, and mammary gland. The function of NIS and the transport of iodine are influenced by both the sodium and potassium-activated adenosine triphosphatase (Na^+-K^+-ATPase) or sodium pump activity (Waugh 2019a, b).

THs can promote growth and development and enhance energy metabolism by improving the metabolism of sugar, protein, and fat. In addition, THs promote the excitability of the sympathetic system and enhance the absorptive capacity of the digestive system. Thyroid function is mainly regulated by hypothalamus and pituitary gland. Three levels of hypothalamus, pituitary, and thyroid gland are closely related and constitute the hypothalamic-pituitary-thyroid axis. The hypothalamus stimulates the release and synthesis of thyroid stimulating hormone (TSH) by secreting thyrotropin-releasing hormone (TRH) in the anterior pituitary gland. TSH can stimulate thyroid follicular hyperplasia, and synthesis and secretion of THs. When the free T_3 and T_4 in the blood reach a certain level, negative feedback effect will be produced to inhibit the secretion of TRH and TSH, forming an automatic control loop of TRH-TSH-THs secretion. Changes or defects in any one of these regulatory systems may cause modification in thyroid synthesis and secretion (Waugh 2019a, b).

Influences of Fluoride Exposure on Function of Thyroid

Fluoride can inhibit function and impact human THs (mainly appearing to hypothyroidism). In our previous investigations, we found that the levels of T_3 and T_4 in serum of the patients with skeletal fluorosis in the area of coal-burning type of endemic fluorosis were significantly decreased, and TSH compensatively increased (Yu et al. 1985). The thyroid function of children in the area of water-type of endemic fluorosis was also studied (Wang et al. 2020a, b). The increase of serum TSH level was positively correlated with the fluoride contents in drinking water and urine of the children living in the area of endemic fluorosis, while the decreased levels of total T_4 and free T_3 was negatively correlated with the fluoride content in drinking water and urine. THs play a crucial role in human growth and intelligence development. Excessive exposure to fluoride may not only affect the changes of TSH, $T_{3,}$ and T_4, but also the growth and development, and intelligence of school-age children through the above changes. The altered thyroid function induced by fluoride exposure in children in the endemic fluorosis area might be involved in the mechanism of the changed intelligence of these children (Wang et al. 2020a, b).

Thyroid cytotoxicity due to fluoride is causing an increase in public concern since fluoride can affect the secretion of THs and induce the changed morphology in histological and ultrastructural levels of the thyroid gland (Abdelaleem et al. 2018). Fluoride and iodide can both have harmful effects on the thyroid. The endocrine toxicity of iodide is generally higher than that of fluoride, but exposure to both elements has the greatest impact on the thyroid (Jiang et al. 2016). In addition to affecting THs, fluoride causes damage to the follicular epithelial cells in thyroid gland and increases the diameter of cell nucleus. Fluoride and iodide have different toxic mechanisms on thyroid cells. Fluoride significantly increased total T_3 and T_4 levels in the absence of iodine, suggesting that the effect of fluoride on the thyroid may be related to iodine concentration, and the combined action of the two can induce apoptosis and even death of thyroid cells. The inositol-requiring enzyme 1 (IRE1) is one of the receptors of endoplasmic reticulum (ER) stress. Drinking water in many areas contains excessive amounts of both fluoride and iodide. Experimental studies showed that after feeding drinking water containing excessive fluoride and/or iodide to rats for 8 months, the morphology of thyroid follicles and THs levels of the animals were obviously modified. Exposure fluoride and iodine by drinking water to rats alone or together induced apoptosis of thyroid cells. However, the expression of IRE1-related factors was increased only in the fluoride-alone and fluoride and iodin combined groups, suggesting that IRE1-induced apoptosis is manly involved in fluoride-induced cytotoxic processes (Liu et al. 2016a, b). One of our previous studies showed that the rats were fed the water containing high fluoride (30 ppm) and normal iodine for a long time, severe damages to function and morphology of thyroid gland were observed, but the rats fed with drinking water with low fluoride (10 ppm) alone only showed slight abnormal changes in the ultrastructure of thyroid epithelium (Guan et al. 1988). The rats with iodine deficiency had goiter accompanied by compensatory function changes. Interestingly, the rats with iodine deficiency and 10 ppm fluoride in drinking water occurred goiter accompanied by obviously morphological and structural damages in thyroid follicular epithelial cells and the inhibited thyroid compensatory function or even decreased the function. It is suggested that iodine deficiency and high fluoride have synergistic effect on thyroid injury.

Furthermore, TH levels from residents living in two different areas with drinking water fluoride concentrations of 0–0.29 and 0.3–0.5 ppm, respectively, were studied. The results showed that the high concentration of fluoride in drinking water (even less than the standard fluoride concentration of 0.5 ppm) also induced an decreased effect on TSH and T_3, suggesting that the patients with hypothyroidism are advised to use standard household water purification equipment to avoid excessive level of fluoride in their drinking water (Kheradpisheh et al. 2018). A large cross-sectional study examining the effects of fluoride on growth and obesity in children showed that low to moderate fluoride exposure was associated with overweight and obesity in children, particularly in girls, which was correlated to the educational level of their parents (Liu et al. 2019).

Mechanisms of Thyroid Damage Induced by Excessive Fluoride

The mechanism of thyroid damage caused by excessive fluoride is complex. At present, the inhibitions of NIS gene expression, Na^+-K^+-ATPase activity and thyroid peroxidase (TPO) activity by fluoride exposure are concerned as the hypothesis of the mechanism.

A number of studies have shown that fluoride can significantly inhibit the expression of NIS gene and the activity of Na^+-K^+-ATPase. The expression of NIS gene was activated by prolactin, but inhibited by fluoride. Fluoride inhibits the NIS and reduces the absorption of iodine. Moreover, these changes are dose-dependent with the concentrations of serum fluoride. NIS is a plasma membrane glycoprotein that mediates the transport of reactive iodine in the thyroid and digestive tract mucosa, respiratory mucosa, placenta, and breast. The transport of iodide by NIS depends on the activity of Na^+-K^+-ATPase.

Furthermore, there is already evidence in that tumor necrosis factor-α (TNF-α), TNF-β1, interleukin (IL)-6 and IL-1β, interferon-γ, insulin-like growth factor 1 (IGF1), phosphatidylinositide 3-kinases, and thyroglobulin can inhibit the expression of NIS. Fluoride also can inhibit the expression of NIS by up-regulating the expression and/or activity of the above inhibitors (Waugh 2019a, b). The activity of Na^+-K^+-ATPase is associated with the pathophysiology of many diseases, such as asthma and allergic disease, metabolic disease, cancer, cardiovascular disease, neurodevelopment, etc. Excessive fluoride reduces the activity of Na^+-K^+-ATPase, which may result in dysfunction of the iodine pump, inducing a decrease in iodine uptake by the thyroid gland, and adverse affect on the accumulation of iodide in the thyroid gland. Fluoride decreased the expression and function of NIS, resulting in the impaired thyroid iodide absorption, the reduced iodide concentration and the iodine deficiency (Waugh 2019a, b). The bioavailability of iodine in children lived in endemic fluorosis area may be reduced, which can affect their development procedure and health in a long term. In particular, maternal iodine deficiency can have adverse effects on the health of offspring, increasing the chances of miscarriages, stillbirths, the risk of attention-deficit/hyperactivity disorder and autism, and the reduced IQ. It is worth noting that lower iodine bioavailability may decrease both the body's capacity of antioxidant and anti-inflammatory, leading to the impaired immunity and an increased risk of cancer in the body (Waugh 2019a, b).

Exposure to high fluoride early in life not only significantly impaired thyroid function and intelligence in children living in endemic fluorosis areas, but also led to overweight. Iodine metabolism is impaired in adults and their offspring in areas with chronic fluorosis. In a word, these changes may be related to the inhibition of NIS gene expression and the decrease of Na^+-K^+-ATPase activities by fluoride toxicity.

Influences of Fluoride Exposure on the Activity of TPO

TPO is a hemoglobin substance containing iron porphyrins and has an effect on the synthesis of THs, which can activate iodine and tyrosine iodination, and promote the coupling of iodinated tyrosine. Fluoride inhibits the activity of TPO and therefor influences many aspects in the synthesis of THs. In addition, fluoride may also affect the uptake of iodine by thyroid, further aggravating thyroid iodine deficiency (Singla and Shashi 2013). Therefore, fluoride may cause thyroid dysfunction by impairing the absorption of iodine and/or reducing its ability to concentrate.

Parathyroid Function and Calcium Homeostasis

Metabolism and Effect of Parathyroid Hormone (PTH)

PTH is an important hormone secreted by parathyroid cells to regulate the metabolism of calcium and phosphorus in the body. Its synthesis and secretion are mainly controlled by calcium-sensitive receptors (CaSR) in the parathyroid glands. The maintenance of calcium homeostasis by PTH could be achieved in part by its ability as a direct action on bone and kidney, and an indirect action on gastrointestinal tract. Under physiologic conditions, a small amount of PTH is secreted into the blood in a pulsed form, and the serum Ca^{2+} concentration is maintained at a normal level by adjusting the renal tubular's reabsorption of Ca^{2+} and activity of renal 1-α-hydroxylase. Excessive secretion of PTH or injection of pharmacological dose can both regulate the function of osteoclasts and stimulate the activity of osteoblasts.

Parathyroid hormone 1 receptor (PTH1R) is a class B multidomain G-protein coupled receptor, which is expressed by target cells (such as osteoblasts and osteocytes in bone, and renal tubule cells), and mediates the cellular response to PTH (Yu et al. 2018). PTH, an important component of the bone-parathyroid-kidney (BPK) feedback loop, is responsible for the dynamic balance of minerals (Bergwitz and Jüppner 2010).

Possible Influences of Fluoride Exposure on the Body through PTH (Including Changes in Serum Free PTH Content, Abnormal Calcium Circulation, and Bone Turnover)

Fluoride, a highly electronegative ion, combines with calcium ions with a positive charge in mineralized tissues such as teeth and bones. High amount of fluoride intake induces pathological changes in mineralized tissues, presenting dental and skeletal fluorosis. In addition, interactions of fluoride absorption with calcium can reduce fluoride uptake and use.

Epidemiological studies in India have shown a higher incidence of sclerosis and secondary hypoparathyroidism in skeletal muscle fluorosis among children in calcium-deficient endemic fluorosis areas than those in calcium-abundant endemic fluorosis area, suggesting that low calcium increases the toxicity of fluoride (Teotia et al. 1998). At the same time, however, excessive fluoride exposure may cause hypocalcemia. Fluoride plays a driving role in the cause of secondary hyperparathyroidism by interfering calcium homeostasis followed elevating serum PTH levels (Koroglu et al. 2011; Xu et al. 2010). The decrease in serum calcium may be due to the reduction of free calcium level caused by the formation of CaF_2 complexes with fluoride and calcium ions, leading to hypocalcemia reactions (Xu et al. 2010).

Studies in vivo and in vitro have confirmed that fluoride can regulate the secretion of intact PTH (iPTH) with a rapid effect. Changes in iPTH can be observed after a single dose of fluoride administration. Fluoride exposure can also mediate iPTH secretion through changes in calcium levels. The overall effect of fluoride is not limited to the parathyroid gland, but also affects the gene expressions of BPK feedback loop, including PTH, CaSR, αKlotho, fibroblast growth factor receptor 1c, vitamin D receptor, and parathyroid hormone like hormone (Puranik et al. 2015).

PTH, PTH-related peptide (PTHrP), and CaSR are involved in maintaining calcium homeostasis. CaSR is a G-protein-coupled receptor that regulates calcium homeostasis after activation by extracellular calcium ion. Structurally, PTHrP is correlated to PTH and affects bone formation by promoting the survival and differentiation of osteoblasts. CaSR is a G protein-coupled receptor that regulates calcium homeostasis after activation by extracellular calcium ion. CaSR maintains calcium homeostasis in the body by regulating PTH secretion and inhibiting the transcription of parathyroid genes and cell proliferation. Receptor-activator of nuclear factor kappa-B ligand (RANKL) and osteopontin (OPN) in osteoclasts play an important role in regulating the differentiation and activation of osteoclasts, and have obvious influence on bone remodeling. RANKL initiates osteoclast formation through multiple downstream signaling pathways including activation of mitogen-activated protein kinase, nuclear factor kappa-B, activated protein 1, cellular oncogene Fos, activated T nuclear factor, and cytoplasmic 1 (Silva and Kousteni 2015).

Excessive fluoride intake may cause the increase of calcitonin, which leads to hypocalcemia in patients with fluorosis, and then increase the level of PTH in the serum to maintain the calcium level. Fluoride up-regulates expression of CaSR by regulating extracellular calcium, which then leads to changed regulation of PTH and PTHrP expression. Moreover, fluoride has a dose-dependent effect on the expression of CaSR, RANKL, and osteoprotegerin (OPG) in MC3T3-E1 cells and rats. These results indicate that high fluoride enhances osteoclast activity and bone remodeling. Fluoride affects CaSR expression by regulating extracellular calcium, thereby affecting the secretion of PTH and PTHrP. Probably, fluoride may be able to affect calcium homeostasis by regulating PTH, PTHrP, and CaSR (Wang et al. 2015).

Low calcium and high fluoride can significantly increase PTH secretion, and expressions of PTHrP, osteocalcin (OCN), and OPN. The target genes for the physiological effects of PTH on bone cells are sclerosteosis (SOST) and RANKL.

PTH provides dual anabolic (SOST/sclerostin pathways) and catabolic (OPG/RANKL/receptor-activator of nuclear factor kappa beta pathways) skeletal actions that are balanced. Fluoride exposure can reduce the expression of SOST/sclerostin and increase RANKL. Interestingly, co-treatment with PTH and low concentrations of fluoride amplified the trend of inhibition, but reversed the effects of high doses of fluoride on SOST and RANKL. In addition, the co-processing of fluoride and PTH greatly stimulated the expression of PTH1R (Yu et al. 2018).

For years, it is common that excessive fluoride exposure reduces the level of free calcium in the body, which may cause CaSR to be activated, thus increases PTH and PTHrP. In turn, more PTH may lead to abnormal calcium homeostasis and bone metabolism. In short, fluoride can affect PTH and PTHrP levels, thereby contributing to active bone turnover.

Influences of Fluoride Exposure on Insulin Metabolism and Diabetes

Insulin, secrete by islet B cells, is a key hormone that can comprehensively promote anabolic metabolism and cause various symptoms and pathological changes in diabetes. In addition, insulin is a systemic hormone that promotes formations of bone, bone matrix and collagen, and mineralization of normal bone.

Fluoride can affect the release, activity, and sensitivity of insulin. In turn, the state of insulin in vivo interferes with osteogenesis, suggesting that there is a close relationship between insulin and the pathogenesis of bone in the mechanism of fluoride toxicity (Yang et al. 2015). In the pathological process of bone injury induced by more fluoride exposure via mediating insulin receptor signaling pathway, insulin plays an important role, and whereas in the abnormal bone turnover influenced by fluoride exposure IGF1 signaling pathway may have an important effect (Liu et al. 2016a, b).

Studies in vitro have shown the dual effects of fluoride on the function of osteoblast. First, in the fluoride-treated rats the serum insulin level was significantly elevated but glucagon decreased, and the insulin-positive area and the islet size were increased with the morphological analyses. Secondly, fluoride significantly promoted the expression of insulin mRNA and insulin receptor in vitro, suggesting a correlation of insulin secretion with exposure of fluoride. Therefore, it indicates that insulin signaling pathway may be involved in the potential occurring procedure of skeletal fluorosis (Hu et al. 2012).

Some studies have shown that fluoride reduces insulin sensitivity eventhough a study indicated that fluoride might enhance glucose homeostasis in diabetes (Lobo et al. 2015). It was shown that long-term treatment with fluoride can increase insulin resistance, reduce insulin signaling, raise plasma insulin, OCN and TNF-α concentrations, decline tibial trabecular bone area, and led to change of metabolism in bone of ovariectomized rats (Pereira et al. 2017). In an investigation concerning the relationship between fluoride and diabetes in 22 states of the USA from 2005 to 2010, the positive relationship between the exposure of fluoride and the incidence of diabetes was observed (Fluegge 2016). In the Newfoundland and Labrador region of Canada, the epidemiological studies found that higher incidence of type 1 diabetes in children was associated with higher level of fluoride in the local drinking water (Chafe et al. 2018). Despite the limitations of the current study, it can be argued that diabetes may be caused by a variety of factors, including epigenetic factor. However, the effect of fluoride on insulin metabolism and diabetes mellitus still needs to be further studied.

Influences of Fluoride Exposure on the Reproductive System

The reproductive toxicity of fluoride has been caused widespread concern. High amount of fluoride exposure significantly influences not only sperm quantity and quality as well as fertilization in males, but also alternates hormone level related with the hypothalamic-pituitary-testicular (HPT) axis, involving sex hormone-binding globulin (SHBG), androgen binding protein (ABP), estradiol, luteinizing hormone, and testosterone, thus interfering with the male reproductive endocrine function (An et al. 2019; Ma et al. 2017).

There is an interaction between fluoride exposure and serum concentrations of follicle-stimulating hormone and estradiol, and the reciprocity among different loci of the estrogen receptor alpha (ESR-α) gene may affect serum testosterone levels. Current research results indicate that long term exposure to fluoride in drinking water may regulate male reproductive hormone levels in area of endemic fluorosis, and the reciprocity between fluoride exposure and ESR-α polymorphisms may influence serum hormone levels related to male HPT axis (Ma et al. 2017). Both ABP and SHBG are essential for the regulation of HPT axis. The concentration of SHBG affects the level of ABP. Long-term fluoride exposure not only reduces the serum SHBG content of male farmers in endemic fluorosis areas, but the combination of fluoride exposure and specific ESR-α polymorphisms (*XbaI* and *rs3798577*) causes a decrease in serum ABP of male residents in the areas (An et al. 2019).

In addition, exposure of fluoride also significantly influences the female reproductive system, changes the structure and function of ovarian cells, and hinders the developmental potential of oocytes (Wang et al. 2017; Li et al. 2020). It has been found that overexposure to fluoride can cause abnormal transcription profiles of ovulation-related genes (*tgfβ1*, *bmp15*, *gdf9*, *mprα*, *mprβ*, and *ptg2β*) and impair ovarian function, in which mechanism may be related to the adverse effects of fluoride on FSH, LH, and VTG levels (Li et al. 2020). High amount of fluoride significantly induced alternation of microRNA, exhibiting the decrease in 40 miRNAs and the increase in 37 miRNAs (Wang et al. 2020a, b). The hormone secretion and efficiency of the hypothalamic-pituitary-ovarian (HPO) axis may be altered by gene–environment interactions. Studies have shown that serum gonadotropin-releasing hormone (GnRH) and estradiol levels associated with the

HPO axis are significantly reduced in women exposed to fluoride. In addition, fluoride exposure interacted with serum estradiol concentrations in women exhibiting the G allele of FSHR -29G > A (Zhao et al. 2015). Fluoride affected glucose transport during porcine oocyte maturation and impaired the development of parthenogenetic activation by disturbing DNA methylation of NNAT (Liu et al. 2018).

In areas of endemic fluorosis, there is a significant association of PRL gene polymorphism to the increased risk of fluorosis. A case-control study found that there was a genetic association between estrogen receptor (ESR) gene and ESR polymorphisms, and dental fluorosis among children exposed to high fluoride in Henan, China. Estrogen and estrogen-related receptor can influence osteoblast activity, which is benefit to deposit calcium and phosphate in bone (Ba et al. 2011). Similarly, studies conducted in Tibetan, Kazakh, and Mongolian subjects in China found that the rs1341239 polymorphism of the PRL gene promoter is related to skeletal fluorosis resulted from the brick tea type of endemic fluorosis. Interestingly, the residents with T allele in PRL rs1341239 exhibited a protective effect on skeletal fluorosis, which is more pronounced in the residents under the age of 45 or with short time of exposure, especially in Kazakh women (Li et al. 2017).

Environmental fluoride exposure is related to lower testosterone and estradiol, associated with others sex steroid hormones, such as SHBG and ABP mainly in male, and GnRH and estradiol in female (Bai et al. 2020). It should be noted that when other factors remain the same, abnormal transcriptional profiles of oogenesis-related genes, such as miRNAs and ESR polymorphisms, may play an important role in the risk of fluoride exposure.

Conclusion

Excessive intake of fluoride can have a toxic effect on the endocrine system or organs (e.g., thyroid gland, parathyroid gland, islets, and gonads). The direct toxicity of fluoride can result in abnormal morphology of these organs or indirectly interfere with the synthesis and secretion of its own hormones, cause disorders of iodine metabolism, calcium homeostasis, bone turnover, hypothalamic-pituitary-testicular axis, and hypothalamic-pituitary-ovarian axis, and reduce insulin sensitivity, thus disturbing bone metabolism, growth and development in the body with intellectual abnormalities. It is worth to note that abnormal changes in genes, miRNAs, and estrogen receptor polymorphisms caused by fluoride exposure may play an important role in the risk of toxic effects of fluoride on endocrine system. Nevertheless, the complexity of the endocrine system makes it difficult to reveal the relationship between fluoride and the endocrine disruption, which needs more in-depth research and exploration to understand the correlation.

References

Abdelaleem MM, El-Tahawy NFG, Abozaid SMM, Abdel-Hakim SA. Possible protective effect of curcumin on the thyroid gland changes induced by sodium fluoride in albino rats: light and electron microscopic study. Endocr Regul. 2018;52:59–68.

An N, Zhu J, Ren L, Liu X, Zhou T, Huang H, et al. Trends of SHBG and ABP levels in male farmers: influences of environmental fluoride exposure and ESR alpha gene polymorphisms. Ecotoxicol Environ Saf. 2019;172:40–4.

Ba Y, Zhang H, Wang G, Wen S, Yang Y, Zhu J, et al. Association of dental fluorosis with polymorphisms of estrogen receptor gene in Chinese children. Biol Trace Elem Res. 2011;143:87–96.

Bai R, Huang Y, Wang F, Guo J. Associations of fluoride exposure with sex steroid hormones among U.S. children and adolescents, NHANES 2013-2016. Environ Pollut. 2020;260:114003.

Bergwitz C, Jüppner H. Regulation of phosphate homeostasis by PTH, vitamin D, and FGF23. Annu Rev Med. 2010;61:91–104.

Chafe R, Aslanov R, Sarkar A, Gregory P, Comeau A, Newhook LA. Association of type 1 diabetes and concentrations of drinking water components in Newfoundland and Labrador, Canada. BMJ Open Diabetes Res Care. 2018;6:e000466.

Fluegge K. Community water fluoridation predicts increase in age-adjusted incidence and prevalence of diabetes in 22 states from 2005 and 2010. J Water Health. 2016;14:864–77.

Guan ZZ, Zhuang ZJ, Yang PS, Pan S. Synergistic action of iodine-deficiency and fluorine-intoxication on rat thyroid. Chin Med J. 1988;101:679–84.

Hu CY, Ren LQ, Li XN, Wu N, Li GS, Liu QY, et al. Effect of fluoride on insulin level of rats and insulin receptor expression in the MC3T3-E1 cells. Biol Trace Elem Res. 2012;150:297–305.

Jiang Y, Guo X, Sun Q, Shan Z, Teng W. Effects of excess fluoride and iodide on thyroid function and morphology. Biol Trace Elem Res. 2016;170:382–9.

Kheradpisheh Z, Mirzaei M, Mahvi AH, Mokhtari M, Azizi R, Fallahzadeh H, et al. Impact of drinking water fluoride on human thyroid hormones: a case-control study. Sci Rep. 2018;8:2674.

Koroglu BK, Ersoy IH, Koroglu M, Balkarli A, Ersoy S, Varol S, et al. Serum parathyroid hormone levels in chronic endemic fluorosis. Biol Trace Elem Res. 2011;143:79–86.

Li BY, Yang YM, Liu Y, Sun J, Ye Y, Liu XN, et al. Prolactin rs1341239 T allele may have protective role against the brick tea type skeletal fluorosis. PLoS One. 2017;12:e0171011.

Li M, Cao J, Zhao Y, Wu P, Li X, Khodaei F, et al. Fluoride impairs ovary development by affecting oogenesis and inducing oxidative stress and apoptosis in female zebrafish (Danio rerio). Chemosphere. 2020;256:127105.

Liu Q, Liu H, Yu X, Wang Y, Yang C, Xu H. Analysis of the role of insulin signaling in bone turnover induced by fluoride. Biol Trace Elem Res. 2016a;171:380–90.

Liu H, Hou C, Zeng Q, Zhao L, Cui Y, Yu L, et al. Role of endoplasmic reticulum stress-induced apoptosis in rat thyroid toxicity caused by excess fluoride and/or iodide. Environ Toxicol Pharmacol. 2016b;46:277–85.

Liu X, Nie ZW, Gao YY, Chen L, Yin SY, Zhang X, et al. Sodium fluoride disturbs DNA methylation of NNAT and declines oocyte quality by impairing glucose transport in porcine oocytes. Environ Mol Mutagen. 2018;59:223–33.

Liu L, Wang M, Li Y, Liu H, Hou C, Zeng Q, et al. Low-to-moderate fluoride exposure in relation to overweight and obesity among school-age children in China. Ecotoxicol Environ Saf. 2019;183:109558.

Lobo JG, Leite AL, Pereira HA, Fernandes MS, Peres-Buzalaf C, Sumida DH, et al. Low-level fluoride exposure increases insulin sensitivity in experimental diabetes. J Dent Res. 2015;94:990–7.

Ma Q, Huang H, Sun L, Zhou T, Zhu J, Cheng X, et al. Gene-environment interaction: does fluoride influence the reproductive hormonesin male farmers modified by ERα gene polymorphisms? Chemosphere. 2017;188:525–31.

Pereira AG, Chiba FY, de Lima Coutinho Mattera MS, Pereira RF, de Cássia Alves Nunes R, et al. Effects of fluoride on insulin signaling and bone metabolism in ovariectomized rats. J Trace Elem Med Biol. 2017;39:140–6.

Puranik CP, Ryan KA, Yin Z, Martinez-Mier EA, Preisser JS, Everett ET. Fluoride modulates parathyroid hormone secretion *in vivo* and *in vitro*. Cells Tissues Organs. 2015;200:413–23.

Silva MC, Kousteni S. Chapter 8 - cellular actions of PTH: osteoblasts, osteoclasts, and osteocytes. In: The parathyroids. London: Elsevier Inc.; 2015. p. 127–37.

Singla S, Shashi A. Thyroid peroxidase activity as toxicity target for fluoride in patients with thyroid dysfunction. Curr Res Microbiol Biotechnol. 2013;1:53–7.

Skórka-Majewicz M, Goschorska M, Żwierełło W, Baranowska-Bosiacka I, Styburski D, Kapczuk P, et al. Effect of fluoride on endocrine tissues and their secretory functions - review. Chemosphere. 2020;260:127565.

Teotia M, Teotia SP, Singh KP. Endemic chronic fluoride toxicity and dietary calcium deficiency interaction syndromes of metabolic bone disease and deformities in India: year 2000. Indian J Pediatr. 1998;65:371–81.

Wang Y, Duan XQ, Zhao ZT, Zhang XY, Wang H, Liu DW, et al. Fluoride affects calcium homeostasis by regulating parathyroid hormone, PTH-related peptide, and calcium-sensing receptor expression. Biol Trace Elem Res. 2015;165:159–66.

Wang HW, Zhao WP, Liu J, Tan PP, Zhang C, Zhou BH. Fluoride-induced oxidative stress and apoptosis are involved in the reducing of oocytes development potential in mice. Chemosphere. 2017;186:911–8.

Wang M, Liu L, Li H, Li Y, Liu H, Hou C, et al. Thyroid function, intelligence, and low-moderate fluoride exposure among Chinese school-age children. Environ Int. 2020a;134:105229.

Wang Z, Dong S, Chen Q, Chen C, Xu Z, Dong Z. Analysis of the microrna profile by sequencing in ovarian granular cells from women suffering fluorosis with reproductive dysfunction. Biol Trace Elem Res. 2020b;197:101–6.

Waugh DT. Fluoride exposure induces inhibition of sodium/iodide symporter (NIS) contributing to impaired iodine absorption and iodine deficiency: molecular mechanisms of inhibition and implications for public health. Int J Environ Res Public Health. 2019a;16:1086.

Waugh DT. Fluoride exposure induces inhibition of sodium-and potassium-activated adenosine triphosphatase (Na^+, K^+-ATPase) enzyme activity: Molecular mechanisms and implications for public health. Int J Environ Res Public Health. 2019b;16:1427.

Xu H, Liu QY, Zhang JM, Zhang H, Li GS. Elevation of PTH and PTHrP induced by excessive fluoride in rats on a calcium-deficient diet. Biol Trace Elem Res. 2010;137:79–87.

Yang C, Zhang M, Li Y, Wang Y, Mao W, Gao Y, et al. Streptozotocin aggravated osteopathology and insulin induced osteogenesis through co-treatment with fluoride. Biol Trace Elem Res. 2015;168:453–61.

Yu YN, Guan ZZ, Liu JL, Liu SZ, Chen YQ. Changes of serum T_4, T_3 and TSH in skeletal fluorosis patients. Chin J Endem. 1985;3:188–90.

Yu X, Yu H, Jiang N, Zhang X, Zhang M, Xu H. PTH (1-34) affects bone turnover governed by osteocytes exposed to fluoride. Toxicol Lett. 2018;288:25–34.

Zhao MX, Zhou GY, Zhu JY, Gong B, Hou JX, Zhou T, et al. Fluoride exposure, follicle stimulating hormone receptor gene polymorphism and hypothalamus-pituitary-ovarian axis hormones in Chinese women. Biomed Environ Sci. 2015;28:696–700.

Chapter 11
Damages to Reproductive System by Coal-burning Type of Endemic Fluorosis

Yue-Hai Xiao and Fa Sun

Abstract Coal-burning type of endemic fluorosis is capable of causing damage to both the male and female reproductive systems. The effects of coal-burning type of endemic fluorosis on the male reproductive system are mainly manifested in the following aspects: changes in testicular tissue structure, disturbances in sex hormone levels, imbalance in the testicular oxidative/antioxidant system, elevated apoptosis rate of spermatogenic cells, and reduced semen quality. The effects of coal-burning type of endemic fluorosis on the female reproductive system are mainly manifested in the following aspects: fluorosis interferes with the synthesis and secretion of sex hormones; coal-burning type of endemic fluorosis destroys the structure and function of reproductive organs such as the uterus and ovaries; fluorosis affects the structure and function of the placenta, and leads to fetal dysplasia, malformation, and even stillbirth.

Keywords Fluorosis · Mechanism · Prevention · Reproductive system · Treatment

Introduction

Living environment is one of the critical factors affecting human fertility. It is known that fluoride is one of the essential trace elements. However, prolonged excessive fluoride intake will cause damage to the reproductive system. Coal-burning type of endemic fluorosis can lead to changes in testicular tissue structure, disturbance of sex

Y.-H. Xiao
Department of Urology at the Affiliated Hospital of Guizhou Medical University, Guiyang, Guizhou, People's Republic of China

F. Sun (✉)
Department of Urology at the Affiliated Hospital of Guizhou Medical University, Guiyang, Guizhou, People's Republic of China

Health Commission of Guizhou Province, Guiyang, Guizhou, People's Republic of China
e-mail: sunfagz@gmc.edu.cn

Z.-Z. Guan (ed.), *Coal-burning Type of Endemic Fluorosis*,
https://doi.org/10.1007/978-981-16-1498-9_11

hormone levels, imbalance of testicular oxidative/antioxidant system, elevated apoptosis rate of spermatogenic cells, and decreased semen quality. Many previous studies mainly focus on the effects of fluorosis on the male reproductive system. In recent years, the damage of fluorosis to the female reproductive system has gradually drawn the universal concern of the entire humankind. Epidemiological investigations have found a potential link between long-term severe fluoride exposure and female primary infertility, and fluoride is an important risk factor for female infertility (Shi et al. 2012). This section will elaborate the effects of fluorosis on the male reproductive system and female reproductive system, respectively, in combination with the latest research progress at home and abroad.

Effect of Fluorosis on Sperm and Semen Quality

Any factor affecting spermatogenesis, maturation, expulsion, capacitation, and fertilization can lead to male infertility (Chen et al. 2018). According to a case report, a young patient in an area of coal-burning type of endemic fluorosis in Guizhou Province, China, has not had children for three years after marriage because of long-term consumption of such food with high fluoride pollution caused by burning coal. The results of semen analysis revealed that sperm liquefaction, sperm survival rate, and sperm motility were significantly reduced. After physical examination, his spouse excluded female reproductive system diseases, suggesting that the patient's low sperm quality may be related to long-term fluoride intake (Shi et al. 2012). Additionally, numerous animal experiments have also confirmed that excessive intake of fluoride is able to reduce sperm motility and sperm density, affect semen quality, and impair reproductive function in rats (Gang et al. 2010).

Effects of Fluorosis on Sperm Morphology and Structure

Long-term fluoride exposure is capable of affecting semen quality, decrease sperm count and motility, and increase the malformation rate of sperm in a dose-dependent manner (Wang et al. 2018). For example, mice that ingest fluoride in drinking water develop dental fluorosis, along with reduced sperm survival and sperm motility. In the study of sperm damage caused by coal-burning type of endemic fluorosis, obvious symptoms of dental fluorosis were found in rats fed with coal-baked corn consumed by people in fluorosis areas. Meanwhile, ultrastructure observation under electron microscope revealed that reduced spermatogenesis, poor head cap formation, double-headed sperm, pyknosis dissolved nucleus, denatured and edematous mitochondria sheath in fluorosis rats.

Effect of Fluorosis on Sperm Acrosome

The sperm acrosome is a cap-like structure that originates from the Golgi and covers the anterior portion of the sperm nucleus. Acrosome is a specialized lysosome that promotes sperm binding to the egg by releasing hydrolytic enzymes (Berruti 2016). Electron microscopy revealed that the acrosome substance of fluoride-exposed rat sperm was deficient, and there was a significant difference in the number of actin required and the proportion of tyrosine phosphorylation during acrosome capacitation. Jiang et al. studied the changes in mRNA expression levels of seven genes associated with the entire acrosome biogenesis process under fluorinated conditions, The mRNA expression levels of three genes (*Zpbp1*, *Spaca1*, and *Dpy19l2*) among the aforementioned genes, showed a decreasing trend during the acrosomal phase, indicating that fluoride may have altered the ultrastructure of the acrosome and nuclear membrane during spermatogenesis and interfered with acrosome biogenesis by down-regulating the expression of *ZPBP1*, *SPACA1*, *DPY19L2*, and *LMNB2* (Jiang et al. 2019).

Effects of Fluorosis on the Testis

The testis is the site of sperm production and the endocrine organ of male sex hormones (Chen et al. 2018). The damage caused by fluoride to the testis is the result of many effects, and excessive fluoride can destroy the morphology and function of testicular germ cells. Animal experimental studies have confirmed that fluoride is able to break through the blood–testis barrier and act directly on testicular tissue, destroying spermatogenic cells, sertoli cells (SC_S), and interstitial cells at all levels of the testis, affecting the function of the testis, and the fluoride content in the testis increases in a dose-dependent manner with the dose of fluoride taken.

Effect of Fluorosis on Testis Weight

Animal experimental studies found that fluorosis could cause a decrease in testicular mass and testicular protein content, and increased cholesterol, leading to delayed development of accessory sexual organs such as the testis (Gupta et al. 2007). The potential reason was speculated that decreased weight of testicular and accessory gonadal organs may be related to hormonal imbalance caused by fluorosis. However, some scholars have found no significant difference in testicular mass changes after fluoride exposure, and even opposite studies have reported that the relative wet weight of the testis increased in rats treated with fluoride, and the authors speculated that the reason may be due to compensatory changes or fluid accumulation resulting in increased wet weight of the testis (Ghosh et al. 2002). More interestingly, the fluoride exposure test was performed on male rats, and the results suggested that there was no correlation between the testicular weight of the experimental group and the control group, but the right testicular weight of the F1 generation control group grouped in the experiment was higher than that of the P generation control group, the left testicular weight of the F1 generation was significantly lower than that of the P generation in the 25 ppm fluoride exposure group, and the right testicular weight of the F1 generation was significantly higher than that of the P generation in the 100 ppm fluoride exposure group. The reasons for the inconsistent results reported in many studies may be due to different experimental factors such as fluorine exposure dose, time, detection method and sample number, which require further study.

Effect of Fluorosis on Testicular Tissue Structure

As early as 1973, it was found that fluoride caused reproductive impairment in male mice when they were fed with fluoridated water and fluoridated diet (Messer et al. 1973). In a study, mice were fed with a fluoride-containing diet. Three months later, degeneration and necrosis of the seminiferous tubules of the testis, cessation of spermatogenesis and the most significant damage to the spermatids at all levels of spermatogenic cells were observed under optical microscope, manifested as poor formation of the head cap, lack of acrosomal material, no condensation of the nucleus, no formation of axonemes in the rear cytoplasm as well as degeneration of the mitochondrial sheath (Kulbir 1980). After that, in an experiment, the significant changes were found in testicular tissue structure of dogs inhaled fluoride air after 5 weeks, and pathological changes such as testicular degeneration also occurred after inhalation of fluorine-containing air in mice (Zahrorohkor and Tarinsky 1981).

In our group, a study on the effect of coal-burning type of endemic fluorosis on the male reproductive system was carried by feeding SD rats with coal-baked corn consumed by people in coal-burning type of endemic fluorosis areas. We found that the rats developed dental fluorosis, successfully replicating the coal-burning type of endemic fluorosis rat model. In the experimental study, the structure of seminiferous tubules in the testis of rats with coal-burning type of endemic fluorosis was damaged to varying degrees, and the observation under a light microscope revealed that loose and disordered structure of spermatogenic cells at all levels, swollen cells, increased size, vacuolated cytoplasm, and partial necrotic cells. Besides, the degree of damage was aggravated with the increase of fluoride dose (Xiao et al. 2015).

Effect of Fluorosis on Testicular Ultrastructure

Transmission electron microscopy revealed that the ultrastructure of testicular tissue in fluoride-exposed rats had a significant effect, and the lesions were aggravated with the extension of fluoride dose and exposure time, and the detailed changes were as follows. (1) Fluoride-exposed rat spermatogonia were pyknotic, mitochondria were swollen in a ring shape, edge centralization of Chromatin, and lysosomes were increased. (2) In primary spermatocytes, the cytoplasm dissolved locally, lysosomes increased, mitochondrial cristae decreased or disappeared, nuclear membrane space widened, and even nuclear pyknosis and dissolution occurred. (3) Effects of fluorosis on ultrastructure of secondary spermatocytes were observed. Extensive widening and dissolution of nuclear membrane space, expansion and vacuolization of endoplasmic reticulum, and even local depression and dissolution of nuclear membrane, nuclear pyknosis and dissolution were observed in secondary spermatocytes. (4) The gap at the junction between SC_S and spermatogonia was widened and vacuolated. The mitochondria of SC_S were swollen and showed pleomorphic changes, commonly annular and dumbbell-shaped, with significantly increased lysosomes, liquefaction of the cytoplasmic matrix, and even cell atrophy. Additionally, the phenomenon of degeneration and pyknosis occurs in the SC_S, the cytoplasm of SC_S was distributed around the spermatogenic cells. The appearance of ring mitochondria, increased smooth endoplasmic reticulum, expansion and increased lysosomes in SC_S are all manifestations of active cell function. With the increase of fluoride intake, SC_S further showed mitochondrial degeneration, cytoplasmic vacuolization, accumulation of super large lipid droplets, and the function tended to be low and even the whole cell degenerated and atrophied. (5) Drinking-water type of fluorosis studies revealed that Leydig cells were found to be significantly damaged under electron microscopy, with the most significant changes in mitochondria and smooth endoplasmic reticulum. Mitochondria decreased or showed polymorphic changes, some mitochondrial cristae dissolved and destroyed, showing vacuolization, smooth endoplasmic reticulum expanded, and even showed bullae, lysosomes and lipid droplets increased, which provided a morphological basis for the decrease of serum testosterone (T) levels in fluorosis. In the study of testicular injury in male rats by coal-burning type of endemic fluorosis, it was found that the smooth endoplasmic reticulum in the cytoplasm of rats in the exposure group was slightly dilated after 90 days; after 120 days, the smooth endoplasmic reticulum was significantly dilated or even vacuolated, mitochondria were reduced, the number of lipid droplets was reduced, yet lysosomes were increased; after 180 days, the cells shrunk, a large number of lipofuscin granules appeared in the cytoplasm, chromatin edge was centralized, and even the nuclei of the cells were pyknotic and dissolved, the cells ruptured, and the cytoplasm escaped. In addition, the lesions also worsened with increasing fluoride dose. (6) After 90 days of fluoride exposure, there are no obvious lesions in seminiferous tubule myoid cells; while at 120 and 180 days, myoid cells showed pyknosis, oblong shape, irregular membrane shrinkage, reduced cytoplasm, and even cell nuclear pyknosis and nuclear membrane thickening.

Effect of Fluorosis on Testicular Blood–Testis Barrier

The content of fluoride in testicular tissue increased in the fluoride-exposed group, indicating that fluoride broke through the blood–testis barrier, and the structure and protective effect of the blood–testis barrier were both damaged. Moreover, the fluoride content in the testicular tissue of the fluoride-exposed group was significantly higher than that of the control group, demonstrating that fluoride broke through the blood–testis barrier and produced damage to spermatogenic cells, SC_S, and interstitial cells in the testis, and SC_S are one of the crucial components of the blood–testis barrier.

Impairment of Testicular Spermatogenic Function by Fluorosis

Effect of Fluorosis on Apoptosis of Spermatogenic Cells in Testis

Apoptosis is an active death process that occurs after cells receive a certain signal or a certain stimulus. Excessive intake of fluoride will disrupt testicular structure and induces apoptosis of spermatogenic cells in male rats, which leads to spermatogenic dysfunction (Zhao et al. 2018). Another animal experiment with fluoride exposure by drinking water, gavage, and intraperitoneal injection have shown that excessive fluoride intake can cause increased apoptosis of spermatogenic cells and affect testicular reproductive function (Zhang et al. 2016). In the study of coal-burning type of endemic fluorosis, it was found that both coal-burning type of endemic fluorosis and sodium fluoride injection through drinking water or intraperitoneal injection could increase the apoptosis rate of testicular spermatogenic cells; coal-burning type of endemic fluorosis could increase the apoptosis of spermatogenic cells in rats and increasing nutrition could effectively reduce the number of apoptotic cells. Apoptotic cells mainly occur in primary spermatocytes and spermatogonia, with scattered distribution, incomplete nuclei, irregular clumps in shape, and inconsistent size (Tian et al. 2018; Xiao et al. 2015).

Effect of fluorosis on pro-apoptotic/anti-apoptotic genes of testis: Apoptosis is regulated by a variety of related genes, and fluoride is capable of inducing apoptosis through both intrinsic and extrinsic pathways. Fluorosis is able to elicit the up-regulation of the expression of pro-apoptotic genes *Bax*, *C-kit*, and *Fas*, and the down-regulation of the expression of the anti-apoptotic gene *Bcl-2* in testicular spermatogenic cells (Liu et al. 2013). Fluorosis may regulate spermatogenic cell apoptosis through apoptosis-related gene regulation pathways, resulting in increased apoptosis of spermatogenic cells in rats, which thereby causes spermatogenic failure. *Bcl-2* and *Bax*, which are anti-apoptotic genes and pro-apoptotic genes, respectively, are a pair of closely apoptosis-related genes but have opposite effects. Bcl-2 interacting mediator (Bim) belongs to the same B-cell lymphoma-2 (Bcl-2) family, and the mutual coordination between anti-apoptotic and pro-apoptotic members of the Bcl-2 protein family plays the switch role of the apoptosis. When there is no apoptotic stimulus from the outside, Bcl-2 anti-apoptotic proteins are sequestered as integral membrane proteins of organelles thereby inhibiting the release of pro-apoptotic factors, and Bax is localized in the cytoplasm as a pro-apoptotic protein in an inactive form. When cells receive apoptosis-stimulating signals, Bax undergoes conformational changes, translocates from the cytoplasm to the mitochondrial membrane, and interacts with anti-apoptotic proteins in and on the membrane, so that anti-apoptotic proteins lose their inhibitory activity against apoptosis, and the permeability pores on the mitochondrial membrane open, resulting in the down-regulation of mitochondrial membrane potential and the release of cytochrome C, which thereby initiates mitochondria-mediated apoptosis (Mikhailov et al. 2001). The ratio of Bcl-2 to Bax expression determines whether the cells survive or apoptosis after receiving apoptotic signal stimulation, and when Bcl-2 is overexpressed, heterodimer Bcl-2/Bax is formed to play a role in inhibiting apoptosis; over-expressed Bax can combine with Bcl-2 to form homodimer Bax/Bax, counteracting the antiapoptotic effect of Bcl-2 and playing the role of promoting apoptosis. The change of Bcl-2 to Bax ratio regulates the balance of cell proliferation and apoptosis, with a low ratio showing high apoptosis, while a high ratio showing low apoptosis rate and high survival rate. Studies have shown that the expression of Bcl-2 protein is decreased and the expression of Bax is increased in the testicular tissue of rats with coal-burning type of endemic fluorosis. With the increase of fluoride dose as well as the prolongation of fluoride exposure time, the symptoms of fluorosis in rats are gradually aggravated, the expression of Bcl-2 in rat testis is gradually decreased, and the expression of Bax is gradually increased (Xiao et al. 2015), which is similar to a relevant report of fluoride exposure via a high-fluoride diet (Liu et al. 2013). The proportional change of Bcl-2 to Bax expression levels determines whether apoptosis as well as the degree of apoptosis occurs in cells at defined sites. Fluorosis down-regulates the expression of apoptosis-related genes *Bcl-2*, up-regulates the expression of Bax, and decreases the expression ratio of Bcl-2/Bax in rat testicular tissue, thereby inducing apoptosis in testicular spermatogenic cells.

Fluoride can activate the oxidative stress-mediated mitogen-activated protein kinase (MAPK) signaling pathway by phosphorylating c-Jun N-terminal kinases (JNKs) and extracellular regulated protein kinases (ERK). It is well-known that JNK and ERK are two family members of MAPK that are essential for the regulation of cell growth and apoptosis. JNKs are transcription factors that regulate pro-apoptotic gene expression and activate apoptotic signaling pathways. There is evidence that reactive oxygen species (ROS) can lead to apoptosis through JNK phosphorylation. ERK can promote apoptosis through multiple mechanisms, which include activation of caspase-8 and stimulation of cytochrome c release.

In conclusion, fluorosis may increase spermatogenic cell apoptosis by affecting the regulation of apoptosis-related genes in rats, which thereby affecting testicular reproductive function.

Effect of Fluorosis on Oxidative Stress in Testicular Tissue

Under normal physiological conditions, the production and clearance of reactive oxygen species in the male reproductive system are maintained in a balanced state; under pathological conditions, if the production of reactive oxygen species exceeds a critical level, it will cause a state of oxidative stress and damage spermatogenic cells resulting in impaired sperm production and maturation, which leads to infertility. Studies at home and abroad suggest that oxidative stress injury is one of the main mechanisms of fluoride toxicity to the body. Fluorosis can cause oxidative stress damage in male reproductive organs such as testis, prostate, seminal vesicle, and epididymis. Coal-burning type of endemic fluorosis can lead to imbalance of oxidative system and antioxidant system in rat testis, and damage of germ cells. Oxidative stress injury may play an important role in reproductive toxicity of male rats induced by coal-burning type of endemic fluorosis.

The mechanism of the effect of fluorosis on the testis is not clear at present. Some scholars have proposed the oxidative stress theory and concluded that oxidative stress is an important link in the mechanism of fluorosis. Ghosh et al. (2002) studied the oxidative stress effect of drinking-water type of fluorosis on testicular tissue and found that the toxic effect of reproductive system caused by excessive fluoride is closely related to testicular oxidative stress. SOD is an important enzyme antioxidant, one of the important substances to remove reactive oxygen species, which plays a very important role in the oxidative and antioxidant balance of the body and can protect the body from oxidative damage, known as the first line of defense of the antioxidant system in the body, and the activity of SOD can reflect the antioxidant capacity of tissues. Our group found that the SOD activity in each fluoride-exposed group was significantly lower than that in the control group in the testicular tissue of rats with coal-burning type of endemic fluorosis, which was similar to the results of Liang et al. 's study on the animal model of fluoride in drinking water (Liang et al. 2001), and the SOD activity in each fluoride-exposed group showed a decreasing trend with the increase of fluoride dose, suggesting that coal-burning type of endemic fluorosis can indeed lead to a decrease in the antioxidant capacity of the testis and is closely related to the fluoride dose. Malondialdehyde (MDA) is the product of lipid peroxide production, and its content can indirectly reflect the level of free radicals and the degree of lipid peroxidation in testicular tissue, but also reflect the degree of damage to testicular germ cells. The research group found that in coal-burning type of endemic fluorosis, the MDA content in rat testicular tissue was significantly increased compared with the control group, and with the increase of fluoride dose on the rise. The results obtained by adding sodium fluoride to drinking water were similar, and it was speculated that coal-burning type of endemic fluorosis could increase the amount of peroxidation products in the testis and damage testicular germ cells. Nitric oxide synthase (NOS) is divided into three subtypes: inducible nitric oxide synthase (iNOS), neuronal nitric oxide synthase (nNOS), and endothelial nitric oxide synthase (eNOS). The ability of testicular tissue to synthesize NO can be reflected by measuring total NOS and iNOS activities. NO is one of the biologically active free radicals closely related to oxygen free radicals. The study found that the NOS activity and iNOS activity in the testicular tissue of rats in each fluorosis group were significantly increased compared with the control group, suggesting that the synthesis ability of NO in the testicular tissue of rats with coal-burning type of endemic fluorosis was enhanced, further indicating that the oxidative stress effect in the testicular tissue of rats with coal-burning type of endemic fluorosis was enhanced (Xiao et al. 2015). Fluoride exposure dose was positively correlated with testicular fluoride content, degree of testicular histopathological damage and degree of increased oxidative activity, but negatively correlated with the decrease of antioxidant capacity, indicating that coal-burning type of endemic fluorosis can cause imbalance between oxidative system and antioxidant system in rat testicular tissue and damage the structure of testicular germ cells.

Effect of Fluorosis on Testicular Cellular Immune Privilege/Autophagy

Autophagy plays a crucial role in cell differentiation, development, homeostasis, cell survival, and death. Regulation of autophagy leads to several pathological processes, including cancer, infection, muscle, and degenerative diseases. Fluoride has been reported to impair the autophagic degradation of rat SC_S, and autophagy may play a key role in the protective response to fluoride-induced oxidative damage (Feng et al. 2019).

Previous studies have demonstrated detrimental effects of fluoride on SC_S, such as oxidative stress and apoptosis, however, the effect of fluoride on Sertoli cell immune privilege is unknown. Different concentrations of sodium fluoride were used to treat mouse SCs with fluoride in vitro to establish a fluoride-treated SCs model. After 48 h of incubation, F-SCs were transplanted under the renal capsule of mice for 21 day or co-cultured with splenic lymphocytes for 48 h. GATA binding protein 4 (GATA4) immunohistochemical analysis of renal subcapsular SCs grafts showed that SCs were less distributed and immune cell infiltration was evident in the F-SCs group. In addition, the levels of Fas ligand (Fas-L) protein and mRNA in non-cocultured F-SCs decreased with increasing fluoride concentration. When co-cultured with F-SCs, the lymphocyte viability was significantly increased and the apoptosis rate was lower in the F-SCs group. FasL protein and mRNA expression was decreased, lymphocyte Fas expression was decreased, and caspase 8 and caspase3mrna expression was also decreased in a dose-dependent manner after co-culture of F-SCs. These results indicate that fluoride affects testicular immune privilege by interfering with the Fas/FasL system. The role of autophagy in fluoride-induced male reproductive toxicity was investigated in vivo and in vitro Leydig cell models. Using transmission electron microscopy, the testicular tissue of fluoride-exposed mice was observed and an increase in the number of autophagy in Leydig cells was found. Further studies showed that fluoride increased the mRNA and protein expression levels of autophagy markers light chain 3, Beclin 1, and autophagy-related protein in primary interstitial cells. The finding highlights the role of autophagy in fluoride-induced testicular and Leydig cell damage (Zhang et al. 2017).

Effects of Fluorosis on Epididymis and Vas Deferens

The vas deferens includes the epididymis, vas deferens, ejaculatory duct, and urethra common for urination. Among them, epididymis has the function of transporting and storing sperm, and is also an important site of sperm maturation, and makes sperm maturation acquire the function of motility and fertilization ability. The epididymis has two barrier effects: the immune barrier and the blood–testis barrier. Studies have confirmed that fluoride ingested by the body can enter the epididymis, resulting in increased fluoride content in the epididymis; fluoride can have damaging effects on the epididymis and vas deferens.

A test of fluoride exposure in mature male rats revealed that the weights of the cauda epididymis and vas deferens were significantly increased (after 30 days of the experiment); the sperm motility of the cauda epididymis was significantly reduced from 15 to 30 days of the experiment; the sperm count of the cauda epididymis was significantly reduced, especially after 30 days of the experiment; and the fertility of the experimental group was significantly reduced compared with the control group. Degeneration occurred in the caudal tubules of the epididymis, vacuolation and nuclear pyknosis occurred in the epithelial cells of the epididymal duct; the lamina propria of the vas deferens endings was thickened, and nuclear pyknosis occurred in the epithelium; the sperm showed "sickle" malformation changes under the microscope. Fluorosis can reduce the protein content of cauda epididymis tissue, and fluoride ions may be related to the synthesis of proteins.

Sodium fluoride significantly decreased the activities of SOD and catalase (CAT) in the epididymis. The mRNA expression of both SOD and CAT in the epididymis was significantly decreased in the fluoride-exposed group. The findings suggest that the epididymis also loses the balance between oxidative stress and antioxidant defense in the presence of fluoride, and that the epididymis is more sensitive to fluoride than the testis (Sun et al. 2018).

Effects of Fluorosis on the Prostate and Seminal Vesicles

Human accessory gonads include seminal vesicle, prostate and urethral bulb glands, among which the prostatic fluid secreted by the prostate is the main component in semen (Chen et al. 2018). There are relatively few studies on fluorosis on the prostate and seminal vesicles. Studies have found that fluorosis could cause different degrees of damage to the epithelial cells and stromal tissues of the rat prostate. Under the electron microscope, the microvilli decreased, mitochondria swelled, cristae disappeared, nucleus pyknosis, rough endoplasmic reticulum decreased and Golgi apparatus degenerated were found in prostatic acinar epithelial cells of the fluoride-exposed group; collagen fibers in the interstitium were released and smooth muscle showed different degrees of degeneration; fluorosis could cause a decrease in the mass of rat prostate and a significant decrease in the content of sialic acid in the prostate. Fluorosis could decrease the weight of seminal vesicle and increase the level of fructose in seminal vesicle. Fluorosis can cause a decrease in the protein content of the prostate.

Effects of Fluorosis on Male Reproductive Endocrine Levels

Effects of Fluorosis on Sex Hormone Levels

The entire male reproductive activity is a regular, sequential, and coordinated physiological process, and male reproductive activity needs to be performed under the control and regulation of the body's neuroendocrine glands, and abnormal reproductive endocrine levels can affect male reproductive function (Chen et al. 2018). Coal-burning type of endemic fluorosis can affect the changes of hormones in the hypothalamic-pituitary-testicular axis, thereby interfering with reproductive endocrine function in male rats. In addition, reducing the intake of coal-baked food and improving nutritional status can reduce the condition of fluorosis in rats and reduce the reproductive endocrine effects in male rats.

Testosterone in serum is mostly generated by Leydig cells and is divided into two forms in the blood, one in the conjugated form, accounting for 98%, and the other in the free form, accounting for only 2%. Some researchers treated Leydig cells with fluoride and found that fluoride-induced changes included decreased cell proliferation, steroidogenesis, and related gene expression, and the ability of cells treated with fluoride to secrete testosterone was significantly reduced (Yu et al. 2018). The non-intervention group (unmodified stove village), intervention group (modified stove village), and control group (healthy adults) in coal-burning type of endemic fluorosis wards in Guizhou province were randomly sampled to collect peripheral venous blood for sex hormone levels. The study found that there was significant difference in serum testosterone level between coal-burning intervention group and non-intervention control group, and serum testosterone in non-intervention group was significantly lower than that in control group; there was no significant difference in serum free testosterone level between non-intervention group and control group; 3 there was significant difference in serum luteinizing hormone (LH) level, and LH level in non-intervention group was higher than that in intervention control group; 3 there was significant difference in serum follicle stimulating hormone (FSH) level, and serum FSH level in non-intervention group was higher than that in control group; there was no significant difference in serum FSH level between non-intervention group, intervention group, and control group.

Fluoride can affect male reproduction by affecting the production of steroid hormones. Fluoride also suppressed the expression levels of genes and proteins critical for steroidogenesis (Orta Yilmaz et al. 2018). Fluoride down-regulates transcription factors involved in the development of infertility by decreasing testosterone and cyclic adenosine monophosphate levels.

The concentration of testosterone in serum and testis of fluoride-exposed rats was lower than that of the control group, especially with the prolongation of fluoride-exposure time and the increase of fluoride-exposure dose. At the same time, the concentration of testosterone in serum and testis could increase with the increase of nutrition. Therefore, it is speculated that coal-burning type of endemic fluorosis may directly destroy the structure of the testis, resulting in a decrease in the ability of interstitial cells to synthesize and secrete testosterone, and may also interfere with the regulatory effect of the hypothalamic-pituitary-gonadal axis, indirectly causing a decrease in testosterone levels inside and outside the testis, and increasing nutrition can improve the effect of fluoride on testosterone.

Estradiol in males mainly comes from the peripheral metabolic process of testosterone, a small amount is secreted by SC_S, and the serum estradiol concentration level is decreased in each exposure group, which may be due to the decrease of testosterone, aromatization is converted to estradiol and then decreased. On the other hand, coal fluorosis directly damages the structure and function of SC_S and reduces the function of SC_S to secrete E2.

The effect of fluoride on LH and FSH in adult males remains controversial. Some scholars have found that the serum LH and FSH of adult males in fluorosis areas are significantly increased; however, some scholars believe that the serum testosterone and LH concentrations of adult males in high fluoride areas are not significantly changed compared with those in control areas. Most reports indicated that LH and FSH concentrations were significantly increased in the fluoride-exposed group compared with the control group (Cheng et al. 2004). Experimental findings of fluorosis caused by coal-burning pollution. The mechanism may concern that LH is positively regulated by gonadotropin-releasing hormone (GnRH) levels and negatively regulated by gonadal hormones testosterone and estradiol 2, involving (1) fluoride is an activator of G protein and can promote the synthesis and release of LH; (2) the long-term effect of high fluoride reduces the GnRH receptor of the pituitary gland and affects the normal conduction of GnRH stimulated signals; (3) the decrease of serum testosterone concentration negatively feeds on the pituitary gland and enhances the ability of the adenohypophysis to synthesize and secrete LH, thereby increasing the control serum LH. The reason for the increase in FSH is unknown, and under physiological conditions, inhibin produced by SC_S negatively feeds back to regulate FSH synthesis and secretion in the pituitary gland. The increase of FSH may be due to the fact that fluorosis reduces the ability of SC_S to produce inhibin, the inhibin level decreases, and negative feedback causes an increase in FSH, and the specific mechanism needs to be confirmed by further experiments.

GnRH secreted by the hypothalamus plays an important role in the regulation of reproductive endocrine. After feeding sodium fluoride in drinking water for 8 weeks,

GnRH levels were found to be increased compared with the control group (Ma et al. 2008), and GnRH levels were decreased in each exposure group compared with the control group, but increased with the fluoride dose in the exposure group (Sun et al. 2011). It may be because chronic fluorosis not only indirectly affects the negative feedback regulation of GnRH by sex hormones, but also directly affects the structure and function of the hypothalamus, so that GnRH synthesis and secretion are affected, and the mechanism needs to be further studied.

Decreased mass of testicular and accessory gonadal organs caused by fluorosis may be related to fluoride-induced changes in sex hormone levels.

Effect of Fluoride on Serum Inhibin b (Inhb) in Adult Men

The secretion of InhB is mainly derived from SC_S, and the level of InhB in male serum is significantly positively correlated with its sperm concentration in semen, which will be used as one of the indicators for direct evaluation of testicular spermatogenic function. When FSH is increased due to abnormal causes in men, normal SC_S will increase InhB secretion, which negatively feeds back on the pituitary gland to reduce FSH excess secretion. Clinically, sex hormone levels can be detected as normal or decreased FSH, while InhB is significantly increased, which indicates that InhB plays a negative feedback role on FSH. Different doses of fluoride exposure can lead to different degrees of decrease in serum InhB levels and a significant increase in FSH in adult males. It was demonstrated that serum InhB was significantly lower in the high-fluoride group than in the low-fluoride group, and that the fluoride contact index was negatively correlated with InhB levels; there was a good correlation between InhB levels and sperm concentration (Ortiz-Pérez et al. 2003). A survey in our group relating adult males in the area of coal-burning type of endemic fluorosis Guizhou province, China, revealed that the serum InhB level in the non-intervention group was lower than that in the intervention control group.

Therapeutic Agents and Prognosis

Fluoride damage to the male reproductive system has been confirmed, but whether the damage caused by fluoride to the male reproductive system is temporary or persistent permanent damage has received much attention from the scientific community. It has been found that whether the damage to the male reproductive system by fluorosis is reversible is closely related to the fluoride dose, and fluoride causes irrecoverable arrest during the haploid stage of spermatogenesis (Han et al. 2019). Fluoride had significant effects on testicular histomorphology, sperm acrosome, and chromosome ultrastructure in rats; 2 weeks after stopping fluoride intervention, it was found that some reproductive function indicators of rats in the low-dose and medium-dose fluorosis groups were significantly recovered, while the relevant reproductive function indicators of rats in the high-dose fluorosis group were not significantly recovered, suggesting that male reproductive function damage caused by severe fluorosis may be difficult to recover and reverse rapidly in the short term. At present, some drugs have been found to improve male reproductive system damage caused by fluorosis, as follows:

Testosterone

Testosterone has a protective effect on reproductive system damage caused by fluorosis. The findings (Ghosh et al. 2002), fluorosis can produce oxidative stress damage to the testis, prostate, and epididymis, causing a significant increase in the levels of conjugated dienes and malondialdehyde in rat testis, prostate, and epididymis tissues; the levels of conjugated dienes and malondialdehyde in rat testis, prostate, epididymis, and other organ tissues were decreased after treatment with testosterone and vitamin E (Vit E); there was no significant difference between the therapeutic effects of testosterone and Vit E. These results suggest that testosterone can reduce the oxidative stress injury caused by fluorosis on sexual organs and has certain clinical significance in protecting the oxidative stress injury of reproductive organs caused by fluorosis. Some experts predict that testosterone is expected to be used in the treatment of patients in fluorosis wards in the future.

Vitamin E and Lycopene

It has been shown that fluorosis induces oxidative stress, which leads to impairment of cellular redox homeostasis and lipid peroxidation. Vit E and lycopene are well-established antioxidants and present good protection under fluoride exposure. Relevant studies further confirmed the antioxidant effects of Vit E and lycopene, and found that they alleviated fluorosis-induced oxidative stress and apoptosis by inhibiting JNK and ERK phosphorylation (Tian et al. 2018). The related research results provide a new idea for the treatment of male reproductive system injury caused by fluorosis.

Choline

Choline is an important nutrient that has been extensively studied for its role in the nervous system. Choline has been found to reduce fluoride damage to the male reproductive system. Male mice were given 150 mg/l sodium fluoride, 5.75 g/kg choline, and both from pregnancy to 15 weeks postpartum. The results showed that fluoride exposure decreased body weight gain, decreased sperm count and survival rate, changed testicular histology, down-regulated the mRNA expression of *NGF*, *Ras*, *Raf*, and *MEK* genes in the testis, and significantly decreased the expression of NGF and MEK proteins in the testis. Examination of the data from choline-treated mice revealed that the choline-treated group ameliorated the relevant changes induced by fluoride. Choline has been suggested to reduce testicular toxicity due to fluorosis (Zhang et al. 2016).

Taurine

Fluorosis can affect the function of the pituitary-testicular gonadal axis, resulting in abnormal levels of sex hormones. Taurine could significantly inhibit the increase of oxidative stress indicators induced by fluoride and enhance the antioxidant enzyme activity and glutathione level in rat brain, testis, and epididymis; in addition, taurine can reverse fluorine-induced elevation of inflammatory markers and caspase-3 activity, and improve the tissue structure of brain, testis, and epididymis in fluoride-treated rats. Taurine significantly reversed the fluoride-induced decrease in testicular testosterone levels and acid phosphatase, alkaline phosphatase, and lactate dehydrogenase activities in rats, accompanied by an enhancement of sperm function. Therefore, taurine may be a protective agent against male reproductive dysfunction caused by fluoride exposure (Adedara et al. 2017).

Extracts of Ginseng and Panama Leaf

Some scholars have established a fluorosis diabetic mouse model, and when it was confirmed that fluorosis diabetic mice showed hyperglycemia and fluoride accumulation, the mice was treated with ginseng extract and Panama leaf extract. Studies have found that ginseng extract and Panama leaf extract significantly improve sperm density, motility, morphology, and testicular biochemical indicators in fluorosis mice after intervention (Sm and Mahaboob 2017).

Insulin-like Growth Factor-I (IGF-I) and Human Chorionic Gonadotropin (hCG)

Some scholars have found that insulin-like growth factors-1 and hCG have protective effects on testicular Leydig cell function (Yu et al. 2018).

Damage to the Female Reproductive System Due to Coal-Burning Type of Endemic Fluorosis

The female reproductive system is very sensitive to fluoride exposure. Fluorosis can damage the structure and function of the ovaries and uterus and affect their fertility. Epidemiological investigations have found that long-term severe fluoride exposure has a potential link with female primary infertility, and fluoride is an important risk factor for female infertility (Shi et al. 2012). The latest animal experimental studies have found that fluorosis decreases the conception rate and pregnancy rate of female rats in a dose-dependent manner. We will analyze the effects of fluorosis on the female reproductive system from the following aspects.

Effects of Fluorosis on the Uterus

Fluorosis can damage the uterus. Animal experiments revealed that the shape and size of uterine vessels changed, endometrial glands became hypertrophic, and endometrial cells became larger in fluoride-exposed female rats.

Effects of Fluorosis on the Ovary

Fluorosis caused by coal-burning pollution is capable of inducing toxic effects on the ovaries of female rats. In the experiment of simulating coal-burning type of endemic fluorosis animals, microscopic observation of ovaries in the fluoride-exposed group showed: progressive edema of ovarian granulosa cells, blurred cell contour and structure, and increased degeneration of atretic follicles and corpora lutea. Coal-burning type of endemic fluorosis have toxic effects on the ovaries of female rats, and the mechanism may be that coal-burning type of endemic fluorosis accelerates the degeneration and atresia of follicles at all levels, causes follicular granulosa cell edema, follicular cell damage, affects the synthesis and secretion of estradiol, reduces the ovulatory function of the ovaries, and ultimately affects reproductive function.

Fluorosis is capable of destroying the structure of ovarian tissue and apoptosis exists in ovarian granulosa cells. Light microscopy revealed that the granulosa cells and interstitial cells of the ovary in the fluoride-exposed group were significantly edematous, the cell outline and structure were blurred, the number of mature follicles and corpora lutea was reduced, and the degenerated follicles were increased. Electron microscopy revealed that the nuclei of ovarian granulosa cells in fluoride-exposed female rats were reduced, the nuclear membrane was irregularly zoned, and the nuclear membrane space was widened; the nucleoli became smaller or disappeared, the chromatin accumulated into lumps, a large number of lipid droplets appeared in the cytoplasm, and mitochondria were widely deformed (Chen et al. 2015).

The number of large follicles, medium follicles, and small follicles in fluoride-exposed female rats was significantly lower than that in the non-fluoride-exposed group, and was closely related to the fluoride-exposed dose.

Effects of Fluorosis on Ovarian Granulosa Cells

Granulosa cells play an important role in the female reproductive system, are the main functional cells that synthesize and secrete estrogen, progesterone, and inhibin, and play an important role in the development of follicles and embryos. Studies have confirmed that fluorosis can cause apoptosis of ovarian granulosa cells, combined with abnormal sex hormone secretion. Coal-burning type of endemic fluorosis can destroy the ultrastructure of ovarian tissue cells, destroy the mitochondria of granulosa cells, affect the energy supply of ovarian cells, lead to apoptosis of granulosa cells in the ovary, and ultimately affect the normal function of the ovary.

Effects of Fluorosis on the Placenta and Fetus

Fluoride may eventually lead to stillbirth, fetal dysplasia, and even fetal malformations by disrupting the structure and function of the maternal placenta. Fluorosis can affect the placenta, and animal experimental studies have found that the number of live fetuses in female mice exposed to fluoride is reduced and the placental mass is increased.

Effects of Fluorosis on Female Sex Hormone Levels

Fluorosis can lead to disturbances in female reproductive endocrine levels. Population surveys in fluorosis wards have found that fluoride exposure can cause a decrease in serum estradiol levels, and estradiol levels are negatively correlated with age in the high fluoride group (Zhou et al. 2013). Animal experiments also revealed that the levels of sex hormones such as estradiol, progesterone, FSH, LH, and testosterone were lower in female rats exposed to fluoride than in the non-exposed group. However, some studies on animal experiments are inconsistent (Chen et al. 2015), and scholars all over the world need to further expand the sample size, remove interfering factors and conduct further in-depth study in the future.

Oxidative Stress Damage to the Female Reproductive System by Fluorosis

Oxidative stress is also one of the mechanisms of female reproductive system damage caused by coal-burning type of endemic fluorosis. Sodium fluoride can induce oxidative stress and promote the development of inflammatory response in rat uterine tissue, and MAPK signaling pathways may be involved. Fluorosis can cause the imbalance of oxidation-antioxidant system in rat uterine tissue, and excessive fluoride damages rat endometrial tissue by activating inflammatory response through oxidative stress. ROS and antioxidant enzymes superoxide dismutase (SOD), catalase, and glutathione were significantly higher in rat uterine tissue. Sodium fluoride treatment leads to the production of a large number of peroxides in rat uterine tissue, and at the same time, it also stimulates the body to produce a large number of antioxidant substances, breaking the balance between the body's oxidative system and antioxidant system. The phosphorylation efficiency of p38 and c-Jun N-terminal kinase was significantly higher in the sodium fluoride treatment group, and the expression and phosphorylation levels of ERK and Phosphatidylinositol-3-kinase (PI3K) were also increased, suggesting that MAPK and PI3K signaling pathways may play an important role in the oxidative stress-induced inflammatory response in uterine tissues induced by sodium fluoride. Studies have confirmed that fluorosis can cause oxidative stress in female rats, activate ERK and JNK signaling pathways in female rats, cause oxidative stress in female rats, induce granulosa cell apoptosis, reduce sex hormone levels and the synthesis and secretion of sex hormones, and finally lead to a decrease in pregnancy rate (Liao et al. 2015).

Conclusion

Coal-burning type of endemic fluorosis can damage both male and female reproductive systems, ultimately affecting human fertility. In recent years, with the decline of overall human fertility all over the world, especially the decline of overall male sperm quality, the impact of fluoride on the male system has renewed interest. Fluoride exposure significantly alters the body's expression of genes related to stress response, detoxification, antibacterial, transport, oxidative phosphorylation, and ribosomes (Tang et al. 2018). However, the specific mechanism of reproductive system damage caused by coal-burning type of endemic fluorosis is not fully understood and needs further in-depth study, which is expected to provide a theoretical basis for the prevention and control of coal-burning type of endemic fluorosis.

References

Adedara IA, Olabiyi BF, Ojuade TD, Idris UF, Onibiyo EM, Farombi EO. Taurine reverses sodium fluoride-mediated increase in inflammation, caspase-3 activity, and oxidative damage along the brain-pituitary-gonadal axis in male rats. Can J Physiol Pharmacol. 2017;95:1019–29.

Berruti G. Towards defining an 'origin'-the case for the mammalian acrosome. Semin Cell Dev Biol. 2016;59:46–53.

Chen XH, Xia SH, Yu SW, Xu JF, Yu MJ, Wang C. Effects of coal-burning type of fluorosis on hypothalamic-pituitary-ovary axis function in female rats. Chin J Endem. 2015;34:93–7.

Chen XP, Wang JP, Zhao JZ. Surgery. 9th ed. Beijing: People's Medical Publishing House; 2018.

Cheng PZ, Yun ZJ, Sun GD, Wang YT, Gao HX, Ma AH, et al. Effects of fluoride on secretion of hormone. Endem Dis Bull. 2004;19:14–6.

Feng Z, Liang C, Manthari RK, Wang C, Zhang J. Effects of fluoride on autophagy in mouse sertoli cells. Biol Trace Elem Res. 2019;187:499–505.

Gang YD, Wang JL, Wang PG, Zhang BZ, Wang YL, Li J. Impact of fluorine poisoning on sperm motility of male rats. Chin J Endem. 2010;29:139–41.

Ghosh D, Das S, Maiti R, Jana D, Das UB. Testicular toxicity in sodium fluoride treated rats: association with oxidative stress. Reprod Toxicol. 2002;16:385–90.

Gupta RS, Khan TI, Agrawal D, Kachhawa JB. The toxic effects of sodium fluoride on the reproductive system of male rats. Toxicol Ind Health. 2007;23:507–13.

Han Y, Yu Y, Liang C, Shi Y, Zhu Y, Zheng H, et al. Fluoride-induced unrestored arrest during haploid period of spermatogenesis via the regulation of DDX25 in rats. Environ Pollut. 2019;253:538–51.

Jiang S, Liang C, Gao Y, Liu Y, Han Y, Wang J, et al. Fluoride exposure arrests the acrosome formation during spermatogenesis via down-regulated Zpbp1, Spaca1 and Dpy19l2 expression in rat testes. Chemosphere. 2019;226.874–82.

Kulbir K. Histological findings of mice testes following fluoride ingestion. Fluoride. 1980;13:160.

Liang G, Sun GF, Li YY, Dai GJ. Effects of fluoride on lipid peroxidation and the capacity of anti-oxidation of rats. Chin J Endem. 2001;20:109–10.

Liao XG, Li YL, Qiu YW, Zhou YJ, Geng YQ, Liu XQ, et al. Sodium fluoride activates P38 and JNK via induction of oxidative stress to promote uterine inflammation in rats. Acta Acad Med Militaris Tert. 2015;37:1744–9.

Liu J, Cui H, Peng X, Fang J, Zuo Z, Wang H, et al. Dietary high fluorine induces apoptosis and alters Bcl-2, Bax, and caspase-3 protein expression in the cecal tonsil lymphocytes of broilers. Biol Trace Elem Res. 2013;152:25–30.

Ma XY, Cheng XM, Li FR, Zhang JS, Li Y, Ba Y, et al. Experimental research on endocrine disturbing effect of fluorine on hypothalamus-hypophysis-testis axis in male rats. J Hyg Res. 2008;37:733–5.

Messer HH, Armstrong WD, Singer L. Influence of fluoride intake on reproduction in mice. J Nutr. 1973;103:1319–26.

Mikhailov V, Mikhailova M, Pulkrabek DJ. Bcl-2 prevents Bax oligomerization in the mitochondrial outer membrane. J Biol Chem. 2001;276:18361–74.

Orta Yilmaz B, Korkut A, Erkan M. Sodium fluoride disrupts testosterone biosynthesis by affecting the steroidogenic pathway in TM3 Leydig cells. Chemosphere. 2018;212:447–55.

Ortiz-Pérez D, Rodrlguea-Martinez M, Martinez F, Borja-Aburto VH, Castelo J, Grimaldo JI, et al. Fluoride-induced disruption of reproductive hormones in men. Environ Res. 2003;93:20–30.

Shi XQ, Liu D, Tan LC. A case of male infertility in high fluorine area. Chin J Endem. 2012;31:462.

Sm S, Mahaboob BP. Fluoride exposure aggravates the testicular damage and sperm quality in diabetic mice: protective role of Ginseng and Banaba. Biol Trace Elem Res. 2017;177:331–44.

Sun F, Li CB, Xiao YH, Shi JQ, Gu J, Xie C, et al. Effects of fluorosis of coal burning on reproductive and endocrinology of male rats. Basic Clin Med. 2011;31:1077–81.

Sun Z, Li S, Guo Z, Li R, Wang J, Niu R, et al. Effects of fluoride on SOD and CAT in testis and epididymis of mice. Biol Trace Elem Res. 2018;184:148–53.

Tang W, Xiao Y, Li G, Zheng X, Yin Y, Wang L, et al. Analysis of digital gene expression profiling in the gonad of male silkworms (Bombyx mori) under fluoride stress. Ecotoxicol Environ Saf. 2018;153:127–34.

Tian Y, Xiao Y, Wang B, Sun C, Tang K, Sun F. Vitamin E and lycopene reduce coal burning fluorosis-induced spermatogenic cell apoptosis via oxidative stress-mediated JNK and ERK signaling pathways. Biosci Rep. 2018;38:1003.

Wang C, Chen Y, Manthari RK, Wang J. Abnormal spermatogenesis following sodium fluoride exposure is associated with the downregulation of CREM and ACT in the mouse testis. Toxicol Ind Health. 2018;34:219–27.

Xiao YH, Tian Y, Sun F, Li CB, Tang KF, Sun C, et al. Role of excessive fluorine on protein expression of the related apoptosis gene in rat testis. Chin J Endem. 2015;34:254–9.

Yu Y, Han Y, Niu R, Wang J, Manthari RK, Ommati MM, et al. Ameliorative Effect of VE, IGF-I, and hCG on the fluoride-induced testosterone release suppression in mice leydig cells. Biol Trace Elem Res. 2018;181:95–103.

Zahrorohkor A, Tarinsky JK. Fluorosis: geographical pathology and some experimental findings. Fluoride. 1981;14:182.

Zhang J, Zhang Y, Liang C, Wang N, Zheng H, Wang J. Choline supplementation alleviates fluoride-induced testicular toxicity by restoring the NGF and MEK expression in mice. Toxicol Appl Pharmacol. 2016;310:205–14.

Zhang J, Zhu Y, Shi Y, Han Y, Liang C, Feng Z, et al. Fluoride-induced autophagy via the regulation of phosphorylation of mammalian targets of rapamycin in mice leydig cells. J Agric Food Chem. 2017;65:8966–76.

Zhao WP, Wang HW, Liu J, Tan PP, Luo XL, Zhu SQ, et al. Positive PCNA and Ki-67 expression in the testis correlates with spermatogenesis dysfunction in fluoride-treated rats. Biol Trace Elem Res. 2018;186:489–97.

Zhou YJ, Zhang HL, He JL, Chen X, Ding Y, Wang Y, et al. Effects of sodium fluoride on reproductive function in female rats. Food Chem Toxicol. 2013;56:297–303.

Part III
Molecular Pathogenesis of Endemic (Chronic) Fluorosis

Chapter 12
Enhanced Oxidative Stress

Long-Yan Ran and Zhi-Zhong Guan

Abstract Accumulating and widely accepted evidence indicates that enhanced oxidative stress is an important cause of the systemic damage induced by endemic (chronic) fluorosis. Extensive investigations on patients with endemic fluorosis, animal models of this disease and cultured cells exposed to high levels of fluoride all show that after a large amount of this ion is deposited in the bones, teeth and soft tissues, the levels of free radicals are elevated, which leads to lipid peroxidation (resulting in production of malondialdehyde), along with carbonylation of proteins and formation of 8-hydroxy-deoxyguanine in DNA. At the same time, chronic fluorosis reduces both the activities of antioxidant enzymes such as superoxide dismutase, glutathione peroxidase and catalase, and the levels of antioxidant substances such as reduced glutathione and coenzyme Q. In addition, chronic fluorosis often damages mitochondria, thereby producing large amounts of reactive oxygen species. This enhanced oxidative stress damages cells and leads to apoptosis, which, of course, impairs the functioning of tissues and organs. Importantly, antioxidant compounds and similar substances used to treat chronic fluorosis clinically have already given positive results. Hopefully, our constantly improving knowledge of the mechanisms underlying this disease will soon lead to the development of effective treatment.

L.-Y. Ran
The Key Laboratory of Endemic and Ethnic Diseases (Guizhou Medical University), Ministry of Education, Guiyang, Guizhou, People's Republic of China

Department of Pathology at the Affiliated Hospital of Guizhou Medical University, Guiyang, Guizhou, People's Republic of China

Z.-Z. Guan (✉)
The Key Laboratory of Endemic and Ethnic Diseases (Guizhou Medical University), Ministry of Education, Guiyang, Guizhou, People's Republic of China

Department of Pathology at the Affiliated Hospital of Guizhou Medical University, Guiyang, Guizhou, People's Republic of China

The Provincial Key Laboratory of Medical Molecular Biology, Guiyang, Guizhou, People's Republic of China
e-mail: zzg@gmc.edu.cn

Z.-Z. Guan (ed.), *Coal-burning Type of Endemic Fluorosis*,
https://doi.org/10.1007/978-981-16-1498-9_12

Keywords Antioxidant · Endemic fluorosis · Free radicals · Mechanism · Oxidative stress

Introduction

Endemic (chronic) fluorosis can cause systemically pathological damages to a multitude of tissues and organs. Understanding the pathogenesis of this complex disease is a key to developing effectively clinical treatments. Over the years, numerous epidemiological and experimental investigations have attempted to unravel this complexity and it has been proposed that changes in proteins or nucleic acids, enzyme activities, collagen, calcium balance, etc. are involved. However, none of these mechanisms can explain the systemic damages due to chronic fluorosis.

Our own early studies revealed elevated levels of lipid peroxide and reduced activities of antioxidant enzymes both in the serum and a variety of organs of animals with chronic fluorosis and in the serum of patients suffering from endemic fluorosis due to burning coal (Guan et al. 1989, 1991, 1992; Guan and Yang 1991). Based on such findings, we first proposed that increased levels of free radicals may be the central link of the different types of damage associated with chronic fluorosis (Guan et al. 1989).

During the nearly 30 years that have since been passed, many other researchers have provided additional evidence for oxidative stress being the underlying mechanism, which may help design effective drug treatment of this condition. Here, we focus on changes in the balance between oxidative and antioxidant systems in connection with the development of chronic fluorosis, including the relationships between oxidative stress and multiple metabolic pathways and studies on experimental and clinical antioxidant therapy.

Elevated Level of Oxidative Stress Induced by Chronic Fluorosis

Free Radicals and Reactive Oxygen Species (ROS)

The term free radicals refer to atoms or chemical moieties with unpaired electrons formed in connection with the splitting of a covalent bond in response to external stimuli, such as hydrogen radicals ($H^{\bullet}$), chlorine radicals ($Cl^{\bullet}$), methyl radicals ($CH3^{\bullet}$), and so on. In biological systems the most commonly occurring are oxygen free radicals, such as superoxide anion ($O^{-\bullet}$), hydroxyl radical (•OH), lipid oxygen radicals ($RO^{\bullet}$), peroxyl radical ($ROO^{\bullet}$), nitric oxide ($NO^{\bullet}$), hydrogen peroxide (H_2O_2), and singlet oxygen ($^1O_2$), referred to collectively as reactive oxygen species or ROS (Schieber and Chandel 2014). Some free radicals are involved in normal metabolism and others are bactericidal, but excessive amounts of free radicals can destroy biological structures and metabolic processes, resulting in damage and dysfunction.

Antioxidant Systems

A number of antioxidant systems, both enzymatic and non-enzymatic, can effectively inhibit the oxidation of cellular components by free radicals, either by acting directly on these radicals or by preventing their production. The main antioxidant enzymes include superoxide dismutase (SOD), glutathione peroxidase (GSH-px), catalase (CAT), and peroxiredoxin (Prx). In addition, antioxidant substances include primarily reduced glutathione (GSH), vitamin E, vitamin C, bioflavonoids, selenium, carotene, and coenzyme Q. Previously believed to function solely as an electron transmitter in the mitochondrial respiratory chain, Coenzyme Q is now believed to be a potent endogenous antioxidant (Guan et al. 2000).

Imbalance Between Oxidative and Antioxidant Systems

In biological systems free radicals are produced continuously and under normal circumstances, cells can maintain a balance between oxidative and antioxidant systems. However, when large quantities of free radicals are produced suddenly in response to endogenous and/or exogenous stimuli, the scavenging capacity of antioxidant systems may be exceeded, resulting in oxidative stress. It is important to remember that oxidative stress not only can damage tissues and cells, but also plays an important regulatory role in cellular communication and homeostasis of the body, e.g., in coordination of cell proliferation, differentiation and apoptosis through the regulation of gene expression or related transcription factors (Görlach et al. 2015).

In a state of oxidative stress, free radicals, and in particular ROS, can attack and oxidize lipids, proteins, and DNA, leading to their crosslinking, breakdown, and/or conversion into metabolites such as the MDA and 4-hydroxynonenal (4-HNE) arising from lipid peroxidation, carbonyl groups on proteins, and 8-hydroxy-2-′-deoxyguanosine (8-OHdG) in DNA. Thus, oxidative stress may play a major role in the pathogenesis of a variety of diseases, such as cardiovascular diseases, neurodegenerative diseases, aging, adiposity, atherosclerosis, type II diabetes, and cancer (Görlach et al. 2015).

Chronic Fluorosis is Associated with an Elevated Level of Oxidative Stress

Fluorine, a relatively reactive element, causes bursts in cellular respiration and reactions that can give rise to free radicals. Excessive fluoride may attack oxygen and/or interfere with oxygen metabolism in other ways that increase the cellular level of ROS (Yu et al. 2001). In 1989, based on our findings with experimental fluorosis, we first proposed that this elevated level of free radicals induced by excessive fluoride may be a common cause of the damage to various organs (Guan et al. 1989). Later, we found that both adults and children living in an area characterized by endemic fluorosis due to coal burning exhibit high levels of oxidative stress (Guan et al. 1991, 1992).

To study the effects of fluoride on oxidative stress in greater detail, we examined residents of regions in Bijie and Shuicheng cities of China where the incidence of coal-burning fluorosis is extremely high. In addition to elevated levels of MDA and lower SOD activity in serum, the activity of myeloperoxidase (MPO) in the blood of patients with endemic fluorosis was increased significantly. Since MPO can promote the production of free radicals, both its activity and polymorphism in the gene encoding this enzyme may influence the level of oxidative stress (Zhang et al. 2013). Furthermore, the activities of copper/zinc SOD (Cu/Zn-SOD), CAT and GSH-Px, as well as the level of MDA in the serum of residents of these patients were determined (Wang et al. 2014). These patients demonstrated much higher levels of MDA and lower expression of Cu/Zn-SOD than healthy individuals, indicating that fluorosis can reduce the activities of antioxidant enzymes.

The plasma total oxidative status (TOS), total antioxidant capacity (TAC), and oxidative stress index (OSI) in patients with endemic fluorosis were investigated (Varol et al. 2013). The results showed that the levels fluoride in urine, and TOS and OSI in plasma from the patients with chronic fluorosis were significantly higher, while TAC in plasma was significantly lower. Correlation analysis indicated that in all the groups TAC was negatively, whereas TOS and OSI were positively correlated with urine fluoride, suggesting that oxidative stress plays an important role in the pathogenesis of the endemic fluorosis. A study was conducted to find out the level of oxidative stress and effect of supplementation of vitamin C, vitamin D, and calcium on levels of SOD and fluoride in serum or urine in children residing in endemic fluorosis area (Ailani et al. 2009). The study indicated an increasing oxidative stress in cases of fluorosis with increasing drinking water fluoride concentration. After treatments by the supplied materials a significant reduction in serum fluoride and an increased urinary fluoride were detected. A high positive correlation was observed in fluoride level in serum and urine between pre-treatment and post-treatment groups.

In our experimental studies, we observed a significant elevation in the total level of oxidation and reduced antioxidant capacity in the blood and certain soft tissues of rats with chronic fluorosis. Moreover, the levels of lipid peroxidation and protein oxidation in cultured cells exposed to high concentrations of fluoride were elevated (Gao et al. 2008, 2009). Further studies revealed that, in addition to the common changes associated with oxidative stress, excessive fluoride lowered the activities of glutathione reductase and glutathione-S-transferase (GSTs) in rat brains (Chauhan et al. 2013). These increased contents of ROS, singlet oxygen, and nitrotyrosine, as well as the raised activity of nitric oxide synthase (NOS) may be related, at least in part, to the activation of microglia cells, which can augment oxidative stress (Xu et al. 2012).

In mice with chronic fluorosis, the frequency of micronucleus in polychromatic erythrocytes was significantly higher than in controls, demonstrating the genotoxic potential of the enhanced oxidative stress associated with this condition (Manivannan et al. 2013). In connection with the three-dimensional culture of human osteoblasts in the presence of fluoride, cell proliferation was either stimulated or inhibited, depending on the fluoride concentration and exposure time, but in all cases there was more oxidative stress, indicating a potential role for such stress in regulating cell proliferation (Liu et al. 2011a). Furthermore, chronic fluorosis impairs the mitochondrial respiratory chain in a manner that can promote the production of ROS (Fina et al. 2012). Finally, fluoride can also attack metal ions with two or three positive charges in the active center of antioxidant enzymes, thereby inhibiting their activity (Goschorska et al. 2015).

Chronic Fluorosis is Associated with Modifications in the Structures of Lipids in Cellular Membranes

The lipids of which cellular membranes are composed consist primarily of phospholipids, dolichol, ubiquinone, and cholesterol. The main species of phospholipids are phosphatidyl ethanolamine (PE), phosphatidylcholine (PC), phosphatidylserine (PS), phosphatidylinositol, cardiolipin, and sphingomyelin. These lipids play a key role in maintaining not only the normal structure of cells, but also functions such as signaling, ion transport, intracellular enzyme activities, and binding of ligands to their receptors.

We found that after treating rats for 7 months with 30 or 100 ppm fluoride in drinking water, the total content of phospholipid in their brains was reduced by 10% and 20%, respectively (Guan et al. 1998). PE, PC, and PS were the phospholipids whose levels were attenuated most. Interestingly, with both levels of exposure after 3 months, the ubiquinone content of the brain was lowered, but after 7 months this content was clearly elevated. Moreover, in rats fed a diet containing a high content of fluoride, the total phospholipid and ubiquinone contents of the kidney were reduced, with PE and PC being the main phospholipids influenced (Guan et al. 2000). In addition, the proportions of polyunsaturated fatty acids in renal PE and PC declined. Furthermore, the contents of PE, PC, and PS in the liver of rats exposed to high-dose fluoride were also lowered (Wang et al. 2000). In molecular structure of PE and PC, the proportion of polyunsaturated fatty acids (20:4 and 22:6) declined, whereas the relative amount of saturated fatty acids (16:0 and 18:0) rose. The total ubiquinone content in liver was reduced 11% and 42% by the diets containing 30 ppm and 100 ppm fluoride, respectively, with decreases in both ubiquinone-9 and ubiquinoine-10.

Interestingly, chronic fluorosis is associated with elevated levels of free radicals, which oxidizes the lipids in biological membranes (and especially polyunsaturated fatty acid moieties in phospholipids), as well as proteins and DNA as targets. These alternations in membrane lipids might play an important role in the pathogenesis of chronic fluorosis.

Effects of the Oxidative Stress Induced by Chronic Fluorosis on Mitogen-Activated Protein Kinase (MAPK)

MAPK, a family of widely occurring cytoplasmic serine/threonine protein kinases, includes the extracellular signal-regulated kinase 1/2 (ERK1/2), c-Jun N-terminal kinase (JNK), and p38 mitogen-activated protein kinase (p38 MAPK) pathways. MAPK signaling is activated by a number of extra- and intra-cellular stimuli, including cytokines, growth factors, and related hormones, as well as oxidative stress and stressors for the endoplasmic reticulum. These pathways in MAPK play key roles in the regulation of numerous cellular processes including proliferation, differentiation, the stress response, motility, growth, survival, and death. For example, MAPK cascades, and in particular ERK1/2, are required for long-term potentiation (LTP) of the hippocampus, thereby regulating cellular responses to extracellular stimuli, neuronal differentiation and plasticity.

In the brains of rats with chronic fluorosis, the levels of both phospho- and total-ERK1/2, as well as phospho-MEK1/2 protein are elevated, but with no change in the level of ERK1/2 mRNA, and activation of phospho-ERK1/2 is attenuated (Liu et al. 2010). This attenuated activation of ERK1/2 may be related to the impaired learning and memory exhibited by these same animals, while the increase in ERK1/2 may be compensatory. Oxidative stress may be involved in these effects of fluoride on MAPK signaling (Cao et al. 2014).

In both the brains of rats with chronic fluorosis and SH-SY5Y cells cultured in the presence of high concentrations of fluoride, expression of the phospho-JNK protein was enhanced and more apoptosis observed (Liu et al. 2011b). Moreover, these two alterations were positively correlated. In addition, expressions of p38 MAPK, NF-kB p65, and caspase-3 proteins in peripheral blood mononuclear cells from patients with high levels of fluoride living in the region of endemic coal-burning fluorosis were elevated (Chen et al. 2010).

Furthermore, fluoride enhanced the phosphorylation of JNK and ERK, in which the effect was attenuated by vitamin E or lycopene. Thus, in patients with coal-burning fluorosis, vitamin E and lycopene prevented apoptosis in spermatogenic cells by suppressing the JNK and ERK signaling stimulated with oxidative stress (Tian et al. 2018).

Effects of the Oxidative Stress Induced by Chronic Fluorosis on Receptors

Acetylcholine Receptors

Acetylcholine receptors, which play a key role in cholinergic signaling, are of two types, i.e., nicotinic acetylcholine receptors (nAChRs) and muscarinic acetylcholine receptors (mAChRs).

Neuronal nAChRs, members of the super-family of ligand-gated ion channels, are involved in a variety of important physiological functions. In the central nervous system α (2-10) and β (2-4) subunits combine to form both hetero- and homopentameric neuronal nAChRs, among which the main subtypes present in the brain are α4β2 and α7. In response to acetylcholine, these nAChRs participate in a wide range of brain activities and functions, including cognitive enhancement, increased arousal and decreased anxiety, regulatory effects on several Ca^{2+}-dependent neuronal processes, the actions of receptors for other neurotransmitters and neuroprotection (Solari and Hangya 2018). Not surprisingly, alterations in the numbers and/or function of these subtypes are characteristic of various pathological conditions.

In our early investigations on rats with chronic fluorosis, we found lowered levels of nAChR proteins in combination with impaired learning and memory. Moreover, high concentrations of either fluoride or ferrous iron ions were found to lower the level of nAChR proteins and promote lipid peroxidation and protein oxidation in cultured cells in a concentration-dependent manner (Gao et al. 2008). Furthermore, the cells exposed to high concentrations of fluoride exhibited binding sites for [^{125}I] α-bungarotoxin (commonly used as a ligand for α7 nAChR) with a lowered B_{max}. Interestingly, the antioxidants vitamin E and GSH could attenuate these effects of fluoride on the receptors. These findings indicate that the oxidative stress associated with fluorosis might cause a deficit in α7 nAChR that could be related to the impairments in learning and memory.

In attempt to unravel the mechanism(s) by which fluoride reduces the activity of nAChRs, PC12 cells were exposed to different concentrations of fluoride for 48 hours and rats administered high doses of fluoride in their drinking water for 7 months (Shan et al. 2004). Significant decreases in the level of nAChR proteins (with no change in the level of the corresponding mRNA) and obvious increases in protein oxidation and lipid peroxidation were detected both in the cells and brains of the rats. These alternations were mostly prevented by pre-treatment with an antioxidant. These results indicate that the deficit in nAChRs associated with fluoride toxicity occurs at the posttranscriptional level and may involve damage of these receptors by oxidative stress.

Furthermore, resveratrol (trans-3,4′,5-trihydroxystilbene, RSV), which regulates the antioxidant and neuroprotective sirtuins (SIRTs), attenuates both the impaired learning and memory in rats with chronic fluorosis and their offspring, as well as the level of oxidative stress in the brains of these rats and in cells cultured in the presence of high concentrations of fluoride (Zeng et al. 2019, 2020). Resveratrol also attenuated the decreases in the levels of the α7 and α4 nAChR subunits caused by fluorosis.

In general, mAChRs, a large family of G-protein-coupled receptors, are involved in the modulation of learning, memory, and neuronal plasticity in the central nervous system, as well as in important physiological functions such as respiration, circulation, movement, and temperature regulation. The five subtypes M1, M2, M3, M4, and M5 all contain a single peptide and are located in the cellular membrane. When activated, these receptors activate the G protein to which they are coupled, causing a variety of physiological and biochemical effects.

The expression of M1 and M3 protein, as well as the corresponding levels of mRNA, was suppressed not only in rats with chronic fluorosis, but also in their offspring (Dong et al. 2015). Notably, these reductions correlated positively with the deficits in learning and memory and were linked to the increases in ROS (such as OH, H_2O_2, and O_2^-) and lipid peroxidation, as well as to the decreases in the activities of antioxidant enzymes (such as SOD and GSH-Px). Thus, the high level of oxidative stress induced by fluorosis might alter the expression of mAChRs.

Advanced Glycation End Products (AGEs)/the Receptor for Advanced Glycation End Products (RAGE)

AGEs, heterogeneous products of non-enzymatic malliard reactions of aldose sugar with proteins and lipids, are important in connection with the pathogenesis of several chronic diseases. The physical, chemical, and functional properties of proteins modified in this fashion can be changed and, in addition, AGEs can act via specific receptors, among which RAGE is the most well-known. The major consequence of AGE/RAGE signaling is induction of oxidative stress through activation of NADPH oxidases. Subsequently, this elevated intracellular oxidative stress stimulates protein kinase and ERK1/2, resulting in the translocation and activation of nuclear factor-kappa (NF-κB) and subsequent up-regulation of NF-κB-dependent genes, the products of which ultimately exert deleterious effects.

To explore the underlying mechanisms by which chronic fluorosis damages the brain, we examined the levels of AGEs/RAGE, NADPH oxidase-2 (NOX2), ROS, and MDA in the brains of rats receiving 5 or 50 ppm fluoride in their drinking water for 3 or 6 months, as well as in SH-SY5Y cells exposed to high concentrations of fluoride in their incubation medium for as long as 48 h (Zhang et al. 2015). In both of these systems, the levels of AGEs/RAGE and NOX2 protein and corresponding species of mRNA were elevated following long-term, high-dose exposure to fluoride. In contrast, no significant changes in these parameters were detected in the rats exposed for only 3 months. In addition, the levels of ROS and MDA in the SH-SY5Y cells exposed to high-dose fluoride were elevated in a manner that correlated positively with the levels of AGE/RAGE, suggesting that excessive fluoride may activate AGE/RAGE signaling by enhancing oxidative stress.

Receptors for N-methyl-D-Aspartic Acid Receptors (NMDARs)

NMDARs mediate the signals in the glutamatergic system, the major excitatory neurotransmitter system in the central nervous system. These receptors play key roles in neuronal maturation and the establishment of synapses during the development of the nervous system and their potential involvement in excitotoxic entry and massive intracellular accumulation of Ca^{2+} has been discussed extensively in connection with several neurodegenerative diseases.

In the brains of adult rats with chronic fluorosis and their pups, as well as in primary neurons exposed to high concentrations of fluoride, the levels of the mRNAs encoding the GluN1 and GluN2B subunits and of the corresponding proteins were elevated, whereas expression of GluN3A was lowered and that of GluN2A unchanged (Wei et al. 2018). In addition, the level of phospho-CaMKII was reduced and Ca^{2+} influx and apoptosis enhanced both in vivo and in vitro. Thus, modifications of this kind of receptor may also be involved in the brain damage induced by chronic fluorosis.

Interestingly, NMDAR-mediated oxidative stress has been implicated in disruption of the blood–brain barrier (BBB) in connection with a variety of neuropathological diseases. Betzen et al. (2009) found that the expression of NR1 significantly increased 72 h after 30 min exposure to superoxide, peroxynitrite or hydrogen peroxide, resulting in increased cellular response to NMDAR stimulation and diminished monolayer impedance. Therefore, they conclude that oxidative stress upregulates NMDAR on cerebrovascular endothelium and thus heightens susceptibility to glutamate-induced BBB disruption.

Excessive release of neuronal nitric oxide synthase (nNOS) with NO dependent during the stimulation of high NMDARs results in productions of hydroxyl (HO•) and peroxynitrite anion radicals, which are responsible for oxidative injury (Guix et al. 2005). Recently, we found that exposure of chondrocytes or neuronal cells to high concentrations of fluoride elevated their content of NO, activity of inducible nitric oxide synthase (iNOS), and levels of iNOS and nNOS protein, MDA, and ROS (Deng et al. 2018). At the same time, the activities of the antioxidant enzymes SOD, GSH, and CAT were reduced and the ROS level and the content of MDA were significantly increased in the cells exposed to fluoride. Interestingly, non-selective inhibition of iNOS with L-NMMA or more selective inhibition with 1400 W attenuated these effects, indicating the close relationship between the NO system and oxidative stress (He et al. 2020).

Effects of the Oxidative Stress Induced by Chronic Fluorosis on Various Enzymes

Fluoride can form insoluble salts with cations such as calcium and magnesium and in this manner long-term intake of excessive fluoride can change the activities of various enzymes in tissues and cells. In addition, the elevated levels of free radicals associated with chronic fluorosis can lower enzyme activities. Furthermore, fluoride may act directly as an agonist or inhibitor of certain types of enzymes, e.g., inhibiting antioxidant enzymes such as SOD, GSH-px and CAT, but stimulating MPO (Zhang et al. 2013). The activity of MPO in the blood of patients with endemic fluorosis is elevated, which can promote oxidative stress by producing more free radicals. Lactate dehydrogenase (LDH) is an important oxidant enzyme located in the cytoplasm and. the total serum activity of the isoenzymes LDH1-LDH5 is also higher in patients with endemic fluorosis, indicating the presence of systemic lesions in skeletal muscles, cardiac muscle, and the lungs (Arulkumar et al. 2017).

Many exogenous chemicals are converted to highly reactive, most often electrophilic intermediates in the first step of their biotransformation and these intermediations can bind covalently to important cellular macromolecules and thereby cause damage. Glutathione-S-transferase (GST) catalyzes the binding of nucleophilic glutathione both to various electrophilic xenobiotics and their electrophilic intermediates, preventing such covalent binding and playing a role in detoxification. Excessive fluoride decreased the activity of GST in rat brains (Chauhan et al. 2013) and, moreover, high levels of both fluoride and GST activity were detected in the saliva of children living in an area with a naturally high level of fluorine in the drinking water. This increase in GST activity may well reflect a need to inactivate free radicals produced by exposure to fluoride (Bonola-Gallardo et al. 2017), providing a supplementary approach to the elimination of high levels of oxidative stress.

As a matter of fact, fluoride can also inhibit enzyme activity by binding to functional groups other than those in the active center of enzymes. The sodium- and potassium-activated adenosine triphosphatase (Na^+, K^+-ATPase) plays vital roles in normal brain development and function and loss of this activity is implicated in numerous disorders. Fluoride appears to influence this activity in a variety of ways. For instance, chronic exposure to fluoride during pregnancy and early infancy may impair Na^+, K^+-ATPase activity, with both short- and long-term implications for disease and other adverse health effects (Waugh 2019).

The erythrocytes of individuals living in an area of endemic fluorosis contained elevated levels of sodium and reduced levels of potassium, as well as lower Na^+, K^+-ATPase activity, and the magnitudes of all of these changes were correlated to the fluoride contents of both the serum of the subjects themselves and in the drinking water in this region (Shashi and Meenakshi 2015). Since the activity of Na^+, K^+-ATPase activity, a membrane-bound enzyme, is inhibited by lipid peroxidation (Namazi et al. 2019), we speculate that these alterations may be related to the increased oxidative stress associated with chronic fluorosis.

Changes in Mitochondria Correlated to the Oxidative Stress Induced by Chronic Fluorosis

Even under normal physiological conditions, consumption of oxygen in connection with the production of ATP by the respiratory chain in mitochondria produces a certain amount of ROS. Consequently, an increase in this mitochondrial activity can be per se a source of oxidative stress. Furthermore, production of ROS is markedly enhanced under many pathological conditions when the respiratory chain is impaired. Mitochondrial DNA, which encodes certain of the proteins essential for oxidative phosphorylation, is located close to the respiratory chain and is thus more extensively oxidized by the ROS produced than is nuclear DNA. Cumulative oxidative damage to mitochondrial DNA is implicated in the negative effects of chronic fluorosis on health.

Mitochondria are continuously undergoing functional fusion and fission and the balance between these processes determine the overall morphology of this organelle. Too much fission leads to fragmented mitochondria, with shortened membranes, while excess fusion produces long mitochondria that are overly interconnected. Fusion involves mitochondrial GTPases, such as the mitofusions (Mfn1 and Mfn2) and optic atrophy 1 (OPA1), whereas fission is promoted by fission 1 (Fis1) and dynamin-related protein 1 (Drp1).

In the cortices of the brains of rats with chronic fluorosis, the level of Mfn1 protein was clearly reduced, whereas those of Fis1 and Drp1 were elevated, with almost the same changes in the levels of the corresponding mRNAs (Lou et al. 2013). The mitochondria were fragmented and located further away from the axons of the cortical neurons. These abnormal mitochondrial dynamics induced by chronic fluorosis might help promote oxidative stress.

In addition, the expression of proteins related to mitochondrial morphology in the kidney and heart of rats with chronic fluorosis was investigated to examine the possibility that mitochondrial injury might contribute to the increased oxidative stress in multiple organs (Qin et al. 2015). In the case of the kidney, the levels of both Mfn1 protein and mRNA were clearly reduced, those of Fis1 elevated, the level of MDA increased and the total antioxidative capacity lowered. At the same time, swollen or fragmented mitochondria were observed under the electronic microscope. Similar changes were observed in the heart. These findings indicate that the effects of chronic fluorosis on mitochondria may help explain the increase in oxidative stress in a variety of tissues.

It has been reported that exposure of F9 embryonic carcinoma cells to fluoride is cytotoxic elevates the production of ROS by mitochondria and induces apoptosis (Peng et al. 2019). Consistent with this finding, stimulation of SOD2 activity in these same cells attenuates this mitochondrial oxidative stress.

Apoptosis Resulting from the Oxidative Stress Induced by Chronic Fluorosis

Apoptosis, a programmed form of cell death actively controlled by gene products, is essential for both homeostasis and development. Many forms of cellular stress stimulate pro-apoptotic signaling pathways that activate the caspase proteases and cause mitochondrial dysfunction. Apoptotic cells undergo characteristic changes in morphology, including rounding up, blebbing of the cellular plasma membrane and fragmentation of the nucleus and DNA it contains. Elevated apoptosis is characteristic of many diseases.

Numerous investigators have concluded that fluorosis induces apoptosis by disturbing signaling that controls gene expression and protein activity. B-cell lymphoma-2 (Bcl-2), an apoptotic signaling factor encoded by one of the most important oncogenes, is involved in fluoride-induced apoptosis. For example, exposure of human gingival fibroblasts to fluoride down-regulates Bcl-2, an event that is followed by activation of a mitochondrial cell death pathway involving the release of cytochrome C from the mitochondria into the cytosol; activation of caspase-3, -8, and -9; cleavage of poly (ADP-ribose) polymerase; and the expression of voltage-dependent anion channels (Lee et al. 2008). In addition, fluoride-induced apoptosis involves increases in the levels of the Fas and Fas ligand (Fas-L) proteins and their mRNAs.

It has been demonstrated that the oxidative stress and consequent modulation of intracellular redox homeostasis, lipid peroxidation, and protein carbonylation, together with altered gene expression caused by chronic fluorosis can lead to mitochondrial dysfunction. In particular, the oxidative stress resulting from the intake of excessive fluoride, which is concentration dependent, is believed to be involved in the induction of apoptosis. For instance, elevated levels of ROS, MDA, protein carbonyl groups, and 8-OHdG, and along with decreased levels of GSH-Px, cytoplasmic thioredoxin reductase 1 (TrxR1), and Bcl-2, as well as apoptosis were observed in the serum. And liver and kidney of broilers exposed to high-dose fluoride (Wang et al. 2018).

In addition, when H9c2 cells were exposed to different concentrations of fluoride, cell proliferation was inhibited and apoptosis augmented (Yan et al. 2015). As the concentration of fluoride increased, so did the incidence of apoptosis and generation of ROS, while the mitochondrial membrane potential declined. These findings indicate that the generation of ROS and perturbation of mitochondria are involved in the apoptosis caused by chronic fluorosis (Yang et al. 2015).

Dietary selenium, which is an essential component of the active sites of antioxidant selenoproteins such as GSH-Px and TrxR1, attenuates the oxidative stress and apoptosis induced by fluoride. Furthermore, lycopene (LYC), an effective antioxidant, attenuates various aspects of fluoride toxicity, including the raised levels of MDA and ROS, increased activities of caspase-3 and -9, more pronounced expression of the Bax and Bcl-2 genes, and reductions in the activities of SOD and GSH-px (Li et al. 2017). Thus, antioxidant compounds like vitamin E, selenium, and LYC may counteract fluoride-induced apoptosis by suppressing the ERK and JNK signaling mediated by oxidative stress (Tian et al. 2018).

Prevention and Treatment of Chronic Fluorosis by Antioxidants

For many years, development of effective drug treatment for endemic fluorosis has been hampered by an inadequate understanding of the pathogenesis of this disease. However, the present belief in that oxidative stress is a key factor in this connection has led to extensive efforts to find treatments based on elimination of this stress. Antioxidant compounds such as vitamins E and C, selenium, and LYC all prevent or reverse oxidative stress and could therefore potentially be used to treat chronic fluorosis (Peng et al. 2019; Tian et al. 2018; Wang et al. 2018; Dong et al. 2015; Gao et al. 2008).

Indeed, protection by vitamin E against the neurotoxicity induced by fluorosis has been investigated extensively. In this context, we found that treatment of neuronal cells exposed to high levels of fluoride with vitamin E or reduced glutathione attenuated the reduction in nAChRs and neurotoxicity (Gao et al. 2008). Moreover, vitamin E prevented the elevated production of $O_2^{-\bullet}$ and MDA in the brains of rats administered a diet high in fluoride and attenuated the impairment of the learning and memory of these animals (Dong et al. 2017), perhaps by restoring nAChRs.

Since oxidative stress is responsible for many of the adverse effects of endemic fluorosis, we have developed a mixture of traditional Chinese medicines which, at least in experimental systems, improves the activity of antioxidant enzymes, counteracts the damage caused by free radicals, and promotes urinary excretion of fluoride. In addition, clinical trials have shown that this mixture improves the self-reported symptoms and bone X-rays of patients with chronic fluorosis, as well as increasing the level of fluoride in their urine and reducing their oxidative stress. However, much work remains in order to determine whether this mixture of traditional Chinese medicines will prove to be effective in preventing and/or treating chronic fluorosis.

Conclusions

Accumulating evidence indicates that the systemic injuries caused by endemic (or chronic) fluorosis may be due to elevated oxidative stress, a conclusion now widely accepted. Such stress, caused by an imbalance between cellular oxidative and antioxidative systems leading to enhanced production of ROS, can impair the lipid structure of cell membranes, signal transduction, receptor functions, and enzyme activities. This oxidative stress theory remains the focus of extensive research and the basis of current attempts to develop more effective treatment of chronic fluorosis.

References

Ailani V, Gupta RC, Gupta SK, Gupta K. Oxidative stress in cases of chronic fluoride intoxication. Indian J Clin Biochem. 2009;24:426–9.

Arulkumar M, Vijayan R, Penislusshiyan S, Sathishkumar P, Angayarkanni J, Palvannan T. Alteration of paraoxonase, arylesterase and lactonase activities in people around fluoride endemic area of Tamil Nadu, India. Clin Chim Acta. 2017;471:206–15.

Betzen C, White R, Zehendner CM, Pietrowski E, Bender B, Luhmann HJ, et al. Oxidative stress upregulates the NMDA receptor on cerebrovascular endothelium. Free Radic Biol Med. 2009;47:1212–20.

Bonola-Gallardo I, Irigoyen-Camacho ME, Vera-Robles L, Campero A, Gómez-Quiroz L. Enzymatic activity of glutathione S-transferase and dental fluorosis among children receiving two different levels of naturally fluoridated water. Biol Trace Elem Res. 2017;176:40–7.

Cao J, Chen J, Wang J, Klerks P, Xie L. Effects of sodium fluoride on MAPKs signaling pathway in the gills of a freshwater teleost, cyprinus carpio. Aquat Toxicol. 2014;152:164–72.

Chauhan SS, Ojha S, Mahmood A. Effects of fluoride and ethanol administration on lipid peroxidation systems in rat brain. Indian J Exp Biol. 2013;51:249–55.

Chen Q, Wang Z, Xiong Y, Zou X, Liu Z. Comparative study of p38 MAPK signal transduction pathway of peripheral blood mononuclear cells from patients with coal-combustion-type fluorosis with and without high hair selenium levels. Int J Hyg Environ Health. 2010;213:381–6.

Deng MF, He WW, Gui CZ, Guan ZZ. Attenuation of 7-nitroindazole on the raised level of nitric oxide and the activity of nitric oxide synthetase as well as the toxic effect in SH-SY5Y cells exposed by high fluoride. Fluoride. 2018;51:328–39.

Dong YT, Wei N, Qi XL, Liu YJ, Chen D, Zeng XX, et al. Attenuation effect of vitamin E on the deficit of learning and memory of rats with chronic fluorosis – the mechanism may involve muscarinic acetylcholine receptors. Fluoride. 2017;50:354–64.

Dong YT, Wang Y, Wei N, Zhang QF, Guan ZZ. Deficit in learning and memory of rats with chronic fluorosis correlates with the decreased expressions of M1 and M3 muscarinic acetylcholine receptors. Arch Toxicol. 2015;89:1981–91.

Fina BL, Brance ML, Brun LR, Ringlli A. Fluoride inhibition of oxygen consumption and increased oxidative stress in rats. Fluoride. 2012;45:343–8.

Gao Q, Liu YJ, Guan ZZ. Decreased learning and memory ability in rats with fluorosis: increased oxidative stress and reduced activity of cholinesterase in the brain. Fluoride. 2009;42:266–74.

Gao Q, Liu YJ, Guan ZZ. Oxidative stress might be a mechanism connected with the decreased alpha 7 nicotinic receptor influenced by high-concentration of fluoride in SH-SY5Y neuroblastoma cells. Toxicol In Vitro. 2008;22:837.

Goschorska M, Gutowska I, Olszewska M, Baranowska-Bosiacka I, Rać M, Olszowski T, et al. Effect of sodium fluoride on the catalase activity in THP-1 macrophages. Fluoride. 2015;48:274–82.

Guan ZZ, Wang YN, Xiao KQ, Dai DY, Chen YH, Liu JL, et al. Influence of chronic fluorosis on membrane lipids in rat brain. Neurotoxicol Teratol. 1998;20:537–42.

Guan ZZ, Xiao KQ, Zeng XY, Long YG, Cheng YH, Jiang SF, et al. Changed cellular membrane lipid composition and lipid peroxidation of kidney in rats with chronic fluorosis. Arch Toxicol. 2000;74:602–8.

Guan ZZ, Yang PS. Changes of lipid peroxidation and antioxidant substances in serum and erythrocytes of rats with chronic fluorosis. Chin J Endem. 1991;9:4–6.

Guan ZZ, Yang PS, Su YS, Wang YN. Changes of lipid peroxide content and anti-oxidant level in blood of the children living in the F-AI combined toxicosis area of Shuicheng of Guizhou. J Guiyang Med Coll. 1992;16:198–200.

Guan ZZ, Yang PS, Wang YN, Wei ZD, Zhou LY, Li FC. Study on antiaging index in plasma and erythrocytes of adult patients with fluorosis and aluminum poisoning. Guizhou Med. 1991;15:265–7.

Guan ZZ, Yang PS, Yu ND, Zhuang ZJ. An experimental study of blood biochemical diagnostic indices for chronic fluorosis. Fluoride. 1989;22:112–8.

Guix FX, Uribesalgo I, Coma M, Muñoz FJ. The physiology and pathophysiology of nitric oxide in the brain. Prog Neurobiol. 2005;76:126–52.

Görlach A, Dimova EY, Petry A, Martínez-Ruiz A, Hernansanz-Agustín P, Rolo AP, et al. Reactive oxygen species, nutrition, hypoxia and diseases: problems solved? Redox Biol. 2015;6:372–85.

He WW, Wang WY, Deng J, Xiang J, Deng MF, Gui CZ, et al. L-NMMA and 1400W, inhibitors of iNOS, attenuate the induction of iNOS and NO in primary rabbit costal chondrocytes by fluoride. Fluoride. 2020;53:220–38.

Lee JH, Jung J, Jeong YJ, Park JH, Yang KH, Choi NK, et al. Involvement of both mitochondrial- and death receptor-dependent apoptotic pathways regulated by Bcl-2 family in sodium fluoride-induced apoptosis of the human gingival fibroblasts. Toxicology. 2008;243:340–7.

Li W, Jiang B, Cao X, Xie Y, Huang T. Protective effect of lycopene on fluoride-induced ameloblasts apoptosis and dental fluorosis through oxidative stress-mediated caspase pathways. Chem Biol Interact. 2017;261:27–34.

Liu H, Sun JC, Zhao ZT, Zhang JM, Xu H, Li GS. Fluoride-induced oxidative stress in three-dimensional culture of OS732 cells and rats. Biol Trace Elem Res. 2011a;143:446–56.

Liu YJ, Guan ZZ, Gao Q, Pei JJ. Increased level of apoptosis in rat brains and SH-SY5Y cells exposed to excessive fluoride - a mechanism connected with activating JNK phosphorylation. Toxicol Lett. 2011b;204:183–9.

Liu YJ, Gao Q, Wu CX, Guan ZZ. Alterations of nAChRs and ERK1/2 in the brains of rats with chronic fluorosis and their connections with the decreased capacity of learning and memory. Toxicol Lett. 2010;192:324–9.

Lou DD, Guan ZZ, Liu YJ, Liu YF, Zhang KL, Pan JG, et al. The influence of chronic fluorosis on mitochondrial dynamics morphology and distribution in cortical neurons of the rat brain. Arch Toxicol. 2013;87:449–57.

Manivannan J, Sinha S, Ghosh M, Mukherjee M. Evaluation of multi-endpoint assay to detect genotoxicity and oxidative stress in mice exposed to sodium fluoride. Mutat Res. 2013;751:59–65.

Namazi G, Asa P, Sarrafzadegan N, Pourfarzam M. Decreased Na+/K+-ATPase activity and altered susceptibility to peroxidation and lipid composition in the erythrocytes of metabolic syndrome patients with coronary artery disease. Ann Nutr Metab. 2019;74:140–8.

Peng W, Xu S, Zhang J, Zhang Y. Vitamin C attenuates sodium Fluoride-induced mitochondrial oxidative stress and apoptosis via sirt1-SOD2 pathway in F9 cells. Biol Trace Elem Res. 2019;191:189–98.

Qin SL, Deng J, Lou DD, Yu WF, Pei J, Guan ZZ. The decreased expression of mitofusin-1 and increased fission-1 together with alterations in mitochondrial morphology in the kidney of rats with chronic fluorosis may involve elevated oxidative stress. J Trace Elem Med Biol. 2015;29:263–8.

Schieber M, Chandel NS. ROS function in redox signaling and oxidative stress. Curr Biol. 2014;24:453–62.

Shan KR, Qi XL, Long YG, Nordberg A, Guan ZZ. Decreased nicotinic receptors in PC12 cells and rat brains influenced by fluoride toxicity--a mechanism relating to a damage at the level in post-transcription of the receptor genes. Toxicology. 2004;200:169–77.

Shashi A, Meenakshi G. Inhibitory effect of fluoride on Na+, K+ATPase activity in human erythrocyte membrane. Biol Trace Elem Res. 2015;168:340–8.

Solari N, Hangya B. Cholinergic modulation of spatial learning, memory and navigation. Eur J Neurosci. 2018;48:2199–230.

Tian Y, Xiao YH, Wang BL, Sun C, Tang KF, Sun F. Vitamin E and lycopene reduce coal burning fluorosis-induced spermatogenic cell apoptosis via oxidative stress-mediated JNK and ERK signaling pathways. Biosci Rep. 2018;38:1003.

Varol E, Icli A, Aksoy F, Bas HA, Sutcu R, Ersoy IH, et al. Evaluation of total oxidative status and total antioxidant capacity in patients with endemic fluorosis. Toxicol Ind Health. 2013;29:175–80.

Wang Q, Cui KP, Xu YY, Gao YL, Zhao J, Li DS, et al. Coal-burning endemic fluorosis is associated with reduced activity in antioxidative enzymes and Cu/Zn-SOD gene expression. Environ Geochem Health. 2014;36:107–15.

Wang YN, Xiao KQ, Liu JL, Dallner G, Guan ZZ. Effect of long term fluoride exposure on lipid composition in rat liver. Toxicology. 2000;146:161–9.

Wang YX, Xiao X, Zhan XA. Antagonistic effects of different selenium sources on growth inhibition, oxidative damage, and apoptosis induced by fluorine in broilers. Poult Sci. 2018;97:3207–17.

Waugh DT. Fluoride exposure induces inhibition of sodium-and potassium-activated adenosine triphosphatase (Na^+, K^+-ATPase) enzyme activity: molecular mechanisms and implications for public health. Int J Environ Res Public Health. 2019;16:1427.

Wei N, Dong YT, Deng J, Wang Y, Qi XL, Yu WF, et al. Changed expressions of N-methyl-d-aspartate receptors in the brains of rats and primary neurons exposed to high level of fluoride. J Trace Elem Med Biol. 2018;45:31–40.

Xu S, Liu Z, Ling Y, Fei W, Sun G. A role of fluoride on free radical generation and oxidative stress in BV-2 microglia cells. Mediat Inflamm. 2012;2012:102954.

Yan X, Yang X, Hao X, Ren Q, Gao J, Wang Y, et al. Sodium fluoride induces apoptosis in H9c2 cardiomyocytes by altering mitochondrial membrane potential and intracellular ROS level. Biol Trace Elem Res. 2015;166:210–5.

Yang Y, Lin X, Huang H, Feng D, Ba Y, Cheng X, et al. Sodium fluoride induces apoptosis through reactive oxygen species-mediated endoplasmic reticulum stress pathway in Sertoli cells. J Environ Sci. 2015;30:81–9.

Yu Y, Wang S, Gao Q, Liu J, Liang R, Jin S, et al. Determine the free radical contents of bone, liver and kidney in the rats exposed to fluoride with electron spin resonance (ESR). Chin J Endem. 2001;20:342–4.

Zeng XX, Deng J, Xiang J, Dong YT, Cao K, Liu XH, et al. Protections against toxicity in the brains of rat with chronic fluorosis and primary neurons exposed to fluoride by resveratrol involves nicotinic acetylcholine receptors. J Trace Elem Med Biol. 2020;60:126475.

Zeng XX, Deng J, Xiang J, Dong YT, Cao K, Liu XH, et al. Resveratrol attenuated the increased level of oxidative stress in the brains and the deficit of learning and memory of rats with chronic fluorosis. Fluoride. 2019;52:149–60.

Zhang KL, Lou DD, Guan ZZ. Activation of the AGE/RAGE system in the brains of rats and in SH-SY5Y cells exposed to high level of fluoride might connect to oxidative stress. Neurotoxicol Teratol. 2015;48:49–55.

Zhang T, Shan KR, Tu X, He Y, Pei JJ, Guan ZZ. Myeloperoxidase activity and its corresponding mRNA expression as well as gene polymorphism in the population living in the coal burning endemic fluorosis area in Guizhou of China. Biol Trace Elem Res. 2013;152:379–86.

Chapter 13
High Level of Inflammation

Wen-Feng Yu and Zhi Huang

Abstract In low concentration, fluoride is considered a necessary compound for human health. Long-term exposure to excessive content of fluoride is the reason for a disease called fluorosis. In more than 25 countries and regions in the world, the excessive fluoride content in the environment has led to the occurrence of chronic endemic fluorosis. Fluorosis not only damages bones and teeth, but also causes extensive pathological damage to human organs/systems. The pathogenesis of endemic fluorosis has not been fully elucidated. Many studies have shown that the mechanism of chronic fluorosis causing extensive pathological damage in the body may be oxidative stress induced by accumulation of fluoride. As an essential intermediate process that causes inflammation, it is such oxidative stress causes extensive damage to the body's tissues. At present, there are few reports about the pathological role of inflammation in the pathogenesis of endemic fluorosis. Therefore, this section outlines the pathological role of inflammation in the pathogenesis of endemic fluorosis and focuses on the effects of inflammation on organs/systems (such as kidney, thyroid, cardiovascular, brain, and reproductive system) during fluoride exposure, together with its possible mechanism of injury.

Keywords Cytokines · Fluorosis · Inflammation · Necrosis · Toxic effect

W.-F. Yu (✉)
Department of Biochemistry and Molecular Biology, School of Basic Medical Science, Guizhou Medical University, Guiyang, Guizhou, People's Republic of China

The Key Laboratory of Endemic and Ethnic Diseases (Guizhou Medical University) of the Ministry of Education, Guiyang, Guizhou, People's Republic of China

The Provincial Key Laboratory of Medical Molecular Biology, Guiyang, Guizhou, People's Republic of China
e-mail: wenfengyu@gmc.edu.cn

Z. Huang
Department of Interventional Medicine at the Affiliated Hospital of Guizhou Medical University, Guiyang, Guizhou, People's Republic of China

Z.-Z. Guan (ed.), *Coal-burning Type of Endemic Fluorosis*,
https://doi.org/10.1007/978-981-16-1498-9_13

Introduction

Endemic fluorosis was an environmental geochemical endemic disease that eroded teeth (dental fluorosis), bones (skeletal fluorosis), and human body's soft-tissue organs (Jha et al. 2011). In China, the cases of endemic fluorosis could be roughly classified into three types: drinking water-type, coal-burning type, and brick-tea type of endemic fluorosis (Guan et al. 2018). Chronic fluorosis is a widespread pathological disease afflicting multiple systems and organs. There were two main kinds of damage to the body: bone damage and non-bone damage. Bone damage were osteosclerosis, osteomalacia, secondary hyperparathyroidism, and osteoporosis, whereas non-bone damage mainly afflicted the central nervous system, liver, kidney, thyroid or other tissues and organs (Wei et al. 2019). The pathogenesis of endemic fluorosis was mainly related to the increase of oxidative stress level and apoptosis, change of signal transduction pathway, cell membrane structure and function, mitochondria as well as the disorder of calcium metabolism (Yadav et al. 2019). As the most widely studied theory, oxidative stress is also a necessary intermediate process to cause inflammatory reaction, which has been reported in many diseases. However, only a few reports tapped the pathological role of inflammatory response in the pathogenesis of endemic fluorosis. Thus, whether it participates in the pathogenesis of endemic fluorosis is a subject worth studying.

An Overview of Chronic Inflammatory Response

Chronic Inflammatory Response

Inflammation refers to the dynamic balance process of self-recovery and the body's defensive response to the stimulation in living tissues, including the vascular system. It is specifically manifested as swelling, heat, redness, pain and even dysfunction. If the inflammation cannot be completely eliminated by immune cells, or the tissue damage cannot be fully repaired, inflammatory response will phase in chronic inflammatory stage. Generally, chronic inflammation is caused by repeated acute inflammation and cannot be cured, but the presence of inflammatory factors and the continuous damage of tissues it brings is the root cause. The cytokines involved in the inflammatory response could be roughly divided into four categories, namely proinflammatory cytokines, inflammatory cytokines, inflammatory cytokines regulatory factors and anti-inflammatory cytokines (Jiang et al. 2019). Proinflammatory cytokines include tumor necrosis factor-α (TNF-α), interleukin (IL)-1β, IL-6, IL-8, granulocyte-macrophage colony-stimulating factor (GM-CSF), and cyclooxygenase-2 (COX-2), etc.; inflammatory cytokines include c-reactive protein (CRP), fibrinogen, serum amyloid A (SAA), cell adhesion molecules, plasminogen activator inhibitor-1 (PAI-1), etc.; inflammatory cytokines regulatory factors include nuclear factor-kappa B (NF-κB); anti-inflammatory cytokines include insulin (INS), α-melanocyte stimulating hormone (α-MSH), endothelial nitric oxide (ENO), IL-4, IL-IO, IL-13, adenosine triphosphate, etc. Various cytokines play different roles in different stages of inflammation.

Proinflammatory Cytokines

A substance capable of inducing the reduction of transplanted tumors in vivo and apoptosis in vitro was found (Carswell et al. 1975), then it was named tumor necrosis factor (TNF). In 1985, macrophages produced TNF officially was named as TNF-α (Shalaby et al. 1985), which is mainly secreted by activated monocytes or macrophages. Astrocytes, microglia, $CD4^+$ lymphocytes, natural killer cells, and neurons can also secrete it. Moreover, excessive TNF-α can not only activate itself, but can also promote the secretion of other cytokines such as IL-1, IL-6, and induce the generation of nitric oxide, endothelin, oxygen free radicals, and other medium.

Mainly produced by activated monocyte macrophages, IL-1β has the functions of regulating immune cells, promoting hematopoiesis, causing fever, and so on. It is an important substance that mediates the inflammatory response, as well as one of the most powerful inflammatory mediators in vivo. In the active phase of inflammatory diseases, non-activated pro-IL-1β is synthesized and then IL-1β activation followed. Therefore, IL-1β could also be regarded as inflammatory cytokines in certain way (Mwantembe et al. 2001).

IL-6 is a multifunctional cytokine that can be synthesized by different cells, including activated T cells and B cells, monocytes macrophages, endothelial cells, epithelial cells, and fibroblasts. It could induce inflammatory response, promote the differentiation and activation of a variety of immune cells, and enhance the effect of other cytokines (Calabrese and Rose-John 2014). However, it also had negative feedback regulation effect. In the case of increased concentration, the synthesis of TNF-α could be inhibited. Therefore, IL-6 could block the IL-1 receptor, thus reducing the inflammatory response (Rose-John 2012).

GM-CSF is a cytokine secreted by endothelial cells, fibroblasts, macrophages, and activated T cells under the stimulation of inflammatory factors (bacterial lipopolysaccharide (LPS), IL-1, and TNF-α). It is an important chemokine and activator of inflammatory cells. In the early stages of inflammation, it plays a role in transmitting external information as a signaling molecule. GM-CSF regulates various functions of immune cells, including cell adhesion, pathogen pattern recognition receptor expression, promotion of secretion of inflammatory factors (IL-12, IL-18, IL-6, and TNF-α, etc.), phagocytosis and kill pathogens. The generation and effect of GM-CSF can be regulated by IL-l, TNF-α, monocyte inflammatory protein, and so on, which was also confirmed in further studies in vivo. Moreover, the newly generated GM-CSF could be used as proinflammatory cytokines to induce the production of IL-1β and TNF-α in inflammatory sites (Fitzgerald et al. 2003).

COX2 is also considered to be a proinflammatory cytokine and an inducible enzyme of "inflammatory response." It is in a very low or non-expression state under normal conditions; when stimulated, it can express and synthesize rapidly on monocyte macrophages, vascular endothelial cells, and other cells closely related to inflammation, so as to participate in a variety of pathological and physiological processes. In the process of inflammatory response, COX2 was induced and expressed in a large amount, which would lead to the amplification and enhancement of inflammatory response. Moreover, prostaglandin (PG) E2, PGD2, PGF2, PGI2, and thromboxane A2 (TxA2) could be generated by COX2 pathway, the blood vessels would be expanded and the edema would be promoted (Tong et al. 2018).

Inflammatory Cytokines

According to recent studies, it was found that the increased level of some proinflammatory mediators and acute phase proteins in blood circulation suggesting the existence of chronic inflammatory response. CRP and SAA were two important APPs in the human body (Gabay and Kushner 1999). They are mainly synthesized by the liver in response to some proinflammatory cytokines. CRP is mainly regulated by circulating IL-6 and TNF-α under most situations. Its serum concentration is sensitive to various inflammatory states. CRP is a sign of cytokine generation under chronic inflammatory conditions, which indicates the chronic inflammatory states. For instance, it is expressed in conditions such as systemic lupus erythematosus, various arthritis, obesity, and sleep disorders. Moreover, in chronic inflammatory responses, the CRP level can remain constant for a long time without any day-and-night changes in its concentration. Therefore, CRP can be considered as a sensitive serological index to evaluate systemically inflammatory diseases with mild-grade. SAA is an acute phase protein, which plays an important role in acute and chronic inflammation. It is currently a commonly used inflammatory index in clinical testing. When bacterial or viral infection occurs, the body's SAA level would increase significantly (Ye and Sun 2015).

Inflammatory Cytokine Regulator

NF-κB was a toll-like receptor protein activated by the bond of inflammatory cytokines (TNF-a, IL-1, and IL-18, etc.) with pathogen-associated molecular patterns. It played a key and core regulatory role in inflammatory response (Lin et al. 2004). The presence of NF-κB in the cytoplasm does not show biological activity and it forms heteropolymer with Inhibitor of NF-κB (IκB). In general, the trimer formed by the binding of NF-κB (p50/p65) and IκB has no transcriptional activation function. When stimulated by external factors, IκB is degraded by protease. Then, the activated NF-κB enters the nucleus from the cytoplasm and binds to the characteristic site (κB site) of DNA to regulate the basics (growth factor, transcription factor, cytokine, chemokine, anti-apoptotic protein, etc.). Such process was termed signal transduction pathway of NF-κB (Li and Verma 2002). The signal transduction pathway of NF-κB can be activated by many factors such as pathogen, stress, and proinflammatory. It is involved in inflammation, apoptosis, and immune response. The abnormal increase of NF-κB activity can be induced by TNF-α, LPS, IL-l, etc., and it would lead to excessive release of a large number of proinflammatory cytokines. If NF-κB is over activated, the sustained expression of NF-κB regulated genes and its consequential generation of chronic inflammatory response will appear.

Anti-inflammatory Cytokines

INS can not only regulate blood glucose and reduce the harm of hyperglycemia to the body, but also directly inhibit the generation of inflammatory mediators, reduce inflammatory response, shorten the course of disease, and improve prognosis. These effects indicate its great clinical application value. Recent studies found that INS owns certain anti-inflammatory effect, which could be manifested by inhibiting the activity of NF-κB. Such anti-inflammatory effect enhanced the activity of anti-inflammatory cytokines or through other indirect ways. With such function, the levels of various inflammatory factors (CRP, adhesion molecules, PAI-1, IL-1) were significantly reduced, while the generation and release of anti-inflammatory cytokines (such as Eno, IL-4, IL-10) were increased (Zand et al. 2017).

α-MSH is a pre-opioid corticotropin derived peptide with anti-inflammatory effects on various acute and chronic inflammation. It can directly act on inflammatory cells in peripheral tissues in the form of endocrine, autocrine, paracrine; it mainly acts as an anti-inflammatory role on monocyte macrophages and neutrophils, inhibiting the secretion of proinflammatory cytokines, as well as promoting the secretion of anti-inflammatory cytokines. Furthermore, α-MSH could also pause the phosphorylation of NF-κB inhibitor IκB in vitro (TNF-α or LPS), thus inhibiting the dissociation of NF-κB and IκB, and the transcription of new genes mediated by NF-κB further (major histocompatibility complex class 1, IL-1, IL-6, G-CSF, nitric oxide synthase, COX2, chemokines, complement receptor, cell adhesion molecules). These genes played important roles in the development of inflammation (Mulero et al. 2019).

However, there is no absolute limit between anti-inflammatory cytokines, as they only play an anti-inflammatory role in certain ranges. For example, changes during the expression of anti-inflammatory cytokines IL-4, IL-10, and IL-13 are in phase similar to that of proinflammatory cytokines IL-6. Nevertheless, when they are overexpressed, not only the inflammation would not be inhibited, but might be promoted to spread. Therefore, many positive and negative feedback loops are formed between the cytokines in its complex network. Thus, the proinflammatory and anti-inflammatory mediators in the response are in the dynamic homeostasis.

Basic Pathological Changes of Inflammatory Response

The basic pathological changes of inflammation include metamorphism, exudation, and proliferation of local tissues with a certain sequence in the whole change. Metamorphism refers to local tissue degeneration and necrosis, which can occur in both parenchyma and stroma. The degeneration of parenchyma cells often includes fatty degeneration, coagulative or liquefying necrosis, while the degeneration of stroma can be manifested as mucoid degeneration, cellulose like degeneration or necrosis. Exudation refers to the process that the fluid, protein, and blood cells in the inflammatory local tissue enter the stroma, body cavity, surface or mucosal surface. Exudative lesions in the center are important signs of inflammation, the inflammatory exudation is caused by the increase of local vascular permeability and active outflow. In the inflammatory exudate, protein content is relatively high, and it contains more cell components and debris. Therefore, the proportion of exudate in serous cavity caused by inflammation is higher than 1.020, with a mixed appearance; the cell content is relatively large, which is different from that caused by the obstruction of venous return. Proliferation indicates that under the stimulation of inflammatory factors, tissue disintegration products or some physical and chemical factors, the macrophages, endothelial cells, and fibroblasts in the inflammatory area can proliferate. It is observed that even and diffuse enlargement of proliferative tissues and organs or the single or multiple proliferative nodules formed in tissues and organs. It is this proliferative reaction that makes the damaged tissue repaired.

Generally, the early stage is dominated by metamorphism and exudation, while the later stage is by hyperplasia. The three of them are closely related. Metamorphism belongs to damage process, while the other two belong to the process of anti-damage process. Inflammation is a unified process of injury, anti-injury, and repair, in which the repair reaction of inflammation can also lead to the dysfunction of the body.

Inflammatory Responses in Endemic Skeletal Fluorosis

Chronic fluorosis is a widespread pathologically damaging disease on multiple systems and organs in the body. The damage is categorized as bone damage and non-bone damage.

The most serious damage of excessive fluoride is on osseous tissue. Endemic fluorosis mainly eroded the teeth and bones, which was manifested as dental fluorosis and skeletal fluorosis in most cases (Perumal et al. 2013). Mild symptom patients usually experience sore pain, mostly in joints and limbs. However, with the aggravation of the disease, the articular function of severe patients will be impaired and even the limbs will be deformed. The characteristics of this disease are similar to that of general senile degenerative arthritis, which is a mild degree with long-term inflammatory state. The changes of endemic fluorosis cases and knee osteoarthritis

(OA) in control group from Turkey were compared (Savas et al. 2001). Subjects were divided according to age and gender. Later, such study proposed that the degree of OA could be significantly increased by endemic fluorosis; moreover, osteoarthritis in the fluorosis area was more frequent. There was a quantitative difference but no essential difference between endemic fluorosis and general osteoarthritis, but symptoms of the former was more severe than the latter. Most patients with osteoarthritis showed the increase of CRP, SAA, α-acid mucin, and haptoglobin. Through the histopathological observations of the knee synovium of fluorosis rats, it was found as compared with the control group, the medium and high fluoride group had obvious tissue congestion and edema, inflammatory cell infiltration, synovium cell layer increased to 3–10 layers, fibrous tissue hyperplasia, and macrophage (synovium cells) could be seen (Zeng et al. 2009). All these suggested that there was inflammatory reaction in the knee joint of fluorosis rats.

Inflammatory Responses of Non-bone Systems in Endemic Fluorosis

Non-bone damage mainly refers to the damage that endemic fluorosis afflicts on brain, heart, liver, kidney, gland, and other systems. Brain damage is the most serious one besides bone organs.

Central Nervous System Damage

As early as 1937, a report indicated that patients with endemic fluorosis had central nervous system symptoms related to spinal cord radiculopathy (Short et al. 1937). Moreover, existing research showed that in addition to the damages mentioned above, changes in brain function and morphology could also be caused by excessive fluoride. The accumulation of fluoride ions will form fat soluble complex, thus penetrating the barrier between blood and brain, and accumulating in brain tissue. Then, it causes changes in the shape, metabolism and function of the brain tissue, thus affecting learning and memory ability. After the examination of fetal brain tissue in fluorosis area, it was found that in addition to the increase of fluoride in brain, the mitochondria of neurons in cerebral cortex were enlarged, the rough endoplasmic reticulum was expanded, and the nuclear membrane was broken (He et al. 1989). Such results indicated that the neurons in fluorosis were denatured. Meanwhile, this study observed in previous experiments under electron microscope that while having chronic fluorosis, a number of mitochondrial cristae disappeared, mitochondria swelled, and mitochondrial cross-sectional area changed (Yu 1999). Excessive fluoride might also damage mitochondria by inducing high levels of reactive oxygen species, resulting in the redistribution of mitochondria in rat cerebral cortex neurons. Such results caused their aggregation in the neurons, reduced the synapse, and influenced its function. It was observed that an increase in advanced glycation end products (AGEs), the receptor for advanced glycation end products (RAGE), NADPH oxidase-2 (NOX2) mRNA and protein level as well as in ROS content in the brain of rats with chronic fluorosis (treated by long-term feeding of 50 ppm fluoride containing water). Thus, it could be concluded that excessive fluoride could activate AGE/RAGE pathway and enhance oxidative stress (Zhang et al. 2015). Through the past research, it was found that fluoride could damage the mitochondria of nerve cells and make them produce oxidative stress. However, previous studies confirmed that inflammation is related to oxidative stress in many diseases. Therefore, whether fluorosis can lead to chronic inflammation of the central nervous system is a subject worth further studying.

In the central nervous system, glial cells are great in number and microglia are in non-activated state under normal physiological conditions. However, when stimulated by exogenous agents, they can be activated to be an immuno-protective role. However, over activation of microglia will lead to the generation of a large number of cytokines and active oxygen products, and inflammatory factors such as TNF-α and IL-1β will be released. Thus, over activation aggravates the inflammatory response of the central nervous system. It was discovered that fluoride could induce microglia to generate inflammatory factors and activate NF-κB signal pathway (Tang et al. 2015). In addition, it was confirmed that the activation of microglia was the main manifestation of neuroinflammation in the brain. Moreover, the release of the inflammatory factors TNFα-and IL-1β was confirmed that excess of both could damage the blood–brain barrier and aggravate the central nervous system damage. So, excessive fluoride can definitely lead to inflammatory reactions in the brain.

Liver Damage

More and more researchers pay much attention to the direct damage of fluoride to liver tissue and its function. Some studies showed that when fluoride entered the body, it would consume the antioxidants in the body, and the free radicals would clear the enzymes. It eventually caused the imbalance between liver oxidation and antioxidant defense mechanism, and caused the liver into oxidative stress state, thus causing cell damage (Wang et al. 1997). Liver contains a lot of unsaturated fatty acids, and it is vulnerable to free radical attack. Earlier studies under light microscope proved that there were degenerative changes in the liver of fluorosis animals, dilation and congestion in the central vein of liver lobule, and punctate necrosis in the hepatocytes (Kolodziejczyk et al. 2004). Observing the rat model with chronic fluorosis under the electron microscope, it was found that the mitochondria of liver cells were swollen, vacuole like degeneration occurred with calcification, cristae were broken or disappeared; white lipid droplets could be found in the cytoplasm, and the increase of volume density and surface area density showed different degrees of swelling (Huang et al. 2010). All above findings indicate the existence of inflammatory reaction.

Renal Damage

Through urine, feces, and sweat, fluoride can be discharged from the body. Thus, the kidney is the most important way to discharge fluorine. The damage of fluoride to kidney is mainly manifested by renal insufficiency, which leads to the decrease of defluorination ability in kidney and the aggravation of fluorosis with fluoride in the body. In early studies, it was found that the kidney of fluorosis rats (fed with 50 ppm high fluoride water for a long time) were observed to have decreased alkaline phosphatase activity, abnormal mitochondrial morphology and structure, and focal necrosis of proximal convoluted tubules (Yu et al. 1988). In subsequent study, it was further confirmed that fluoride could cause kidney damage, degeneration, and edema of renal tubular epithelial cells. In scrious cases, these results could lead to cell necrosis, which was one of the basic pathological changes of inflammation (Gao et al. 2005). This further indicated that there was inflammatory reaction in kidney in fluorosis. Syndecan-4 is a kind of transmembrane heparan sulfate proteoglycan. Its extracellular domain can interact with many ligands, such as growth factors, and can regulate a variety of cell biological effects. At the same time, it also mediates the inflammatory response. Studies showed that NF-κB could mediate the increase of syndecan-4 expression. The expression level of syndecan-4 and NF-κB protein in the kidney of rats exposed to fluoride were increased (Zhang et al. 2013). It could be seen that the long-term ingestion of excessive fluoride might change the expression of syndecan-4 and NF-κB, thus participating in the process of kidney damage caused by endemic fluorosis. Therefore, it was believed that the pathogenesis of endemic fluorosis might be related to the chronic inflammatory response.

Damage on Heart and Blood Vessels

Fluoride is an active non-metallic element, which easily combines with blood calcium and magnesium. Such combination causes calcium metabolism disorder and in turn leads to functional and organic damage of cardiovascular system. Through the electrocardiogram (ECG) examination of clinical fluorosis patients, it could be found that their rhythm is abnormal, especially sinus arrhythmia and sinus bradycardia. T wave depression or biphasic patients also accounted for a large proportion, P-R and Q-T interval changes were not obvious (Wang 1995). Through the investigation of residents' ECG in the high fluoride area, it was found that their abnormal ECG changes were significantly higher than those in the control group, and the myocardial damage in the residents of high fluoride area was diffuse (Su et al. 1995). In animal experiments, it was also found that the amplitude of T wave and Q-T interval decreased in rats treated with high fluorine water (Song et al. 2000). Besides, fluorosis could cause damage to heart, which was mainly manifested in the elevation of myocardial enzyme spectrum level, the shortening of Q-T period, and the change of myocardial morphology (Liu et al. 2018). All the data above proved that excessive fluoride did damage on heart, affected the systolic function of the heart, reduced the pumping capacity of the heart, and lead to abnormal ECG. Combined with the pathological changes of myocardial cells, it could be inferred that these damages might be caused by chronic inflammation of the heart in endemic fluorosis patients. The toxic effect of fluoride on cardiovascular system was a complex process of multiple pathological reactions that promoted and interacted with each other. Therefore, whether the cardiovascular damage is related to inflammatory reaction should be further explored.

Thyroid Damage

Thyroid has a strong ability to absorb fluoride since it is an important endocrine organ. Excessive fluoride will cause direct damage to thyroid cells. This kind of cytotoxic damage will increase with time and accumulation of fluorine in the body, so that the thyroid function can be inhabited. In previous studies (Yu et al. 1998), it was proved that increase was detected and observed with stereological method in average circumference, average cross-sectional area, bulk density, and average volume of mitochondria of thyroid follicular epithelial cells of fetuses in fluorosis area. While in other experiments, it was observed that the number of thyroid gland follicular cells, colloid concentration, interstitial capillary hyperplasia, and thyroid tissue were increased in fluorosis rats, and there were obvious hyperplastic nodules (Liu et al. 2002). Some studies found that after observations of the epithelial structure of thyroid follicles in fluorosis rats under light microscope, an increase in number, level and disorder arrangement were observed in follicular epithelium hyperplasia. Besides, columnar or high columnar cells could protrude from the follicular cavity forming papilla. In addition, some of the follicles had less glia, empty cavities, and red stained homogeneous cavities; some of the thyroid tissues had necrosis, structural destruction, unclear cell structure, or even disappearance of nuclei. Some of the interstitial blood vessels were also dilated and congested, the lumen was enlarged, even congested or infiltrated by inflammatory cells; some of the interstitial fiber tissues were focal or diffuse hyperplasia. From these studies, it could be confirmed that the inflammatory reaction was of great significance in the thyroid injury of endemic fluorosis.

Reproductive System Damage

Fluoride can cause significant changes in the testis and epididymis of male animals in tissue morphology and even cells. The changes mainly manifested in the decrease of sperm count, activity, deformity rate, and fertility. The mechanism of fluorosis on testis remained unclear, but some scholars proposed oxidative stress. Oxidative stress was an important link in the mechanism of fluorosis. In a study on oxidative stress effect of fluorosis on testis, it was found that the toxic effect on reproductive system caused by excessive fluoride was closely related to testis oxidative stress (Ghosh et al. 2002). Another experiment indicated that the testicular seminiferous tubules in the fluoride exposed groups were damaged to varying degrees, and the degree of damage was aggravated with the increase of fluoride dose (Xiao et al. 2011). The structure of spermatogenic cells at all levels was loose and disordered, the cells were swollen, the volume was increased, the cytoplasm was vacuolated, some cells were necrotic, and the degree of damage was increased with the exposure of fluoride. All these indicated that there was inflammatory reaction in testicular tissue.

Ovary is the main sex organ of female, which has the function of generating mature eggs and secreting steroid hormone. At present, most toxicity research of fluoride to reproductive system focused on male, with a few relative researches on female. It has been reported that fluoride can produce toxic effects on the reproductive system of female animals (Zou et al. 2014). In the model of coal-burning type of fluorosis rats, when the group was exposed to fluoride for 60 days, the nuclear membrane of granular cells slightly shrank, the distribution of chromatin in the nucleus was normal. The mitochondria slightly deformed, and a small amount of lipid droplets could be observed. With the prolonged fluoride exposure time, the nuclear membrane shrinkage gradually increased, and the edge of chromatin became more obvious. After 180 days of fluoride exposure, it was found that the nuclear membrane shrinkage was serious, endoplasmic reticulum and ribosome were few, a large number of mitochondria were slightly swollen, and there were more lipid droplets in the cytoplasm. While in the high fluoride group, with the increase of fluoride exposure time, the following were observed: nuclear membrane of granular cells shrank into lace shape, the chromatin was massive, there was edge aggregation, the nucleolus became smaller or disappeared, there were more lipid drop like structures in the cytoplasm, a large number of mitochondria were seriously deformed, and the cells showed apoptosis like changes. The results showed that fluorosis could affect the ultrastructure of rat ovarian granulosa cells leading to apoptosis. As an important type of ovarian cells, granulosa cells play an important role in the process of oocyte maturation in vitro and in vivo, and they are both important cells for estrogen synthesis. The damage to them will affect the development of oocytes and lead to the disorder of estrogen synthesis, thus affecting the normal reproductive system function. It was proved that in the early stage of fluoride exposure, the expression of insulin-like growth factor-I (IGF-I) in the ovaries of female rats with endemic fluorosis had no significant change compared with the normal level, but the content of serum estradiol was significantly increased; while in the middle and late stage of fluoride exposure, the expression of IGF-I in the ovaries and the content of serum estradiol were gradually decreased, which indicated that endemic fluorosis might lead to the damage of ovarian granulosa cells (Li et al. 2018).

Conclusion

In conclusion, there may be a link between inflammatory response and endemic fluorosis, and chronic fluorosis may be accompanied by inflammatory response. Currently, despite the lack of relative established research, the inflammatory response does provide a new approach to study the pathogenesis, treatment, and prevention of chronic fluorosis. Such approach also deepens the understanding of the basic pathological changes with various studies on the change of inflammatory factors in chronic fluorosis. Past research all contributes to the further understanding of the pathogenesis of endemic fluorosis, as well as provides a new theoretical basis for drug treatment and prevention of this disease. Therefore, more efforts need to be made in such kind of research.

References

Calabrese LH, Rose-John S. IL-6 biology: implications for clinical targeting in rheumatic disease. Nat Rev Rheumatol. 2014;10:720–7.

Carswell EA, Old LJ, Kassel RL, Green S, Fiore N, Williamson B. An endotoxin-induced serum factor that causes necrosis of tumors. Proc Natl Acad Sci U S A. 1975;72:3666–70.

Fitzgerald SM, Chi DS, Hall HK, Reynolds SA, Aramide O, Lee SA, et al. GM-CSF induction in human lung fibroblasts by IL-1beta, TNF-alpha, and macrophage contact. J Interf Cytokine Res. 2003;23:57–65.

Gabay C, Kushner I. Acute-phase proteins and other systemic responses to inflammation. N Engl J Med. 1999;340:448–54.

Gao Q, Wang SL, Yu YN, Liu JL, Xiao KQ. Morphological changes of free radical in kidney of rats with chronic fluorosis. Guizhou Med J. 2005;29:213–5.

Ghosh D, Das S, Maiti R, Jana D, Das UB. Testicular toxicity in sodium fluoride treated rats: association with oxidative stress. Reprod Toxicol. 2002;16:385–90.

Guan ZZ, Wu CX, Qi XL. Evaluation of health status of population in coal-burning endemic fluorosis area in Guizhou province after comprehensive treatment. J Guizhou Med Univ. 2018;43:1124–8.

He H, Chen ZS, Liu WQ. Effects of fluoride on human fetuses. Chin J Control Endem Dis. 1989;3:136–8.

Huang DY, Wang D, Zhang XH, Yu MJ, Liu XL, Chang AS. Ultrastructure observation of liver in rats with fluorosis caused by coal burning under the high aluminum condition. Mod Pre Med. 2010;37:3104–7.

Jha SK, Mishra VK, Sharma DK, Damodaran T. Fluoride in the environment and its metabolism in humans. Rev Environ Contam Toxicol. 2011;211:121–42.

Jiang N, Li Y, Shu T, Wang J. Cytokines and inflammation in adipogenesis: an updated review. Front Med. 2019;13:314–29.

Kolodziejczyk L, Kuźna-Grygiel W, Myśliwiec Z. Protective effect of chrysin in rats subchronically exposed to sodium fluoride. Fluoride. 2004;37:209–20.

Li Q, Verma IM. NF-kappaB regulation in the immune system. Nat Rev Immunol. 2002;2:725–34.

Li XH, Xia SH, Liu YM, Yu MJ. Effect of low protein intake on ovary IGF-I expression of rats with coal-burning fluorosis. J Environ Health. 2018;35:402–4.

Lin L, Spoor MS, Gerth AJ, Brody SL, Peng SL. Modulation of Th1 activation and inflammation by the NF-kappaB repressor Foxj1. Science. 2004;303:1017–20.

Liu GY, Zhang WD, Gu JH, Chai CY. Effects of fluoride on ultrastructure of thyroid mitochondria in chick. J Shanghai Jiaotong Univ. 2002;03:224–7.

Liu ZM, Liu L, Xie Y. Study on the mechanism of heart damage in endemic fluorosis. Chin J Control Endem Dis. 2018;33:44–86.

Mulero MC, Huxford T, Ghosh G. NF-κB, IκB, and IKK: integral components of immune system signaling. Adv Exp Med Biol. 2019;1172:207–26.

Mwantembe O, Gaillard MC, Barkhuizen M, Pillay V, Berry SD, et al. Ethnic differences in allelic associations of the interleukin-1 gene cluster in South African patients with inflammatory bowel disease (IBD) and in control individuals. Immunogenetics. 2001;52:249–54.

Perumal E, Paul V, Govindarajan V, Panneerselvam L. A brief review on experimental fluorosis. Toxicol Lett. 2013;223:236–51.

Rose-John S. IL-6 trans-signaling via the soluble IL-6 receptor: importance for the pro-inflammatory activities of IL-6. Int J Biol Sci. 2012;8:1237–47.

Savas S, Cetin M, Akdoğan M, Heybeli N. Endemic fluorosis in Turkish patients: relationship with knee osteoarthritis. Rheumatol Int. 2001;21:30–5.

Shalaby MR, Aggarwal BB, Rinderknecht E, Svedersky LP, Finkle BS, Palladino MA, et al. Activation of human polymorphonuclear neutrophil functions by interferon-gamma and tumor necrosis factors. J Immunol. 1985;135:2069–73.

Short HE, Robert GR, Bernard TW, Mannadiyar AS. Endemic fluorosis in the Madras Presidency. Indian J Med Res. 1937;25:553–61.

Song SL, Bian JC, Xiang YZ, Ye P. Effect of fluorosis on electrophysiology of myocardial cells and electrocardiogram of the rats. Bull Endem Dis. 2000;1:11–2.

Su L, Zhang HM, Ba TL. Electrocardiogram observation of residents in high fluoride area. Neimenggu Pre Med. 1995;01:34–5.

Tang TT, Yu WF, Guan ZZ. Enhanced expression of inflammatory cytokines and nuclear factor-κB in microglia by overdose fluoride. Chin J Endem. 2015;34:785–9.

Tong D, Liu Q, Wang LA, Xie QP, Huang Y, Jiang J. The roles of the COX2/PGE2/EP axis in therapeutic resistance. Cancer Metastasis Rev. 2018;37:355–68.

Wang JL. Effects of fluoride on the cardiovascular system. Guiyang: Guizhou Science and Technology Press; 1995.

Wang Y, Zhao BL, Li X, Su Z, Xi W. Spin trapping technique studies on active oxygen radicals from human polymorphonuclear leukocytes during fluoride-stimulated respiratory burst. Fluoride. 1997;30:5–15.

Wei W, Pang S, Sun D. The pathogenesis of endemic fluorosis: Research progress in the last 5 years. J Cell Mol Med. 2019;23:2333–42.

Xiao YH, Sun F, Li CB, Shi JQ, Gu J, Xie C, et al. Effect of endemic fluoride poisoning caused by coal burning on the oxidative stress in rat testis. Acta Acad Med Sin. 2011;33:357–61.

Yadav KK, Kumar S, Pham QB, Gupta N, Rezania S, Kamyab H, et al. Fluoride contamination, health problems and remediation methods in Asian groundwater: a comprehensive review. Ecotoxicol Environ Saf. 2019;182:1–23.

Ye RD, Sun L. Emerging functions of serum amyloid A in inflammation. J Leukoc Biol. 2015;98:923–9.

Yu YN. Somatopic study on mitochondrial injury during fluorosis. Chin J Public Health. 1999;10:879.

Yu YN, Guan ZZ, Zhang WY, Jiang SF, Tang H, Yang WQ, et al. Changes of renal histochemistry, light microscopy and electron microscopy in rats with chronic fluorosis. Chin J Endem. 1988;7:2–4.

Yu YN, Dong Z, Liu JL. Determination of stereological parameters of fetal thyroid in fluorosis area, Guizhou Province. Chin J Control Endem Dis. 1998;03:148–9.

Zand H, Morshedzadeh N, Naghashian F. Signaling pathways linking inflammation to insulin resistance. Diabetes Metab Syndr. 2017;11:307–9.

Zeng JS, Li L, Mo YM, Yu MJ, Zhu CL. Expression of bone morphogenetic protein-2 and bone morphogenetic protein-7 in the knee joint synovial of fluorosis rats. Chin J Endem. 2009;28:28–31.

Zhang KL, Lou DD, Guan ZZ. Changes of transmembrane proteo-glycosaminoglycan-4 and nuclear factor-B in kidney of rats with chronic fluorosis. Chin J Endem. 2013;32:133–5.

Zhang S, Zhang X, Liu H, Qu W, Guan Z, Zeng Q, et al. Modifying effect of COMT gene polymorphism and a predictive role for proteomics analysis in children's intelligence in endemic fluorosis area in Tianjin, China. Toxicol Sci. 2015;144:238–45.

Zou WB, Xia SH, Xu JF, Yu SW, Wang FQ, Wang SJ, et al. Effects of fluorosis caused by coal burning on ovarian granular cells in female rats. J Environ Health. 2014;31:124–7.

Chapter 14
Effect of Fluoride on Signal Transduction Pathways

Xiao-Lan Qi

Abstract With cytotoxicity in nature, fluoride is the essential micro-element for human body. Excessive or deficient intake would invariably cause lesions. Meanwhile, signal transduction is a highly complicated network in human body, which plays a key regulation role for cellular proliferation, differentiation, metabolism, and apoptosis. A substantial number of studies indicated that fluoride worked physiologically and pathologically via regulating multiple signal transduction pathways. Such results shed light on further understanding of fluoride mechanism and screening medicine targets for fluorosis prevention and treatment.

Keywords Drug target · Fluorosis · Signal transduction · Toxicity mechanism

Introduction

Signal transduction refers to the process that cells respond accordingly to extracellular signals, the binding between signal molecules in extracellular environment (primary messenger) and target cell receptor is then turned into perceptible signal (secondary messenger) via signal conversion. There are multiple signal transduction methods and pathways, which then form multilayer cross-regulation into a complex network. It aims to make the overall body respond to external environmental changes in most appropriate way. Excessive fluoride intake causes chronic fluorosis and poses damages on multiple organs and systems of the body, which is related with the pathological and physiological function of fluoride-regulated signal transduction pathways. Further exploring the influence of fluoride on signal transduction

X.-L. Qi (✉)
The Key Laboratory of Endemic and Ethnic Diseases (Guizhou Medical University) of the Ministry of Education, Guiyang, Guizhou, People's Republic of China

The Provincial Key Laboratory of Medical Molecular Biology, Guiyang, Guizhou, People's Republic of China
e-mail: xlq@gmc.edu.cn

Z.-Z. Guan (ed.), *Coal-burning Type of Endemic Fluorosis*,
https://doi.org/10.1007/978-981-16-1498-9_14

pathways will surely assist better understanding of fluoride effect mechanism and screening drug targets for fluorosis prevention and treatment. The following content focuses on relevant study of fluoride and cellular signal transduction.

Effect of Fluoride on G-protein Coupled Receptor (GPCR) Signal Transduction Pathway

In terms of structure, GPCR is monomeric protein, with seven transmembrane α helixes in its three-dimensional structure; binding site of signal transmitting molecule, guanosine triphosphate (GTP) binding-protein (G-protein in short), can be found at both the carboxyl end of its peptide chain and cytoplastic loop linking of 5th and 6th transmembrane helix. G-protein is the first signal molecule of this signal pathway, found in the inside of cytoplasmic membrane in the form of α, β, γ subunit trimer. α subunit has multiple active sites, including the site binds with receptor and to be activation-regulated by the receptor, the site binds with β and γ subunits, the site binds with guanosine diphosphate (GDP) or GTP, and the site interacts with downstream effector molecules. The α subunit also has the activity of GTP enzyme. G-protein functions as the switch of cell in signal transduction process. In most cases, the α subunit of G-protein binds with GDP in shut-down state; when extracellular ligand binds with GPCRs, intracellular protein domain of GPCRs couples with α subunit and replaces α subunit bind-GDP with GTP; α subunit decouples with β or γ subunit, then turns into active state and involves in next signal transmitting process; the hydrolysis of GTP reverses it back to inactive state. There are various α, β, and γ subunits, which then bind with a variety of G-protein coupled receptors in different combinations and form the diversified signal transduction systems.

The activation of G-protein can act on effector molecules like cyclic adenosine monophosphate (cAMP) and phospholipase C (PLC). Such action can transmit extracellular signal into intracellular one, and then intracellular signal will act on downstream molecule to regulate cell proliferation and differentiation. Fluoride is a specific G-protein activating agent, which can trigger the activation of various G-proteins. It was discovered in the 1980s that fluoride stimulates cell proliferation and is related to G-protein (Sternweis and Gilman 1982). Early study considered that fluoride acted on GPCR, later study found that fluoride acted on α subunit of G-protein and fluoride-activated cAMP involved with micro-amount of aluminum (Susa 1999). Other studies further identified that the mechanism of fluoroaluminate-activated G-protein was based on structural and spatial similarity of (aberrant lateral root formation 4 (AlF_4)) molecule and phosphate (Nagasaka et al. 1993; Park et al. 2013), aluminum fluoride penetrated membrane to bind inactive α subunit of G-protein directly and formed Gα-GDP-AlF_4 compound; conformation of such compound was similar with activated Gα-GTP compound, then α subunit was in activated state and transferred extracellular signals.

cAMP Signal Transduction Pathway

cAMP is the key effector molecule after G-protein activation. The activation of cAMP transduction pathway includes the activation of Gαs/Gαi/o protein, adenylyl cyclase (AC), cAMP, and protein kinase A (PKA) in sequence. The rising level of cAMP indicates the activation of Gαs, while decreasing level of cAMP indicates the activation of Gαi. Fluoride can induce apoptosis of non-small cell lung cancer cell (NSCLC, A549) by activating cAMP/PKA pathway; the pretreatment of H89, a specific inhibitor of PKA, can obviously inhibit fluoride-induced A549 cell apoptosis; increased generation of cAMP and activated PKA will not affect A549 cell apoptosis. These findings reveal that increased level of cAMP can enhance other fluoride-induced apoptosis signals, but such signals will not be strong enough to trigger apoptosis. Another explanation proposed that fluoride induced continuous PKA activation and then triggered apoptosis (Refsnes et al. 2003). Fluoride can also increase cAMP level of resident macrophage and inflammatory human peritoneal macrophages by inducing AC activation (Gutowska et al. 2012). Pre-treated mouse embryonic osteoblast precursor cells (MC3T3-E1) with AC activator forskolin, fluoroaluminate can lower cAMP level by 90% by activating Gαi protein and pertussis toxin can reverse such effect (Susa et al. 1997), while fluoride can also lower cAMP level in rat hepatocytes. This indicates that fluoride's effect on cAMP signal pathways differs on different organs. Further research on osteoblast and hepatocyte is required to determine whether fluoride regulation on cAMP pathway is related with fluoride-induced apoptosis or not.

Protein Kinase C (PKC) Signal Transduction Pathway

Extracellular signal molecule binds with GPCR Gαq/11 on cell surface, then it activates PLC on plasmalemma for hydrolysis of phosphatidylinositol-4,5-bisphosphate (PIP2) into two secondary messengers: inositol 1,4,5-triphosphate (IP_3) and diacylglycerol (DAG).

Fluoride triggers increased concentration of intracellular Ca^{2+} through multiple pathways, including triggering the release of intracellular calcium stores, inhibiting calcium pump, and activating calcium channel. The activation of $G\alpha_{q/11}$ protein-PLC signal transduction pathway is one of the major pathways. The IP_3 produced by such pathway can bind with the membrane receptor on rough endoplasmic reticulum (ER), enabling IP3-gated Ca^{2+} channel to open; hence, Ca^{2+} will be released from rough ER to cytoplasm and triggers increased Ca^{2+} concentration; however, Ca^{2+} overload is the common feature of cytotoxicity and apoptosis. In both human pulmonary artery endothelial cells and hepatocyte cultured in vitro, fluoride can trigger elevated IP3 and DAG to cause the increase of intracellular free Ca^{2+} concentration. Increased Ca^{2+} concentration will bind with calmodulin (CaM) and then activate calcium/calmodulin dependent protein kinase (CaMK) successively to regulate cellular function and induce apoptosis. CaM and CaMK inhibitor can obviously inhibit A549 cell apoptosis induced by fluoride, while calcium ionophore can promote the fluoride-induced apoptosis (Refsnes et al. 2003). In the study of fluoride effect on nervous system, it was discovered that chronic fluoride stimulation could trigger Ca^{2+} overload in mouse hippocampus synaptosomes; Ca^{2+} bound with CaM to activate calcium/calmodulin dependent protein Kinase II (CaMKII) and triggered elevated c-fos expression, which finally induced neuron apoptosis (Teng et al. 2018). In addition, renal cell apoptosis was obviously increased since fluoride could increase CaM expression in such cells (Shao et al. 2020). All these studies suggested that Ca^{2+}-CaM-CaMKII pathway was one of the major pathways of toxic effect of fluoride.

Fluoride can trigger increased mouse synaptosome Ca^{2+} concentration obviously, then the expression of following factors in calcium ion signal transduction pathway was found to be abnormal, including CaMKII, c-fos, nuclear transcription factor-kappa B (NF-κB), Bcl-2, and Bax (Zhang et al. 2011a). Such results indicated that fluoride-induced hippocampus synaptosome Ca^{2+} concentration increase might be the promoter of fluoride-induced neuron apoptosis; on the other hand, fluorine ion promoted NF-κB expression and regulated pro-apoptosis gene expression so it could induce and aggravate neuron apoptosis. Fluoride can also induce the early and continuous increase of Ca^{2+} levels in human neuroblastoma SH-SY5Y cells, which may be related to fluoride-induced reactive oxygen species (ROS) generation and cytotoxicity (Xu et al. 2013). In a study on the kidneys of rats with chronic fluorosis, it was found that fluoride exposure significantly increased the expression of CaM in renal cells and decreased the expression of CaMKII (Shao et al. 2020). Furthermore, the expression of anti-apoptotic Bcl-2 protein in renal cells was significantly reduced, while the expression of pro-apoptotic Bax protein increased significantly. With the increase of fluoride exposure concentration and time, renal cell apoptosis increased significantly. The Ca^{2+}-CaM-CaMKII pathway may also be an important pathway for fluoride-induced nephrotoxicity.

DAG activates a variety of PKC under the synergism of Ca^{2+} and then causes a cascade reaction through PKC. The activation of PKC was also involved in fluoride-induced A549 cell apoptosis process. Fluoride induced a continuous increase in PKC activity of A549 cells and induced cell apoptosis, pretreatment with terephthalic acid (TPA), an activator of PKC, for 20 h almost eliminated fluoride-induced and basic PKC activity (less than 5% of the level of the untreated group), and almost prevented the apoptosis. It indicated that PKC actively involved in fluoride-induced apoptosis. Further study suggested that TPA could obviously enhance fluoride-induced A549 cell apoptosis within certain sodium fluoride (NaF) concentration (0–5 mmol/l). As the fluoride concentration reached 7.5 mmol/l, apoptosis level obviously lowered, while the proportion of fluoride-induced propidium iodide (PI) staining positive cells was obviously increased and almost all cells showed PI staining positive. Such results suggested that the combined effect of TPA and fluoride was to induce cellular necrosis rather than apoptosis under relatively high fluoride level. The intriguing fact was that 5 mmol/l NaF induced apoptosis was enhanced by pretreatment A549 cells with PKC specific inhibitor GF109203X (10 μmol/l). It could be explained that the ability of PKC to induce apoptosis or anti-apoptosis in varied types of cell and toxic injury was highly dependent on the activation of different PKC subtypes.

Hence, the activation of G-protein by fluoride could stimulate various G-protein signal transduction pathways, such as PKA and PKC related systems. These pathways could directly or indirectly promote or inhibit fluoride-induced apoptosis in a coordinated manner. Considering the key function of GPCR in fluorosis development, it would be of great importance to research and demonstrate the relationship between GPCRs and fluorosis for the development of new anti-fluorosis medicine.

Effect of Fluoride on Mitogen-Activated Protein Kinases (MAPKs) Signal Transduction Pathway

MAPKs were the major messenger to send signals from cell surface to nucleus. They are a group of serine-threonine kinase which could be activated by a variety of extracellular stimulations. By third degree phosphorylation, MAPKs were activated by the signals of upper stream molecule, mitogen-activated protein kinase kinase (MAPKK), and mitogen-activated protein kinase kinase kinase (MAPKKK); then MAPKs regulated multiple key cellular physiological/pathological processes (cellular growth, differentiation, stress adaptation to environment, inflammation response). MAPKs family included three key members: extracellular signal-regulated protein kinases (ERK1/2), c-Jun N-terminal kinases (JNK_S), and p38. A substantial number of studies indicated that MAPK signal transduction pathway was involved in various cellular injury processes caused by fluoride.

It was found that 5 nM NaF induced human odontoblast mineralization of a rat odontoblast-like cell line (MDPC-23 cells) apoptosis (Karube et al. 2009); meanwhile, phosphorylated JNK level increased in 1 h and peaked at 2 h and then decreased slightly. Phosphorylated p38 level increased in 1h and peaked at 4–6 h and remained elevated at 24 h. However, fluoride has a biphasic effect on the phosphorylation of ERK. 30 m after fluoride treatment showed increased phosphorylated ERK level, 2–6 h later decreased to control level; 8 h later, phosphorylated ERK level increased again and rose obviously with the duration of fluoride treatment. JNK inhibitor SP600125 treatment could obviously inhibit fluoride-induced apoptosis; while MAPK kinase 1/2 (MEK1/2) inhibitor U0126 treatment could only partially inhibit NaF induced MDPC-23 apoptosis. Inhibition of p38 activity with SB203580 has no significant effect on fluoride-induced apoptosis. Such results indicated that fluoride-induced MDPC-23 apoptosis was largely determined by JNK, while ERK pathways showed less contribution. The MAPK pathway involved in A549 cell apoptosis induced by fluoride is different. p38 and JNK participate in fluoride-induced A549 cell apoptosis, while phosphorylation activation of ERK can resist apoptosis. Such findings suggested that the role of MAPK in fluoride-induced apoptosis might vary due to different cellular types. Previous study recorded that

Smad protein involved in transforming growth factor-β1 (TGFβ1) induced apoptosis in the process of fluoride-induced MDPC-23 apoptosis. Interference between MAPK and Smad protein might be related with fluoride-induced MDPC-23 cell apoptosis. There was also research on replicated germ cell from rat's testicle with coal-burning type of fluorosis, it discovered that fluoride could upregulate levels of phosphorylated p-JNK, p-Jun, and pERK to induce its apoptosis. It suggested that fluoride might play a major role in reproductive system injury.

There were studies focused on the effect of varied dosages of fluoride (0, 12, 24, and 48 mg/kg) on liver inflammation in mice (Chen et al. 2019). The results suggested that fluoride could trigger histopathological injury on liver, including degeneration and necrosis of hepatocyte, as well as inflammatory cell infiltration. Such changes were in line with the inflammatory factor changes in liver. Further study discovered that ERK, JNK, and p38 pathway were all activated in hepatocyte of mice with fluorosis. Such manifestations were obvious with the increase of dosage and extension of fluoride exposure time, since changes were dose- and time-dependent. This study suggested that the activation of MAPK signal pathway could increase inflammatory agent, such as the expression of interleukin-1β (IL-1β), interleukin-6 (IL-6), interleukin-8 (IL-8), and cyclooxygenase-2 (COX-2), and it played a major role in fluorosis liver injury mechanism. Another study proved that MAPK signal pathway was a participant of fluorosis-induced inflammation response: exposing A549 to 5 mmol/l NaF discovered that COX-2 mRNA and protein expression level obviously increased in 4 h, peaked in 8 h and 12 h, then decreased mildly in 24 h (Ridley and Matsuoka 2009). Phosphorylated level of ERK and p38 began to increase 30 m after being treated by NaF. MEK1/2 inhibitor U0126 or p38 inhibitor SB203580 was added for pretreatment, which could all obviously inhibit upregulation of COX-2 protein expression level induced by NaF. It suggested that MAPK pathway involved in the upregulation of COX-2 protein expression induced by fluoride. Activation of ERK and p38 promoted the COX-2 generation.

There was a correlation between the injury of fluoride-induced pulmonary artery endothelial barrier and activation of ERK pathway. 20 mmol/l NaF was used to treat cattle pulmonary artery endothelial cells for 2 h could obviously increase the permeability of albumin and lower its trans-epithelial electrical resistance, and then caused endothelial barrier damage. Meanwhile, such process came along with the activation of ERK, ERK2 was activated at 4 min, then peaked at 10 min and remained there for 60 min; the activation of ERK1 also peaked at 10 min and then soon decreased. In addition, fluoride shortly increased phosphorylated level of Raf-1. The selective Raf-1 inhibitor ZM-336372 could obviously lower fluoride-induced ERK and phosphorylated level of caldesmon. Fluoride would not trigger the activation of ERK after the interference of MEK1/2 inhibitor U0126, and phosphorylated level of CaM obviously decreased at the meantime (Bogatcheva et al. 2006). The above results suggested that fluoride induced activation of Rho/Raf-1/MEK successively, then activated ERK and CaM; the activated CaM could trigger myoglobin ATP enzyme to lose the ability to inhibit or cause the actin filament to rearrange, then ultimately lead to pulmonary artery endothelial barrier injury.

Fluorosis caused the learning and memory decrease in rats. Such mechanism might be associated with the activation of ERK-MAPK signal pathway in brain tissue by fluoride. The test on brain tissue from rats with coal-burning type of endemic fluorosis and drinking water type of endemic fluorosis revealed that the expression level of phosphorylated ERK 1/2 protein and gene expression level of ERK 1/2 in brain of fluorosis group were increased significantly; such increase was proportional to the fluoride dosage and showed a dose-dependent relationship with the learning and memory reduce in rats. This may be due to the fact that fluorine caused abnormal development and differentiation of nerve cells and increased apoptosis through this pathway, affecting cell function, resulting in a decrease in brain learning and memory functions (Liu et al. 2010). There was also a study showed that 0.8 mg/l fluoride or 0.8 mg/l fluoride with MEK1/2 inhibitors treated primary cultured hippocampal cells could both obviously decrease pERK1/2 level. This might be associated with the difference of fluoride concentration and experiments in vitro/vivo.

This part summarized the key regulating feature of MAPKs signal transduction pathway in fluorosis caused cytotoxicity and apoptosis. MAPKs signal transduction pathway was involved in cellular pathological and physiological process (growth, development, differentiation, and apoptosis) and associated with multiple signal transduction pathways. Therefore, it is pressing to conduct in depth research on the role of MAPKs signal transduction pathway in fluorosis-induced physiological processes together with its correlation and effect on other signal transduction systems.

Effect of Fluoride on Apoptosis Signal Transduction Pathway

Cell apoptosis refers to the independent and orderly death process resulting from the programmed cell death, which was activated by various intrinsic and extrinsic cellular signals under certain physiological or pathological conditions. Although the detailed mechanism of apoptosis process remained unclear, caspase played an irreplaceable role in the apoptosis process. Such process was the cascade amplified response of caspase irreversible limited hydrolysis substrate. Apoptosis was strictly under control, and multiple apoptosis related regulating proteins were involved in regulating apoptosis process. There were two major apoptosis pathways, extrinsic death receptor pathway and intrinsic mitochondrial pathway. In addition, ER stress signal transduction pathway also played a major role in cell apoptosis. These pathways were all involved in the process of fluoride caused cellular injury.

Death Receptor Mediated Apoptosis Signal Transduction Pathway

The toxic effect of fluoride on teeth may be related to the apoptosis of ameloblasts, gingival fibroblasts and cementoblasts induced by fluoride. Twenty-four hours treatment by 3.2 mmol/l NaF could significantly induce ameloblast apoptosis, obvious increase in protein level was found in FasL and Fas, upregulation of cleaved caspase-8, caspase-3, and poly adenosine diphosphate-ribose polymerase (PARP). Such results further proved that fluoride could induce ameloblast apoptosis by activating FasL/Fas death receptor pathway (Wang et al. 2016). Meanwhile, blocked FasL/Fas signal transduction could partially counter fluoride-induced apoptosis, which indicated that other pathways might involve in fluoride-induced apoptosis. It was reported that treatment of human gingival fibroblasts with 10 mmol/l and 20 mmol/l fluoride for 6 h can lead to a significant increase in the level of apoptosis (Lee et al. 2008), and the expression level of FasL was increased in a dose-dependent manner, enzyme activity of caspase-8 and -3 was obviously increased. Such results indicated that fluoride could induce human gingival fibroblast apoptosis via death receptor pathways and then cause gingiva damage. Similar study discovered that fluoride could induce cementoblast OCCM-30 cell apoptosis via death receptor pathway (Ni et al. 2018).

In neuroblastoma SH-SY5Y cells, exposure to 40 and 80 mg/l fluoride for 24 h obviously lowered cellular activity and significantly increased apoptosis rate (Xu et al. 2011). Fas, FasL, caspases-3 and -8 mRNA expression level and caspase-3 activity were significantly higher than the control group. Fas agonist CH-11 obviously enhanced fluoride-inducing apoptosis with synergism. While Fas blocking antibody ZB4 partially inhibited fluoride-induced SH-SY5Y cell apoptosis. These results indicated that exposure to fluoride could induce apoptosis and stimulate the Fas/FasL death receptor of neuronal cells, which might be one of the mechanisms of fluoride caused neural system damage.

A study on fluoride treated sertoli cells (SCs) of mice discovered that the protein and mRNA levels of FasL and the mRNA level of caspase-8 and -3 in the fluoride treated group were significantly decreased with the increase of fluoride concentration (Sun et al. 2017). When splenic lymphocytes were co-cultured with fluoride treated SCs cells, it was found that the co-cultured lymphocytes showed obvious high cell viability, low apoptosis rate, and reduction of Fas, which indicated that fluoride interfered with Fas/FasL death receptors pathway affects the immunity of the testis, which may play a role in the damage of the fluoride on reproductive system.

Mitochondrial Mediated Apoptosis Signal Transduction Pathway

Fluoride could cause abnormal changes in apoptosis level, since it acted on more than signal molecules in extrinsic apoptosis pathway, but also on intrinsic ones. Multiple studies proved that fluoride could induce apoptosis via mitochondrial pathway by increasing cytochrome C (Cyt C), caspase-3 and -9 mRNA or protein level of multiple cells, including cytogenesis, osteoblast, HL-60, Leydig cells, H9C2 cardiomyocyte, oocyte, human pulmonary BEAS-2B cells, rat's liver and thymus.

As the most common early symptoms of chronic fluorosis, dental fluorosis is triggered by excessive intake of fluoride in enamel development period and then causes increased ameloblast apoptosis in both secretory and mature phases. Such increase ultimately causes abnormal enamel matrix structure due to the decreased enamel matrix protein secretion and leads to mineralization disorders. Mice ameloblast LS8 was treated with fluoride for 48 h and discovered that 2 mmol/l NaF treatment caused an increase in ameloblast apoptosis rate; Bcl-2 gene and protein expression level were all lowered (Yang et al. 2013). The study suggested that Bcl-2 was involved in regulation of fluoride-induced ameloblast apoptosis and affected the development of enamel, which could explain the development of dental fluorosis. Furthermore, fluoride reduced the cell viability of human gingival fibroblast in a dose- and time-dependent manner; induced apoptosis with concomitant chromatin condensation and (deoxyribonucleic acid) DNA fragmentation (Lee et al. 2008). In addition, fluoride increased the level of Cyt C released from the mitochondria into the cytosol; enhanced the caspase-9, -8, and -3 activities, and the cleavage of PARP; and upregulated the voltage-dependent anion channel 1 and FasL, however, downregulated Bcl-2. These results suggest that the mitochondrial pathway is involved in the apoptosis of human gingival fibroblasts induced by fluoride, thus causing gum damage.

In a study (Yan et al. 2009) osteoblast was isolated from neonatal rat's cranium and then treated with NaF (0.012 mmol/l, 0.12 mmol/l, 0.24 mmol/l, 0.48 mmol/l, and 0.72 mmol/l) for 24–72 h. The results showed that fluoride induced osteoblast apoptosis in dose-dependent way; gene expression levels of caspase-9 and -3 in 0.12 mmol/l dose group were obviously increased, while those in 0.48 mmol/l dose group were not obviously increased, which suggests that certain dosage of fluoride could cause the development of skeletal fluorosis by triggering osteoblast apoptosis via caspase-dependent mitochondrial pathway. There were also possibilities that other caspase non-dependent pathways might be involved in the apoptosis process. Fluoride could also increase permeability of mitochondrial membrane in osteoblast and activate mitochondrial apoptosis signal transduction pathway by reducing mRNA expression of Bcl-2 and increasing mRNA expression of Bcl2-associated X protein (Bax) (Yang et al. 2011).

Fluoride showed distinct reproductive toxicity. Studies have shown treated primary cultured rat's testicle Leydig cell with different concentrations of fluoride, and it was found that fluoride treatment caused increased apoptosis level of testicle

Leydig cell, which was dose-dependent and time-dependent. Bcl-2 protein expression was downregulated, while protein expression of Bax, caspase-9 and -3 was upregulated by fluoride. It suggested that fluoride could cause reproductive system injury by triggering testicle Leydig cell apoptosis via mitochondrial pathway (Song et al. 2014). Another research established fluorosis model by adding 100 mg/l NaF to drinking water to mice for 90 days. Research on the ovaries found that compared with the control group, ovarian cell apoptosis in fluorosis group obviously increased, while mRNA and protein expression level in caspase-3, -9, Bax and Cyt C of fluorosis group obviously upregulated, expression level of Bcl-2 obviously downregulated (Wang et al. 2017). Such results suggested that fluoride decreased growth potential of ovarian cells by inducing ovarian cell apoptosis via mitochondrial pathway.

ER Stress Mediated Apoptosis Signal Transduction Pathway

ER is the major site for folding and maturing of transmembrane protein, secretory protein, and ER retention protein. When the cell received certain extrinsic stimulations, ER will initiate a series of regulation mechanisms. Accumulation of misfolded protein and unfolded protein will trigger ER stress (ERS), then ERS will trigger the aggregation of intracavitary misfolded protein and unfolded protein as well as balance disorder of Ca^{2+}. The body can maintain the milieu interne stability by unfolded protein response (UPR) to protect the cells from ERS caused injury and restore cellular functions. For mammals, UPR is mediated by three transmembrane sensory molecules on protein kinase R (PKR)-like endoplasmic reticulum kinase (PERK), activating transcription factor 6 (ATF6), and inositol-requiring enzyme 1 (IRE1). It plays the role of restoring balance state of ER and promoting cell survival by inhibiting nascent protein translation, enhancing protein folding ability and accelerating misfolded protein degradation. However, when ERS lasts for too long or UPR adaptive functions are deficient, severe or chronic UPR activation can regulate relevant molecules to mediate apoptosis. A large number of researches discovered that fluoride triggered apoptosis by inducing ERS was closely related with its cellular toxicity.

There is a correlation between ERS and fluorosis caused skeletal and cartilage disease. Relevant molecules in UPR signals (PERK and X-box binding protein-1 (Xbp1)) play a major regulation role in skeletal and cartilage development, and ERS was also involved in skeletal fluorosis development. The effects of fluoride on osteoblast (MC3T3-E1 cells) were investigated (Li et al. 2019). The results indicated that the proliferation of MC3T3-E1 cells was inhibited by exposure to fluoride. In addition, apoptosis was induced by fluoride, as caspase-associated proteins showed a higher level of expression and apoptotic bodies were formed. Furthermore, ERS induced by fluoride activated the unfolded protein response and upregulated the expression of the glucose-regulated proteins 94 (GRP94) and 78 (BiP). Therefore, ERS plays a vital role in fluoride-induced autophagy and apoptosis. Furthermore, apoptosis is promoted following the inhibition of fluoride-induced autophagy. Such data indicate under fluoride treatment, the ERS-signaling pathway is activated, leading to apoptosis and autophagy and affecting the proliferation and survival of MC3T3-E1 cells.

The results from treating rats with 150 mg/l NaF and different concentrations of $CaCO_3$ indicated that fluoride-induced pathological and ultra-microstructure damage on bone tissues; DNA fragmentation and an increase in osteocalcin (OCN) were found on 120 days; fluoride exposure upregulated mRNA and protein expression as well as downregulated Bcl-2 expression of Bax, calpain, caspase-12, -9, -7, -3, chronic actinic dermatitis (CAD), PARP, and apoptosis-inducing factor (AIF) (Wang et al. 2019a). It might be related with fluoride-induced ERS, while calcium supplement could counter such effect of fluoride.

Ameloblast has vital function in enamel generating process, since its proteinase secretory function is closely related with enamel formation and mineralization. Exposure to high content fluoride can impair ameloblast function in tooth development. It leads to intracellular protein accumulation and decreased mineralization. All these results are inseparable from fluoride's effect on ERS signal transduction pathway. Low concentration of fluoride (1.9–3.8 mg/l) could obviously inhibit ameloblast growth cultured in vitro, while high concentration fluoride (38 mg/l) could upregulate ERS relevant genes in ameloblast, such as molecular chaperone glucose-regulated protein 78 (GRP78), Xbp1, and protein expression level of growth inhibition and GADD153/C/EBP homologous protein (CHOP). These findings suggested that fluoride could induce ERS in ameloblast and UPR activation. Such process was also accompanied with caspase-dependent DNA fragmentation changes in ameloblast (Zhang et al. 2016b). Since CHOP is an important regulatory factor in the process of ERS-mediated apoptosis, it suggests that the apoptosis of ameloblasts induced by high fluoride is related to ERS. The above-mentioned changes were also proved solid in animal test (in vivo). These results suggest that exposure to high fluoride during the development of tooth enamel can damage the ER function of ameloblasts, cause ERS and UPR activation in ameloblasts, interfere with normal protein synthesis and secretion of ameloblast.

Apart from bone damage, ERS was also involved in fluorosis caused soft tissue injury and closely related to apoptosis. In SH-SY5Y cells exposed to fluoride, the expressions of GRP78, IRE1, apoptosis-inducing factor CHOP, and cleaved-caspase-3 obviously increased. Such results suggested that fluoride could induce neural cells apoptosis by activating ERS/IRE1/CHOP pathways in ER, and the same results were got in vivo study (Niu et al. 2018). It was observed that exposure to fluoride could enable eIF2α phosphorylation and CHOP expression in pancreatic cells (Ito et al. 2009). It could be reasoned that PERK/eukaryotic initiation factor-2α (eIF2α) signal pathways might be involved in fluoride caused pancreatic cell damage and be related to fluoride caused digestive system damage. In rats with fluorosis, the gap between the glomerulus and the wall of the renal capsule is obviously enlarged, and there was uneven distribution of epithelial cells and cellular necrosis. In addition, excessive fluoride intake caused GRP78, XbpP1, and CHOP mRNA and protein expression increase in renal tissues. It suggested that excessive fluoride intake could induce renal damage via IRE1/CHOP pathway. Excessive fluoride exposure would also lead to reproductive system function disorder and ultimately lead to cellular pathological damage and apoptosis. Fluorosis would not only cause histopathological changes and abnormal ultra-microstructure in testicles. Excessive fluoride would enable upregulated expression in GRP78, PERK, and CHOP of Sertoli cells as well as eIF2α phosphorylation. It could be concluded that fluorosis was involved in ERS-mediated PERK/CHOP pathway to promote testicles cellular apoptosis and then trigger fluoride damage on reproductive system.

Phosphatidylinositol 3-Kinase/Protein Kinase B (PI3K/PKB) Signal Transduction Pathway

PI3K/Akt signal pathway is one of the major pathways involving in cellular proliferation. It functions as anti-apoptotic as its major role and pro-apoptotic as well. PI3K is an intracellular phosphatidylinositol kinase. As a heterodimer, it is composed by p110 catalytic subunit and p85 regulator subunit with lipid kinase activity and protein kinase. It could be activated directly by receptor tyrosine kinase or Ras and p110. The products (PIP_2 and phosphatidylinositol 3, 4, 5-trisphosphate) are key secondary messengers which can further activate Akt. Then, the activated Akt can regulate apoptosis in multiple pathways, including inhibiting caspase-9 phosphorylation directly to inactivate it, inhibiting mitochondrial release apoptosis relevant factors like Cyt C and AIF to inhibit apoptosis.

100 nmol/l Wortmannin, specific inhibitor of PI3K, could obviously enhance A549 cell apoptosis induced by fluoride (0–3.75 mmol/l) (Refsnes et al. 2003); fluoride treated C2C12 myoblast of mice could upregulate protein level of Bcl-2-associated death promoter, Bcl-2 and Bax, while expression of 3-phosphoinositide dependent protein kinase-1 and p-Akt1 would be significantly lowered (Zhou et al. 2018). Such studies indicated that PI3K- Akt pathway was the major part in fluoride-induced apoptosis. Another study suggested that PI3K- Akt signal pathway also played a vital role in pathogenesis of fluorosis triggered liver damage. In the liver tissue of rats with chronic fluorosis, mRNA and protein expression of PI3K and Akt 1 all increased obviously, while levels of p-PI3K and p-Akt obviously decreased (Fan et al. 2015). Such results indicated that fluorine ion could inhibit activation of downstream protein of PI3K-Akt pathway by inhibiting the activation of p-PI3K. This might be related with the increased hepatocyte apoptosis in fluoride-induced chronic fluorosis.

NF-κB Signal Transduction Pathway

NF-κB signal transduction pathway shoulders major function in apoptosis regulation. This pathway is widely found in multiple histocytes and a major nuclear factor in nucleus. NF-κB family of mammals' cell includes five members: NF-κB1 (p50), NF-κB2 (p52), RelA (p65), RelB, and c-Rel (P75). NF-κB refers to the NF-κB1 heterodimer protein formed by p65/p50; RelB/p52 subunit forms NF-κB2 heterodimer protein. Inhibitor of NF-κB (IκB) is the inhibit factor of NF-κB, it can form compound with NF-κB heterodimer in cytoplasm and then inhibit NF-κB from entering nucleus to play its transcription function; conversely, it can inhibit NF-κB from binding with DNA in the nucleus directly. Studies have found that the target proteins regulated by NF-kB include pro-apoptotic factors (Bax, caspase-11, Fas, and FasL), apoptosis inhibitory factors (Bcl-2, B-cell lymphoma-XL, apoptosis-inducing protein inhibitor of apoptosis protein), and cell cycle regulatory factors. NF-κB is also involved in multiple responses like immune, stress, inflammation by regulating adhesion molecule, growth factor, angiogenic factors, and so on.

Multiple studies found that abnormal relevant molecules of NF-κB signal transduction pathway shared close relationship with the apoptosis regulation and occurrence/development of chronic fluorosis organ damage. NF-κB is a major component in the formation and differentiation of osteoclast. Research on rats with chronic fluorosis discovered that low content fluoride could stimulate the expression of NF-κB in osteoclasts, reduce the apoptosis, and increase the activity of osteoclasts and at the same time promote the production of osteoblasts, stimulate cell proliferation, and lead to hyperostosis; high fluoride content caused severe cellular function damage, inhibited NF-κB expression, increased osteoclast apoptosis, and decreased activity, which in turn aggravated bone sclerosis. In osteoclast-like cell treated with different concentrations of fluoride for 7d, the expression level of Fas and FasL was all obviously increased with the increase of fluoride dosage; however, protein expression level of NF-κB was obviously lowered. It suggested that fluoride promoted osteoclast-like cell apoptosis by inhibiting NF-κB expression (Sun et al. 2002). These findings all suggested that NF-κB has apoptosis inhibiting effect in fluoride-induced osteoclast apoptosis process.

NF-κB signal transduction pathway was also involved in renal damage process which was caused by chronic fluorosis. In renal rats with chronic fluorosis, the degree of renal damage deteriorated with the increase of fluoride dose, and expression level of NF-κB in all fluoride groups showed linear increase. In addition, such process also came with abnormal changes of antioxidant enzyme activity in renal tissues as well as the accumulation of lipid peroxide malondialdehyde (MDA). The results suggested that oxidative stress could be the major trigger for fluoride caused abnormal expression of NF-κB in kidney, hyperactive NF-κB could cause the abnormal expression of relevant downstream gene, then triggered immune-inflammatory response in renal tissues, ultimately caused renal structure change and lead to renal damage (Oyagbemi et al. 2017). Another study fed Sprague-Dawley (SD) rats with 10 mg/l and 50 mg/l drinking water for 6 m. The study discovered dose-dependent increase in gene and protein expression levels of NF-κB p65 and membrane transporter polysaccharide-4 (syndecan-4) while inhibiting the translocation and expression of NF-κB could decrease the protein expression of syndecan-4. This study suggested that renal damage caused by upregulation of syndecan-4 might be regulated by NF-κB. In contrast with the above research findings, fluoride was found to decrease expression of NF-κB in kidney of similar animal models with chronic fluorosis. SD rats (100–120 g) were exposed to 5 and 50 mg/l NaF drinking water for 8 months. Later, the test found that with the increase of fluoride dosage, gene and protein expression levels of renal tissue NF-κB p50 and NF-κB p65 were gradually decreasing compared with control group, while protein expression level of IκBα was gradually increasing; apoptotic cells were distributed in renal tubule epithelial cells; apoptosis related protein also showed consistent changes with the gradual increase of fluoride dosage. It is suggested that fluoride may inhibit the activation of NF-κB p50 and NF-κB p65, so that NF-κB is in a resting and non-functional state, resulting in abnormal expression of NF-κB related downstream genes, thereby affecting the structure and function of the kidney. Increased expression of IκBα could bind more p50/p65 to hold NF-κB in cytoplasm and weaken its anti-apoptotic effect, increase apoptosis of renal tubule epithelial cells and induce nephrosis. The inconsistent findings of above-mentioned researches might be related with fluoride dosage, fluoride exposure time, and feeding method. NF-κB was involved in renal inflammatory response caused by fluoride, in addition to its function in regulating renal tissue apoptosis. This suggested that NF-κB mediated inflammatory response was also involved in fluoride caused renal damage process.

NF-κB is also one of the target molecules of neural injury, liver injury, and cardio-vascular injury caused by chronic fluorosis. The study of chronic fluorosis rats discovered sporadic distribution of hippocampal neurons with disappeared cell outline and vacuole in fluorosis rat group. Such changes would be even more obvious with the increasement of fluoride exposure time and concentration. Protein expression of NF-κB p65 in hippocampal tissues of fluorosis group showed increase in different degrees, while protein expression of Bcl-2 and Bax, and Bcl-2/Bax ratio showed decrease in varying degrees. The above results indicated that fluoride can promote NF-κB activation and regulate expression of apoptosis related gene. Hence, fluoride can induce and aggravate neurons apoptosis. Other studies have also proved that the increased expression of synaptosome Ca^{2+} and NF-κB r65 may be the molecular basis of chronic fluorosis-induced central nervous system damage (Zhang et al. 2011a). Rats were given intragastric administration of 12, 24, and 48 mg/kg NaF for 48 days to replicate fluorosis model. The study discovered that fluoride could trigger liver histopathological injury, including liver cellular degeneration and necrosis, inflammatory cell infiltration, obvious upregulation of expression of pro-inflammatory cytokines interleukin-1 (IL-1), IL-6, IL-8, monocyte chemoattractant protein-1 and COX-2 in hepatocyte and downregulation of anti-inflammatory cytokines interleukin-4 (IL-4) and transforming growth factor-β (TGF-β). In this way, fluoride could induce hepatic inflammatory response. Meanwhile, fluoride could upregulate mRNA and protein expression level of NF-κB and

IKK-β in liver and downregulate that of IκB. Such results indicated that fluoride could activate NF-κB signal pathway, which could be related with fluoride-induced hepatitis response and promote liver injury (Chen et al. 2019). There was also study proved that fluoride triggered hypertension was due to induced reactive oxygen generation and activated NF-κB expression of kidney and heart.

To sum, apoptosis is a major part and key feature of fluorosis process. In addition to the typical mitochondrial pathway and death receptor pathway, fluoride works its toxicity out of inducing apoptosis by ERS and can affect apoptosis by regulating PI3K/Akt and NF-κB pathways. The piling studies on fluoride and apoptosis pathways enabled people to grasp a better understanding of fluorosis pathogenesis. Therefore, regulating apoptosis signal transduction pathway as target point should be explored for early stage prevention and control of fluorosis as well as research and development of anti-fluorosis medicine.

Effect of Fluoride on TGF-β/Smad Signal Transduction Pathway

TGF-β superfamily is formed by polypeptide growth factor subfamilies with related structure and functions, including 30 proteins like TGF-βs, bone morphogenic proteins (BMPs), growth differentiation factors, glial line-derived neurotrophic factors, etc. By regulating a series of downstream gene activity, this superfamily is involved in many vital biological processes, like nervous system differentiation, development of teeth and bones, as well as carcinogenesis. In the process of TGF-β signal transmitting from membrane to nucleus, Smad protein is the key part as TGF-β binds with its receptor and further phosphorylates Smad protein (Smad1, Smad5, and Smad8); then Smad molecule is separated from membrane receptor and translocated to nucleus after binding with Smad4 molecule in cytoplasm. In nucleus, Smad compound works on specific target gene with the involvement of other DNA-binding protein and then regulates the transcription of target gene.

TGF-β is indispensable for the differentiation of osteoblast and osteoclast. A study analyzed rats' bone mineral density treated by different concentrations of fluoride with or without TGF-β1 inhibitor SB431542, along with TGF-β signal transduction and related gene expression of bone turnover. The study found that SB431542 treatment would affect rat's bone density and gene expression, which indicated that TGF-β1 and activin receptor-like kinase 5 (ALK5) might be involved in fluoride-induced bone turnover and bone formation. In vitro experiment indicated that low concentration fluoride could improve bone marrow mesenchymal stem cell cellular activity, activity of alkaline phosphatase, and expression of osteocalcin protein, but high concentration fluoride could inhibit these expressions. Such findings might be related to the increased gene expression of *Runx2* and *ALK5* in low concentration fluoride treated cell, since high concentration fluoride also inhibited gene expression of *Runx2* and *ALK5* (Yang et al. 2017). Another study discovered that fluoride treatment could induce upregulated expression of osteoblast TGF-β receptors (TβRs), Smad3, and MAPK. Treating osteoblast with TGF-β could reverse the inhibition of high dose fluoride on osteogenic differentiation in bone marrow mesenchymal stem cells (MSCs) (Zhang et al. 2019). It suggested that TGF-β receptor mediated Smad3 and MAPK signal transduction was indispensable for regulating viability and differentiation of fluoride on osteoblast while activating TβR2-smad3 signal transduction pathway could reverse inhibition of high dosage fluoride on osteoblast differentiation.

The study on mice osteoblast cells with TGF-β1 overexpression and silencing in vitro indicated that when osteoblast exposure to fluoride would cause proliferation rate of rat's osteoblast to decrease obviously in dose-dependent way, at the same time, expression level of TGF-β1 and autophagy would rise accordingly (Zhao et al. 2018). Overexpression of TGF-β1 obviously increased fluorescent intensity of multidisciplinary diagnostic center staining cells, mRNA and protein expression level of homologen of yeast autophagy gene *Atg6/Vps30* (Beclin1) and light chain 3-II/I (LC3-II/I), and lowered p62 at the meantime, and vice versa for TGF-β1 silencing. All these results indicated that TGF-β1 functioned as mediator in fluoride-induced mice osteoblast autophagy and affected osteoblast differentiation. Studies on osteoclasts have also shown that different doses of fluoride have a biphasic effect on the viability, differentiation, formation, and function of osteoclasts. Low dose of fluoride has a significant stimulating effect on osteoclast activity, differentiation, and function, while high dose fluoride could inhibit such stimulant function after a certain period of time and inhibit its differentiation and function. Fluoride treated osteoblast-like cells enhanced the expression of TβR1 and Smad3. With the increase of fluoride concentration, Smad3 and protein expression of phosphorylated Smad3 showed obvious dose-dependent features. For osteoclast-like cells treated by SB431542, fluoride stimulation on Smad3 and expression of phosphorylated Smad3 were blocked. These findings indicated that TβR1/Smad3 were involved in the dual regulation effect of fluoride on osteoclast. It also suggested that fluoride had such stimulant effect via TβR1-phosphorylated Smad3 signal pathway in TGF-β1.

Fluoride could lead to dental fluorosis by affecting TGF-β/Smads signal transduction pathway. As its obvious inhibition on TGF-β expression of incisor enamel organ, fluoride could obviously inhibit Smad2/3 expression in rat's incisor development and expression of Smadl and Smad5 in inner enamel epithelial (Zhang et al. 2011b). This suggested that fluoride could inhibit the expression of TGF-β and its downstream signal molecule Smads at the same time and magnify such inhibition. Fluoride may affect the normal signal transduction between epithelium and mesenchyme by interfering with the TGF-β/Smad signal transduction pathway, thereby inhibiting the differentiation of ameloblasts and odontoblasts and subsequent matrix synthesis and secretion.

To sum, current study on fluoride toxicity in TGF-β/Smad signal transduction pathway was primarily focusing on bone damage. Its effect in fluorosis caused non-bone or soft tissue damage required further research.

Effect of Fluoride on Nuclear Factor Erythroid 2-Related Factor 2 (NRF2) Signal Transduction Pathway

Chronic fluorosis causes comprehensive damage on the body, possibly due to oxidative stress resulting from fluoride accumulation induced free radicals' level to increase. Nrf2/antioxidant response element (ARE) signal pathway might play a major role in this development.

Nrf2 and its cytoplasmic adaptor Keap1 are the central control of cellular antioxidant response. In physiological state, Nrf2 exists in cytoplasm and Nrf2 signal transduction pathway remains in relatively silent state. Nrf2 binds with Keap1 closely to anchor Keap1 protein on actin constructed cytoskeleton, prohibiting Keap1 from entering nucleus to develop its transcriptional activity. When the body has oxidative stress or is stimulated by activator signal, Keap1 and Nrf2 were decoupled to transfer Nrf2 into nucleus. Then, it binds with Maf protein in gene into heterodimer, then being recognized and binds with ARE. Such binding would activate downstream antioxidant protein gene for transcription and increase cellular antioxidant ability. Due to NADPH quinone oxidoreductase-1 (NQO1) and heme oxygenase-1 (HO-1), which are the major antioxidant proteinase in Nrf2/ARE pathway, their inducing expression is indispensable in enhancing body's antioxidant ability.

In the early stage of fluorosis, fluoride ions attack the cysteine of Keap1 and change the conformation of Keap1. The increased Nrf2 drift away into nucleus and activate Nrf2/ARE signal transduction pathway. Such change compensates early-stage fluorosis damage on the body by increasing the content of superoxide dismutase, catalase, and glutathione S-transferase in body, which can enhance the antioxidation of body and scavenge free radicals. In the late stage of fluorosis, activity and content of antioxidant enzyme decrease. Fluorine ion penetrates membrane and nuclear membrane successively and finally triggers oxidative stress by damaging ARE (disenable the binding between Nrf2 and ARE), downregulating the expression of antioxidant enzyme, lowering antioxidant system function, and passively accumulating free radicals. Excessive reactive oxygen free radicals caused lipid peroxide in membrane phospholipid, which would damage the biological membrane and let in more fluorine ions. Lipid peroxides such as MDA have direct damage to DNA and protein, destroy the ARE sequence, and further downregulate the expression of antioxidant enzymes. A study proved that animal tissues with Nrf2 gene knockout were more prone to the damage of various toxics.

The research on mice ameloblast cultured in vitro discovered that fluoride could induce the expression of Nrf2 gene and protein after fluoride treated ameloblast, which would then enhance the expression of antioxidant enzyme HO-1 and NQO1 in downstream of Nrf2 signal transduction pathway (Zhou et al. 2019b). Nrf2 agonist tBHQ could induce enhanced gene expression of ameloblast HO-1, NQO1, Nrf2, which had protection on cells and could antagonize toxicity of fluoride on ameloblast to a certain degree. This suggested that Nrf2 signal transduction pathway was vital in fluoride-induced oxidative stress damage. Plus, in test of fluoride treated osteoblast (cultured in vitro) for 24 and 48 h, it was found that relatively high concentration (1.00–4.00 mmol/l) could continuously induce gene expression of osteoblast Nrf2; relatively low concentration of fluoride (0.50 mmol/l) after 72 h treatment could also induce gene expression of Nrf2. In cytoplasm, Nrf2 protein showed a decreasing tendency with the increase of fluoride dosage, while Nrf2 protein content in nucleus showed an increasing tendency with the increase of fluoride concentration. Such findings suggested that fluoride could induce the increase of gene expression level of osteoblast Nrf2, hence, it could trigger the expression increase of HO-1 and NQO1 of Nrf2-ARE downstream target gene, which could then enhance antioxidant defense ability of osteoblast.

Nrf2 signal transduction pathway was also involved in fluoride neurotoxicity process. After treating PC12 cells for 24 h with different concentrations of fluoride, it was found that the production of intracellular ROS increased. 20 mg/l and 40 mg/l NaF could obviously upregulate the expression of Nrf2 protein; but its expression decreased with the increase of fluoride dosage, which suggested that low dosage of fluoride could activate protein expression of Nrf2; in high dosage fluoride condition, a large amount of ROS generation led to cellular decompensation, downregulated expression level of glycine cleavage system protein H and HO-1, decreased glutathione content, and increased apoptotic rate. Treating cells with Nrf2/ARE pathway inducer tBHQ could obviously reduce the toxicity of fluoride on cells, which indicated that activating such pathway could partially antagonize the damage of fluoride-induced oxidative stress on cells (Wu et al. 2015). In the study of microglia, 25 mg/l NaF obviously stimulated the expression of microglial cell Nrf2 and enhanced its accumulation in nucleus while decreased the expression of Keap1 protein. Activating microglial cell Nrf2 could antagonize fluoride-induced oxidative stress and inflammation and also silence Nrf2 for counteraction (Song et al. 2020). In the study on zebrafish, fluoride treatment could cause enhanced cerebral oxidative stress level and increased Nrf2 expression, with hyperactivity, frequent surface activity, tumbling, and vertical swimming. It also proved that fluoride could regulate oxidative stress via Nrf2 pathway (Mukhopadhyay et al. 2015).

In the study of animal model with chronic fluorosis, it was discovered that fluoride could activate Nrf2 signal transduction pathway in cerebral tissues. The SD rats were exposed to drinking water containing 50 mg/l NaF for 10 months. It found that Nrf2, Keap1, NQO1, HO1 of cerebral cortex (including frontal lobe, parietal lobe, and occipital lobe) and hippocampus (CA3, CA4, and dentate gyrus) all obviously increased. Adding antioxidant vitamin E for interference could obviously inhibit fluoride caused protein expression increase of Nrf2 pathway. The above-mentioned results suggested that chronic fluorosis could trigger obvious increase in related factor's expression of Nrf2/ARE signal transduction pathway in rat brain. This was related with changes and regulation of oxidative stress state and might be involved in pathogenesis of chronic fluorosis brain damage.

Another study differed from the above-mentioned study (fluoride active cerebral tissue Nrf2 signal transduction pathway). Wistar male rats (weighing 160–180 g) were given intragastric administration of NaF (5 mg/kg) for 4 weeks. The results suggested that fluoride could trigger pathological changes of renal tissues and cause apoptosis and inflammation of renal tissues; fluoride could also obviously decrease protein expression level of renal tissue Nrf2, HO-1, γ-GCS, and GST and upregulate protein expression level of Keap1. The findings suggested that fluoride had inhibition on Nrf2 signal transduction pathway of kidney, since fluoride could damage its normal antioxidant defense and lead to oxidant injury in kidney (Thangapandiyan and Miltonprabu 2014). In the study on fluoride caused lung injury, same method was applied (subjects were given intragastric administration of 5 mg/kg NaF for 4 weeks), and similar results were obtained, which caused lung inflammation, increased oxidative stress, pulmonary edema, and decreased antioxidant ability, meanwhile, the levels of Nrf2 and HO-1 were decreased and Keap1 protein expression was upregulated. Epigallocatechin gallate (EGCG), which had antioxidant and anti-inflammation effect, could obviously improve the body's antioxidant state by activating Nrf2 transmit to nucleus, inhibiting oxidative stress and expression of inflammatory cytokines and Keap1. It indicated that EGCG could relieve rat's oxidative lung injury induced by fluoride by activating Nrf2/Keap1 pathway (Shanmugam et al. 2016).

Oxidative stress is the major mechanism of fluorosis damage. Nrf2 signal transduction pathway is a major pathway in regulating ROS of fluorosis process. A substantial amount of study proved that it is a functional target which could serve as entry for fluorosis prevention and treatment.

Effect of Fluoride on Wnt/β-catenin Signal Transduction Pathway

Wnt signal transduction pathway is a pathway with multi-downstream channels stimulated by the binding of ligand protein Wnt and membrane protein receptor. Wnt pathway is involved in regulating various cellular physiological activities, including proliferation, differentiation, transfer and gene expression. Wnt/β-catenin signal transduction pathway is the typical pathway of Wnt signal transduction pathway, the extracellular Wnt signal would trigger Wnt protein, Frz protein and lipoprotein receptor-related protein (LRP) LRP5/6 protein on membrane to form trimer and move the signal into cell, then the signal could inhibit the activity of glycogen synthase kinase-3β (GSK3β) to prevent GSK3β from phosphorylating β-catenin. In this way, it could inhibit β-catenin degradation and accumulate β-catenin in cytoplasm steadily. Later on, β-catenin could enter nucleus and work with transcription factor TCF/lEF to activate and regulate the expression of target gene. Recent studies indicated that Wnt/β-catenin signal transduction pathway was indispensable in inducing the directed differentiation from mesenchymal stem cells of bone marrow to osteoblast and regulating bone metabolism as well as physiological activities of osteoblast. Runx2 was also termed as core binding factor-al, which was the downstream target gene of β-catenin. Wnt/β-catenin could affect osteoblast and bone formation by regulating the expression of Runx2. Piling evidences indicated that pathogenesis of fluorosis shared a close relationship with Wnt/β-catenin signal transduction pathway.

The study on rat model with sub-chronic fluorosis revealed that the increase of bone trabecula led to increase of bone mass in fluorosis group. Meanwhile, the mRNA level of LRP5, β-catenin, and Runx2 rose with the increase of fluoride dosage in fluorosis group, so did the protein expression of β-catenin. It suggested that Wnt/β-catenin signal transduction pathway was involved in pathological changes of skeletal fluorosis in rats with sub-chronic drinking water type of fluorosis. Another study on rats with chronic fluorosis also discovered that increased bone trabecula density of rat's thigh bone in fluorosis group came with osteosclerosis (high fluoride group was obviously higher than low fluoride group). At the same time, activity of osteoblast and content of bone alkaline phosphatase (BALP), specific marker of bone formation, obviously increased in serum with the increase of fluoride dosage. Taking osteosclerosis into account, it could prove that fluoride simulated enhanced ossification (Pan et al. 2014). The above-mentioned results suggested that fluoride could enhance expression of Wnt3α and β-catenin in osteoblast; together with the results of BALP, it could be reasoned that fluoride could induce ossification by stimulating Wnt3α to upregulate the expression of BALP.

As the major target of fluoride, osteoblast is involved in bone matrix apposition, bone remodeling, calcium homeostasis, and collagen synthesis. Existing study indicated that 10^{-6} mol/l NaF could obviously promote the proliferation and differentiation of primary osteoblast of mice; meanwhile, it also significantly increased the expression level of Wnt3a, LRP5, Dv1, β-catenin, LEF1, cMyc, and Cyclin D1 while decreased mRNA and protein expression level of GSK3β. The results suggested that the fluoride-induced activation of typical Wnt /β-catenin signal pathway was dependent on the activity of GSK3β. The pathway also inhibited the degradation of β-catenin protein and promoted β-catenin's entering from cytoplasm into nucleus (Wang et al. 2019b). Similar studies also proved that fluoride could upregulate gene expression of primarily cultured osteoblast β-catenin and also inhibit β-catenin degradation by promoting phosphorylation of GSK3β (Ser9) and Akt (Ser473). Such process would accumulate β-catenin in the cell and then moved to nucleus to induce the expression of downstream target gene Runx2. Adopting DKK-1 (Wnt antistatic agent) could completely inhibit phosphorylation of fluoride-induced GSK3β-Ser9, protein expression of β-catenin as well as proliferation and differentiation of osteoblast (Pan et al. 2014).

In summary, fluoride regulates downstream target gene expression by activating Wnt/β-catenin signal transduction pathway. In this way, fluoride could promote osteoblast differentiation, enhance osteoblast functions, break the balance of ossification and bone resorption, tilt the balance from bone turnover to ossification, and ultimately cause sclerotic skeletal fluorosis with increased bone mass. Since fluorosis is a chronic systemic disease affecting multiple organs and systems, whether Wnt/β-catenin signal transduction pathway has effect in other parts caused by fluorosis need to be further studied.

Effect of Fluoride on Autophagy Signal Transduction Pathway

When the cell responds to extrinsic stimulation, autophagy can effectively scavenge misfolded protein and damaged cells to protect the cells. Autophagy effect could be reflected by changes of autophagy gene expression. For mammals, autophagy-related genes are termed as autophagy-related (ATG) genes, including microtubule related globin *LC3*, *beclin1*, *Atg5* and *Atg7*, etc. Beclin1 is the mammal ortholog of Atg6, which interacts with class III phosphoinositide 3-kinase /VPS34 and enables other Atg protein to reposition on phagophore assembly site for autophagosome formation. As UBL, LC3 is capable to bind with phosphatidyl ethanolamine (PE) into LC3-PE conjugate and position on pre-autophagosomal structure, which is the prerequisite for autophagosome extension. In the formation of autophagy, cytoplasmic LC3 (i.e. LC3-I) will enzymatically decompose a small portion of peptides and transform into autophagosome membrane type (ie LC3-II) Hence, LC3-II/lC3-I ratio could reflect the autophagy condition. Deficient autophagy is harmful for the body, as it abnormally accumulates misfolded protein in the body, while excessive autophagy is also detrimental, since it can trigger autophagic cell death (type II programmed cell death). With the widespread attention to fluorosis, the study of autophagy mediated by fluorosis has made new progress, but further research is still needed.

According to previous studies, autophagy was involved in fluoride caused bone damage. In a study, 6w old SD rats were exposed to 50, 100 mg/l, and 125 mg/l NaF drinking water for 6 weeks. The results showed that the expression of silent information regulator factor 2-related enzyme 1 (sirtuin 1, Sirt1) protein, autophagy-related proteins Atg5, Atg7, and Atg8/lC3 in mature ameloblasts of the fluoride group increased in a dose-dependent manner, while the expression of the above proteins in the secretion stage of ameloblasts did not change significantly (Suzuki and Bartlett 1842). In addition, exposing rat ameloblast LS8 with different concentrations of NaF (0.5 mg/l, 1.0 mg/l, and 3.0 mmol/l) for 4 h discovered that protein expression of SIRT1 showed increased dose-dependent feature, and NaF exposure group (3.0 mmol/l) showed obvious upregulation of expression level of Atg5, Atg7, and LC3II compared with control group. To further interfere with resveratrol (SIRT1 activating agent), it was found that resveratrol could obviously enhance fluoride-induced autophagy and significantly lower it mediated shear activation of caspase-3 and PARP to inhibit apoptosis (Suzuki et al. 2015). These results indicated that resveratrol could protect fluoride caused ameloblast injury by inducing the activation of SIRT1/Autophagy signal transduction pathway.

Reproductive toxicity of fluoride is also related with the activation of autophagy pathway. A study treated male mice with 0, 0.6, 1.2, and 2.4 mM NaF for 90 days. The increase of mitochondrial autophagy in testis tissue was observed by transmission electron microscope, especially in Leydig cells (Liang et al. 2020). In addition, after treating TM3 mice testicle Leydig cell with fluoride for 24 h, it was discovered that mitochondrial membrane potential was obviously decreased and the number of lysosomes was obviously increased. At the meantime, exposure to fluoride also increased expression of PTEN induced putative kinase 1 (PINK1) and prohibitin 2 in TM2 Leydig cells. These results indicate that fluoride can induce mitochondrial damage and PINK1/Parkin-mediated mitochondrial hypertrophy in testicular cells, especially testicular stromal cells, which may help elucidate the mechanism of fluoride-induced male reproductive toxicity. It was reported that exposed SD rats to fluoride in drinking water before pregnancy and removed testicles of the offspring on the 56th day after birth. The results showed that exposure to fluoride enhanced the induced increased apoptosis of testicular cells, the level of cell surface death receptor Fas was obviously increased, and Cyt C increased in parallel. Such findings indicated that two apoptotic pathways both involved in this process. The intriguing fact was that fluoride treatment also increased the number of autophagosome and level of autophagy marker LC3-II, but level of Beclin1 was not increased. Expression of substrate p62 (autophagy degraded) obviously increased, which indicated that the accumulated autophagy was due to damaged autophagy degradation rather than increased formation (Zhang et al. 2016a). The finding further proved that fluoride-induced male reproductive toxicity was related with autophagy pathway.

Another study used SD rats exposed to sodium fluoride from 60 days before pregnancy to 6 months after delivery as the research object, it was found that fluoride would impair the learning and memory of rat's offspring, reduce the number of neurons, inhibit autophagy of hippocampal cells, and enhance apoptosis (Zhou et al. 2019a). These results were verified in neuroblastoma SH-SY5Y cells. Concerning the mechanism, mammalian/mechanistic target of rapamycin (mTOR) signal (responsible for autophagy guidance in vivo) is activated in both vivo and vitro; rapamycin targeted inhibition on mTOR can protect SH-SY5Y from autophagy defect and excessive apoptosis (Zhou et al. 2019a). In this way, survival rate of neuron is increased. It indicates that autophagy defect is vital in fluoride neural toxicity, and mTOR might be the promising target of fluoride neural toxicity prevention and treatment.

Conclusion

Cellular signal transduction had a vital regulation role in cellular proliferation, differentiation, metabolism, and death. In other words, occurrence and development of various diseases directly or indirectly originates from abnormal cellular signal transduction molecules and signal transduction processes. These GPCRs signal transduction pathways concerning GPCRs, MAPKs, TGF-β/Smad, Nrf2, Wnt/β-catenin, and autophagy are all involved in physiological and pathological effect of fluoride. Despite the differences in dosage and time of fluoride exposure on signal transduction pathway in various organs and system, it is clear that the fluoride toxicity mechanism is closely related with the regulation of signal pathway. Such results provide new target for fluorosis treatment and drug target, which has a great significance in fluorosis prevention and treatment.

References

Bogatcheva NV, Wang P, Birukova AA, Verin AD, Garcia JG. Mechanism of fluoride-induced MAP kinase activation in pulmonary artery endothelial cells. Am J Physiol Lung Cell Mol Physiol. 2006;290:1139–45.

Chen L, Kuang P, Liu H, Wei Q, Cui H, Fang J, et al. Sodium fluoride (NaF) induces inflammatory responses via activating MAPKs/NF-kappaB signaling pathway and reducing anti-inflammatory cytokine expression in the mouse liver. Biol Trace Elem Res. 2019;189:157–71.

Fan B, Yu Y, Zhang Y. PI3K-Akt1 expression and its significance in liver tissues with chronic fluorosis. Int J Clin Exp Pathol. 2015;8:1226–36.

Gutowska I, Baranowska-Bosiacka I, Siennicka A, Telesinski A, Stanczyk-Dunaj M, Wesolowska T, et al. Activation of phospholipase A2 by low levels of fluoride in THP1 macrophages via altered Ca^{2+} and cAMP concentration. Prostaglandins Leukot Essent Fatty Acids. 2012;86:99–105.

Ito M, Nakagawa H, Okada T, Miyazaki S, Matsuo S. ER-stress caused by accumulated intracistanal granules activates autophagy through a different signal pathway from unfolded protein response in exocrine pancreas cells of rats exposed to fluoride. Arch Toxicol. 2009;83:151–9.

Karube H, Nishitai G, Inageda K, Kurosu H, Matsuoka M. NaF activates MAPKs and induces apoptosis in odontoblast-like cells. J Dent Res. 2009;88:461–5.

Lee JH, Jung JY, Jeong YJ, Park JH, Yang KH, Choi NK, et al. Involvement of both mitochondrial- and death receptor-dependent apoptotic pathways regulated by Bcl-2 family in sodium fluoride-induced apoptosis of the human gingival fibroblasts. Toxicology. 2008;243:340–7.

Li X, Meng L, Wang F, Hu X, Yu Y. Sodium fluoride induces apoptosis and autophagy via the endoplasmic reticulum stress pathway in MC3T3-E1 osteoblastic cells. Mol Cell Biochem. 2019;454:77–85.

Liang C, Gao Y, He Y, Han Y, Manthari RK, Tikka C, et al. Fluoride induced mitochondrial impairment and PINK1-mediated mitophagy in Leydig cells of mice: *In vivo* and *in vitro* studies. Environ Pollut. 2020;256:113438.

Liu YJ, Gao Q, Wu CX, Guan ZZ. Alterations of nAChRs and ERK1/2 in the brains of rats with chronic fluorosis and their connections with the decreased capacity of learning and memory. Toxicol Lett. 2010;192:324–9.

Mukhopadhyay D, Priya P, Chattopadhyay A. Sodium fluoride affects zebrafish behaviour and alters mRNA expressions of biomarker genes in the brain: Role of Nrf2/Keap1. Environ Toxicol Pharmacol. 2015;40:352–9.

Nagasaka Y, Okubo M, Larner J. Mechanisms of aluminum fluoride- and insulin-stimulated p33 mRNA accumulation in rat hepatoma cells: involvement of a G protein and kinase action and demonstration of effects on mRNA turnover. Cell Signal. 1993;5:633–41.

Ni J, Li Y, Zhang W, Shu R, Zhong Z. Sodium fluoride causes oxidative stress and apoptosis in cementoblasts. Chem Biol Interact. 2018;294:34–9.

Niu Q, Chen J, Xia T, Li P, Zhou G, Xu C, et al. Excessive ER stress and the resulting autophagic flux dysfunction contribute to fluoride-induced neurotoxicity. Environ Pollut. 2018;233:889–99.

Oyagbemi AA, Omobowale TO, Asenuga ER, Adejumobi AO, Ajibade TO, Ige TM, et al. Sodium fluoride induces hypertension and cardiac complications through generation of reactive oxygen species and activation of nuclear factor kappa beta. Environ Toxicol. 2017;32:1089–101.

Pan L, Shi X, Liu S, Guo X, Zhao M, Cai R, et al. Fluoride promotes osteoblastic differentiation through canonical Wnt/beta-catenin signaling pathway. Toxicol Lett. 2014;225:34–42.

Park B, Yang YM, Choi BJ, Kim MS, Shin DM. Activation of G proteins by aluminum fluoride enhances RANKL-mediated osteoclastogenesis. Korean J Physiol Pharmacol. 2013;17:427–33.

Refsnes M, Schwarze PE, Holme JA, Lag M. Fluoride-induced apoptosis in human epithelial lung cells (A549 cells): role of different G protein-linked signal systems. Hum Exp Toxicol. 2003;22:111–23.

Ridley W, Matsuoka M. Fluoride-induced cyclooxygenase-2 expression and prostaglandin E2 production in A549 human pulmonary epithelial cells. Toxicol Lett. 2009;188:180–5.

Shanmugam T, Selvaraj M, Poomalai S. Epigallocatechin gallate potentially abrogates fluoride induced lung oxidative stress, inflammation via Nrf2/Keap1 signaling pathway in rats: an in-vivo and in-silico study. Int Immunopharmacol. 2016;39:128–39.

Shao D, Zhang J, Tang L, Yu Q, Hu X, Ruan Q, et al. Effects and molecular mechanism of L-type calcium channel on fluoride-induced kidney injury. Biol Trace Elem Res. 2020;197:213–23.

Song G, Wang RL, Chen ZY, Zhang B, Wang HL, Liu ML, et al. Toxic effects of sodium fluoride on cell proliferation and apoptosis of Leydig cells from young mice. J Physiol Biochem. 2014;70:761–8.

Song C, Heping H, Shen Y, Jin S, Li D, Zhang A, et al. AMPK/p38/Nrf2 activation as a protective feedback to restrain oxidative stress and inflammation in microglia stimulated with sodium fluoride. Chemosphere. 2020;244:125495.

Sternweis PC, Gilman AG. Aluminum: a requirement for activation of the regulatory component of adenylate cyclase by fluoride. Proc Natl Acad Sci U S A. 1982;79:4888–91.

Sun YM, Yang FJ, Li YM, Lu B, Zhu M, Qiu MC. Expression of Fas, FasL, and NF-kappa B in the process of osteoclast-like cell apoptosis effected by sodium fluoride. Acta Acad Med Sine. 2002;24:491–4.

Sun Z, Nie Q, Zhang L, Niu R, Wang J, Wang S. Fluoride reduced the immune privileged function of mouse Sertoli cells via the regulation of Fas/FasL system. Chemosphere. 2017;168:318–25.

Susa M. Heterotrimeric G proteins as fluoride targets in bone. Int J Mol Med. 1999;3:115–26.

Susa M, Standke GJ, Jeschke M, Rohner D. Fluoroaluminate induces pertussis toxin-sensitive protein phosphorylation: differences in MC3T3-E1 osteoblastic and NIH3T3 fibroblastic cells. Biochem Biophys Res Commun. 1997;235:680–4.

Suzuki M, Bartlett JD. Sirtuin1 and autophagy protect cells from fluoride-induced cell stress. Biochim Biophys Acta. 1842;2014:245–55.

Suzuki M, Bandoski C, Bartlett JD. Fluoride induces oxidative damage and SIRT1/autophagy through ROS-mediated JNK signaling. Free Radic Biol Med. 2015;89:369–78.

Teng Y, Zhang J, Zhang Z, Feng J. The effect of chronic fluorosis on calcium ions and CaMKIIalpha, and c-fos expression in the rat hippocampus. Biol Trace Elem Res. 2018;182:295–302.

Thangapandiyan S, Miltonprabu S. Epigallocatechin gallate supplementation protects against renal injury induced by fluoride intoxication in rats: Role of Nrf2/HO-1 signaling. Toxicol Rep. 2014;1:12–30.

Wang L, Zhu Y, Wang D. High-dose fluoride induces apoptosis and inhibits ameloblastin secretion in primary rat ameloblast. Biol Trace Elem Res. 2016;174:402–9.

Wang HW, Zhao WP, Liu J, Tan PP, Zhang C, Zhou BH. Fluoride-induced oxidative stress and apoptosis are involved in the reducing of oocytes development potential in mice. Chemosphere. 2017;186:911–8.

Wang J, Yang J, Cheng X, Xiao R, Zhao Y, Xu H, et al. Calcium alleviates fluoride-induced bone damage by inhibiting endoplasmic reticulum stress and mitochondrial dysfunction. J Agric Food Chem. 2019a;67:10832–43.

Wang J, Yang J, Cheng X, Yin F, Zhao Y, Zhu Y, et al. Influence of calcium supplementation against fluoride-mediated osteoblast impairment *in vitro*: involvement of the canonical Wnt/beta-catenin signaling pathway. J Agric Food Chem. 2019b;67:10285–95.

Wu J, Cheng M, Liu Q, Yang J, Wu S, Lu X, et al. Protective role of tert-butylhydroquinone against sodium fluoride-induced oxidative stress and apoptosis in PC12 cells. Cell Mol Neurobiol. 2015;35:1017–25.

Xu B, Xu Z, Xia T, He P, Gao P, He W, et al. Effects of the Fas/Fas-L pathway on fluoride-induced apoptosis in SH-SY5Y cells. Environ Toxicol. 2011;26:86–92.

Xu Z, Xu B, Xia T, He W, Gao P, Guo L, et al. Relationship between intracellular Ca^{2+} and ROS during fluoride-induced injury in SH-SY5Y cells. Environ Toxicol. 2013;28:307–12.

Yan X, Feng C, Chen Q, Li W, Wang H, Lv L, et al. Effects of sodium fluoride treatment *in vitro* on cell proliferation, apoptosis and caspase-3 and caspase-9 mRNA expression by neonatal rat osteoblasts. Arch Toxicol. 2009;83:451–8.

Yang S, Wang Z, Farquharson C, Alkasir R, Zahra M, Ren G, et al. Sodium fluoride induces apoptosis and alters bcl-2 family protein expression in MC3T3-E1 osteoblastic cells. Biochem Biophys Res Commun. 2011;410:910–5.

Yang T, Zhang Y, Li Y, Hao Y, Zhou M, Dong N, et al. High amounts of fluoride induce apoptosis/cell death in matured ameloblast-like LS8 cells by downregulating Bcl-2. Arch Oral Biol. 2013;58:1165–73.

Yang C, Wang Y, Xu H. Correction: fluoride regulate osteoblastic transforming growth factor-beta1 signaling by mediating recycling of the type I receptor ALK5. PLoS One. 2017;12:e0176772.

Zhang J, Zhu WJ, Xu XH, Zhang ZG. Effect of fluoride on calcium ion concentration and expression of nuclear transcription factor kappa-B rho65 in rat hippocampus. Exp Toxicol Pathol. 2011a;63:407–11.

Zhang XL, Xi SH, Cheng GY, Zhang Y. The effect of different concentrations of fluoride on the expression of Smad2/3 in ameloblast of rat incisor. Shanghai J Stomatol. 2011b;20:230–3.

Zhang S, Niu Q, Gao H, Ma R, Lei R, Zhang C, et al. Excessive apoptosis and defective autophagy contribute to developmental testicular toxicity induced by fluoride. Environ Pollut. 2016a;212:97–104.

Zhang Y, Zhang K, Ma L, Gu H, Li J, Lei S. Fluoride induced endoplasmic reticulum stress and calcium overload in ameloblasts. Arch Oral Biol. 2016b;69:95–101.

Zhang J, Jiang N, Yu H, Yu X, Guo F, Zhao Z, et al. Requirement of TGFbeta signaling for effect of fluoride on osteoblastic differentiation. Biol Trace Elem Res. 2019;187:492–8.

Zhao Y, Li Y, Gao Y, Yuan M, Manthari RK, Wang J, et al. TGF-beta1 acts as mediator in fluoride-induced autophagy in the mouse osteoblast cells. Food Chem Toxicol. 2018;115:26–33.

Zhou BH, Tan PP, Jia LS, Zhao WP, Wang JC, Wang HW. PI3K/AKT signaling pathway involvement in fluoride-induced apoptosis in C2C12cells. Chemosphere. 2018;199:297–302.

Zhou G, Tang S, Yang L, Niu Q, Chen J, Xia T, et al. Effects of long term fluoride exposure on cognitive ability and the underlying mechanisms: Role of autophagy and its association with apoptosis. Toxicol Appl Pharmacol. 2019a;378:114608.

Zhou X, Chen Z, Zhong W, Yu R, He L. Effect of fluoride on PERK-Nrf2 signaling pathway in mouse ameloblasts. Hum Exp Toxicol. 2019b;38:833–45.

Chapter 15
Changes of Receptors

Yang-Ting Dong and Zhi-Zhong Guan

Abstract Chronic fluorosis is an endemic disease that severely impairs human health. Studies have proved that the change of various receptors in the body is attributed to excessive fluoride intake. For instance, it will reduce the expression of acetylcholine receptor that affects the release of neurotransmitters and will cause changes of N-methyl-D-aspartic acid receptor that impacts cellular Ca^{2+} influx and of advanced glycation end product receptor, augmenting thc inflammatory response of the body, etc. These changes are the key mechanism of multi-system injuries caused by fluorosis. With the increased attention on more and more receptors in recent years, the role they played in the pathogenesis of chronic fluorosis is becoming ever more essential. Therefore, an in-depth study of its mechanism of action is possibly vital to uncover the injury of chronic fluorosis and can also provide a theoretical basis for prevention and clinical treatments.

Keywords Acetylcholine receptor · N-Methyl-D-aspartic acid receptor · Peroxisome proliferators-activated receptor · Receptor for advanced glycation end products · Vitamin D receptor

---

Y.-T. Dong (✉)
The Key Laboratory of Endemic and Ethnic Diseases (Guizhou Medical University) of the Ministry of Education, Guiyang, Guizhou, People's Republic of China

The Provincial Key Laboratory of Medical Molecular Biology, Guiyang, Guizhou, People's Republic of China
e-mail: ytd1988@gmc.edu.cn

Z.-Z. Guan
The Key Laboratory of Endemic and Ethnic Diseases (Guizhou Medical University), Ministry of Education, Guiyang, Guizhou, People's Republic of China

The Provincial Key Laboratory of Medical Molecular Biology, Guiyang, Guizhou, People's Republic of China

Department of Pathology at the Affiliated Hospital of Guizhou Medical University, Guiyang, Guizhou, People's Republic of China

Z.-Z. Guan (ed.), *Coal-burning Type of Endemic Fluorosis*,
https://doi.org/10.1007/978-981-16-1498-9_15

Introduction

Cell receptor is a biomacromolecule on the cell membrane or within the cell that can recognize and bind with ligand, a bioactive molecule specifically binding with receptors, in the cell communication. Most receptors are proteins, a few are glycolipids (Wang and Zhong 2012), and some are compounds of the two. At present, the receptor is divided into cell membrane receptors and intracellular receptors according to its location in the cell. Cell membrane receptor is a type or class of molecule on the cell surface, which can recognize and bind with a specific bioactive substance (known as a ligand). The compound of the receptor and ligand is able to activate and initiate a series of physical and chemical changes, leading to the ultimate biological effect of the substance. The changes of various factors in the cellular environment are connected with the corresponding changes that affect the physiological processes in the cell by the effect of cell membrane receptors. The intracellular receptor refers to the receptor located in the cytoplasm or nucleus within the cell membrane. This type of receptor serves as a transcription factor since its corresponding hormones have to pass through the membrane and enter the cell before binding with it. Membrane receptors are found on the cell membrane, and most of them are glycoproteins, they are divided into three categories according to their structure and function.

Classification and Function of Cell Membrane Receptors

Membrane receptors are found on the cell membrane, and most of them are glycoproteins, they are divided into three categories according to their structure and function.

Ligand-Gate Ion Channel Receptors

Ligand-gate ion channel receptors are also called the ligand-gated receptor. Cell membrane receptors are activated by such extracellular ligands as neurotransmitter receptor, while the intracellular receptors are activated by intracellular ligands. Currently, 13 kinds of ligand-dependent ion channel receptors have been discovered, each of which is comprised of a ligand-binding site and an ion channel. When the ligand binds with the ligand-gated receptor, the ion channel can be opened or closed, altering the permeability of the membrane. Although different receptors contain different numbers and types of subunits, their basic structures are similar. Therefore, information transduction does not depend on any diffusible factors within the cell or of the membrane, making it possible to transmit information quickly.

G-protein Coupled Receptors

G-protein coupled receptors (GPCRs), also known as a serpentine receptor, are composed of a peptide chain, whose N-terminal is outside the cell and C-terminal forms the intracellular in the tail. The middle part of the receptor is the serpentine α-helical structure, three extracellular loops and three intracellular loops. The receptor of this type has been studied most extensively and thoroughly. At present, there are more than 1000 known GPCRs characterized by the fact that the second and third loops of the cytoplasmic surface can be coupled to G proteins, able to respond to many hormones and neurotransmitters including ligand-binding domains and intracellular domains. When the ligand binds with the receptor on the cell surface, their interaction causes a change in the conformation of the receptor, allowing it to interact with the G protein of the cell membrane to form a compound ligand-receptor-G protein with high affinity.

Single Transmembrane α-helical Receptor

This type of receptor often exists in monomer form in the absence of ligands and binds with the cell membrane with a single transmembrane α-helix domain with polypeptide chain as a ligand. When receptor and ligand are binding together (or couple with), protein tyrosine kinase activity or protein serine/threonine kinase activity will present. Due to the difference in the structure of receptors, they are divided into four types: (1) Protein tyrosine kinase receptors, including insulin receptors, and various growth factor receptors. The insulin receptor is a $\alpha 2\beta 2$ tetramer formed by two pairs of αβchains through disulfide bonds. (2) Protein serine and threonine kinase receptors. Transforming growth factor β (TGF-β) family is divided into two subfamilies-type I (TβR-I) and type II (TβR-II), of which type I is more conservative than type II receptor. TGF-β receptor family transduces information through the receptor's serine/threonine protein kinase. (3) Non-catalytic single transmembrane receptor—cytokine receptor with its ligands such as cytokines and growth hormone. Although the receptor presents no enzymes catalytic activity, it has a domain that binds with non-receptor tyrosine kinases in the juxtamembrane domain. When the receptor binds with the ligand, it can be coupled to the kinase to exhibit enzymatic activity, phosphorylating the tyrosine residues of the receptor and substrate. (4) Receptors with guanine cyclase activity, whose ligands include atrial natriuretic peptide and guanylin. The receptors consist of homotrimers and homotetramers, of which each subunit includes an extracellular receptor domain of N-terminal, a transmembrane domain, a protein kinase-like domain within the membrane, and a C-terminal bird glycosyl cyclase catalytic domain. The protein kinase-like domain has no kinase activity with its function currently unknown.

Classification and Function of Intracellular Receptors

Intracellular receptors share a similar basic structure because of their great homology. They usually have two different domains. One is an intermediate domain that binds with DNA and the other is an N-terminal domain that activates gene transcription. In addition, there are two binding sites. One is a site for binding to lipid ligands at the C-terminal and the other is for binding to inhibitory proteins. The intracellular receptors are all trans-acting factors with zinc finger structure as their DNA binding domain, which is usually a monomeric protein composed of 400–1000 amino acid residues and includes five regions: (1) the high variable region determining the specificity of the gene transcription; (2) the DNA binding domain that is rich in cysteine and can rotate along the DNA helix and bind to it to dimerize the receptor; (3) the hormone binding domain, which is a short sequence and may interact with transcription factor as well as push the receptor to move into the nucleus; (4) the hormone hinge domain at the C-terminal of the domain, which can bind to ligands and heat shock proteins to dimerize the receptor and activate transcription; (5) the carboxyl terminal domain with its function unknown. When the hormone binds with the receptor, the conformation of the receptor changes, exposing the nuclear transfer site of the receptor and the DNA binding site. The hormone–receptor complex transfers into the nucleus and binds with the hormone response element (HRE) adjacent to the specific gene on the DNA, which in turn changes the gene expression profile of the cell and changes in cell function. For example, steroid hormone receptors are mostly trans-acting factors. After binding to ligands, they bind to cis-acting elements of DNA to regulate gene transcription. According to the different locations of receptor molecules on cells, they are divided into two categories.

The Cytoplasmic Receptor

It is located in the cytoplasm of the target cell with its ligands like steroid hormones, vitamin D3, thyroxine, and retinoic acid. And its intracellular receptor is an inducible transcription factor, which manifests transcription activity after binding with its ligands to promote such process. Signals of extracellular information molecules are transmitted to the cell or nucleus through the cell information pathway, which results in many biological effects and examples are the opening or closing of ion channels and changes in ion concentration, changes in enzyme activity, substance metabolism, and gene expression and response, and the effects of cell growth, development, differentiation as well as proliferation.

The Nuclear Receptor

It is located in the nucleus of the target cell. The receptor of steroid hormones, retinoic acid, vitamin D (Vit D), and thyroid hormone is in the nucleus. Such signaling molecules bind with serum proteins and transport to target tissues and permeate the plasma membrane into the cell through the nuclear pore and specific nuclear receptor to form a hormone–receptor compound and change the receptor conformation. Hormone–receptor compound binds with a gene-specific regulatory domain, also known as HRE, which affects gene transcription. Gene activation induced by steroid hormone is usually divided into two stages: (1) rapid primary reaction stage, which directly activates transcription of a few special genes; (2) delayed secondary reaction stage. During this stage, the gene product of the primary reaction activates the transcription of other genes, magnifying the primary reaction. Thyroxine is also a small lipophilic molecule with the same mechanism of action as steroid hormones. However, there are also individual lipophilic molecules (such as prostaglandins) whose receptors are on the plasma membrane.

The Impact of Chronic Fluorosis on Acetylcholine Receptors

Acetylcholine (ACh) that plays an important role in the peripheral and the central nervous system (CNS) is the earliest and most standard neurotransmitter. When cholinergic excitation occurs, ACh is released to the synaptic space as a neurotransmitter, which transmits information to the interior of the target cell through the receptor on the post-synaptic membrane, causing the cell to respond to the neurotransmitter rapidly. In the early twentieth century, it was found that the pharmacological response of ACh receptor to natural alkaloids muscarine and nicotine was different. According to this, they are divided into nicotinic acetylcholine receptors (nAChRs) and muscarinic acetylcholine receptors (mAChRs).

The Structure, Function, and Distribution of nAChRs

The receptor is a ligand-gated ion channel receptor with five subunits on the cell membrane, which belongs to a gene superfamily with a common origin called the cys-loop receptor. It is a pentamer with a pentagonal shape in cross-section with an ion channel formed in the center and can be activated by ACh and nicotine. The activated nAChR contributes to Na^+, K^+, and Ca^{2+} plasma flow (Wang et al. 2020).

The nAChRs are divided into central and peripheral types. Types and combinations of peripheral nAChR subunits are relatively fixed, which is found at the neuromuscular junction, responsible for the neurotransmitter conversion between nerves and muscles. Central nAChR consists of brain nAChR and ganglion nAChR with many possibilities for subunit types and combination, which is the molecular basis of the complicated function of nAChRs in the nervous system.

At present, it is believed that the central nAChRs contain at least 12 subunits from α2-10 to β2-4, each of which is encoded by different genes. Different combinations of these subunits exhibit different pharmacological and physiological characteristics. Therefore, different nAChR subtypes can present different affinity and ion permeability when binding with the same ligand. The subunits of nAChRs are the functional ones with binding site to the ligands, and the subunits are the structural subunits, maintaining the structural stability of the receptors. nAChRs are distributed in pre- and post-synapse. After the pre-synaptic nAChRs are activated, the inflow of Na^+ depolarizes the membrane, therefore opening the voltage-gated Ca^{2+} channel. This process is known as positive feedback regulation (Lewis and Picciotto 2020).

The nAChR α4β2 and α7 are the receptor subtypes most widely distributed in the CNS, of which nAChR α4β2 is distributed practically in the entire nervous system, while nAChR α7 is found in the cerebral cortex, hypothalamus, and hippocampus and some brainstem nuclei (Wang and Lindstmm 2018). The α7 nAChR is expressed in the hippocampus and cortical neurons and plays a vital role in the formation of memory. α7 nAChR mediates neurotransmitter transmission as a post-synaptic receptor and can also be used as a pre-synaptic modulator to regulate the release of neurotransmitters such as glutamate, γ-aminobutyric acid, dopamine, and norepinephrine (Baranowska and Wisniewska 2017) when the agonist binds with nAChRs, the structure of the receptor changes with channel opening to allow more Ca^{2+} to pass through the channel. The closing, opening, and desensitization of Ca^{2+} channels after binding with the ligand realize the structure–activity state transition of nAChRs. There are usually three structure–activity states for nAChRs, and the conversion between them constitutes the complexity of nAChRs. nAChRs are related to brain functions such as brain learning and memory, intellectual development, and recognition ability. In addition, they regulate the functions of many other receptors and have neuroprotective effects against cytotoxicity.

The Effect of Fluoride on nAChRs

Professor Guan's group studied the changes of neuronal nAChRs in the brain after excessive fluoride exposure. Firstly, the group used different concentrations of fluorine to observe neuroblastoma cells of human brains (SH-SY5Y) and detect changes in the levels of protein and gene expressions of neuronal nAChRs in the cultured cell. The 3-(4, 5-dimethylthiazol-2-yl)-2, 5-diphenyl tetrazolium bromide reduction rate of SH-SY5Y cells after fluorine exposure was significantly lower than that of the control group, indicating that the cell function was damaged, impacting the survival status. Forty-eight hours after fluorine treatment of the cells, nAChR α3 and α7 protein levels were reduced, resulting in impaired receptor function and nerve cell function. The study observed mRNAs of two nAChRs subunits and the results showed there was no significant difference and therefore excluded the possibility of abnormal gene expression caused by excessive fluorine attacking DNA or RNA. The study also applied antioxidants to antagonize the high concentration of fluorine on nAChRs injury and achieved satisfactory results. It explained that the neurological injury caused by fluorosis might be demonstrated, to a certain extent, by the reduction of the expression of neuronal nAChRs. According to the researchers' analysis, radicals could damage the structure of the cell membrane and attack receptor proteins. Since the nicotine receptor is the membrane intrinsic receptor protein, susceptible to the attack of radical, the radical injury can be considered as an essential factor causing structural and functional changes of nAChRs. This study linked the oxidative stress theory of fluoride injury to changes in neurotransmitter receptors, providing further in-depth research directions for the mechanism of nerve damage caused by fluorosis. The group found that the expression level of receptor subunit protein of nAChR $\alpha4\beta2$ was reduced in the brain tissue of experimental drinking-water chronic fluorosis rats. The decreased binding rate of nAChRs with the radioligand [^{3}H]Epibatidine or [^{125}I]-bungarotoxins compared to the control group indicated that under excessive fluoride exposure, the expression of neuronal nAChRs was inhibited, which might be related to changes in nervous system function (Guan et al. 2011).

Coal-burning fluorosis is prevalent, in Southwest China, typically in Guizhou province. To explore the nervous damage mechanism of coal-burning fluorosis and compare it with drinking-water fluorosis, the research group has conducted related studies of rats in experimental coal-burning fluorosis. The study found that the expression levels of nAChR $\alpha3$, $\alpha4$, and $\alpha7$ of receptor subunit protein in the brain tissue of rats exposed to fluorine were significantly lower than those in the control group. The mRNA expression of nAChR $\alpha3$, $\alpha4$, and $\alpha7$ of subunit in brain tissues of rats exposed to high-concentration fluorine was significantly lower than that in the control group. Moreover, these changes were related to the concentration of negative ions in their brain tissue and to the changes in the learning and memory abilities of experimental animals (Gui et al. 2011).

Structure, function, and Distribution of mAChRs

The mAChRs belong to GPCRs that is a family of receptors mediated by second messenger receptors. They are a glycoprotein on the cell membrane and have a specific spatial structure and stereoselectivity. mAChRs participate in learning and memory in the CNS and modulate neuroplasticity. They are involved in critical physiological functions such as respiration, circulation, exercise, and body temperature regulation. The receptor has a molecular weight of about 50,000 kDa, which is composed of a peptide chain. The N-terminus of the peptide chain is located outside the membrane and has one or more carbonylation sites. The C-terminus of the peptide chain extends into the cytoplasm. The amino acid sequence of mAChRs forms seven transmembrane segments (TM I-VII), three intracellular loops (i1-i3) and three extracellular loops (o1-o3). On the cell membrane, seven transmembrane domains are arranged in a spoke structure, forming a transmembrane hollow facing out of the cell, which is the site where mAChRs bind to agonists or antagonists. The ligand recognition site is located outside the transmembrane domain. Asp residue in the third transmembrane domain is the specific site H for the alkylation of M-receptor agonist Ach and antagonist phenyl choline. Other studies have suggested that the Asn residue in the sixth transmembrane domain of M-receptor is an important group that binds to antagonists, especially atropine. The cysteine residues between the 1st and 3rd loops of the extracellular loops of M-receptor form disulfide bonds; amino acid triplets (Asp, Arg, Tyr) between the second intracellular loop and the plasma membrane are of great significance for the expression and function of the receptor.

At present, the five subtypes of mAChRs are, respectively, expressed as M1, M2, M3, M4, and M5. It is distributed in the effector cells dominated by postganglionic parasympathetic fibers and coupled with different G proteins in the periphery, causing a series of physiological and biochemical effects, such as vascular smooth muscle relaxation, bradycardia, decreased myocardial contractility, the increased contractility of the smooth muscle of gastrointestinal tract and bladder, the changes in adenylyl cyclase activity, the activation of the creatine phospholipid system, and the opening and closing behavior of ion channels. At the same time, its various subtypes are distributed in the CNS. These subtypes play a role in the transmission or the inhibition of neuronal impulse, including in various functions of the brain such as arousal, attention, emotion response, and motility regulation, especially in the regulation of advanced cognitive activities such as learning and memory (Dong and Guan 2014).

The Effect of Fluoride on mAChRs

In the case of cholinergic excitation, ACh as a neurotransmitter is released to the synaptic space and then combines with nAChRs or mAChRs to produce corresponding biological activities.

As mentioned before, mAChRs are a member of the G-protein coupled receptor family, which are located on the cell membrane and consist of a peptide chain. After being activated, they transduce subsequent signals through the G-protein coupled to them, causing various physiological and biochemical effects in the cell. mAChRs have a more extensive range of functions than nAChRs. In the peripheral nervous system, mAChRs can regulate smooth muscle contraction, glandular secretion, heart rate, etc., in the CNS, and they participate in advanced cognitive functions such as learning and memory and motion control. There are common subtypes (M1, M2, and M3) in the brain. The existing studies have shown that M1 and M2 are closely related to the ability of learning and memory. At present, a subject research in our group has proved that the levels of protein and gene expression of M1 and M3 in the brain tissues of rats with chronic fluorosis and their offspring significantly reduced. Through the Morris water maze experiment, the declined ability of learning and memory was observed in these rats with chronic fluorosis and their progeny (Dong et al. 2015a).

ACh is an excitatory transmitter of many preganglionic nerve fibers and parasympathetic postganglionic nerve fibers. Studies have shown that the activity of acetylcholinesterase in the synaptic plasma membrane of fluorosis rats and their progeny significantly decreases with the elevating dose of fluoride and extending time, causing the increase of the Ach content in brain tissue. In the case of chronic fluorosis, the oxidative stress level in the brain tissue increased with the elevating dose of fluoride. Excessive free radicals can attack macromolecular substances such as cell membrane lipid structure and protein molecules and destroy the structure and function of cholinergic neurotransmitter receptors with cell membrane mosaic, causing the change of neurotransmitter-related cholinesterase activity in cells and a decrease in learning and memory function of rats with chronic fluorosis (Dong and Guan 2014). Certain factors such as the declined expression of mAChRs in the brain, the weakening coupling effect of mAChRs and G protein, the decreasing ability to mediate the generation of second messengers, and the reduced Ca^{2+} influx affected the inhibition of neurotransmitter release. At the same time, mAChRs and nitric oxide synthetase (NOS)/nitric oxide had a change, which influenced NOS activity. The cells failed to function normally, directly or indirectly altering the activity of some signaling pathways and synaptic plasticity, then leading to a further reduction in the ability of learning and memory (Dong et al. 2015b). Besides, in case of chronic fluorosis, there were increases in the oxidative stress level in brain tissue and reactive oxygen species (ROS), and there was a decrease in antioxidant level, which resulted in the degeneration or disfunction of mAChRs, and the decrease of the number of cholinergic neurons causing neuronal damage. Chronic fluorosis can cause abnormal changes in the biomembrane lipid structure of nerve cells in brain tissue. The damage to the cell membrane will inevitably affect the structure and function of mAChRs on the cell membrane, leading to a reduction in its expression, and further to brain damage. As a result, the ability of learning and memory reduce (Dong et al. 2016).

The Impact of Chronic Fluorosis on N-methyl-D-Aspartate Receptors (NMDARs)

Structure, Function, and Distribution of NMDARs

NMDARs are specific receptors for excitatory amino acid systems (EAAs), forming the family of ionotropic glutamate receptors with a-amino-3-hydroxy-5-methyl-4-isoxazolepropionate (AMPA) subtypes and kainite (KA) subtypes. The NMDAR is a ligand-gated ion channel that has many different allosteric regulatory sites and is highly permeable to Ca^{2+}. In particular, Ca^{2+} influx by the opening of NMDAR channels is both ligand-gated and voltage-dependent. In static status, Mg^{2+} was in the channel to prevent Ca^{2+} influx; once the receptor was activated, its receptor protein conformation changed and the ion channel opened so that cations such as K^+, Na^+, and Ca^{2+} could be in and out of the cell, as a result, which caused membrane depolarization and neuronal excitation. NMDARs can regulate the survival of neurons, the development of dendrite and axon and synaptic plasticity. They can affect the formation of neuronal circuits and the processes of learning and memory. The receptors are mainly composed of three types of subunits, namely NR1, NR2, NR3. NR2 has four subtypes (NR2A, NR2B, NRC, NR2D) encoded by independent genes. NR3 acquires NR3A and NR3B through different splicing. The NR1 subunit is widely distributed in the brain. Its expression has been throughout the entire developmental stage of the brain and has an upward trend in the developing hippocampus, reaching a peak at three weeks after birth and then gradually decreasing. The NR1 subunit is a functional unit of the NMDARs and is an essential part of the receptor channel. Its expression level represents the total amount of the NMDARs. Thus, the NR1 subunit has a close bearing on learning and memory. NR2 is an NMDA regulatory subunit, and its distribution in the CNS is specific in a tissue region. It changes with the development of the CNS, especially the brain. NR2A and NR2B of NR2 affect synaptic plasticity. After the removal of NR2A C-terminus, although the mice can survive, it has obstacles on synaptic plasticity and learning and memory. The loss of NR2A gene leads to damage to the eyelid conditioned reflex and learning ability of mice; the site-directed mutation (F637A) of NR2A can even cause nonfunctional NMDARs composed of it. The two subunits of NR3 are also distributed in the brain, while the distribution of NR3A is wider than NR3B. From the analysis of immunocytochemistry, NR3B is abundantly expressed in motoneurons. The NMDAR is mainly composed of NR1 and NR2 subunits. NR2 alone does not respond to agonists and binds only to NR1 to be active.

The Effect of Fluoride on NMDARs

There are a large number of NMDARs on the neuronal synapses in the hippocampus. Therefore, NMDARs are considered to be the essential substances in learning and memory. A large number of animal experiments have shown that NMDARs play an important role in short-term memory, spatial memory, and passive avoidance memory. In addition, NMDARs are closely related to the development of the CNS and to many complex physiological and pathological mechanisms, such as synaptic plasticity, long-time potentiation (LTP), excitatory neurotoxicity, neurodegenerative denaturation, etc. the activation of NMDARs can lead to an increase in neuron excitotoxicity, and thereby to a variety of neurological disorders.

Post-synaptic NMDARs are activated in the processes of learning and memory in animals. Ca^{2+} influx activates nNOS with calmodulin and catalyzes the production of NO by L-arginine. NO can pass the cell membrane to enter the pre-synaptic component and activate guanylate cyclase. Increase of cyclic guanosine monophosphate promotes the synthesis of glutamate and the release of the transmitter, and then the glutamate acts on the post-synaptic NMDARs to achieve LTP induction. Glutamate is the maximum content and the strongest excitatory amino acid in the CNS. Under normal condition, glutamate mainly exists in the vesicles of nerve endings. In the case of the depolarization of nerve endings to make itself excited, glutamate is released into the synaptic cleft through Ca^{2+}-dependent methods and acting on specific receptors in the post-synaptic membrane, which completes excitatory synaptic transmission and other physiological functions. It is then rapidly degraded by enzymes and is reuptaken by neurons and glial cells. In our previous study (Wei et al. 2018), we found that there were significant increases in expression levels of protein of NR1, NR2A, and NR2B, mRNA, and phospho-NR2B protein, but there was a decrease in the level of NR3A. The activation of NMDARs can lead to increased neurotoxicity. Chronic fluorosis may trigger an increased release of pre-synaptic glutamate or the increased release and/or decreased reuptake of EAAs in the CNS due to the activation of NMDARs, resulting in excessive excitability of post-synaptic EAAs receptors for a long time, finally which will cause excitotoxic neuronal damage or death. The activation and excitation of NMDARs can also promote the formation of reactive nitrogen intermediates with cytotoxicity and reactive oxygen intermediates. They facilitate the influx of extracellular Ca^{2+} to activate phosphatase, protease, nitric oxide synthase, etc., which causes necrosis or apoptosis of nerve cells, and then the nerve functional disturbance in the brain. In contrast, the inhibition of the expression level of phosphorylated NMDARs and their declined activity to reduce Ca^{2+} influx can alleviate CNS damage by excessive fluoride, which is consistent with previous cell experiments (Bi et al. 2019; Ge et al. 2018).

The Impact of Chronic Fluorosis on Advanced Glycation End Products (AGEs) and Their Receptors

Structure, Function, and Distribution of AGEs

AGEs are a heterogeneous group of compounds that are formed by non-enzymatic reactions between the free amino groups of proteins, lipids, or nucleic acids and the carbonyl groups of glucose or reducing sugars (Wang et al. 2016), in correlation with the receptor for advanced glycation end products (RAGE). The human encoding RAGE gene is on chromosome 6, including 11 exons and 10 shorter introns. Any other subtypes of RAGE have not been found in humans. RAGE is highly homologous, the cDNA sequence homology of which is about 90% among humans, cattle, adult rats, and pups. In amino acid sequence analysis, RAGE is comprised of over 400 amino acids, including larger extracellular fragment, the transmembrane domain, and shorter intracellular domain. RAGE is a multi-ligand receptor, which explains its complexity in biological activity. RAGE can activate various signal transduction pathways when combined with different ligands, resulting in diverse corresponding effects. For instance, nuclear factor-kappa B (NF-κB) may regulate the transcription of RAGE by binding to the sites on the 5′ non-coding DNA sequences. AGEs can directly destroy many proteins in organisms via glycosylation and alter downstream signaling pathways when combined with RAGE, resulting in a series of biological effects.

The Impact of Fluoride on AGEs and Their Receptors

In recent years, studies have been conducted on the mechanism of the altered AGEs and RAGE signal transduction pathways in nerve injuries induced by chronic fluorosis. Among the downstream factors associated with AGE/RAGE signal transduction pathways, NF-κB is the primary focus. NF-κB is a multi-transcription regulator of many injury-responsive genes, the activation of which can promote the expression of downstream damage factors, leading to the injuries of gene organization. The enzyme-linked immunosorbent assay indicated that in comparison to the controls, the levels of AGEs in the brains of rats in two groups after 6 months of exposure to fluoride were elevated. In contrast, no significant changes in the parameter were detected in the rats exposed for 3 months. Immunohistochemical results proved that AGEs were located in intercellular substance and cytoplasm, and RAGE was located in the membrane. Compared to the controls, the levels of AGEs and RAGE significantly increased in the rats following 6 months of exposure to fluoride, while for the rats exposed for 3 months, no significant changes in the parameter were detected. Additionally, the elevated RAGE protein expression and the transcription level in the brains of rats after 6 months of exposure to fluoride were measured by RAGE western blotting and mRNA expression, which correlated positively with the immunohistochemical results (Zhang et al. 2013). RAGE expression was found to increase with time after 24-h fluoride exposure. The alteration of AGE/RAGE signal transduction pathways can activate neuron apoptosis. Besides, the activation of downstream NF-κB and reduced nicotinamide adenine dinucleotide phosphate oxidase 2 can be significantly inhibited along with the decreased apoptosis rate of nerve and oxidative stress level after treating with RAGE blocker and antioxidant (Zhang et al. 2015). The study indicated that the alteration of AGE/RAGE signal transduction pathway might play an important role in the occurrence mechanism of nerve injuries induced by fluoride.

The Impact of Chronic Fluorosis on Vit D Receptor

Structure, Function, and Distribution of Vit D

Vit D is a fat-soluble vitamin that plays a vital role in human health, especially children. Active Vit D is mainly combined with vit D receptor to exert a biological effect (Hanel and Carlberg 2020). In 1998, Baker et al. cloned human vit D receptor (VDR), with a relative molecular weight of 45 × 103, consisting of 427 amino acid residues. VDR is a member of the steroid hormone/thyroid hormone receptor family, consisting of nuclear receptor and membrane receptor. Different species of VDR vary enormously in relative molecular weight but are similar in amino acids and structural composition. VDR has six functional regions, ranging from the N-terminal to the C-terminal, each of which functions differently but cooperates. The N-terminal short region is a transcription regulatory domain, which has the A/B domain and 24 amino acid residues. The region is ligand-independent and has tissue and cell specificities. Region C is the VDR binding domain, encoded by VDR gene exons II and III, and is mainly involved in the vit D responsive element (VDRE) (Belorusova et al. 2020). Region D is the antigen-determining domain and has strong immunogenicity which may be related to nuclear localization. Region E is a ligand-binding domain, encoded by exons V and IX of VDR gene, and is the main site binding with active Vit D. The free active vit D is quickly transferred to the nucleus after entering cells to be combined with VDR, which makes VDR undergo phosphorylation and conformational change and then bind to retinoid X receptor (RXR). VDR-RXR heterodimers promote the combination of CDR and VDER, formulating VDR-RXR-VDER complex. VDR-RXR functions as a switch to play a role in the function of activated vit D and the conformational switch can be stabilized by activated Vit D in return. VDR can be induced by co-inhibitor factors at the C-terminal of VDRE to prohibit target genes. VDR can be expressed in multiple organ systems, suggesting a wide use of Vit D, and is associated with diseases such as tuberculosis, HBV chronic infection, and osteoporosis.

The Impact of Fluoride on VDR

Fluorides can cause a series of bone changes such as sclerosis, osteoporosis, osteophyte, ostealleosis, and fracture, and tissue calcification and ossification in joint capsule, tendon, and ligament, along with fluorosis arthropathy. Vit D is a regulatory factor for the metabolism, the lack of which may cause bone metabolism-related diseases. It is found that VDR can mediate 1,25(OH)2 D3 to regulate osteoblast, osteoclast, chondrogenic differentiation and proliferation to enhance calcium deposition, and facilitate the synthesis of osteoclast-activating factor via osteoblast to promote the growth of epiphysis. Fluoride can disrupt the balance of bone formation and resorption to influence the regulation of bone metabolism. The VDR gene is located on chromosome 12, 75 kb in length, with 9 exons and 8 introns. Multiple restriction sites have been found by studying the VDR gene polymorphisms associated with diseases, the three most common types of which are BsmI, FokI, and TaqI. In the study of fluorosis group in Bijie City, Guizhou province, the frequencies of genotypes and alleles in VDR gene BsmI and FokI of fluorosis group contrasted sharply with that of the controls. While the difference between the two groups in VDR-TaqI gene polymorphisms showed no statistical significance, but those carrying with VDR bb, ff, Tt are relatively susceptible to fluorosis. Due to the rare reports on the correlation between VDR and fluorosis, it remains elusive in the relationship between the dental fluorosis susceptibility of both children and adults exposed to fluoride and the VDR gene Apa I, Bsm I, Fok I, and Taq I polymorphisms (Zhang et al. 2012). The elucidation of the relationship requires further study.

The Impact of Chronic Fluorosis on Peroxisome Proliferators-Activated Receptors (PPARs)

Structure, Function, and Distribution of PPARs

PPAR is a member of the nuclear hormone receptor superfamily and a ligand-dependent nuclear transcription factor (Schubert et al. 2020). PPARs mainly exist in three subtypes encoded by different genes, PPAR-α, β, and γ, the last of which can be further divided into PPAR-γ1 and PPAR-γ2 according to the different N-terminals. The amino acid residues of human PPAR-α, β, and γ are 468, 442, and 479, respectively (Derosa et al. 2018). PPARs are widely distributed in organisms, the distribution of which varies in nerve system, liver, heart, kidney, spleen, muscle, and adipose tissue. PPARs are mostly co-expressed but different in levels to a large degree since they are species- and tissue-specific. PPAR-α, β, and γ vary in distribution and expression in the CNS. PPARs expression can be detected by neurons and glial cells that PPAR-β is widely distributed, while PPAR-α and γ are regional (Kushwaha et al. 2019). Studies have shown that the expression of PPAR-γ in the brain is mainly concentrated in microglia and astrocyte, and these cells exert vital regulatory effects on the CNS inflammatory response and immune response.

The Impact of Fluoride on PPAR

PPAR-γ is found to be widely expressed in the CNS and plays a significant role in the nerve system development. The activation of PPAR-γ may exert a direct influence on the expressions of superoxide dismutase (SOD), catalase, and glutathione peroxidase 3 (GPx3) (Mahmoud et al. 2017), which shows that PPAR-γ plays a prominent role in anti-oxidative stress system. The research group found that the elevated fluoride content was accompanied with the decreased PPAR-γ protein and mRNA expression, along with the reduction in the protein expression levels for peroxisome proliferator-activated receptors-gamma coactivator-1α (PGC-1α). Besides, relevant analysis revealed that PPAR-γ expression in the hippocampal and cortical regions exhibited a negative correlation with fluoride concentration in the brain of adult rats, but positively correlated with serum SOD activity and negatively correlated with malondialdehyde (MDA) level. Besides, the protein expression of PGC-1α was negatively correlated with fluoride levels in the blood but positively related to the study and memory ability of rats. It shows that longer fluoride exposure with higher concentration can increase the fluoride accumulated in the body and the oxidative stress level, resulting in the reduction of PPAR-γ and the function impairment of its downstream activation factor, PGC-1α, which eventually cause functional disorder of nervous system in adult rats and the impairment of their study and memory ability (Liu et al. 2018a, b). Besides, as a nuclear receptor, PPAR-γ can regulate mesenchymal stem cells to transform from osteocyte to adipocyte, prohibiting bone formation to promote bone adipogenesis, which may serve as an important mechanism to reduce the bone mass by the activation of PPAR-γ. Currently, the correlation between the single nucleotide polymorphism of PPAR-γ and diseases of the skeletal system has become a concern for researches, while there is no apparent correlation between the polymorphism Rs1801282 of the PPAR-γ gene and fluorosis caused by tea drinking (Yan et al. 2018). It requires further research on the impacts of other sites of PPAR-γ gene or other genes influencing bone mass and their polymorphism on skeletal fluorosis.

Conclusion

The impacts of chronic fluorosis on receptors have been confirmed, mainly including the alternations in expression and function of these receptors, such as nAChRs, mAChRs, NMDA, AGE/RAGE, VDR, and PPAR. The impacts are primary on the genes related to intellectual development and abilities of learning and memory, such as nAChRs, mAChRs, NMDA, and PPAR-γ, the changes of which may exert a pivotal impact on fluoride inducing mechanism of brain injury. A molecular conformational change occurs when a receptor binds to a ligand, resulting in cellular reactions, such as mediated signal transduction, intercellular adhesion, and endocytosis, which can exert effects on metabolism and function of cells, such as the enzyme activity and ion channel. Based on the results, a conclusion can be made that chronic fluorosis may play a significant role in the abnormal expressions of the receptors.

References

Baranowska U, Wisniewska RJ. The 7-nACh nicotinic receptor and its role in memory and selected diseases of the CNS. Postepy Hig Med Dosw. 2017;71:633–48.

Belorusova AY, Chalhoub S, Rovito D, Rochel N. Structural analysis of VDR complex with ZK168281 antagonist. J Med Chem. 2020;63:9457–63.

Bi F, Wei N, Wan L. Effect of fluoride on expression of phosphorylated NMDAR protein in rat hippocampal neurons. Chin J Endem. 2019;38:4–7.

Derosa G, Sahebkar A, Mamoli P. The role of various peroxisome proliferator activated receptors and their ligands in clinical practice. J Cell Physiol. 2018;233:53–161.

Dong YT, Guan ZZ. Function of mascarinic acetylcholine receptors and their effects in the brain damages of rats with chronic fluorosis. Foreign Med Sci. 2014;2:97–103.

Dong YT, Wang Y, Wei N, Zhang QF, Guan ZZ. Deficit in learning and memory of rats with chronic fluorosis correlates with the decreased expressions of M1 and M3 muscarinic acetylcholine receptors. Arch Toxicol. 2015a;89:1981–91.

Dong YT, Wang Y, Wei N, Guan ZZ. Expression of muscarinic acetylcholine receptors in the brain of rats with chronic fluorosis. Chin J Endem. 2015b;34:84–8.

Dong YT, Wang Y, Wei N, Guan ZZ. Levels of oxidative stress in brain and serum of rats with chronic fluorosis and antagonistic effects of vitamin E. Chin J Endem. 2016;35:178–81.

Ge Y, Chen L, Yin Z, Song X, Ruan T, Hua L, et al. Fluoride-induced alterations of synapse-related proteins in the cerebral cortex of ICR offspring mouse brain. Chemosphere. 2018;201:874–83.

Guan ZZ, Gao Q, Gui CZ, Liu YJ, Ran LY. Study on brain injury mechanism of chronic fluorosis. Chin J Endem. 2011;30:352–4.

Gui CZ, Ran LY, Guan ZZ. Expression levels of brain nicotinic acetylcholine receptor mRNA and protein in coal-burning type of fluorosis rats. Chin J Endem. 2011;30:239–42.

Hanel A, Carlberg C. Vitamin D and evolution: pharmacologic implications. Biochem Pharmacol. 2020;173:113595.

Kushwaha R, Mishra J, Gupta AP, Gupta K, Vishwakarma J, Chattopadhyay N, et al. Rosiglitazone up-regulates glial fibrillary acidic protein via HB-EGF secreted from astrocytes and neurons through PPARγ pathway and reduces apoptosis in high-fat diet-fed mice. J Neurochem. 2019;149:679–98.

Lewis AS, Picciotto MR. Regulation of aggressive behaviors by nicotinic acetylcholine receptors: animal models, human genetics, and clinical studies. Neuropharmacology. 2020;167:107929.

Liu XH, Chen D, Zeng XX, Dong YT, Deng J, Qi XL, et al. Correlation between expression of peroxisome proliferator-activated receptor and oxidative stress in brains of rats with chronic fluorosis. Chin J Endem. 2018a;37:450–4.

Liu XH, Chen D, Zeng XX, Dong YT, Deng J, Qi XL, et al. Correlation study on the PGC-1α level in cortex tissue and hippocampus, serum with learning and memory in the brain of rats with chronic fluorosis. J Guizhou Med Univ. 2018b;43:1154–68.

Mahmoud AM, Hussein OE, Hozayen WG, Abd El-Twab SM. Methotrexate hepatotoxicity is associated the oxidative stress, and down regulation of PPARγ and Nrf2: protective effect of 18β glycyrrhetinic acid. Chem Biol Interact. 2017;270:59–72.

Schubert M, Becher S, Wallert M, Maeβ MB, Abhari M, Rennert K, et al. The peroxisome proliferator-activated receptor (PPAR)-γ antagonist 2-chloro-5-nitro-N-phenylbenzamide (GW9662) triggers perilipin 2 expression via PPAR δ and induces lipogenesis and triglyceride accumulation in human THP-1 macrophages. Mol Pharmacol. 2020;97:212–25.

Wang J, Lindstmm J. Orthosteric and allosteric potentiation of heteromeric neuronal nicotinic acetylcholine receptors. Br J Pharmacol. 2018;175:1805–21.

Wang JE, Zhong WY. Types and functions of cell receptors. Biol Teaching. 2012;37:80–1.

Wang JQ, Jiang WY, Zhang LJ, Wang JH, Zhao X, Wang Y. Hypoglycemic effects of saponins, flavonoids, and Polysaccharides of fenugreek with different component compatibilities on type l diabetic rats. J Jilin Univ. 2016;42:1081–6.

Wang XL, Deng YX, Gao YM, Dong YT, Wang F, Guan ZZ, et al. Activation of α7 nAChR by PNU-282987 improves synaptic and cognitive functions through restoring the expression of synaptic-associated proteins and the CaM-CaMKII-CREB signaling pathway. Aging. 2020;12:543–70.

Wei N, Dong YT, Deng J, Wang Y, Qi XL, Yu WF, et al. Changed expressions of N-methyl-d-aspartate receptors in the brains of rats and primary neurons exposed to high level of fluoride. J Trace Elem Med Biol. 2018;45:31–40.

Yan HZ, Li BY, Huo SM, Fan YM, Li Y, Li JJ, et al. Relationship between Rs1801282 polymorphism of peroxisome proliferator activated receptor γ gene and brick-tea type fluorosis. Chin J Endem. 2018;37:107–11.

Zhang C, ShanKR HY, An D, Xu SQ, Zhou SG, et al. Association of vitamin D receptor gene polymorphisms with susceptibility to coal-burning borne fluorosis in Guizhou province. Chin J Endem. 2012;31:130–3.

Zhang KL, Lou DD, Guan ZZ. Expression of receptor for advanced glycation endproducts and nuclear factor κB in brain hippocampus of rat with chronic fluorosis. Chin J Endem. 2013;32:2095–4255.

Zhang KL, Lou DD, Guan ZZ. Activation of the AGE/RAGE system in the brains of rats and in SH-SY5Y cells exposed to high level of fluoride might connect to oxidative stress. Neurotoxicol Teratol. 2015;48:49–55.

Chapter 16
Alteration of Endoplasmic Reticulum Stress

Hui Xu, Feng-Yang Guo, and Zhong-Yuan Zhang

Abstract The endoplasmic reticulum (ER) stress is induced by protein misfolding and nutritional deprivation in many eukaryotic cells. Meantime, an imbalance between protein synthesis and folding capacity of ER also caused the unfolded protein response (UPR). The UPR included three different protein family signals: double-stranded RNA-activated protein kinase like ER kinase, inositol-requiring kinase 1, and activating transcription factor 6, all of which are triggered during ER stress. Malfunction of UPR is induced by aging, genetic mutations, or environmental factors and results in various diseases, especially metabolic disease. The main adverse effects of chronic fluorosis are skeletal, dental, or enamel damages. The soft tissue lesions of fluorosis have been observed and discussed. The underlying mechanisms that mediated these effects remained poorly understand. This section is to review the role of ER stress in mechanism that overdosal fluoride exposure caused body lesion.

Keywords Bone · Endoplasmic reticulum stress · Fluorosis · Unfolded protein response

Endoplasmic Reticulum Stress and Unfolded Protein Response (UPR)

Endoplasmic Reticulum Stress

Endoplasmic reticulum (ER) serves as the major component for protein folding, Ca^{2+} storage, and protein synthesis, which plays an irreplaceable role for the survival and function of cell, as well as normal physiological function (Wang and Kaufman

H. Xu (✉) · F.-Y. Guo · Z.-Y. Zhang
School of Pharmaceutical Sciences, Jilin University, Changchun, Jilin, People's Republic of China
e-mail: xu_hui@jlu.edu.cn

Z.-Z. Guan (ed.), *Coal-burning Type of Endemic Fluorosis*,
https://doi.org/10.1007/978-981-16-1498-9_16

2016). ER is vesicle or tubular-shaped structure and can be further divided into rough ER and smooth ER. Physiological functions of ER range from coordinating synthesis folding of protein in eukaryotic cell, transmitting, signal peptide identifying, glycosylation modification, calcium ion regulation, and intracellular calcium redistribution (Schwarz and Blower 2016).

The ER maintains the steady state of protein folding synthesis, including protein which is transmitted to various cell organelles and extracellular part. Protein synthesis initiates from ribosome, while polypeptide to secretory pathway often comes with an N-terminal signal sequence. As the polypeptide enters ER, the signal sequence is removed by signal peptidase and the new polypeptide acquires the native conformation from being assembled and folded N-linked glycosylation. Hence, protein folding environment of ER distinguishes itself from that of cytoplasm due to its oxidizing property. Therefore, protein folding is the most error-prone step in its expression. The properly folded protein is then transmitted to Golgi apparatus and other cellular parts, while unfolded and misfolded proteins remain in endoplasmic reticulum and cause endoplasmic reticulum stress (ERS). The ERS would stimulate a series of signal transduction to response to the cell, including enhancing the protein folding, terminating the majority protein interpretation, speeding up the protein degradation. Such response is generally termed as UPR.

Three response sensor proteins were found to be the leading part in UPR, including protein kinase R (PKR)-like endoplasmic reticulum kinase (PERK), the activating transcription factor 6 (ATF6), and inositol-requiring enzyme l (IRE1). The binding immunoglobulin protein (BiP) is also named as glucose-regulated protein78, could integrate with the inner end of these three molecules in ER to maintain the inactive state of signal transduction factors. As an energy-consuming process, protein folding demands synergism between trans-organelle and protein from multiple cellular compartments as well as energy from various metabolic pathways. The steady-state shift of ER would lead to misfolded protein accumulation. Oxidative stress, ischemia, imbalanced Ca^{2+} state could lead to such unfolded protein to accumulate within ER. Although the molecular chaperone in ER could active the UPR to protect the molecule, the intensive ERS could induce the signal activation of cell apoptosis and cause molecular death through mitochondrion relied or non-mitochondrion relied pathway. Recent studies proved that ERS played a major role in determining cell survival and apoptosis.

Binding immunoglobulin Protein (BiP)

BiP is a member of the heat shock protein 70 kDa family. The protein folding process refers to the competing process of right and wrong pathway through folding intermediate product. The molecular chaperone functions as improving the protein synthesis efficiency and a competing mechanism of assisting the right pathway. During ERS, the protein expression is upregulated in BiP, calreticulin, and calnexin. BiP is regarded as the symbolic molecular chaperone in ER and plays major regulating role in ERS process. The three protein molecules in ER (IREI, PERK, and ATF6) are the kind of transmembrane sensor protein to response to stress signals. In general condition, these three sensor proteins form composite with BiP from ER and stay inactive. As the unfolded protein in ER accumulates, BiP molecule begins to integrate with the increasingly fresh yet unfolded protein in ER. Such integration causes three sensor proteins to be dissociated from its BiP composite. With the integration between BiP and unfolded or misfolded protein together with endoplasmic reticulum stress associated degradation unfolded or misfolded protein accumulation in ER would be reduced. Hence, protein metabolic pressure of ER is alleviated to preserve the survival of cells. As the ERS stays active or lasts for relatively long time, the dissociation of BiP and PERK could activate PERK/eukaryotic initiation factor 2α (eIF2α) pathway, which is one of the branch pathway of unfolded protein response. In this way, it could activate the interpretation expression increase of downstream activating transcription factor 4 (ATF4). The ATF4 and CCAAT/enhancer-binding protein homologous protein (CHOP) initiate subsegment combination to promote transcription expression of CHOP. Then, such increase would promote the release of cytopigment c to initiate the caspase cascade response and ultimately induce apoptosis. Previous study indicates that intra-ER imbalance-induced stress response could activate calcium protein kinase. The active calcium protein kinase cleaves the cysteine aspartate specific proteinase 12 (caspase 12) on ER membrane to produce active caspase12 segment. Meanwhile, calcium protein kinase could cut Bcl-XL to inactive it and cause apoptosis. Conversely, with the physical separation between BiP and other apoptosis-related protein, BiP could initiate ER-related apoptosis or other apoptosis pathways to cause cell apoptosis. Research findings discovered the differentiation of chondrocyte, which synthesized large amount of extracellular matrix protein, would induce ERS. In such process, BiP not only integrates with unfolded protein and assists it to form correct conformation, but also involves in the activation of three signal pathways of unfolded protein response. It indicates that BiP is a major regulating factor in differentiation of chondrocyte (Couasnay et al. 2019).

UPR Signal Pathway

UPR utilizes four pathways to maintain the steady state in ER: (1) Temporarily reducing the general protein synthesis to prevent the accumulation of unfolded protein; (2) Upregulating the expression of molecular chaperone (BiP and so on) which assists protein folding in ER to enhance its protein folding function; (3) Speeding up unfolded or misfolded protein degradation by ubiquitin-proteasome system; (4) Initiating apoptosis procedures of the affected cells to ensure the survival of body.

PERK Pathway

PERK falls into the category of serine/threonine protein kinase. ERS leads to the aggregation of misfolded or unfolded protein in ER to activate PERK. PERK would activate the downstream eIF2α to block further synthesis of the wrong protein. In this way, PERK would reduce the burden of ER and maintain the balance state. The active protein domain of PERK is homologous counterpart of eIF2α family kinases. The active PERK actives eIF2α to alleviate the burden of ER by terminating the translation of mRNA to reduce the accumulation of misfolded protein accumulation. Certain mRNA transcription factors, such as ATF4, acquire selective advantage of translation under the condition of eIF2α phosphorylation. As one of the transcription factors of basic leucine zipper (bZIP) family, ATF4 can regulate a handful of UPR promoter. Previous studies have identified that ATF4 relative genes include: ER molecular chaperone BiP, glucose-regulated protein94, glutathione biosynthesis, anti-oxidative stress genes, metabolism of amino acid, and transmitting genes. Target of ATF4 can upregulate molecular chaperone level and maintain the balance state of intracellular oxidation–reduction. Furthermore, ATF4 can promote the protein folding process and degradation of protein which cannot be folded. As a matter of fact, it is believed that PERK acts as a sensor of calcium ion level in ER. Certain signals such as glucose starvation and high fatty acid level would negatively regulate BiP to activate PERK level to maintain its steady state.

Inositol-Requiring Kinase 1 (IRE-1) Pathway

IRE-1 is featured with serine/threonine protein domain and endoribonuclease (RNase) protein domain. With unfolded protein response, IRE-1 intracavitary protein domain dissociates from BiP, then causes IRE1 oligomerization and stimulates its autophosphorylation to activate its enzymatic activity. The activated IRE-1 RNase protein domain cleaves X-box binding protein 1 (Xbp1) mRNA to make it spliced Xbp-1. Hence, the translated Xbp-1 is transmitted to the nucleus and integrated with ERS components and unfolded response elements. Such integration would promote the expression of unfolded protein target genes, such as molecular chaperone BiP and so on. In addition, IRE-1 encodes secretory protein by decomposing the majority of mRNA and pre-transcription decomposing, which would reduce the burden of ER in time of unfolded protein response. As IRE-1 is similar with most of tumor necrosis factor family members, it can integrate with E3 ubiquitin ligase tumor necrosis factor receptor and jointly activate the protein kinases which are related with immunization, inflammation, and apoptosis. Apoptosis signal-regulating kinase 1 in particular would lead to the activation of c-Jun N-terminal kinases (JNKs).

ATF6 Pathway

As a member of the bZIP family transcription family, ATF6 is endowed with an ER-related hydrophobic sequence which enables it to link with ER membrane. Compared with PERK and IRE-1 pathway, the mechanism of ATF6 is different from the rest of unfolded protein responses. The dissociation of ATF6 and BiP does not cause the oligomerization of ATF6, rather, it causes the metastasis of ATF6 to Golgi apparatus. Then, the dissociation will release the transcription factor to cytoplasm and enter the nucleus to regulate the gene expression. ATF6α activates ERS-related gene and converts it into homodimer or heterodimer with specific bZIP structure. Meanwhile, ATF6 can induce Xbp-1 expression together with IRE-1. ATF6 first induces the transcription of Xbp-1 mRNA, then the complete Xbp-1 is edited by the activation of IRE1 endoribonuclease. ATF6 integrates with targets like BiP, protein-disulfide isomerase (PDI), and ER degradation-enhancing α-mannosidase-like proteins (EDEMs) 1 to activate the expression of ER chaperone and degrade the misfolded protein.

ER-Related Protein Degradation

ER-related protein degradation refers to misfolded protein on its membrane or within its cavity is released into cytoplasm and then degraded by ubiquitin-proteasome. Such degradation involves four steps: substrate identification, unfolded free protein target positioning, cross-ER membrane transmitting, entering cytoplasm and then being degraded. The ER-related protein degradation is jointly participated by multi-ubiquitin-conjugating enzymes, including Ubc1p, Ubc6p, Ubc7p, and membrane anchor combination, Cue1p. In addition, the degradation also involves joint action of some proteins and ubiquitination pathway. ER-related protein degradation carried out synergistic action with unfolded protein response: on the one hand, unfolded protein response can upregulate ER-related protein degradation function; on the other hand, misfolded protein cannot be eliminated from the ER as it would cause unfolded protein response pathway to be specifically activated. It triggers the feedback regulation of ER-related protein degradation function and unfolded protein response to maintain the balance state of ER.

Relationship between Endoplasmic Reticulum Stress and Oxidative Stress

The new protein folding in ER falls into the category of oxidative folding. The folding process is one of the sources of intracellular oxidative stress as it generates reactive oxygen. Oxidative stress refers to imbalance between oxidation system and antioxidant system that leads to the tissue damage. When the body suffers from adverse stimuli, such imbalance triggers highly reactive molecules, such as reactive oxygen species (ROS) and reactive nitrogen species (RNS), that are generated too much and the degree of oxidation is too severe to remove the oxides. The results of ERS lead to the reactive oxygen accumulation and cause oxidative stress. For one thing, ERS can generate excessive ROS molecules, such as $O_2^{-\bullet}$ and $OH^{\bullet}$, to induce local tissue and molecule oxidative stress by UPR and interaction with mitochondrion; for another, ERS can resist anti-oxidative stress and maintain intracellular balance between oxidation and antioxidation. With ERS activated transmembrane protein related pathway, it can activate antioxidation transcription factors and upregulate the expression of multiple antioxidant enzyme. High level deposition of oxidative stress can affect protein folding and even ERS. If the biological system is exposed to oxidative stress conditions, it will cause oxidation on cellular level to modify the protein, increase lipid and nucleic acid. Then diseases will be induced and cause damage to cognitive function and whole metabolism. It was found that 6-hydroxydopamine caused oxidation will trigger ERS or UPR and ultimately result in apoptosis. The oxidized protein folding in ER and protein folding will change the redox state; in contrast, the generation of reactive oxygen will affect the steady state of ER and protein folding directly or indirectly (or both). The signal pathway of PERK can activate nuclear factor erythroid 2-related factor 2 (NRF2) and ATF4 transcription factors. Miscellaneous oxidative invasion induces the expression of antioxidant gene which contains antioxidant response element (ARE) in a variety of conditions. Transcription factor Nrf family regulates the expression of ARE genes in normal and pathological condition. Nrf2 knockout rat is prone to high level oxidative stress: its glutathione (GSH) level is obviously lowered compared with wild type cell, which indicates Nrf2 plays certain role on anti-oxidative signal pathway (Schmoll et al. 2017). Existing studies indicated that PERK could regulate multiple aspects of cellular oxidative stress and restore equilibrium state of ER protein folding by reducing protein translation with Ire1 and ATF6. Harding (Harding et al. 2003) discovered that ATF4 and Nrf2, as downstream transcription factor of PERK increase metabolism of amino acid to resist intracellular ERS by means of increasing GSH generation. Furthermore, ATF4 could stimulate the expression of oxidation/reduction balance-related gene to protect cells from damage. In less obvious oxidative stress condition, the expression of Nrf2 in ER obviously increases and enhances cellular resistance against oxidative stress (Bellezza et al. 2018). Meanwhile, previous research found that PERK and IRE-1, together with other signal molecules, activated a series of transcription and translation regulation process. Its results lead to the intracellular oxidative stress triggered by reactive oxygen accumulation. It was reported that sodium fluoride (NaF) affects the stress by improving ROS level and induce Nrf2 nucleus metastasis (Yang et al. 2015). Therefore, ERS has a close link with the oxidative stress on signal pathway.

Role of ERS in the Mechanism Underlying Fluoride-Induced Non-bone Damage

Chronic fluorosis is a non-lethal disease with a long progression. Hence, the clinical pathophysiology research on non-bone damage is faced with multi-difficulties. Of which, the major difficulty lies in establishing positive connection between early non-specific and minor symptom with fluorosis. Insufficient autopsy data for pathological morphology observation also poses as another challenge. The existing research has identified its damage on multiple systems and organs, such as heart, liver, kidney, skeletal muscle, immune system, endocrine system, and neuro-system. The mechanism of chronic fluorosis non-bone damage shows the different characteristics of damage level and changes of organs. Current study has proven that the ERS and UPR-induced apoptosis were involved in pathophysiology process of many diseases, including liver, neuro-system, cardiovascular system, endocrine system, and tumor diseases (Cubillos-Ruiz et al. 2017). Since the 1937 endemic fluorosis patient was reported with neurologic manifestation, researchers reported that fluorosis can endanger the nervous system. Niu's vivo experiment (Niu et al. 2018) indicated that exposure of NaF will damage the learning ability and memory of rat, even lead to abnormal histological and ultramicro-structure in rat's hippocampus. In vitro experiment, it further proved NaF induced excessive ERS and autophagy damage in SH-SY5Y cell. This experiment explained the key role of ERS induced autophagy damage in NaF induced apoptosis, which shed light on better understanding fluoride-induced neurotoxicity.

Fluorosis induced non-bone damage is often accompanied with ERS. Current study covers its lesion on kidney, pancreas. and reproductive system. Researchers examined the expression change of ERS marker gene (*BiP*) and key factors (Xbp-1 and PDI) of protein folding in renal tissue of fluoride-treated rat. It was found that low calcium diet exacerbated the fluoride damage on kidney and ERS also played a role in such mechanism. Concerning the study on pancreatic damage, Hu (Hu et al. 2012) found that the insulin-like positive products increased in pancreas islet, pancreatic shape factor significantly reduced, while insulin level in serum obviously increased in one-year fluoride-treated rat. These findings proved fluoride stimulated the synthesis and secretion of insulin. Nevertheless, Ito (Ito et al. 2009) injected NaF (20 mg/kg) on rats and observed 4 days later that granule accumulated in endoplasmic reticulum of pancreatic exocrine cell companied with aggravated autophagy and ribosome degranulation. All these results indicated that the effect of fluoride on pancreatic endocrine and exocrine function was closely related to vigorous protein synthesis or it induced ERS. Deng (Deng et al. 2016) discovered that NaF induced ERS mediated pancreatic cell apoptosis in rat. Zhang (Zhang et al. 2013) observed the whole process of rat pregnancy, infant rat till 8 weeks after adulthood all exposed to fluoride. The research found that such rats showed sperm disorder and reduced sperm cells. It was also found that inflammation change of male rats' reproduction system also triggered significantly increased expression of marker gene (*BiP*, *IRE-1*, and *CHOP*) in ERS, which explained its involvement in excessive fluoride-induced changes in male reproductive system. Relevant study of ERS on fluoride-induced non-bone tissue and organ damage mechanism falls into scarcity. As oxidative stress interacts with ER on various levels, therefore, it is of vital importance to correlate these two and explore their roles in fluorosis non-bone damage mechanism.

Role of ERS in the Mechanism Underlying Fluoride-Induced Bone Damage

ERS and Dental Fluorosis

Dental fluorosis is a chronic endemic fluorosis. With excessive fluoride injuring growing enamel of permanent teeth, which are glazed with chalky stripes or plaque, cinnamon stain, and dent. Basic lesion shows porous feature incomplete mineralization, detained amelogenin of stroma below dental enamel. The majority believes that enamel properties of dental fluorosis should be categorized as maturation defects, which would sabotage the maturation and mineralization of enamel as amelogenin cannot be hydrolyzed and removed in time. Early experiment proved that early stage ameloblast is extremely sensitive to fluoride concentration and excessive fluoride exposure will affect the secretion of ameloblast. However, its mechanism is not yet clear.

Certain amount of fluoride is possible to trigger dental fluorosis by ERS pathway, especially the incisor is sensitive to fluoride. Research found out that fluoride affects ameloblast synthesis and secretion by stimulating ERS (Wei et al. 2013). Subcutaneous injection of sodium fluoride on rats revealed changes of ameloblast during secretory phase of incisor dental germ and proved that excessive fluoride interfered with matter transmitting within ameloblast and hindered synthesis and secretion pathway. It was proven that certain dosage of fluoride stimulated the ERS and unfolded protein response to hinder the synthesis of enamel matrix proteinase. This interference affects the enamel matrix protein metabolism and then exacerbates the lesion of dental fluorosis (Suzuki and Bartlett 2014). Some scientists used gene chip to analyze rat's oral epithelial cells exposed to fluoride in vitro (2 mmol/l NaF) and found out that the expression of ER-related genes, such as DNA damage-inducible transcript 3, ATF4, and Hapa5, was significantly increased (Tabuchi et al. 2014). Zhang (Zhang et al. 2016) proved that exposure to high level of fluoride will cause ameloblast ER in rat, which indicated that calcium overload and high concentrate fluoride triggered ERS played an important role in dental fluorosis. In recent years, scientists began to take interest in the biological and pathological function of ameloblast UPR in enamel, especially on enamel hypoplasia and fluorosis. They found that accumulation of protein in pathologic cell drives UPR to apoptosis (Brookes et al. 2017). Current studies proved that fluoride stimulates ERS and UPR in ameloblast and disturbed matter transmitting leading to interference on the protein synthesis, secretion, and hydrolysis pathway, all of which proves that ERS plays an important role in dental fluorosis development.

ER and Skeletal Fluorosis

Pathogenesis of bone damage of endemic fluorosis (skeletal fluorosis) troubled the medical academia for years, as it diversifies in its pathological change (Mohammadi et al. 2017). In recent years, a substantial animal experiments and cell culture in vitro observations proved that active osteoblast and accelerated bone turnover are the leading characteristic lesions in skeletal fluorosis development and also the pathologic basis for diverse bone lesion. It then served as the specific target to explore mechanism of skeletal fluorosis. Previous studies reported that excessive fluoride intake into body or cell exposed to fluoride in vitro would show hyperplasia and hypertrophy of osteoblast and its precursor companied with abundant rough ER and vigorous protein synthesis and secretion function. The new protein folding in ER falls into the category of oxidative folding. Its folding process will generate reactive oxygen and become one of the sources of intracellular oxidative stress.

The research on osteoblast activity and oxidative stress under fluoride treatment inevitably involves ERS. If the speed of intracellular protein synthesis is too fast and exceeds the folding capacity of ER, it will cause ERS and trigger UPR. Scientists were able to observe the expression of ERS-related gene in bone tissue by adopting molecular biological technology (Liu et al. 2015). Gene and protein expression level of ERS marker BiP increased and significantly increased in low calcium combined with fluoride condition. This study found that toxic effect of low calcium combined with fluoride treatment obviously enhanced the BiP protein expression in osteoblast. The mRNA expression level of Xbp1 and CHOP was also significantly increased in low calcium combined with excessive fluoride condition. The expression level of PDI gene, which was closely related with ER protein oxidative folding, showed a relatively increased trend in low calcium control experiment and low calcium combined with excessive fluoride condition. Sun (Sun et al. 2014) adopted intragastric way to feed fluoride to rats. With the measure on bone pathological figure, bone density, and changes of gene expressions of bone formation and bone resorption-related genes in fluoride-treated rats, they proved that excessive fluoride-treated rats showed characteristic bone turnover from active bone resorption to bone formation. Fluoride-feeding at early stage caused ERS and triggered activation of PERK signal pathway in unfolded protein response. With the fluoride-feeding time being prolonged, not only the expression of Nrf2 enhanced, but also the other two signal pathways of unfolded protein response, IRE-1 and ATF6, were activated. It proved that the development of skeletal fluorosis was often accompanied with ERS and UPR. In addition, the change of PERK signal factor shared consistency with bone resorption of fluorosis, and the change of downstream factor Nrf2 was consistent with the change of osteogenic factors. It was also noticed that oxidative stress level in rats was in the range of compensation at early stage of fluoride-feeding, while decompensation of oxidative stress in rats was recorded in later stage of higher dosage of fluoride. According to the timing of the ERS and oxidative stress, it is speculated that oxidative stress induced by fluorosis in rats is more likely to be a subsequent response induced by ERS, which implied that PERK signal pathway took part in bone resorption regulated by fluoride treatment. The markers of other two ERS response signal pathways, Xbp1 and Nrf2, were most likely to be involved in osteogenic improvement process in later stage of fluoride treatment.

Wei (Wei et al. 2008) discovered PERK gene knocked-out mice lead to bone damage because of impaired osteoblast differentiation, maturation and reduced type I collagen secretion. With the reduced expression of core-binding factor 1 (cbf1) and osterix in osteoblasts and ruled out the possibility that such changes came from downregulated expression of ATF4, PERK could be a key factor to regulate bone development and osteoblast activity. These research findings suggested that low calcium had synergism on toxicity of fluoride. It proved that exposure to certain concentration of fluoride would stimulate the ER in fluorosis rats' tissue. With the increase of fluoride toxicity, ERS in bone tissue cells was also aggravated. Scientists also adopted high-throughput screening to study the fluoride osteoblast proteomics after fluoride exposure. With two-dimensional gel electrophoresis and mass spectrum analysis, BiP, PDI, proteasome 26S ATP enzyme and thioredoxin expressed higher in fluoride-exposed osteoblasts. It implied that obvious ERS occurred in fluoride-exposed osteoblast (Xu et al. 2008). The ATF4 combined with cbf1 stimulated the expression of osteocalcin in osteoblast. Therefore, it was believed that ATF4 played a major role in osteoblasts differentiation, maturity, and bone development.

Using Thapsigargin in ERS to treat osteoblast in vitro and generates ER state from low level to high level could reveal the bi-directional function of ERS in bone tissue formation. Low level ERS can stimulate bone formation, while high level one would induce the osteoblast apoptosis. Subsequently, gene chip analysis proves ERS-stimulated bone formation could trigger upregulation of gene expression level in ATF4, basic fibroblast growth factor, and Fos. Hino (Hino et al. 2010) observed upregulated expression of BiP and PDI in active osteoblast in vitro, while expression of BiP and PDI in osteoblast from osteoporosis patients downregulated. It could be inferred that ERS could be stimulating osteogenic action. Xu (Xu et al. 2010) carried out OS732 cell experiment by adopting certain fluoride concentration which could stimulate osteoblast activity. Their results indicated that fluoride treatment activated ERS and PERK signal pathway of unfolded protein response and PERK downstream factor Nrf2 gene significantly increased in OS732 cells. The small interfering RNA (siRNA) PERK effectively inhibited the PERK gene and protein expression in osteoblast, while expression of ATF4 and Nrf2 was not obviously affected. Certain dosage of fluoride obviously stimulated the expression of ATF4 and Nrf2. The siRNA PERK did not inhibit the fluoride-stimulating osteogenic function, but it inhibited osteoclastogenesis induced by osteoblast (Lü et al. 2014). In order to deeply explore ERS, especially the function of PERK signal factor in fluoride-induced osteoblast, our team utilized MC3T3-E1 cells for a series of experiments to explore the role of ERS in osteogenic function in fluoride exposure osteoblasts. First of all, bi-direction effect of fluoride on osteoblast was observed, of which low dosage of fluoride stimulated cell activity, while high dosage of fluoride inhibited the activity (Li et al. 2014; Zhou et al. 2013). Exposure to fluoride obviously stimulated expression of BiP and mildly promoted the expression of PERK and ATF4 signal pathway. Both low and high dosage of fluoride obviously activated the Xbp1 and ATF6 pathway of unfolded protein response. It indicated that ERS and UPR participated in the bi-directional effect mechanism underlying fluoride exposure induced osteoblasts. The siRNA BiP was effective to lower the expression of BiP gene and protein in osteoblast, which caused downregulation expression of UPR pathway factors (PERK, Xbp1, and ATF6) in fluoride-exposed osteoblasts and decreased expression of osteogenic and osteoclastogenic key transcription factors in fluoride-exposed osteoblast. Fluoride did not activate UPR in osteoblast after BiP expression was interfered. Meantime, osteoblastic activity was lowered. It showed that ERS triggered UPR played a major part in regulating fluoride-exposed osteoblasts differentiation and function. To further identify the role of PERK in osteogenesis of skeletal fluorosis, siRNA PERK was used to reduce the PERK and protein expression in osteoblasts and then observed changes of a series of osteogenic and osteoclastic indexes. PERK siRNA combined with fluoride exposure inhibited osteoblast markers, such as ALP, OCN, and key transcription factor like Runx2 and osterix expression. The combination clearly hindered the gene level of nuclear factor-kappa B ligand and osteoprotegerin—in osteoblast but had limited influence on macrophage colony-stimulating factor (M-CSF).

It proved the key PERK/Nrf2 signal pathway played a major part in fluoride-induced osteoblast activity and function and also a major role in the mechanism that excessive fluoride-induced active osteogenesis and osteoclastogenesis. PERK signal pathway probably functioned in skeletal fluorosis pathogenesis through downstream factor Nrf2 instead of ATF4. The same experiment also observed the GSH/L-glutathione oxidized (GSSG) level of key oxidation–reduction system. It was found that low dosage and medium dosage fluoride exposure could cause oxidation–reduction imbalance and possibly caused activate oxidative stress in osteoblasts. Moreover, high dosage of fluoride can trigger GSH/GSSG buffer system to be inoperative. Considering the changes of ERS after fluoride exposure, it indicated that different dosages of fluoride were more capable to induce ERS and UPR in osteoblasts, triggered the imbalance of GSH/GSSG, and activated oxidative stress in cells.

Fluoride shares closer affinity with calcium. Wang (Wang et al. 2019) reported that adding 1% $CaCO_3$ to the rat feed could significantly lower the fluoride content in bone and reverse the fluoride-induced bone disorder; while adding 2% $CaCO_3$ could significantly improve the fluoride content in bone. They inferred that the over-accumulated calcium in cell triggered the ERS pathway and damaged osteoblast and osteoclast. In vitro experiment also indicated that exposure to fluoride not only lowered the proliferative activity of osteoblasts, but also improved the Ca^{2+} level, expression level of ERS, and apoptosis-related factors in osteoblasts. It indicated that NaF treated released Ca^{2+} accompanied with the possibility of triggering ERS in osteoblast (Li et al. 2019). Previous findings proved that ERS is the major medium for NaF to mediate apoptosis. The relationship among intracellular Ca^{2+}, ERS, and apoptosis in the mechanism of skeletal fluorosis must be further identified.

The balance of osteogenesis and osteoclastogenesis was shattered in the process of skeletal fluorosis and thus caused diversified lesions. Scientists were devoted to study on the crucial part of skeletal fluorosis and trigger mechanism of osteogenesis, osteoclastogenesis, and bone turnover. The upregulated expression of ERS-related gene in bone tissue of skeletal fluorosis rats provided new clues. Scholars primarily focused on the osteogenic function of ERS, but study on relationship between UPR pathway and osteoblast less explored (Tohmonda et al. 2013). Jang (Jang et al. 2012) focused on the regulation of IRE1α-Xbp1 and ATF6 signal factors on osteoblast proliferation and differentiation. This research adopted RNA interference technique to inhibit the genetic expression of BiP and PERK and clarified the role of ERS and PERK pathway in mechanism underlying osteogenesis, osteoclastogenesis, and bone turnover during development of skeletal fluorosis. Changes in bone tissue pathology, bone density, and osteogenesis/osteoclastogenesis related genes expression in rats proved transition from enhanced bone resorption to accelerated bone formation, which is characterized by accelerated bone turnover of skeletal fluorosis. The accompanied alteration of endoplasmic reticulum stress markers and unfolded protein response pathways suggested a close correlation of ERS and the occurrence and development of skeletal fluorosis.

Conclusion

There is a complete mechanism in the cells to response to ERS through a series of signal transduction pathways, including enhancing protein folding capacity, suspending interpretation of most proteins, and accelerating degradation of protein. Activating UPR by molecular chaperone in ER could protect the cell, but the drastic ERS would also activate apoptosis to cause cell death. Therefore, endoplasmic reticulum plays a major role in determining the survival and apoptosis of cell. The previous studies have already proven the status of ERS and UPR in skeletal fluorosis bone and non-bone damage mechanism, as well as their correlation with related transcription factor, biological factors, and cytokine. Current study findings endorse the proposal of ERS, especially UPR pathways, has an inseparable relation with skeletal fluorosis and dental fluorosis. The role of UPR in skeletal fluorosis osteoblast activation has further identified the importance of PERK pathway in different stages of osteoblast proliferation, differentiation, and mineralization. These findings also illustrated the feature of PERK/Nrf2 signal pathway in fluoride-induced osteoblast activity and function, as well as its role in mechanism of active osteogenesis and osteoclastogenesis induced by excessive fluoride. Such findings lend support for future research on fluorosis pathogenesis.

References

Bellezza I, Giambanco I, Minelli A, Donato R. Nrf2-keap1 signaling in oxidative and reductive stress. Biochim Biophys Acta, Mol Cell Res. 2018;1865:721–33.

Brookes SJ, Barron MJ, Dixon MJ, Kirkham J. The unfolded protein response in amelogenesis and enamel pathologies. Front Physiol. 2017;8:653.

Couasnay G, Bon N, Devignes CS, Sourice S, Bianchi A, Véziers J, et al. PiT1/Slc20a1 is required for ER homeostasis, chondrocyte survival, and skeletal development. J Bone Miner Res. 2019;34:387–98.

Cubillos-Ruiz JR, Bettigole SE, Glimcher LH. Tumorigenic and immunosuppressive effects of ERS in cancer. Cell. 2017;168:692–706.

Deng H, Kuang P, Cui H, Chen L, Luo Q, Fang J, et al. Sodium fluoride (NaF) induces the splenic apoptosis via ER (ER) stress pathway *in vivo* and *in vitro*. Aging. 2016;8:3552–67.

Harding HP, Zhang Y, Zeng H, Novoa I, Lu PD, Calfon M, et al. An integrated stress response regulates amino acid metabolism and resistance to oxidative stress. Mol Cell. 2003;11:619–33.

Hino SI, Kondo S, Yoshinaga K, Saito A, Murakami T, Kanemoto S, et al. Regulation of ER molecular chaperone prevents bone loss in a murine model for osteoporosis. J Bone Miner Metab. 2010;28:131–8.

Hu CY, Ren LQ, Li XN, Wu N, Li GS, Liu QY, et al. Effect of fluoride on insulin level of rats and insulin receptor expression in the MC3T3-E1 cells. Biol Trace Elem Res. 2012;150:297–305.

Ito M, Nakagawa H, Okada T, Miyazaki S, Matsuo S. ER-stress caused by accumulated in triacontanol granules activates autophagy through a different signal pathway from unfolded protein response in exocrine pancreas cells of rats exposed to fluoride. Arch Toxicol. 2009;83:151–9.

Jang WG, Kim EJ, Kim DK, Ryoo HM, Lee KB, Kim SH, et al. BMP2 protein regulates osteocalcin expression via Runx2-mediated Atf6 gene transcription. J Biol Chem. 2012;287:905–15.

Li X, Meng L, Wang F, Hu X, Yu Y. Sodium fluoride induces apoptosis and autophagy via the ERS pathway in MC3T3-E1 osteoblastic cells. Mol Cell Biochem. 2019;454:77–85.

Li XN, Lv P, Sun Z, Xu H. Role of unfolded protein response in affecting osteoblast differentiation induced by fluoride. Biol Trace Elem Res. 2014;158:113–21.

Liu L, Zhang Y, Gu H, Zhang K, Ma L. Fluorosis induces ERS and apoptosis in osteoblasts *in vivo*. Biol Trace Elem Res. 2015;164:64–71.

Lü P, Li X, Ruan L, Xu H, Liu Q. Effect of siRNA PERK on fluoride-induced osteoblastic differentiation in OS732 cells. Biol Trace Elem Res. 2014;159:434–9.

Mohammadi AA, Yousefi M, Yaseri M, Jalilzadeh M, Mahvi AH. Skeletal fluorosis in relation to drinking water in rural areas of West Azerbaijan. Iran Sci Rep. 2017;7:17300.

Niu Q, Chen J, Xia T, Li P, Zhou G, Xu C, et al. Excessive ERS and the resulting autophagic flux dysfunction contribute to fluoride-induced neurotoxicity. Environ Pollut. 2018;233:289–99.

Schmoll D, Engel CK, Glombik H. The Keap1-Nrf2 protein-protein interaction: a suitable target for small molecules. Drug Discov Today Technol. 2017;24:11–7.

Schwarz DS, Blower MD. The ER: Structure, function and response to cellular signaling. Cell Mol Life Sci. 2016;73:79–94.

Sun F, Li X, Yang C, Lv P, Li G, Xu H. A role for PERK in the mechanism underlying fluoride-induced bone turnover. Toxicology. 2014;15:52–66.

Suzuki M, Bartlett JD. Sirtuin1 and autophagy protect cells from fluoride-induced cell stress. Biochim Biophys Acta. 2014;1842:245–55.

Tabuchi Y, Yunoki T, Hoshi N, Suzuki N, Kondo T. Genes and gene networks involved in sodium fluoride-elicited cell death accompanying ERS in oral epithelial cells. Int J Mol Sci. 2014;15:8959–78.

Tohmonda T, Yoda M, Mizuochi H, Morioka H, Matsumoto M, Urano F, et al. The IRE1α-XBP1 pathway positively regulates parathyroid hormone (PTH)/PTH-related peptide receptor expression and is involved in PTH-induced osteoclastogenesis. J Biol Chem. 2013;288:1691–5.

Wang J, Yang J, Cheng X, Xiao R, Zhao Y, Xu H, et al. Calcium alleviates fluoride-induced bone damage by inhibiting ERS and mitochondrial dysfunction. Agric Food Chem. 2019;67:10832–43.

Wang M, Kaufman R. Protein misfolding in the ER as a conduit to human disease. Nature. 2016;529:326–35.

Wei J, Sheng X, Feng D, McGrath B, Cavener DR. PERK is essential for neonatal skeletal development to regulate osteoblast proliferation and differentiation. J Cell Physiol. 2008;217:693–707.

Wei W, Gao Y, Wang C, Zhao K, Sun D, et al. Excessive fluoride induces ERS and interferes enamel proteinases secretion. Environ Toxicol. 2013;28:332–41.

Xu H, Ling J, Li GS. Proteomic analysis of osteoblasts exposed to fluoride *in vitro*. Biol Trace Elem Res. 2008;123:91–7.

Xu H, Zhou YL, Zhang XY, Lu P, Li G. Activation of PERK signaling through fluoride-mediated ERS in OS732 cells. Toxicology. 2010;277:1–5.

Yang Y, Lin X, Huang H, Feng D, Ba Y, Cheng X, et al. Sodium fluoride induces apoptosis through reactive oxygen species-mediated ERS pathway in Sertoli cells. J Environ Sci. 2015;30:81–9.

Zhang S, Jiang C, Liu H, Guan Z, Zeng Q, Zhang C, et al. Fluoride-elicited developmental testicular toxicity in rats: roles of ERS and inflammatory response. Toxicol Appl Pharmacol. 2013;271:206–15.

Zhang Y, Zhang K, Ma L, Gu H, Li J, Lei S, et al. Fluoride induced ERS and calcium overload in ameloblasts. Arch Oral Biol. 2016;69:95–101.

Zhou YL, Shi HY, Li XN, Lv P, Li GS, Liu QY, et al. Role of ERS in aberrant activation of fluoride-treated osteoblasts. Biol Trace Elem Res. 2013;154:448–56.

Chapter 17
Mitochondrial Changes

Di-Dong Lou and Zhi-Zhong Guan

Abstract Mitochondrion is an important organelle in eukaryotes to provide adenosine triphosphate as energy for cell activities. It plays a significant role in cell apoptosis, cell signal transduction, ion transfer, and steroid generation. The latest researches showed that the function of mitochondria was consistent with the fission, fusion, and morphological structure of mitochondria. When there is a mitochondrial fusion barrier, mitochondria will constantly split into a single pellet tubule or other shapes or mitochondria will overfuse into a tubular membrane network. The balance between fission and fusion is vital for mitochondria to maintaining cell function. Some scholars reported that the mitochondria of various systems of rats with fluorosis showed vacuolated changes, suggesting that fluorosis could cause dysfunction of mitochondrial fusion. This section will introduce the mechanism of mitochondrial fusion and the related studies of mitochondrial morphology and function in fluorosis.

Keywords Dynamic equilibrium · Endemic fluorosis · Fission · Fusion · Mitochondria

Introduction

The complex organelle mitochondria exist in all eukaryotic cells except for mature red blood cells. They are wrapped by the external mitochondrial membrane and the internal mitochondrial membrane. Through fusion and division, the two membranes

D.-D. Lou (✉)
School of Basic Medicine Sciences, Guizhou University of Traditional Chinese Medicine, Guian, People's Republic of China
e-mail: luoruixi058@gzy.edu.cn

Z.-Z. Guan
The Key Laboratory of Endemic and Ethnic Diseases (Guizhou Medical University), Ministry of Education, Guiyang, Guizhou, People's Republic of China

Department of Pathology at the Affiliated Hospital of Guizhou Medical University, Guiyang, Guizhou, People's Republic of China

Z.-Z. Guan (ed.), *Coal-burning Type of Endemic Fluorosis*,
https://doi.org/10.1007/978-981-16-1498-9_17

worked in coordination to maintain the mitochondrial homeostasis (Farmer et al. 2018). Embedded with membrane proteins which contribute to electron transfer and transport, mitochondrial membrane was mainly composed of phospholipid. Besides synthesizing (adenosine triphosphate) ATP for cell energy, mitochondria also participated in intracellular calcium homeostasis and cell signal transduction (Liang et al. 2002). Mitochondrial membrane permeability increased during cell apoptosis, then cytochrome C stored in the membranous space and apoptosis inducing factor, cysteine protease activator, endonuclease G, and other molecules were released into the cytoplasm. Finally, cells died because of the activation of cysteinyl aspartate specific proteinase (caspase) or the direct action on chromosomal deoxyribonucleic acid (DNA) (Adrain et al. 2001).

The endemic fluorosis prevalence was serious in Shanxi, Xinjiang, Inner Mongolia, Guizhou, and so on (Ma et al. 2017). Not only did it affect the skeletal system, but it could also cause damage to other non-osseous systems or tissues, such as the immune system, nervous system, reproductive system, and cardiovascular system (Sun et al. 2016). It was found from continuous researches that the cause of the pathological changes in tissues of various systems and organs induced by excessive fluorine intake was attributed to the dysfunction of mitochondria. Such cause was the result of intrinsic changes, such as morphological and structural, ultrastructural, and mitochondrial dynamics.

Mitochondrial

Structure and Function of Mitochondrial DNA (mtDNA)

The mtDNA, the only extranuclear DNA, is ringed with the length of 16,569 bp. The two chains of mtDNA, which are termed H chains and L chains, are in reverse parallel. The mtDNA mainly encodes 2 rRNA, 22 tRNA, and 13 polypeptides. The common damage of mtDNA is fragment deletion, single base damage, and small fragment repetition. The mtDNA could be duplicated and delivered monotonously to the offspring as it was inherited in a specific maternal line, so there came mutations when damaged (Adrain et al. 2001).

It was reported that mitochondria were the core target organelle of cell ischemia and hypoxia damage. The activity of key enzymes in oxidative phosphorylation such as cytochrome C oxidase, ATPase, and reduced form of nicotinamide adenine dinucleotide (NADH) reductase was significantly reduced in the early stage of metabolism, and the deletion of mtDNA fragment in these enzyme coding regions was confirmed (Liang et al. 2002). From the analysis of brain striatum of the patients with Parkinson disease (PD), three subunits with lengths of 302, 252, and 242 kDa in the mitochondrial genome complex I subunit were missed, and the missing parts were encoded by mtDNA. It was also reported that the deletion of mtDNA4977 fragment was increased in the striatum and cortex of PD patients. Other studies confirmed that the deletion of mtDNA 4.8 kb was influenced by heavy metals such as manganese. It was widely reported that the deletion of mtDNA fragments was one of the manifestations of mitochondrial damage (Chu et al. 2007).

Mitochondrial Respiratory Chain Function in Oxidative Phosphorylation

It is known that most ATP is provided for various life activities only by an effective mitochondrial oxidative phosphorylation system. The respiratory chain of mitochondria refers to the gradual transfer of paired hydrogen atoms from metabolites through a chain reaction catalyzed by various enzymes and coenzymes and eventually combines with oxygen to produce water and ATP. It has shown that enzymes or coenzymes can be divided into hydrogen transporters and electron transporters, which are arranged on the inner membrane of mitochondria in a certain order. Enzymes and coenzymes mainly consist of five enzyme complexes, namely: complex I (NADH-quinone reductase, prosthetic group-FMN, FeS), complex II (succinic acid-quinone reductase, agon FAD, FeS), complex III (ubiquinone-cytochrome C reductase, prosthetic group-iron porphyrins, FeS), complex IV (cytochrome C oxidase, auxiliary base iron porphyrins, Cu), and complex V (an ATP synthetase). The I to IV enzyme complexes mainly involve in electron transport and V major concerns the synthesis of ATP from ADP and Pi.

Complex I passes electron from nicotinamide adenine dinucleotide to ubiquinone; complex II from the succinic acid to the ubiquinone; complex III from the ubiquinone to cytochrome C, which is water soluble and easily separated from the inner membrane of mitochondria; complex IV passes electrons to oxygen. In the above mentioned four processes, the auxiliary group, FeS, and copper ion (Cu) all participate in the catalytic transfer process. When proton is pumped from the stroma of the mitochondrial inner membrane to the outer membrane, a proton electrochemical gradient is generated (H^+ concentration gradient and transmembrane potential difference) inside and outside the membrane, which is the critical to ATP synthesis. ADP and Pi are driven to generate ATP when protons are reflux along the concentration gradient. Experiments proved that complex I, III, and IV had the effect of proton pump.

There are other oxidative systems in mitochondria as well. Superoxide ions (O_2^-) can be generated during electron transfer in the respiratory chain, and O_2^- can further generate H_2O_2 and hydroxyl radical (•OH). They were collectively known as reactive oxygen group (Mailloux et al. 2013). O_2^- can oxidize unsaturated fatty acid in phospholipid molecule to peroxide lipid to damage biofilm. The formation of lipid peroxidation–protein complex causes loss of protein function. Superoxide dismutase (SOD) was an antioxidant enzyme, which could catalyze the transformation of superoxide dismutase into hydrogen peroxide (Kim et al. 2007). There are three isoenzymes: extracellular-superoxide dismutase (EC-SOD), manganese SOD (Mn-SOD), and SOD with copper/zinc (Cu/Zn-SOD). It was shown that manganese

Mn-SOD was a kind of mitochondrial enzyme, which could convert toxic O_2^- into H_2O_2. Therefore, SOD is an important enzyme for human body to defend against superoxide ion damage in internal and external environment.

Mitochondrial Spatial Structure and Regulatory Mechanism

Mitochondrial dynamics are essential for cell health (Rocha et al. 2018), including mitochondrial division, fusion, transport, autophagic degradation (mitosis), and biogenesis. The dynamic homeostasis of mitochondria depends on its normal spatial structure and orderly regulatory mechanism.

Under transmission electron microscope, the mitochondrial morphology was observed with two-dimensional image with rod-shaped and a highly folded inner membrane, which protruded inward to form ridges and a simple outer membrane, regardless of the three-dimensional morphological characteristics of mitochondria in the cell. In hela cells, three-dimensional image of mitochondria showed a membrane network tubular structure with continuous dynamic changes (Rizzuto et al. 1998). The mitochondrial morphology was revealed by mitochondria-green fluorescent protein probe. It was found that the fluorescence was immediately restored after the partial removal of the mitochondrial tubular network structure, and a connected network was formed in the cell by mitochondria. Recently, some scholars proposed that mitochondria were three-dimensional, membrane-connected tubes in the cell. As mitochondrial fusion and fission frequently occur during the cellular life activity and under the stimulus of external environment, theirs shapes and quantity are variable, maintain a dynamic balance to ensure the integrity of membrane structure and preventing mitochondrial dysfunction and cell apoptosis caused by membrane interstitial molecular leakage (Karbowski and Youle 2003).

In addition, after labeling mitochondria with a membrane potential-sensitive fluorescent probe, the local depolarization of mitochondria was stimulated. The mitochondrial membrane potential in the distance of the network was also decreased, indicating the high connectivity of mitochondria (Skulachev 2001). Numerous cell fusion experiments demonstrated that mitochondria were highly dynamic. It was also found that the morphology of mitochondria changed with the cell cycle. The G1 phase of mitochondria was mainly netlike, while the S phase was fragmented. The morphology of mitochondria was not completely the same under different culture conditions (Margineantu et al. 2002).

In a normal cell, the rate of mitochondrial fusion and fission is comparable, and the quantity and network of mitochondria are stable and balanced. Once damaged, mitochondrial fusion is inhibited so mitochondrial fragmentation occurred. While cell function is enhanced, the process of mitochondrial fission is inhibited, and the degree of mitochondrial networking is enhanced. Under normal conditions, when mitochondrial fusion pathway is disrupted, mitochondrial morphology can be stabilized to a certain extent by inhibiting the fission pathway correspondingly and vice versa.

Table 17.1 Main molecules involved in mitochondrial fusion and fission

Mammalian	Yeast	Bioactivity
Drp1/DVLP/DLP1	Dnm1	Fission
hFis1	Fis1/Mdv2	Fission
Unknown	Mdv1/Fis2/Gag3/Net2	Fission
Mfn1, Mfn2	Fzo1	Fusion
OPA1	Mgm1	Fusion
Unknown	Ugo1	Fusion

Compared to static condition, tubular membrane network dynamic change of mitochondria is more advantageous to the energy transfer and information communication. Moreover, the mtDNA can be exchanged and complemented through the mitochondria dynamic fusion, effectively repairing the damage of mtDNA. In addition, mitochondrial fission can generate different levels of calcium storage, membrane potential, and permeability in mitochondria, while the S phase presents a fragmented structure, which is conducive to the distribution of mitochondria to daughter cells. In recent years, it was discovered that many genes regulate mitochondrial fusion and fission. The main fission and fusion proteins in mitochondria are members of the Dynamin family (Table 17.1), which play an important role in regulating the morphology of mitochondria. Therefore, it is necessary for in-depth studies to describe the morphology and function of mitochondria (Van der Bliek et al. 2013).

At present, the main known mitochondrial fusion execution molecules include the outer membrane fusion protein mitochondrial fusion protein gene (MITofusin gene, Mfn1/2), the intimal fusion protein mitochondrial optic atrophy1 (OPA1), and the yeast molecule Ugo1. Fuzzy Onions (*Fzo*) gene family, separated from the spermatogenesis in Drosophila, was the earliest discovered mitochondrial fusion protein (Lee and Yoon 2016). Subsequently, it was confirmed that mitofusion protein and Fzolp protein had similar functions in mammals and yeast. Both of them were critical proteins in mitochondrial fusion mechanism, while mammalian mitochondrial fusion proteins contained Mfn1 and Mfn2 (Ishihara et al. 2004). Research results confirmed that the deletion of the heat-sensitive Fzolp allele in yeast caused mitochondrial fusion barrier. Then, the tubular membrane network was broken and mitochondria presented fragmented network structure in small spherical or vesicular shape. As a result, mtDNA was lost and respiratory activity was reduced. Similarly, mouse embryonic fibroblasts with defective Mfn1 or Mfn2 alleles also showed mitochondrial fusion disorder. The tubular membrane network was broken and the mitochondrial ATP synthesis function was decreased (Chen et al. 2003).

It was revealed that the homology of protein Mfn1 and Mfn2 was predicted to be 60% and 77%, while the GTPase domain (G-domain) was predicted to be the most homologous and the n-terminal and C-terminal conserved regions were the lowest. The MEFs cells with Mfn1 or Mfn2 absence only maintained partial mitochondrial fusion activity. Overexpression of Mfn2 could promote the reconstruction of mitochondrial tubular connection membrane network in Mfn1-deficient cells, while overexpression of Mfn1 could promote the restoration of wild-type mitochondria in Mfn2 deletion cells. In addition, the mitochondrial morphology could be restored. Either Mfn1 or Mfn2 was deficient or one of them was overexpressed, it demonstrated that partial fusion activity of Mfn1 and Mfn2 was identical and interchangeable. In vitro experiments, it was found that Mfn1 had higher fusion activity than Mfn2. Once the mitochondrial dynamic homeostase-associated protein OPA1 overexpresses, mitochondrial tubular fusion was induced in MEFs cells with Mfn2 deficiency. However, the experimental results of Mfn1 cells were different (Santel and Fuller 2001).

Each fusion protein plays a key role in the regulatory mechanism of mitochondrial fusion. In yeast, strict regulation of Fzolp fusion protein can ensure the normal morphology of mitochondria. Currently, there are two regulatory mechanisms to maintain the normal morphology. One is quantity regulation and the other is control of receptor pairing. Fzolp degradation was controlled by Mdm30p (an F-box protein), which could be ubiquitinated to degrade Fzolp protein by 26S protease. In Mdm30 mutant cells, Fzolp accumulation resulted in excessive fusion and aggregation of mitochondria (Escobar-Henriques et al. 2006). In mammals, the apoptotic factors Bax and Bak not only induced apoptosis, but also formed perforation points on mitochondria to recruit Mfn2 aggregation. The results showed that the mitochondrial fusion activity was injured and presented fractured in cells with Bax/Bak gene knockout (Brooks et al. 2007).

Mitochondrial fission is mainly mediated by Fis1 and Drp1 in mammal cells, and yeast molecule Mdv1/Fis2/Gag3 is also involved. Drp1 was a cytoplasmic protein that moved to the surface of the mitochondria to fission (Zhang et al. 2016). It was a member of the dynamin proteins family with membrane remodeling properties, including G-domain, intermediate domain, variable domain, and GTPase effect domain (Lee and Yoon 2016). Drp1 only participated in the fission of mitochondrial outer membrane, not inner membrane (Labrousse et al. 1999). During the fission process, the mitochondrial outer membrane protein Fis1 absorbed GTPase Drp1 from the cytoplasm to the cytoplasmic surface of mitochondria, where oligomeric Drp1 was assembled into a spiral structure, and the mitochondrial membrane was cut off after GTP hydrolysis. Moreover, overexpression of Drp1K38A could inhibit block cell by inhibiting mitochondrial fission. Overexpression of Fis1 resulted in punctured mitochondria, suggesting that both Drp1 and Fis1 regulate mitochondrial fission (Zhang et al. 2016). In Drp1 RNAi cells, cytoplasmic Bax almost completely disappeared, indicating the protective effect of Drp1 RNAi on apoptosis (Lee et al. 2004). Fis1 could also interact with yeast Mdv1 molecules, suggesting homologous effect with yeast Mdv1 molecules (Tieu et al. 2002).

Recent studies found that the process of mitochondrial fission was related to apoptosis, while fusion was related to anti-apoptosis. Overexpression of molecule mediated mitochondrial fission could induce apoptosis, such as Drp1 or Fis1, while Drp1K38A could inhibit or slow down caspase activation and apoptosis if overexpressed. In contrast, inhibition of Drp1 or Fis1 inhibited apoptosis (Frank et al. 2001; Karbowski et al. 2002). The expression of Fzo1 delayed the activation of Bax/Bak, thus inhibited the release of cytochrome c and apoptosis. Mfn1/2 silence showed enhanced Bak activation, while Drp1 silence showed resistance to Bak activation (Sugioka et al. 2004). Overexpression of molecule Mfn1/2 or OPA1 mediated mitochondrial fusion could promote mitochondrial fusion and inhibit apoptosis. Overexpression of mitofusins or OPA1 could enhance the stimulation of anti-apoptosis, while inhibition of Mfn1/2 or OPA1 could promote apoptosis (Suen et al. 2008). Therefore, cells are more vulnerable when fusion proteins is absence. Mitochondrial fusion is pivotal in the growth and development of the organism. Many studies confirmed that the absence of Mfn1 and Mfn2 simultaneously resulted in complete loss of mitochondrial fusion activity, which not only led to complete fracture of mitochondrial tubular connection membrane network, but also serious decline in mitochondrial respiratory activity, resulting in cell growth stagnation. Embryo of Mfn2 mutants died after germination. It was possible that the mitochondrial fusion function of trophoblast cells was impaired during embryo germination (the cells provided embryonic nutrients). Such results led to the decrease of mitochondrial respiratory activity and the apoptosis of trophoblast giant cells. Mice with Mfn1 or Mfn2 deficiency died at an early age.

It can be inferred that mitochondrial fission is an intermediate link in certain cell apoptosis. The reasons are as follows: (1) mitochondrial fragmentation occurs in many cell apoptosis processes and is earlier than caspase activation. (2) Overexpression of Drp1 or Fis1 can induce apoptosis, while overexpression of Mfn1/2 can inhibit apoptosis. (3) Drp1 and Fis1 are necessary for apoptosis, and inhibition of their activity can prevent or delay apoptosis. (4) During apoptosis, Bax and Drp1 synergistically work on potential mitotic site of mitochondria.

Mitochondrial Changes in Chronic Fluorosis

Morphological Changes of Various Organs Caused by Chronic Fluorosis

The experimental studies showed that long-term excessive fluoride intake could cause myelin destruction and mitochondrial swelling in the spinal cord and cerebellum of rats. It was observed that the myelin sheath of the neocortical cortex was cleaved, axons were with thin myelin sheath, hippocampal myelin fibers were broken, and mitochondria were vacuolated (Reddy et al. 2020). The sciatic nerve of rats showed vacuolation of Schwann cells, enlargement of axons, and destruction of myelin sheath. Researchers also reported that long-term excessive intake of fluoride led to demyelinating changes in cerebral cortex and subcortical areas, changes including: brain hippocampus neural cells, nerve fibers, nerve synapses, injury of the ultrastructure of the blood–brain barrier (mainly for Presynaptic edema), severe degeneration of mitochondria, increase of electron density in the synaptic gap, fusion of presynaptic and posterior membranes, increase of presynaptic active regions, integration of the presynaptic membranes, synaptic gaps, and postsynaptic membranes. From the studies on brain tissue in vitro, it showed that excessive fluoride could also lead to degeneration and apoptosis of nerve cells, such as inhibiting the growth of hippocampal nerve cells. The number of neurons in the dorsal raphe nucleus of rats with fluorosis was slightly lower and the ratio of nucleus plasma increased. The heterochromatin in the nucleus gathered round the edge with increased numbers. The rough endoplasmic reticulum cisterna and Golgi complex squamous sac also expanded. The lipofuscin in neuronal soma, especially in neurite, was significantly increased (Chen et al. 1995). More autophagosomes were observed, suggesting that high fluoride intake had direct damage to neurons in the dorsal raphe nucleus. Fluorosis could also cause direct damage to the spinal cord, mainly to gray matter (Mu et al. 1995). Other experimental results showed that fluorosis could cause the degeneration of nerve cell vacuole in mouse spinal cord, and Nissl body in nerve cells was dissolved, decreased, or even disappeared. More changes were also observed: neuronal shrinkage, neuronal phagocytosis, neuronal cells, and interstitial edema influenced by fluorosis, the decreased number of neurons, and the hemorrhage of spinal cord. Animal study showed that the myocardial cells in the fluorinated group were irregularly arranged with a large number of infiltrated inflammatory cell and nuclear shrinkage (Wang et al. 2017).

Ultramorphological Changes Induced by Fluorosis

In an experiment on endemic fluorosis, experimental objects were the heart, brain, liver, adrenal gland, thyroid. and other tissues of fetuses with fluorosis induced by excessive fluoride intake in Guizhou province, while subjects from the non-fluorosis areas were chosen as the control group. According to pathological histology observation, the researchers found that the mitochondria of experimental group, especially in heart, brain, liver, adrenal gland, thyroid gland viscera essence cell, showed different degrees of degeneration and swelling. All the volume density, average perimeter, area, and increased volume suggested that mitochondria had been affected by fluoride, similar to other causes of mitochondrial damage. The fetus' mothers lived in the ward for a long time, they were diagnosed as patients with chronic fluorosis. They suffered from dental fluorosis with significantly higher urinary fluoride content than that of healthy pregnant women at the same period. The content of skeletal fluoride in fetuses of pregnant women living in endemic fluorosis areas was significantly higher than that of fetuses without endemic fluorosis, indicating that there was also excessive fluoride deposition in fetuses. The results indicated that fluoride may cause fetal fluorosis and abnormal changes in mitochondria (Yu 1999). Some scholars reported that mitochondrial damage of non-bone cells could be observed in animal models with chronic fluorosis. Under scanning electron microscope, hepatocytes of the rats fed with different doses of sodium fluoride in drinking water for 3 months were detected with unclear boundaries. The mitochondria were significantly swollen, their cristae were fractured or disappeared and the density was increased. The endoplasmic reticulum was lamellar dilated. The capillary bile duct dilated, whose microvilli on the lumen decreased with more white lipid droplets.

Observation under electron microscope in the control group showed that the morphology and structure of hepatocytes were normal and clear, the cristae of mitochondria were normal, lipid droplets were rare, and the capillary bile duct and microvilli were visible at the interface of hepatocytes. It suggested that the damage of fluoride to rat liver cells was mainly manifested by the destruction of membrane structure. Free radicals and lipid peroxidation intermediates attacked the sub-organelle membrane and caused peroxidation damage, which might be one of the major mechanisms of liver lesions caused by excessive fluoride (Guo and Sun 2005). Some scholars also found that many organs of fluorosis animals had ultrastructural changes. In animal models with drinking water type of fluorosis, it was found that the heart, liver, lung, kidney, and ulnar nerve had ultrastructural pathological changes to various degrees, and the myocardial fiber bundles were parallel and neatly arranged. The Z-line was blurred, and a few mitochondrial cristae were not obvious between the muscles. The central vein of the liver dilated. Venous congestion in portal areas, hepatocyte swelling, vacuolar degeneration, and lipid droplets could be seen in the cytoplasm. The mitochondrial membrane was partially incomplete and the membrane nucleus was broken. The spleen consisted of a large number of lymphocytes (B cells, T cells round nuclei, and cleft nuclei) where in cytoplasm a few of rough endoplasmic reticulum could be seen. In the splenic sinus, there were red blood cells and increased red pulp. The epithelial cells on the alveolar walls were irregularly arranged with few mitochondria in the cytoplasm and interstitial edema, and hyperemia was observed. In the renal tubular epithelial, the nucleus was irregular, the base membrane thickened, the interstitial edema changed, mitochondria swollen, vacuolar degenerated, and cristae were decreased. The ulnar nerve showed thickened myelin sheath, abundant microfilaments, and damaged mitochondrial structure with decreased number and vacuolated changes. The muscle fiber bundles were in neat arrangement, where sarcomere, Z and M lines showed fuzzy structure with slight degeneration changes. The muscle wall fibers of the iliopopliteal artery were thickened, lumen was narrowed, the endometrium was uneven, and calcification was obvious. The popliteal vein cavity was large, filled with red blood cells. The muscle fiber of skeletal muscle was edema or atrophy, where the sarcomere and adhesion lines are not obvious; the intermuscular capillary, endothelial cells, and internal red blood cells were abundant. The distance between intermuscular mitochondria increased and that between cristae decreased obviously. Immature bone cells were found in teeth with unclear nuclear structure, empty cytoplasm, dysplasia of enamel, disordered arrangement, and loose space (Mo et al. 2005).

In addition, researchers found regular myofilaments, clear mitochondrial cristae, and intervertebral discs in the control group. However, in the sodium fluoride group, myocardial fibers were fractured, mitochondrial cristae were dissolved and fractured, and mitochondria were slightly vacuolated (Wang et al. 2017). The destruction of mitochondrial morphology and structure led to the decrease of ATP production, which explained how fluoride led to mitochondrial dysfunction in cardiomyocytes. Other research found that the length and width of mitochondria in liver tissues of the high fluoride group were 63.87% and 71.85% shorter than those of the control group

(Zhou et al. 2020). With the increase of fluoride dose, the maximum length-width ratio of mitochondria decreased and the vacuolation area expanded. It was observed under electron microscope that the nuclei of rats' ovarian granulosa cells in the control group were normal with clear nucleoli and double layers of nuclear membrane. There were a large number of mitochondria around the nucleus with clear cristae. There was no significant change in the 25 mg/l low fluoride group compared with the control group. The ultrastructure of mitochondria of the 50 mg/l and 100 mg/l medium and high fluoride groups was damaged. Such as mitochondria vacuoles, mitochondrial cristae lysis and fracture could be found. Also, the granulocytes showed obvious apoptotic characteristics. Followed with the increase of fluoride intake, the number of mitochondria increased (Zhao et al. 2019).

In the experiment of fluorosis antagonism, it was found that mitochondrial damage could be reduced by manual intervention to alleviate fluorosis. Investigation was carried out on the effects of selenium and iodine on the mitochondrial damage of mouse kidney cells induced by fluoride. The morphological changes of mitochondria of mouse kidney cells with different iodine content and certain selenium content of fluoride were observed with JM-1200EX transmission electron microscope. As a result, the mitochondria in mice renal cells from the control group showed clearly visible bilayer membranes and cristae; the mitochondria of high fluoride group exhibited indistinct bilayer membrane and cristae; such two features were relatively clear in the low iodine and high fluoride group and in the medium iodine and high fluoride group, and they were fuzzy in the high iodine and high fluoride group. The mitochondria bilayer membrane and cristae in high iodine and high fluorine and selenium group were clearly visible. In conclusion, a certain amount of iodine had an antagonistic effect on the mitochondrial damage of mouse renal cells caused by fluoride, while an excessive amount of iodine had a synergistic effect. A certain amount of selenium also had an antagonistic effect on the mitochondrial damage of mouse renal cells caused by high iodine and high fluorine (Zhang et al. 2002).

Changes of Dynamic Equilibrium of Mitochondrial Fusion and Fission

The mitochondria of cerebral cortical nerve cells in rats with chronic fluorosis were observed by electron microscope. It was found that the mitochondria of cerebral cortical nerve cells in the normal group at 3 and 6 months presented oval rod-shaped structures (sections of three-dimensional membrane network tubular structures), with clear and orderly mitochondrial cristae. However, in the fluorinated group, the mitochondrial rod-like structure was reduced, the mitochondria showed spherical or vacuolar shape with overfused membrane. The mitochondrial cristae were blurred, broken, or disappeared. When Mfn1 was tested by immunohistochemistry, it was found that the expression level of Mfn1 in each group was basically the same as that of Western blot. That was to say, there was no significant difference between the expression level of Mfn1 in the low fluoride group and that of the normal group and no significant change in the mRNA transcription level of Mfn1. The expression of Mfn1 protein was significantly decreased and the mRNA transcription level of Mfn1 was decreased in the high fluoride group. At sixth month of the experiment, the expression of Mfn1 protein in the fluorosis group was significantly lower than that of the normal group, and the mRNA transcription level of Mfn1 was significantly reduced. According to immunohistochemical analysis of Fis1 protein, it showed that there were no significant differences in Fis1 protein expression and no significant changes in Fis1 and Drp1 mRNA transcription level at third month. The expression of Fis1 protein was significantly increased in the high fluoride group, and the transcription level of Fis1 and Drp1 mRNA was increased. At sixth month of the experiment, Fis1 protein expression was significantly higher in the fluorosis group than in the normal group, and transcription level of Fis1 and Drp1 mRNA was significantly higher. The results of Drp1 immunohistochemical localization and protein quantitative analysis were similar to that of Fis1 (Lou et al. 2013). RT-PCR and real-time fluorescence quantitative PCR were used to predict the production of Fis1 and Drp1 mRNA specific PCR products. At third month of the experiment, there were no significant changes in the transcription level of Fis1 and Drp1 mRNA in the low fluoride group, while the transcription level of Fis1 and Drp1 mRNA in the high fluoride group was increased. At sixth month, the level of Fis1 and Drp1 mRNA was significantly increased in the fluorosis group. Compared with the normal group, the level of mtDNA deletion of 4.8 kb in the nerve cells of chronic fluorosis rats was significantly increased; the level of deletion was 80% in the fluorinated group at third month and nearly 90% in the fluorinated group at sixth month. The outline of the cells and mitochondria was revealed by antibody tubulin-α and cyclooxygenase IV. The results showed that the mitochondria of nerve cells in the normal group were evenly distributed, while the concentration of mitochondria in the neurons in the fluorinated group decreased in the synapse. Oxygen free radical fluorescence staining was performed on the neurons in the cerebral cortex of rats with chronic fluorosis. Researchers found that at third month of the experiment, a small amount of fluorescence was observed in the neurons of the low fluoride group, while more and brighter fluorescence was observed in the neurons of the high fluoride group. At sixth month, more and brighter fluorescence was observed in nerve cells of the fluorinated group than in the normal group (Lou et al. 2013).

In addition, it was observed in animal experiments that the protein expression of Mfn1 in the ovary of mice treated with fluoride was upregulated, but the OPA1 expression did not change significantly (Zhao et al. 2019). Moreover, compared with the control group, Mfn1 mRNA expression was increased, but OPA1 expression did not change significantly. Zhou's studies showed that protein expression level of Drp1 and Fis1 in hepatocytes was increased after fluoride treatment (Zhou et al. 2020). However, Fis1 protein expression level was significantly lower in the 100 ppm fluoride group than in 25 ppm and 50 ppm fluoride groups. The expression level of Mfn1 protein was significantly lower in the high fluoride group than in the control group, while the expression level of Fis1 protein was significantly higher in the high fluoride group than in the control group. The mRNA levels of Mfn1 and

Fis1 in renal tissues of rats in the fluorosis group were significantly decreased and increased, respectively, indicating a dose-dependent relationship (Lou et al. 2013).

To sum, chronic fluorosis can cause abnormal morphology in a variety of cell mitochondrial, fusion, and fission disorders, even mitochondrial ultrastructure abnormalities (suggesting mitochondrial damage). The increased deletion of mtDNA fragment of 4.8 kb in nerve cells leads to the loss of mitochondrial coding function. Oxidative phosphorylation is affected by this damage, resulting in incomplete electron respiratory chain and increased production of superoxide free radical. The accumulation of mitochondria and decrease of synapses in fluoride-stained neurons may affect the function of synapses. In the development of chronic fluorosis, mtDNA damage and morphological changes are one of the links that produce high oxygen free radicals and cause cell damage.

Conclusion

Long-term intake of fluoride can cause the accumulation of fluoride in the body, causing changes in bone and non-bone phases. With the development of research on the chronic fluorosis damage to body, the fluoride damage to mitochondria attracted growing attention of researchers. Through electron microscopy, three-dimensional imaging technology, and molecular biology technology, it was found that chronic fluorosis was an abnormal change in the ultrastructure of mitochondria in many organs, which was often manifested as the disappearance and swelling of mitochondrial cristae and imbalance of mitochondrial fusion and mitotic protein. As an important "power factory" of the body with its own DNA and genetic system, mitochondrion is closely related to energy conversion, tricarboxylic acid cycle, oxidative phosphorylation, apoptosis, and other processes. Therefore, the morphological and functional changes of mitochondria play a vital role in the pathogenesis of chronic fluorosis.

References

Adrain C, Creagh EM, Martin SJ. Apoptosis-associated release of Smac/DIALOG from mitochondria requires active caspases and is blocked by BBC-2. EMBO J. 2001;20:6627–36.

Brooks C, Wei Q, Feng L, Dong G, Tao Y, Mei L, et al. Bak regulates mitochondrial morphology and pathology during apoptosis by interacting with mitofusins. Proc Natl Acad Sci USA. 2007;104:11649–54.

Chen H, Detmer SA, Ewald AJ, Griffin EE, Fraser SE, Chan DC, et al. Mitofusins Mfn1 and Mfn2 coordinately regulate mitochondrial fusion and are essential for embryonic development. J Cell Biol. 2003;160:189–200.

Chen X, Song K, Liu D, MA T, Zhao Y. Ultrastructure of dorsal raphe nucleus neurons in rats with experimental fluorosis. Chin J Endem Dis Pre Ctrl (Chinese). 1995;05:260–2.

Chu J, Zhang S, Gen R, Wu H, Wu P, Deng Y, et al. Effects of manganese on deletion of mitochondrial DNA "4834" in the substantia Nigeria striatum of the brain in rats of different age. J Toxicol (Chinese). 2007;2:115–7.

Escobar-Henriques M, Westermann B, Langer T. Regulation of mitochondrial fusion by the F-box protein Mdm30 involves proteasome-independent turnover of Fzo1. J Cell Biol. 2006;173:645–50.

Farmer T, Naslavsky N, Caplan S. Tying trafficking to fusion and fission at the mighty mitochondria. Traffic (Copenhagen, Denmark). 2018;19:569–77.

Frank S, Gaume B, Bergmann-Leitner ES, Leitner WW, Robert EG, Catez F, et al. The role of dynamin-related protein 1, a mediator of mitochondrial fission, in apoptosis. Dev Cell. 2001;1:515–25.

Guo XY, Sun GF. Observation on the ultrastructure of liver in fluorosis rats. Chin J Ctrl Endem Dis (Chinese). 2005;04:210–2.

Ishihara N, Eura Y, Mihara K. Mitofusin 1 and 2 play distinct roles in mitochondrial fusion reactions via GTPase activity. J Cell Sci. 2004;117:6535–46.

Karbowski M, Lee YJ, Gaume B, Jeong SY, Frank S, Nechushtan A, et al. Spatial and temporal association of Bax with mitochondrial fission sites, Drp1, and Mfn2 during apoptosis. J Cell Biol. 2002;159:931–8.

Karbowski M, Youle RJ. Dynamics of mitochondrial morphology in healthy cells and during apoptosis. Cell Death Differ. 2003;10:870–80.

Kim H, Morimoto Y, Ogami A, Nagatomo H, Hirohashi M, Oyabu T, et al. Differential expression of EC-SOD, Mn-SOD and CuZn-SOD in rat lung exposed to crystalline silica. J Occup Health. 2007;49:242–8.

Labrousse AM, Zappaterra MD, Rube DA, van der Bliek AM. C. elegans dynamin-related protein DRP-1 controls severing of the mitochondrial outer membrane. Mol Cell. 1999;4:815–26.

Lee H, Yoon Y. Mitochondrial fission and fusion. Biochem Soc Trans. 2016;44:1725–35.

Lee YJ, Jeong SY, Karbowski M, Smith CL, Youle RJ. Roles of the mammalian mitochondrial fission and fusion mediators Fis1, Drp1, and Opa1 in apoptosis. Mol Biol Cell. 2004;15:5001–11.

Liang WY, Tang LX, Yang ZC, Huang YS. Calcium induced the damage of myocardial mitochondrial respiratory function in the early stage after severe burns. Burns. 2002;28:143–6.

Lou DD, Guan ZZ, Liu YJ, Liu YF, Zhang KL, Pan JG, et al. The influence of chronic fluorosis on mitochondrial dynamics morphology and distribution in cortical neurons of the rat brain. Arch Toxicol. 2013;87:449–57.

Ma Y, Ma Z, Yin S, Yan X, Wang J. Arsenic and fluoride induce apoptosis, inflammation and oxidative stress in cultured human umbilical vein endothelial cells. Chemosphere. 2017;167:454–61.

Mailloux RJ, McBride SL, Harper ME. Unearthing the secrets of mitochondrial ROS and glutathione in bioenergetics. Trends Biochem Sci. 2013;38:592–602.

Margineantu DH, Gregory Cox W, Sundell L, Sherwood SW, Beechem JM, Capaldi RA, et al. Cell cycle dependent morphology changes and associated mitochondrial DNA redistribution in mitochondria of human cell lines. Mitochondrion. 2002;1:425–35.

Mo ZY, Lu ZM, Qiao GX, Wang L. Observation on the ultrastructure of different systems of experimental fluorosis rat. Chin J Endem (Chinese). 2005;24:55–7.

Mu X, Zhang Q, Liu X, Zhao J, Xi D, Zheng S, et al. Photoelectron microscopy of spinal cord injury in rats with chronic fluorosis. J Shihezi Med Coll (Chinese). 1995;2:71–2.

Reddy YP, Tiwari S, Tomar LK, Desai N, Sharma VK. Fluoride-induced expression of neuroinflammatory markers and neurophysiological regulation in the brain of wistar rat model. Biol Trace Elem Res. 2020;199(7):2621–6.

Rizzuto R, Pinton P, Carrington W, Fay FS, Fogarty KE, Lifshitz LM, et al. Close contacts with the endoplasmic reticulum as determinants of mitochondrial Ca^{2+} responses. Science. 1998;280:1763–6.

Rocha EM, De Miranda B, Sanders LH. Alpha-synuclein: pathology, mitochondrial dysfunction and neuroinflammation in Parkinson's disease. Neurobiol Dis. 2018;109:249–57.
Santel A, Fuller MT. Control of mitochondrial morphology by a human mitofusin. J Cell Sci. 2001;114:867–74.
Skulachev VP. Mitochondrial filaments and clusters as intracellular power-transmitting cables. Trends Biochem Sci. 2001;26:23–9.
Suen DF, Norris KL, Youle RJ. Mitochondrial dynamics and apoptosis. Genes Dev. 2008;22:1577–90.
Sugioka R, Shimizu S, Tsujimoto Y. Fzo1, a protein involved in mitochondrial fusion, inhibits apoptosis. J Biol Chem. 2004;279:52726–34.
Sun L, Gao Y, Zhang W, Liu X, Li B, Cui X, et al. Mechanisms underlying endothelin-1 level elevations caused by excessive fluoride exposure. Cell Physiol Biochem. 2016;40:861–73.
Tieu Q, Okreglak V, Naylor K, Nunnari J. The WD repeat protein, Mdv1p, functions as a molecular adaptor by interacting with Dnm1p and Fis1p during mitochondrial fission. J Cell Biol. 2002;158:445–52.
Van der Bliek AM, Shen Q, Kawajiri S. Mechanisms of mitochondrial fission and fusion. Cold Spring Harb Perspect Biol. 2013;5:a011072.
Wang HW, Zhao WP, Liu J, Tan PP, Tian WS, Zhou BH, et al. Atp5j and atp5h proactive expression correlates with cardiomyocyte mitochondrial dysfunction induced by fluoride. Biol Trace Elem Res. 2017;180:63–9.
Yu Y. Stereological study of mitochondrial injury during fluorosis. Public Health Chin (Chinese). 1999;10:3–5.
Zhang C, Shi Z, Zhang L, Zhou Z, Zheng X, Liu G, et al. Appoptosin interacts with mitochondrial outer-membrane fusion proteins and regulates mitochondrial morphology. J Cell Sci. 2016;129:994–1002.
Zhang Z, Xu X, Shen X, Hu Y. Effects of iodine-selenium on mitochondrial injury of mouse kidney cells induced by fluoride. J Zhejiang Normal Univ (Natural Science edition) (Chinese). 2002;04:76–8.
Zhao WP, Wang HW, Liu J, Zhang ZH, Zhu SQ, Zhou BH, et al. Mitochondrial respiratory chain complex abnormal expressions and fusion disorder are involved in fluoride-induced mitochondrial dysfunction in ovarian granulosa cells. Chemosphere. 2019;215:619–25.
Zhou BH, Wei SS, Jia LS, Zhang Y, Miao CY, Wang HW, et al. Drp1/Mff signaling pathway is involved in fluoride-induced abnormal fission of hepatocyte mitochondria in mice. Sci Total Environ. 2020;725:138192.

Chapter 18
Calcium Metabolism Disorder in Chronic Fluorosis

Zi-Gui Zhang

Abstract The biological effect of microelement, calcium (Ca^{2+}), and the effect of endemic fluorosis on Ca^{2+} metabolism have been involved. It also covers the mechanism of endemic fluorosis, existing problems of the disease caused Ca^{2+} metabolism disorder together with its research perspectives.

Keywords Biological effect · Calcium metabolism · Endemic fluorosis

Biological Effect of Calcium

Calcium (Ca^{2+}) is a metal element with a molecular weight of 40.078. It is often found in calcium containing minerals, like limestone, in combined forms in nature. In human body, calcium accounts for 2.0% of body weight in healthy adult body. The majority of Ca^{2+} is found in bone organs in the form of skeletal salt, which also makes calcium the largest proportion of inorganic salt in human body. Very few Ca^{2+} is found in soft tissues, intracellular and extracellular fluid, thus, it maintains a dynamic balance with bone calcium. The human body has different demand on calcium at various growth stages, and calcium deficiency would pose differed impacts at different stages. For example, Ca^{2+} deficiency would cause convulsion and asthma at infant period, abnormal skeletal development at preschool stage, and affect health status of later skeletal development.

Ca^{2+} is the key secondary messenger of regulating physiological functions of cells (including cell survival and signal transduction). In normal physiological status, the calcium content in human body is relatively stable: free Ca^{2+} concentration in cytoplasm at resting state maintains at relatively low level (around 10^{-7} mol/l); Ca^{2+} concentration in intracellular and extracellular organelles (such as mitochondrion and endoplasmic reticulum) is around 10^{-3} mol/l; Ca^{2+} concentration in both inside

Z.-G. Zhang (✉)
College of Chemistry and Life Science, Zhejiang Normal University, Jinhua, Zhejiang, People's Republic of China
e-mail: zzg@zjnu.cn

Z.-Z. Guan (ed.), *Coal-burning Type of Endemic Fluorosis*,
https://doi.org/10.1007/978-981-16-1498-9_18

and outside of cytomembrane and organelles has a difference of 10^4 times. Once the cell is stimulated, the intracellular Ca^{2+} concentration rapidly increases and causes a series of physiological biochemical reaction. This is the signal transduction role of Ca^{2+}. Normally, intracellular and extracellular Ca^{2+} concentration will restore to resting state to maintain the stability of normal physiological function of cells. Once the stability is damaged, it would cause calcium metabolism disorder and lead to functional disorder in the body. Hence, intracellular and extracellular free Ca^{2+} concentration changes play an essential role in cellular physiological functions. Regulating intracellular and extracellular free Ca^{2+} concentration is the key part of maintaining normal functions of human body.

Effects of Endemic Fluorosis on Calcium Metabolism

Effects of coal-burning type of endemic fluorosis on cell Ca^{2+} metabolism. Fluorosis can change the Ca^{2+} balance and affect the cell Ca^{2+} metabolism. Epidemiology study discovered that low Ca^{2+} diet was commonly found in endemic fluorosis areas. Ca^{2+} deficiency in human body was the trigger and exacerbating factors for fluorosis. However, Ca^{2+} overload was obviously found in cells of target organs like liver, kidney, and brain of patients with fluorosis. Fluorosis was also termed as "Ca^{2+} paradox," since a large amount of Ca^{2+} deposited and accumulated in the form of phosphate, then it caused cellular damage and damage to the body. The fluorosis-triggered cellular Ca^{2+} overload mechanism was diversified. The study indicated that fluoride-caused abnormal level of lipid peroxidation could trigger Ca^{2+} overload in hippocampus synaptosomes. As a result, such abnormally caused Ca^{2+} signal pathway disorder (Teng et al. 2017); conversely, it would ultimately cause damage on brain structure and functions, since abnormal expression of Ca^{2+} ion regulated nuclear factor kappa-B (NF-κB) would further affect the apoptosis related protein expression of Bax and Bcl-2 and then enhance apoptosis signals (Zhang et al. 2011). According to a study, N-methyl-D-aspartic acid receptor was excessively activated when affected with fluorosis, the substantial Ca^{2+} inflow further activated a series of reactions to induce the excessive neuron apoptosis or necrosis and ultimately affect the cranial nerves (Wan et al. 2018). A study in vitro showed that 0.5–1.0 mmol/l Ca^{2+} reversed the increase of Ca^{2+} influx caused by the over-opening of Ca^{2+} channel in osteoblasts exposed to fluoride (Wang et al. 2018).

Effects of Endemic Fluorosis on Ca^{2+} Metabolism of Skeletal Organs

To date, study had identified the effect of coal-burning type of endemic fluorosis-caused Ca^{2+} metabolism disorder in pathogenesis of skeletal fluorosis: (1) fluoride is a skeletal element, the blood absorbed fluoride combined with Ca^{2+} into Ca^{2+} fluoride and then lowered the blood Ca^{2+}; (2) fluoride infiltrated into the bone salt crystal, thus increased the bone crystallinity, lowered the solubility; hence it slowed down the osteocalcin release and lowered the Ca^{2+} concentration; (3) fluorosis could reduce the transport of Ca^{2+} through renal tubule endoplasmic reticulum and plasmalemma, and fluorosis was also related to the reduction of SERCA count of individual renal cell membrane; therefore, it resulted in the overload of Ca^{2+} in overall low Ca^{2+}-osteoblast in the body, causing the abnormal activation of osteoblast proliferation, speeding-up the bone turnover and leading to skeletal fluorosis.

The study on dietary Ca^{2+} level affecting fluorosis also served as a proof of the fluorosis effect on Ca^{2+} metabolism of skeletal organs. The past study proved that consuming high Ca^{2+} containing milk for a period would relieve the symptoms of patients with skeletal fluorosis (Rango et al. 2012), while using Ca^{2+} citrate containing toothpaste would relieve the symptoms of child patients with dental fluorosis (Delbem et al. 2012). Low Ca^{2+} could enhance the absorption of fluoride in intestinal tract of mice with fluorosis, thus, it would lead to severe Ca^{2+} deficiency and exacerbate skeletal fluorosis (Li et al. 2003). High Ca^{2+} intake in the long run would reduce the accumulation of fluoride in the body to a certain degree (Yu et al. 2006). All in all, scientific diet plan and improved food pyramid are key parts of fluorosis prevention and control.

L-type Ca^{2+} channel was the major pathway for Ca^{2+} to flow into cells (Robin and Allard 2015), and it played a major role in regulating Ca^{2+} level in cytoplasm. The study showed that exposure to fluoride could affect the expression of osteoblastic transcription factors in MC3T3-E1 while regulating the expression or activity of L-type Ca^{2+} channel could also regulate the Ca^{2+} signal transduction in osteoblasts and bone turnover of patients with fluorosis (Duan et al. 2014).

Effects of Endemic Fluorosis on Ca^{2+} Metabolism of Brain

Ca^{2+} plays a major role in regulating the pathophysiological process, like neuron synaptic transmission, synaptic plasticity, and neurotoxicity, which makes it an important messenger of neurons. It was test proven that Ca^{2+} overload in hippocampus synaptosome was closely correlated with damage in brain function of mice with fluorosis (Teng et al. 2017). The mechanism of brain damage triggered by Ca^{2+} metabolism disorder in brain cell due to fluorosis was a complex one, since the release of intracellular Ca^{2+} ion and regulating mechanism recovery were equally diversified and complex. Such mechanisms included Ca^{2+} channel, Ca^{2+}-binding +

protein, Ca^{2+} storage organelle (endoplasmic reticulum and mitochondrion), sodium-Ca^{2+} exchange, Ca^{2+} pump, downstream Ca^{2+} signal pathway, and apoptosis regulation relevant molecules. In the early stage of the research, serial study was conducted on: effect of fluoride exposure on cell Ca^{2+} ion level of hippocampus, on Ca^{2+}-binding protein of downstream Ca^{2+} signal pathway, on expression level of apoptosis regulation molecule. The findings suggested that abnormal level of fluorosis-caused lipid peroxidation would trigger Ca^{2+} overload in hippocampus synaptosome (Zhang et al. 2011). Neurotoxin would affect Ca^{2+} signal transduction system in neurons in two major pathways: protein kinase C (PKC) pathway and non-PKC pathway (calmodulin-calmodulin kinase [CaM-CaMK] pathway). CaM-CaMK pathway entered the core-cREB-c-fos for expression through pathway cAMP-CaMK/PKA. On the one hand, Ca^{2+} overload triggered abnormal expression of certain study and memory related key proteins (such as CaMKII and c-fos) of non-PKC pathway, CaM-CaMK pathway, and then triggered disorders of Ca^{2+} ion signal pathway. On the other hand, disorders of Ca^{2+} ion signal pathway also caused the abnormal expression of Ca^{2+} ion regulated NF-κB, the activated-abnormal expression of NF-κB would further affect the expression of apoptosis protein Bcl-2 and Bax (Zhang and Zhang 2013). Then it would induce neuron apoptosis to be enhanced and finally impair the brain function (Liu et al. 2011). Therefore, Ca^{2+} overload in neuron, abnormal expression of Ca^{2+}-binding protein and the triggered series of Ca^{2+} metabolism disorder response in downstream Ca^{2+} signal pathway were part of the brain damage mechanism by fluorosis. The study on the effect of different Ca^{2+} levels of feed on study and memory behavior of offspring mice indicated that high Ca^{2+} (70 g/kg) was potent in antagonizing the damage of study and memory of offspring mice caused by mother mice who were exposed to fluoride (Yu et al. 2018), while low Ca^{2+} (0.63 g/kg) would exacerbate the brain function damage of the offspring mice of the same condition. The molecular mechanism of Ca^{2+} intervened fluorosis could be upregulating the expression of Bcl-2 and downregulating the expression of caspase12 to inhibit excessive apoptosis of brain cells of SR mice with fluorosis (Sun et al. 2017). This further proved the effect of fluorosis on brain Ca^{2+} metabolism.

A study (Yu et al. 2019) by using patch clamp to investigate the effect of fluoride exposure on L-type Ca^{2+} ion channel of cone neuron in hippocampus CA_1 region indicated that fluoride exposure lowered the balance state activation voltage of L-type Ca^{2+} channel; the peak current was obviously increased, depolarization time and tail current time of Ca^{2+} current significantly increased. These proved that fluoride could lower the activation voltage of L-type Ca^{2+} channel and prolong its opening time, which would enable excessive Ca^{2+} inflow within unit time and Ca^{2+} ion overload in neurons. All these would lead to Ca^{2+} disorder, triggering apoptosis and ultimately the damage on neurons.

Ca^{2+} channel antistatic agent was a medicine to reduce the intracellular concentration of Ca^{2+}, which would block the Ca^{2+} into cell selectively. Such agent would primarily work on the Ca^{2+} channel of cell membrane. Ca^{2+} channel antistatic agent could be divided into three categories according to its chemical structure: (1) dihydropyridines (DHPs), such as nifedipine and nicardipine;

(2) benzothiazepines BTZs, such as diltiazem; (3) phenylalkylamine, such as verapamil. There were four categories concerning L-type Ca^{2+} ion channel: Cav1.1 (α1S), Cav1.2 (α1C), Cav1.3 (α1D), and Cav1.4 (α1F). Of which, Cav1.2 was the most common combining target spot on the membrane for Ca^{2+} channel antistatic agent and the channel for Ca^{2+} inflow. Liao (Liao et al. 2017) found that, in PC12 (with neuron behavior), expression of Cav1.2, CaMKII, and Bax mRNA increased, while expression of Bcl-2 mRNA lowered after fluoride exposure; L-type Ca^{2+} channel antistatic agent nifedipine could lower the expression level of Cav1.2. The research also suggested that Ca^{2+} channel Cav1.2 might be the key molecular target spot for fluorosis treatment, and nifedipine might be effective medicine to treat fluorosis. Therefore, L-type Ca^{2+} channel could be the major part for investigation on coal-burning type of endemic fluorosis-caused neuron Ca^{2+} metabolism disorder.

Mechanism of Endemic Fluorosis Caused by Ca^{2+} Metabolism Disorder

Explanations on mechanism of coal-burning type of endemic fluorosis never fell short: oxidative stress hypothesis, receptor altering, signal transduction channel altering, enhanced apoptosis, Ca^{2+} metabolism disorder, endoplasmic reticulum, and mitochondrion altering. Of all these explanations, oxidative stress hypothesis was overwhelmingly recognized, and Ca^{2+} metabolism disorder hypothesis also won the support of more scientists in recent years. This hypothesis concluded that fluorosis was a Ca^{2+} metabolism disorder disease, resulted from the Ca^{2+} deficiency in the body while the intracellular Ca^{2+} was overload. Ca^{2+} deficiency/low Ca^{2+} in the body would not only enhance the absorption of intestine tract on fluoride, but also triggered the secretion surge of parathyroid hormone and promoted enhancement of osteoblasts and osteoclastic. Research findings discovered that fluoride could bind with Ca^{2+} ion in the blood for complex compound and caused temporary decrease in the Ca^{2+} concentration in the serum; conversely, fluoride could stimulate the activity of osteoblasts to be increased, then the bone formation would increase the demand of Ca^{2+}, which would further exacerbate Ca^{2+} deficiency. In the meantime, intracellular Ca^{2+} metabolism disorder would trigger a series of harmful reactions in the cell and cause damage to the body.

In recent years, multiple studies provided proof for the mechanism of Ca^{2+} metabolism disorder triggered by coal-burning type of endemic fluorosis from aspects of cell, sub-cell, and molecular level. Oxidative stress hypothesis suggested that the imbalance between reactive oxygen species (ROS) and antioxidant level (such as superoxide dismutase, glutathione peroxidase, and glutathione) triggered such disorder (Lu et al. 2017). Another research suggested that fluoride could inhibit the activity of antioxidant enzyme, trigger over lipid peroxidation, and promote the accumulation of reactive oxygen; excessive reactive oxygen would trigger the modification of macromolecule and hence cause apoptosis by oxidative stress

(Shivarajashankara et al. 2003). The accumulation or over inflow of Ca^{2+} in mitochondrion was another factor causing ROS (Upadhyay 2018). High level of Ca^{2+} in mitochondrion would cause the opening of mitochondrial permeability transition pore, which would cause the escape of electron from electron transport chain and form ROS; such high level of Ca^{2+} could also cause apoptosis by releasing Cyt c from damaging mitochondrial membrane potential (Santo-Domingo and Demaurex 2010). Steep ROS and Ca^{2+} metabolism disorder were all triggers of apoptosis.

The excessive ROS generated by fluorosis could mediate apoptosis in three pathways: intrinsic mitochondrion pathway, endoplasmic reticulum pathway, and extrinsic death receptor pathway. The first two pathways could be triggered by abnormal concentration of Ca^{2+}: the generation of ROS would stimulate the intracavitary Ca^{2+} release of endoplasmic reticulum and lead to endoplasmic reticulum stress. Then, the released Ca^{2+} could be easily absorbed by nearby mitochondrion and trigger its swelling. The results would be the depolarization of mitochondrial membrane and would lead to apoptosis (Zeeshan et al. 2016). Death receptor pathway referred to extrinsic pathway of death domain (which was the combination between death receptor and ligand) combined with Fas associated death domain protein (FADD) activating downstream apoptosis protein to execute cell apoptosis. There were eight existing known death receptors, and the major signal transduction pathways were the tumor necrosis factor receptor (TNFR), TNF-related apoptosis-inducing ligand (TRAIL), and Fas/FasL (Lee et al. 2008); in which, Fas/FasL apoptosis pathway was the most important pathways of death receptor pathways (Miao et al. 2013). It was found that increased intracellular Ca^{2+} level could lead to the combination of Ca^{2+} sensor calmodulin and target protein to form Ca^{2+}/calmodulin compound (Marshall et al. 2015). Ca^{2+}/calmodulin compound could promote FADD collection to participate extrinsic apoptosis pathway by DAP kinase (Papoff et al. 2010). The activated FADD would start caspase8, release it to cytoplasm to activate downstream caspase cascade response and then induce apoptosis (Adams 2003). Past research indicated that above mentioned three pathways mediated apoptosis in oxidative stress triggered by fluoride exposure were exceptionally correlated with fluoride exposure caused abnormal Ca^{2+} concentration.

In recent years, the effect fluorosis had on cell cycle was raising attention. The cell cycle referred to the proliferation of cell, dividing into phase $G1$, phase S, phase $G2$, and phase M. DNA replication was found in phase S, while chromosome segregation was found in phase M. When there was no cell proliferation, the cycle would pause at phase $G0$. The DNA content of phase $G0/G1$ was $2n$, that of phase $G2/M$ was replicated to $4n$, and that of phase S was between $2n$ and $4n$. Two phases in cell cycle were essential: $G1$ to S and $G2$ to M. These two phases were at active transformation of molecular level, which were prone to environment conditions. Research reported that fluoride exposure blocked osteoblast-like cells MC3T3-E1 (Gu et al. 2019) and osteoblast of rat cultured in vitro in phase S, thus cells at phase S abnormally increased (Liu et al. 2018). The molecular mechanism of fluoride exposure affecting cell cycle was yet to be explained. It was also reported that change of intracellular Ca^{2+} level had certain effect on cell cycle, like inducing stasis of cell cycle, affecting DNA replication and mitosis, and lead to apoptosis (Lv et al. 2020). It was suggested that fluoride exposure caused intracellular Ca^{2+} overload could serve to explain fluoride-caused abnormal cell cycle.

Domestic and foreign studies as well as existing research findings suggested that the pathogenesis of endemic fluorosis might involve with cascade response of oxidative stress, change of mitochondrion, endoplasmic reticulum stress, change of signal transduction pathway, apoptosis, and Ca^{2+} metabolism disorder. Of which, Ca^{2+} metabolism disorder was the key part of such cascade response.

Problems and Prospects

Study on mechanism of Ca^{2+} disorder of coal-burning type of endemic fluorosis had made remarkable progress in recent years. However, the complexity of fluorosis made the study on this mechanism reveals more problems: (1) There were studies focused on early process of pathogenesis of fluorosis-caused Ca^{2+} metabolism disorder, such as cause and mechanism of fluorosis-caused intracellular Ca^{2+} overload. However, apart from study on the effect of fluoride exposure on skeletal and non-skeletal organs Ca^{2+} inflow through membrane L-type Ca^{2+} channel, there was limited study on the effect of fluorosis on transportation of intracellular Ca^{2+} to outside plasmalemma, release of Ca^{2+} from intracellular Ca^{2+} stores, and Ca^{2+} stores absorbing Ca^{2+} from cytoplasm. (2) There was no conclusive explanation on occurrence, progress, and some key parts of apoptosis enhancement of fluorosis-caused cellular Ca^{2+} metabolism disorder. Certain key apoptosis signal pathway's role was yet to be defined in the pathogenesis of neuron damage, such as death receptor pathway in fluoride-caused cell damage. (3) There were studies reported individually that the fluoride exposure-caused hippocampus intracellular Ca^{2+} overload triggered abnormal downstream Ca^{2+} signal pathway as well as abnormal apoptosis related regulating molecules, and the molecular mechanism of balance state change of upstream fluoride exposure-caused hippocampus intracellular Ca^{2+}. So far, there was an absence of relevant study combining both researches.

Concerning the previously mentioned problems, it was of vital importance to combine epidemiology study on-site and animal test and adopt advanced molecular biological technology and assays to explore the Ca^{2+} metabolism disorder mechanism caused by coal-burning type of endemic fluorosis.

Conclusion

Epidemiology study and animal test results have all proven that chronic fluorosis-caused Ca^{2+} metabolism disorder may be one of the pathogenesis of coal-burning type of endemic fluorosis. Considering its complexed pathogenesis, which also involves cascade response, such as oxidative stress, apoptosis signal transduction pathway, and abnormal apoptosis, the correlation between fluorosis-caused Ca^{2+} metabolism disorder and these mechanisms, as well as the specific molecular mechanism of such disorder could be worthy of further investigation.

References

Adams JM. Ways of dying: multiple pathways to apoptosis. Genes Dev. 2003;17:2481–95.

Delbem AC, Bergamaschi M, Rodrigues E, Sassaki KT, Vieira AE, Missel EM. Anticaries effect of dentifrices with calcium citrate and sodium trimetaphosphate. J Appl Oral Sci. 2012;20:94–8.

Duan XQ, Zhao ZT, Zhang XY, Wang Y, Wang H, Liu DW, et al. Fluoride affects calcium homeostasis and osteogenic transcription factor expressions through L-type calcium channels in osteoblast cell line. Biol Trace Elem Res. 2014;162:219–26.

Gu XL, Wang Z, Gao J, Han DD, Zhang LM, Chen P, et al. SIRT1 suppresses p53-dependent apoptosis by modulation of p21 in osteoblast-like MC3T3-E1 cells exposed to fluoride. Toxicol In Vitro. 2019;57:28–38.

Lee JH, Jung JY, Jeong YJ, Park JH, Yang KH, Choi NK, et al. Involvement of both mitochondrial- and death receptor-dependent apoptotic pathways regulated by Bcl-2 family in sodium fluoride-induced apoptosis of the human gingival fibroblasts. Toxicology. 2008;243:340–7.

Li GS, Zhang WL, Hua K, Li HY, Xu H. Endemic fluorosis belongs to calcium paradox disease. Petrol Geochem. 2003;22:93–5.

Liao QX, Zhang R, Wang XY, Nian WW, Ke LL, Ouyang W, et al. Effect of fluoride exposure on mRNA expression of cav1.2 and calcium signal pathway apoptosis regulators in PC12 cells. Environ Toxicol Pharmacol. 2017;54:74–9.

Liu YJ, Guan ZZ, Gao Q, Pei JJ. Increased level of apoptosis in rat brains and SH-SY5Y cells exposed to excessive fluoride--a mechanism connected with activating JNK phosphorylation. Toxicol Lett. 2011;204:183–9.

Liu H, Luo Q, Cui HM, Deng HD, Kuang P, Lu YJ, et al. Sodium fluoride causes hepatocellular S-phase arrest by activating ATM-p53-p21 and ATR-Chk1-Cdc25A pathways in mice. Oncotarget. 2018;9:4318–37.

Lu YJ, Luo Q, Cui HM, Deng HD, Kuang P, Liu H, et al. Sodium fluoride causes oxidative stress and apoptosis in the mouse liver. Aging. 2017;9:1623–39.

Lv XG, Zheng QQ, Li M, Huang ZW, Peng M, Sun J, et al. Clioquinol induces S-phase cell cycle arrest through the elevation of the calcium level in human neurotypic SH-SY5Y cells. Metallomics. 2020;12:173–82.

Marshall CB, Nishikawa T, Osawa M, Stathopulos PB, Ikura M. Calmodulin and STIM proteins: two major calcium sensors in the cytoplasm and endoplasmic reticulum. Biochem Biophys Res Commun. 2015;460:5–21.

Miao KK, Zhang L, Yang SY, Qian W, Zhang ZG. Intervention of selenium on apoptosis and Fas/FasL expressions in the liver of fluoride-exposed rats. Environ Toxicol Pharmacol. 2013;36:913–20.

Papoff G, Trivieri N, Crielesi R, Ruberti F, Marsilio S, Ruberti G. FADD-calmodulin interaction: a novel player in cell cycle regulation. Biochim Biophys Acta. 2010;1803:898–911.

Rango T, Kravchenko J, Atlaw B, McCornick PG, Jeuland M, Merola B, et al. Groundwater quality and its health impact: an assessment of dental fluorosis in rural inhabitants of the main Ethiopian rift. Environ Int. 2012;43:37–47.

Robin G, Allard B. Voltage-gated Ca^{2+} influx through L-type channels contributes to sarcoplasmic reticulum Ca^{2+} loading in skeletal muscle. J Physiol. 2015;593:4781–97.

Santo-Domingo J, Demaurex N. Calcium uptake mechanisms of mitochondria. Biochim Biophys Acta. 2010;1797:907–12.

Shivarajashankara YM, Shivashankara AR, Bhat PG, Rao SH. Lipid peroxidation and antioxidant systems in the blood of young rats subjected to chronic fluoride toxicity. Indian J Exp Biol. 2003;41:857–60.

Sun Y, Ke LL, Zheng XR, Li T, Ouyang W, Zhang ZG. Effects of different levels of calcium Intake on brain cell apoptosis in fluorosis rat offspring and its molecular mechanism. Biol Trace Elem Res. 2017;176:355–66.

Teng Y, Zhang J, Zhang ZG, Feng J. The Effect of chronic fluorosis on calcium ions and caMKIIα, and c-fos expression in the rat hippocampus. Biol Trace Elem Res. 2017;182:1–8.

Upadhyay RK. Mitochondrial Ca^{2+} levels lower down rate of metabolic diseases and cardiomyopathies. J Stem Cell Res. 2018;4:82–7.

Wan L, Yang WX, Wei N. Effects of experimental fluorosis on NMDA receptor, Ca^{2+} and apoptosis of SH-SY5Y cells. Chin J Ctrl Endem Dis. 2018;33:131–206.

Wang JM, Zhao YF, Cheng XF, Li YY, Xu HM, Manthari RK, et al. Effects of different Ca^{2+}, level on fluoride-induced apoptosis pathway of endoplasmic reticulum in the rabbit osteoblast *in vitro*. Food Chem Toxicol. 2018;116:189–95.

Yu J, Sun DJ, Wang W, Xu CB, Wu Y, Shi YX, et al. Changes of parathyroid hormone in the serum of fluorosis rats taking different levels of calcium. Chin J Endem. 2006;25:625–8.

Yu QL, Shao DD, Wang XY, Nian WW, Ouyang W, Zhang ZG. Effects of calcium on autonomic activity, learning and memory in neonatal offspring rats induced by fluoride exposure of mother rats before pregnancy. J Toxicol. 2018;32:345–450.

Yu QL, Shao DD, Zhang R, Ouyang W, Zhang ZG. Effects of drinking water fluorosis on L-type calcium channel of hippocampal neurons in mice. Chemosphere. 2019;220:169–75.

Zeeshan HM, Lee GH, Kim HR, Chae HJ. Endoplasmic reticulum stress and associated ROS. Int J Mol Sci. 2016;17:327.

Zhang J, Zhang ZG. Effects of chronic fluorosis on CAMKIIα, c-fos, Bax, and Bcl-2 channel signaling in the hippocampus of rats. Fluoride. 2013;46:135–41.

Zhang J, Zhu WJ, Xu XH, Zhang ZG. Effect of fluoride on calcium ion concentration and expression of nuclear transcription factor kappa-B ρ65 in rat hippocampus. Exp Toxicol Pathol. 2011;63:407–11.

Chapter 19
Altering of Carbohydrate Metabolism

Kai-Lin Zhang and Zhi-Zhong Guan

Abstract Carbohydrate presents itself in the form of glucose and glycogen in human body. Carbohydrate is transported in the form of glucose in blood. Metabolism centers on that of glucose. The metabolism pathways of glucose in human body are: anaerobic glycolysis, aerobic oxidation, glycogen synthesis, glycogenolysis, and gluconeogenesis. Yet, fluoride can inhibit the key enzyme of carbohydrate metabolism pathway to cause its disorder, which then triggers multiple diseases (hyperglycemia as the most commonly diagnosed disease). Meanwhile, fluoride exposure can lower the insulin secretion. Thus, the decrease of insulin mRNA and its β cell secretion is the potential cause of chronic fluorosis-triggered blood glucose level to rise. In addition, carbohydrate forms glycoprotein by bonding with covalent bond and peptide chain amino acid residues, such glycoprotein can transmit information and exert influence on cell recognition, proliferation, and differentiation accordingly. Fluorosis can trigger the altering of carbohydrate metabolism (like collagen and so on) and affect glycoprotein's function in human body. Such influence can cause damage to multiple systems in the body directly or indirectly and lead to diseases ultimately.

Keywords Carbohydrate metabolism · Fluorosis · Glycoprotein · Hyperglycemia

K.-L. Zhang (✉)
Basic Medical College, Guizhou Medical University, Guiyang, Guizhou, People's Republic of China
e-mail: zhangkailin@gmc.edu.cn

Z.-Z. Guan
The Key Laboratory of Endemic and Ethnic Diseases (Guizhou Medical University), Ministry of Education, Guiyang, Guizhou, People's Republic of China

Department of Pathology at the Affiliated Hospital of Guizhou Medical University, Guiyang, Guizhou, People's Republic of China

The Provincial Key Laboratory of Medical Molecular Biology, Guiyang, Guizhou, People's Republic of China

Z.-Z. Guan (ed.), *Coal-burning Type of Endemic Fluorosis*,
https://doi.org/10.1007/978-981-16-1498-9_19

Introduction

Carbohydrate is the compound of polyhydroxyaldehyde or polyhydroxyketone. According to its composition, it can be divided into monosaccharide, oligosaccharide, and polysaccharide. Physiological role of carbohydrate is the energy supply for oxidation. Glucose (Glc) is the essential yet most widely found monosaccharide in nature and the transported form of sugar in blood. Glycogen (Gn) is the polymer of glucose (including liver glycogen, muscle glycogen, and renal glycogen) and storage form of carbohydrate in human body. Glucose and glycogen provide energy for the oxidation in human body. Fluoride exposure will decrease insulin secretion and increase blood glucose level. Long term fluoride exposure will trigger carbohydrate metabolism disorder and cause hyperglycemia. Apart from being the energy supply, glucose plays a major role in information transmitting among cells; it bonds with protein into glycoprotein and forms cell surface receptor and ligand. Existing study considers that glycoprotein is involved in the damage process of chronic fluorosis.

Carbohydrate Metabolism

Carbohydrate metabolism is the essential metabolism pathway, involving a series of complicated chemical reactions among glucose, glycogen, etc. Carbohydrate metabolism, the core of material metabolism, produces various intermediate metabolite via multiple metabolism pathways of carbohydrate. Glucose is the transportation form of carbohydrate in blood, while glycogen is the polymer of glucose and the storage form of carbohydrate in human body. Glucose and glycogen can generate a substantial amount of adenosine triphosphate (ATP) by aerobic decomposition for the physical activities. The carbohydrate metabolism in human body includes anabolism and catabolism of glucose.

Synthesis of Glucose

Being the basic structural unit of glucose, glycogen is the branched macromolecule polysaccharide linked by glycosidic bond. Glycogen synthesis begins with the transformation of glucose into "active glucose donor"-uridine diphosphate glucose (UDPG); then the glucosyl of UDPG is transferred to glycogen primer under the effect of glycogen synthase and extends the glucose chain of glycogen continually; finally, branching enzyme catalyzes the formation of branches. The glycogen stored in human body can be consumed within a dozen of hours, which is the reason the body will transform non-carbohydrates into glucose within 24 h fasting to maintain the normal blood glucose level. Gluconeogenesis is the process of histocyte transforming non-carbohydrate (amino acid, lactic acid, and glycerine) into glucose. Most of gluconeogenesis reactions are the reverse process of glycolysis pathway.

Decomposition of Glucose

Aerobic oxidation of carbohydrate refers to the generating process of ATP and oxidative process of glucose into CO_2 and H_2O under combined effect of cytoplasm and mitochondrion in aerobic condition. It includes three phases: pyruvic acid is produced by anaerobic decomposition of carbohydrate on glucose; pyruvic acid forms Acetyl CoA through oxidative decarboxylation; acetyl CoA enters tricarboxylic acid cycle and undergoes oxidative phosphorylation by coupling. Aerobic oxidation is the major energy supply of glucose oxidation and also the major pathway of energy supply of the majority histocytes in human body.

Glycolysis of carbohydrate refers to the process, in which glucose or glycogen is gradually decomposed into lactic acid and generates limited energy under the condition of oxygen deficiency or failure of oxygen utilization. Research has discovered that metabolism disorder of many microelements (including fluorine) can seriously affect the Mg^{2+} level. Then, it will cause the decreased activity of hexokinase, interfere with normal glycolysis process, and then cause histologic function disorder.

Glycoprotein

Glycoprotein is the bonding protein formed by the bond of single or multiple carbohydrate and peptide chain amino acid residues by covalent bond. The majority extracellular proteins in human body are glycoprotein (including soluble glycoprotein and membrane glycoprotein), and some exist in extracellular matrix protein. The essential biological function of glycoprotein is the biometric identification for cell or molecule, for example, the ligand requires to identify the corresponding glycoprotein receptor on membrane.

Collagen

Collagen is the fundamental structural protein in human body. It serves as the support and stuffing of tissues, which can be widely found in skin, tendon, bone, and cartilage. In addition, collagen molecule and fiber play a vital role in growth and development of organisms, cellular differentiation and adhesion, and antibody conjugation.

Osteocalcin (OCN)

OCN is a special non-collagen bone matrix protein, which is synthesized and secreted by mature osteoblast. OCN falls into the category of non-collagen acidoglycoprotein, a vitamin K dependent calcium-bonding protein. Osteocalcin presents in both carboxylated and uncarboxylated forms: fully carboxylated osteocalcin is often found in bone matrix, shows no bioactivity but increases affinity with calcium ion; it is easy to bond with hydroxyapatite and promotes bone generation. Uncarboxylated osteocalcin (ucOCN) is often found in blood circulation, shows relatively low bone affinity but strong bioactivity, which is the major form of regulating energy metabolism.

P-glycoprotein (P-gp)

P-gp is widely expressed in normal tissues of human body (such as epithelial cells of gastrointestinal tract, liver, kidney, and brain) and plays a vital role in pharmacokinetics (e.g., medicine absorption, distribution, and excretion). P-gp is the expression product of multidrug resistance multidrug resistance gene1, which is the most important membrane transferring protein to prevent the accumulation of anti-cancer drugs in cell (Wu et al. 2019). P-gp involves itself in more than just multiple pathophysiological processes, but also mediates various medicine transfer, excretes a variety of medicine and toxins, and protects brain from toxics. It was proven that the high expression of P-gp can speed up the intestine metabolism (Gwak et al. 2020).

Advanced Glycation End Products

Advanced glycation end products (AGEs) are widely found in blood and multiple tissues (Zeng et al. 2019). They are the type of end-product generated by the reaction between amino-group of protein lysine and carbonyl of reducing sugar under enzyme free condition. Kuhla et al. (2015) reported that AGEs positive staining was widely found in brain of patients with AD. This research suggested that accumulation of AGEs in central nervous system was correlated with degeneration death of neurons. The study on correlation between AGEs level and Alzheimer disease also suggested that AGEs might be related with the occurrence and development of brain injury (Chou et al. 2019). The cross-linking between AGEs and macromolecules (protein, nucleic acid, and so on) will modify glycation of structural protein and then damage its normal structure and function. In the end, such process will trigger neural function disorder. Furthermore, AGEs can bond with specific receptor to activate downstream signal pathway and cause corresponding indirect damage.

Chronic Fluorosis and Disorder of Carbohydrate Metabolism

Chronic fluorosis can inhibit the activity of key enzyme in carbohydrate metabolism and interfere with the normal metabolism. High fluoride content will also trigger the secretion reduction of insulin and cause hyperglycemia.

Chronic Fluorosis Caused Carbohydrate Metabolism Disorder

In the process of glucose anaerobic decomposition, fluoride can inhibit the enolase of glycolysis (Manome et al. 2019). Enolase can catalyze 2-phosphoglycerol to be dehydrated, causes electron redistribution and energy round-up to generate phosphoenolpyruvate (the whole reaction is reversible). Phosphoenolpyruvate and 6-phosphate glucose accumulate in the cell since the enolase is inhibited. In addition, fluoride can bond with metal to inhibit multiple enzymes during carbohydrate metabolism: fluoride can affect the anaerobic glycolysis of carbohydrate by inhibiting the activity of hexokinase, 6-phosphohexokinase, phosphoglycerate kinase, and pyruvate kinase, since fluoride bond with Mg^{2+} during the glycolysis process. It was proven that fluoride in vitro can inhibit glycolysis in erythrocyte: 11.4% of glycolysis in erythrocyte was inhibited when fluoride concentration was 0.1 mM, while 22.7% and 40.1% of glycolysis in erythrocyte were inhibited when fluoride concentration was increased to 1 mM and 10 mM.

In the aerobic oxidation process of glucose, chronic fluorosis inhibits aconitase and succinate dehydrogenase of tricarboxylic acid cycle directly to interrupt the cycle. Hence, citric acid accumulates, pyruvate metabolism is damaged and the normal oxidative phosphorylation is affected. The overall manifestation is insufficient energy, metabolism disorder, and malnutrition. Fluoride can also activate certain enzymes to increase the level of cyclic adenosine monophosphate (cAMP) (such as adenylate cyclase and phosphoglucomutase). In this way, fluoride activates certain protein kinases (adenylate cyclase and phosphoglucomutase) which are dependent on cAMP and cause enhanced glycogenolysis; then glucose is released to blood and causes hyperglycemia. Apart from the above-mentioned pathways, chronic fluorosis can affect other major carbohydrate metabolism pathways, such as phosphoric acid carbohydrate free pathway.

Chronic Fluorosis and Hyperglycemia

Exposure to high fluoride content can cause damage to sugar tolerance: reduced insulin secretion, inhibited glycolysis, consumed glycogen, hyperglycemia and insulin resistance. It has been proven that feeding high fluoride content water to rat for 4 weeks revealed mild damage on hyperglycemia and sugar tolerance. Meanwhile, the study also discovered that insulin secretion was reduced and mRNA expression of insulin in pancreatic β cell of rat was decreased when rats were exposed to 1.35 and 2.26 mM fluoride concentration under the condition of stimulant glucose concentration. Such results suggested that high fluoride concentration could lower the secretion of insulin mRNA and pancreatic β cell of rat, which could ultimately affect the oral glucose tolerance test. This could be the possible explanation of increased level of blood glucose in animals with fluorosis (García-Montalvo et al. 2009).

The study on polycystic ovary syndrome (PCOS) discovered that sex hormone-binding globulin (SHBG) level of PCOS patients in low fluoride content group showed negative correlation with fasting plasma glucose level, while level of high-density lipoprotein (HDL) in high fluoride content group showed positive correlation with fasting plasma glucose level (Szczuko et al. 2019). Phenotype of relatively higher androgen level was negatively correlated with triglyceride level, but positively correlated with HDL. Fluoride can increase the insulin resistance, and high insulin resistance can cause the reduction of SHBG transmitted protein synthesis. Fluoride might enhance the developing mechanism of PCOS by increasing insulin resistance. The study on the effect of insulin and insulin receptor on the development of skeletal fluorosis revealed that: skeletal fluoride concentration of chronic fluoride rat increased, alkaline phosphatase (ALP) activity and OCN level obviously increased. All these indicated that fluoridation showed facilitating effect on the ossification of rats. Conversely, chronic fluorosis rats showed obviously increased serum insulin level and overall decrease of glucagon level. Histology analysis revealed that 1 year after high fluoride content treatment, insulin positive part of rats increased and the size of islet enlarged; fluoride had obvious facilitating effect on the mRNA expression of insulin receptor. The results indicated that insulin secretion and fluoride intake shared close relationship, and insulin signal pathway might involve in the occurrence or development of skeletal fluorosis (Hu et al. 2012).

Chronic Fluorosis and Altering of Glycoprotein

Fluoride accumulation leads to an increase in free radical content, which reduces the role of the antioxidant system and increases oxidative stress levels. This process is considered as the possible comprehensive damage mechanism in human body of chronic fluorosis. Protein glycosylation might be an indicator for oxidative damage when ROS acts on cell protein. With more research on the damage mechanism of chronic fluorosis, the role of glycoprotein in such mechanism grasped more attention that the location and count of certain glycoprotein could change in the pathogenesis of chronic fluorosis.

Collagen

Susheela, Indian scientist, discovered that collagen in bone tissue and non-bone tissue of animals with chronic fluorosis was severely damaged. Hence, she proposed earlier than others that collagen in bone tissue of patients with fluorosis was damaged the worst, and she considered that collagen is the target tissue of fluoride action (Susheela and Mukerjee 1981). The research on osamine polysaccharide metabolism of patients with chronic fluorosis showed that the amount of osamine polysaccharide excreted by urine showed a decreasing trend with the worsening of illness. The results of excessive fluoride on osamine polysaccharide in cartilage matrix and prostate content in lab rats revealed that osamine polysaccharide content in cartilage matrix of rats with chronic fluorosis obviously increased. Excessive fluoride caused obvious increase in hydroxyproline excretion amount, which is exclusively owned by collagen. As collagen was decomposed, its tissue content fell. It explained that excessive fluoride intake can inhibit the biosynthesis of collagen and diminish collagen content by accelerating its decomposition. Cranium osteoblast of rat cultivated by different concentrations of NaF was used to promote the expression of collagen type I (Col I) in mRNA and protein (Li et al. 2006). 90 days after female lab rats (8 weeks old) were fed with fluoridated water (200 ppm), fluoride concentration in plasma, bones, urine, and feces of rats obviously increased; meanwhile, acid solubility of skeletal muscle and concentration of pepsinogen soluble collagen all obviously increased. Expression level of collagen type I α 1 (*COL1A1*) decreased 54%, which proved chronic fluorosis caused collagen in skeletal muscle to be degraded and dissolved, as well as downregulating of *COL1A1* gene expression (Gupta et al. 2013). Electron microscope was adopted to observe ultra-microstructure of skeletal tissue and collagen morphological form of rats with chronic fluorosis. It was discovered that excessive fluoride intake caused the altering of collagen arrangement and ultra-microstructure of skeletal tissue, as well as the expression downregulating of mRNA level (Yan et al. 2015).

Osteocalcin

OCN is the skeletal-specific protein synthesized and secreted by osteoblast; ucOCN can promote pathways of regulating carbohydrate metabolism: proliferation of insulin cell B, secretion of insulin, and improve surrounding tissues' sensitivity towards insulin. There was a study that chose fluoride-exposed workers as fluoride group: subjects were divided into high fluoride load group and low fluoride load group according to urine fluoride concentration and serum fluoride concentration. Radio immune assay was adopted to detect the content of OCN and calcitonin (CT) in serum. Test results showed that OCN and CT content in serum increased with the rise of fluoride load. It indicated that high fluoride exposure would increase the level of serum osteocalcin. OCN and CT in serum might be the early effect indicator of skeletal fluorosis damage (Li et al. 2009). Animal test also showed that fluoride-added group rats with vigorous osteoblast activity, enhanced osteogenesis, increased OCN with the added amount of fluoride exposure. OCN is prone to bond with hydroxyapatite, which is mainly accumulated in bone matrix. The ucOCN is less likely to bond with hydroxyapatite, since it will directly enter the blood circulation once it is secreted. In acid environment, Gla of OCN molecule can be decarboxylated into ucOCN. If the skeletal tissues are in relatively high concentration of fluoride in long term, fluoride mainly acts on osteoblast and osteoclastic to alter the osteogenesis and osteoclastic activity, promotes bone resorption, and secrets acidic components (which can also decarboxylate OCN into ucOCN and enters the blood circulation later) (Dong et al. 2016).

With the increase of ucOCN in patients with chronic fluorosis, it will promote the secretion of insulin, enhance sensitivity of insulin, and improve sugar tolerance. Hence, it was proven that ucOCN is the key participation factor of glucose metabolism regulation. It was test proven that high dose of fluoride could exacerbate weight loss and hyperglycemia caused by diabetes and obviously lower the sensitivity of insulin of rats (Li et al. 2014). Therefore, ucOCN plays a major role in carbohydrate metabolism disorder caused by chronic fluorosis, speeding up and exacerbating the development of diabetes.

P-gp and Advanced Glycation End Products (AGEs)

AGEs are a type of macromolecular substance with complicated structure catalyzed in non-enzymatic condition by protein, nucleic acid, lipid, and carbohydrate. AGEs can directly damage the normal structure of various proteins in human body by glycation; it can also bond with specific receptor for advanced glycation end products (RAGE) to trigger altering of multiple downstream signal pathways and cause a series of biological effects. It has been proven that altering of AGEs/RAGE signal transduction pathway plays a significant role in chronic fluorosis-triggered neural injury mechanism (Zhang et al. 2015).

This research group adopted animal models of chronic fluorosis and fluoride-exposed cell model for research. Immunohistochemistry results showed that AGEs were pinned on neuron cytoplasm, and RAGE was pinned on membrane (Zhang et al. 2015). After 3 months of fluoride exposure, cerebral tissue of rats showed no obvious change in the content of AGEs/RAGE, while such content obviously increased 6 months after fluoride exposure. Western blot and real time quantitative PCR all showed that cerebral tissue of rats in fluoride exposure group with increased AGEs/RAGE protein and mRNA expression. The content of AGEs in cerebral tissues could increase with the added fluoride staining dosage and exposure time. Large amount of fluoride treated SH-SY5Y cell cultured in vitro showed the same results. Positive expression of P-gp was discovered on cerebral neuron membrane of fluoride-exposed rats in chronic fluorosis group. Positive cell count was obviously higher than control group.

The bond between AGEs and its specific receptor RAGE can activate downstream signal transduction pathways. Existing studies focus on AGE/RAGE/mitogen-activated protein kinase (MAPK)/nuclear factor-kappa B (NF-κB). The bond between AGEs and RAGE triggers oxidation and cellular factor cascade response by the phosphorylation of mitogen-activated protein kinase p38 (p38 MAPK) and activation of NF-κB (Chen et al. 2020). NF-κB is the multi-transcription factor of multiple "injury response gene." It can activate the expression of downstream damaged factors and lead to histological injury. It was discovered that inhibitor of NF-κB, IκB, enzyme Ras, and tumor necrosis factor-α (TNF-α) in cerebral tissue of rats with chronic fluorosis all decreased significantly, while expression level of NF-κB in protein and mRNA all increased obviously.

The mechanism of RAGE generated oxidative stress might be activation of nicotinamide adenine dinucleotide phosphate (NADPH) oxidase of different factors (human aortic endothelial cells, human aortic smooth muscle cells, and human corneal endothelial cells) to cause apoptosis (Zhong et al. 2015) and the generation of superoxide. This study found that expression level of NADPH oxidase-2 (NOX_2) in cerebral tissue of chronic fluorosis rats increased. Twelve hours cultivation of fluoride staining cell showed increased expression of NOX_2; the increased expression was positively correlated with fluoride staining time, together with the increase of ROS content. In the meantime, B-cell lymphoma-2 (Bcl-2), Bcl-2 associated X protein (Bax) and cysteine aspartate specific protease-3 (caspase-3) in cerebral tissue of chronic fluorosis rats showed obvious change: apoptosis rate of fluoride staining cultivated neurons increased. Increased protein expression of RAGE, NF-κB, and NOX_2 triggered by high fluoride content could be inhibited to different degrees after using RAGE blocker and antioxidants to treat fluoride exposed cells. Such treatment could also lower the apoptosis rate of fluoride-exposed cells and ROS level.

Study results revealed that chronic fluorosis could activate signal transduction pathway of AGE/RAGE, trigger the activation of factors (NOX2, MAPK, and NF-κB) in downstream of such pathways, cause the rise of oxidative stress level. It could also promote the release of inflammatory factors and cause neuron damage. Altering of AGE/RAGE signal transduction pathway could also activate the process of neuron apoptosis. Adopting RAGE blocker and antioxidant for treatment could obviously inhibit the fluorosis caused activation of NF-κB and NOX_2 in downstream; such treatment could also downregulate neuron apoptosis rate and oxidative stress level. It proved that altering of AGE/RAGE signal transduction pathway played a major role in pathogenesis of neural damage in chronic fluorosis.

Conclusion

The accumulation of fluoride causes free radical content to increase, which then lowers the effect of antioxidant system and increases the level of oxidative stress; then protein is damaged by ROS over time and protein glycosylation is forced to occur. This process is considered as the possible comprehensive damage mechanism in human body of chronic fluorosis. Protein glycosylation might be an indicator for oxidative damage. In the research of glycoprotein's role in chronic fluorosis pathogenesis, obvious altering was discovered in glycoproteins (such as collagen, P-gp, and AGEs) in the development of chronic fluorosis caused damage. It served to support that glycoprotein altering might be a key part in the pathogenesis of chronic fluorosis.

References

Chen SH, Yuan KC, Lee YC, Shih CK, Tseng SH, Alexey A, et al. Iron and advanced glycation end products: emerging role of iron in androgen deficiency in obesity. Antioxidants. 2020;9:261–72.

Chou PS, Wu MN, Yang CC, Shen CT, Yang YH. Effect of advanced glycation end products on the progression of Alzheimer's disease. J Alzheimers Dis. 2019;72:191–7.

Dong X, Hou JQ, Liang XY, Wang CN, Zheng JD, Ma P. The research progress ofosteocalcin. J Jilin Med Univ. 2016;37:67–9.

García-Montalvo EA, Reyes-Pérez H, Del Razo LM. Fluoride exposure impairs glucose tolerance via decreased insulin expression and oxidative stress. Toxicology. 2009;263:75–83.

Gupta AR, Dey S, Swarup D, Saini M. Effects of excessive fluoride ingestion on collagen protein and expression of type I collagen gene in skeletal muscles of rats. Fluoride. 2013;46:149–55.

Gwak EH, Yoo HY, Kim SH. Effects of diabetes mellitus on the disposition of tofacitinib, a janus kinase inhibitor, in rats. Biomol Ther. 2020;28:361–9.

Hu CY, Ren LQ, Li XN, Wu N, Li GS, Liu QY. Effect of fluoride on insulin level of rats and insulin receptor expression in the MC3T3-E1 cells. Biol Trace Elem Res. 2012;150:297–305.

Kuhla A, Ludwig SC, Kuhla B, Munch G, Vollmar B. Advanced glycation end products are mitogenic signals and trigger cell cycle reentry of neurons in Alzheimer's disease brain. Neurobiol Aging. 2015;36:753–61.

Li D, Li YH, Fan Z, Li GS. Influence of fluoride on collagen type I mRNA and protein expression in rat osteoblasts: an *in vitro* study. Chin J Endem. 2006;2:6–12.

Li MF, Ma JX, Song YE. Detection and analysis of serum osteocalcin and serum calcitonin level among different fluoride burden groups. Chin J Ind Hyg Occup Dis. 2009;27:761–2.

Li F, Liu BC, Lu P, Qi XC, Li XN. Toxicity effect of fluoride on rats with type 1 diabetes mellitus induced by streptozotocin and its influence in process of diabetes mellitus. J Jilin Univ. 2014;401:55–9.

Manome A, Abiko Y, Kawashima J, Washio J, Fukumoto S, Takahashi N. Acidogenic potential of oral bifidobacterium and its high fluoride tolerance. Front Microbiol. 2019;10:1099.

Susheela AK, Mukerjee D. Fluoride poisoning and the effect on collagen biosynthesis of osseus and non-osseus tissues of rabbit. Toxicol Eur Res. 1981;3:99–104.

Szczuko M, Splinter J, Zapałowska-Chwyć M, Ziętek M, Maciejewska D. Fluorine may intensify the mechanisms of polycystic ovary syndrome (PCOS) development via increased insulin resistance and disturbed thyroid-stimulating hormone (TSH) synthesis even at reference levels. Med Hypotheses. 2019;128:58–63.

Wu Y, Yao Y, Yun Y, Wang M, Zhu R. MicroRNA-302c enhances the chemosensitivity of human glioma cells to temozolomide by suppressing P-gp expression. Biosci Rep. 2019;39: BSR20190421.

Yan XY, Hao XH, Nie QL, Feng CP, Wang HW, Sun ZL. Effects of fluoride on the ultrastructure and expression of type I collagen in rat hard tissue. Chemosphere. 2015;128:36–41.

Zeng C, Li YY, Ma JZ, Niu LN. Clinical/translational aspects of advanced glycation end-products. Trends Endocrinol Metab. 2019;30:959–73.

Zhang KL, Lou DD, Guan ZZ. Activation of the AGE/RAGE system in the brains of rats and in SH-SY5Y cells exposed to high level of fluoride might connect to oxidative stress. Neurotoxicol Teratol. 2015;48:49–55.

Zhong Y, Cheng CF, Luo YZ, Tian CW, Yang H, Liu BR. C-reactive protein stimulates RAGE expression in human coronary artery endothelial cells *in vitro* via ROS generation and ERK/NF-κB activation. Acta Pharmacol Sin. 2015;36:440–7.

Chapter 20
Changes in Gene Expression

Jie Deng and Zhi-Zhong Guan

Abstract As in many other areas of research, studies on the pathogenesis of fluoride toxicity are involving genetic and epigenetic analyses to an increasing extent. Samples from large numbers of individuals living in areas characterized by endemic fluorosis have allowed exploration of potential relationships between polymorphisms in genes related to bone metabolism and dental and skeletal fluorosis. To date, single nucleotide polymorphisms in osteocalcin, calcitonin receptors, collagen type I α 2, estrogen receptor, vitamin D receptor, and some other genes have been found to be associated with susceptibility to fluoride toxicity. In addition, there is much on-going and in-depth research into the potential significance of epigenetic factors (e.g., DNA methylation, histone modification, non-coding RNA, etc.) with respect to endemic fluorosis. These novel approaches are rapidly improving our understanding of the pathogenesis of endemic fluorosis and should provide a basis for more effective prevention and treatment.

Keywords DNA methylation · Endemic fluorosis · Epigenetics · Free radicals · Gene polymorphism

J. Deng (✉)
The Key Laboratory of Endemic and Ethnic Diseases (Guizhou Medical University), Ministry of Education, Guiyang, Guizhou, People's Republic of China

The Provincial Key Laboratory of Medical Molecular Biology, Guiyang, Guizhou, People's Republic of China
e-mail: mbl2000@gmc.edu.cn

Z.-Z. Guan
The Key Laboratory of Endemic and Ethnic Diseases (Guizhou Medical University), Ministry of Education, Guiyang, Guizhou, People's Republic of China

Department of Pathology at the Affiliated Hospital of Guizhou Medical University, Guiyang, Guizhou, People's Republic of China

Z.-Z. Guan (ed.), *Coal-burning Type of Endemic Fluorosis*,
https://doi.org/10.1007/978-981-16-1498-9_20

Introduction

The development of endemic fluorosis, a geochemical disease caused by environmental factors, is also influenced by genetic factors, lifestyle, the nature access to health services, socio-economic status, and certain additional factors. Thus, although the cause of endemic fluorosis is evident, different individuals living in the same environment and exposed to the same level of fluoride react differently, possibly due to their different genetic make-ups (Ba et al. 2009). Consequently, more and more researches are focusing on genetic factors that may influence the pathogenesis of endemic fluorosis, including single nucleotide polymorphisms (SNPs), an approach that will hopefully provide information that can revolutionize the diagnosis, treatment, and prevention of this disease. At present, this research is centered primarily around genes related to the generation of free radicals and bone metabolism.

In addition, epigenetic changes (i.e., inheritable alterations with independent changes in DNA sequence, including changes in DNA methylation, histone modification, and non-coding RNAs such as miRNAs) in response to environmental factors can alter regulatory mechanisms in the body that lead to homeostatic imbalance. It is becoming more and more evident that appropriate regulation of epigenetic factors is not only of vital importance for the healthy development of organisms and their defense against harmful environmental factors (such as physical and chemical insults, viruses, bacteria, etc.), but also appears inappropriate regulation leading to the development of disease. In recent years, the potential relationship between epigenetic factors and fluorosis has been extensively investigated and accumulating evidence indicates that epigenetic changes play an essential role in the development of this disease.

Genes Associated with Enhanced Susceptibility to Chronic Fluorosis

Genes Related to the Metabolism of Free Radicals

Numerous investigations, both on patients with chronic fluorosis and animal models exposed to fluoride, have revealed the increases in the levels of free radicals and lipid peroxides in body tissues and an accompanying reduction in the activities of antioxidant enzymes, leading to a metabolic imbalance. Moreover, this imbalance becomes more pronounced as the level and period of exposure to fluoride. Thus, a major current hypothesis is that excessive intake of fluoride is harmful because of the resulting production of free radicals.

Fluoride can stimulate a respiratory burst in neutrophils that leads to the release of superoxide anion radicals, hydroxyl radical, nitric oxide, and other free radicals, thereby causing peroxidation of many macromolecules. At the same time, fluoride can attack trace elements that form part of the structure of oxidases, as well as combine with the divalent or trivalent metal ions in the active sites of antioxidant enzymes, thereby inhibiting their activities. As a result, the levels of free radical increase and these electrophilic, reactive radicals attack, among other compounds, the polyunsaturated fatty acids in the phospholipids that form the backbone of biological membranes, changing the structure of these membranes in a dysfunctional manner. Furthermore, free radicals can damage DNA, leading to mutations (e.g., in connection with DNA repair) or other forms of permanent damage.

Our research group at the Guizhou Medical University examined potential genetic polymorphisms in two enzymes related to free radical metabolism, i.e., myeloperoxidase (MPO) and glutathione-S-transferases (GSTs), in the residents living in the areas of coal-burning type of endemic fluorosis, Guizhou of China. Among them, two groups were divided as that the stoves used in their home had been improved to reduce the contamination of fluoride from burning coal and those not improved as comparison. The occurrence frequency of an active gene encoding *GSTP1 AA* (*105Ile*) in both of these groups with chronic fluorosis was significantly lower than that in control individuals without fluorosis living in this same region. In addition, there was no significant difference in this respect between the two groups with improved stoves or not. We speculate that in the presence of excessive intake of fluoride, carrying a *GSTP1 Ile105Val* genotype may reduce the antioxidant capacity and therefore enhance the risk of developing fluorosis.

Furthermore, by utilizing adapter ligation-mediated allele-specific amplification (ALM-ASA) to detect SNPs at five different locations (−1276 C/T, −1228 G/A, −585 T/C, −463 G/A, −163 C/T) in the promoter region of the *MPO* gene, the frequencies, with the exception of the first, differed between the individuals with and without fluorosis. The frequency of 463 G/A was higher in those with than without fluorosis, indicating that high transcription efficiency that allows synthesis of excessive amounts of MPO might help protect individuals in this region from chronic fluorosis. The other three SNPs might also be related to predisposition to fluorosis. We speculated that the three haplotypes constructed successfully in the MPO promoter, i.e., 163C-463G-585T-1228G-1276C, 163C-463G-585T-1228A-1276T, 163T-463G-585T-1228G-1276T, might be associated with chronic fluorosis (Zhang et al. 2013).

Osteocalcin (OCN), Calcitonin (CT), and Calcitonin Receptors (CTR)

The deleterious effects of high level of fluoride on the skeletal system are mediated primarily through alterations in the actions of hormones that regulate calcium. OC, a mediator of osteoblast differentiation and maturation, also plays a key role in bone mineralization. The levels of both OC and CT, which stimulate bone formation and resorption, respectively, in the serum of patients with dental fluorosis were elevated to different extents in comparison to the corresponding levels in control individuals living in the regions without fluorosis, suggesting an imbalance of bone metabolism in the patients with endemic fluorosis (Ba et al. 2009). The same investigation revealed that although the frequency of OC Hind III HH genotypes in children with dental fluorosis was higher than those of children without dental fluorosis living in the same region, the distribution frequency of the three OC Hind III genotypes did not show any statistical significant. Thus, the predisposition of children living in areas with endemic fluorosis to develop dental fluorosis cannot yet be related to this type of polymorphism.

CTR not only activates adenylate cyclase, thereby increasing the level of cyclic adenosine monophosphate, but also mediates the elevation of intracellular calcium levels in response to CT. CTR plays a vital role in the maintenance of the calcium–phosphorus balance in the body and its mutation may alter the binding of its ligands and thereby influence receptor signal transduction. Application of conditional logistic regression models revealed that individuals living in Chongqing of China, an area of coal-burning type of endemic fluorosis, carried CT + TT genotypes of CTR AluI and appeared more susceptible to this disease. Moreover, the presence of these genotypes was correlated with high levels of fluoride ($\geq$1 ppm) in the urine. These investigators speculated that this polymorphism in the CTR gene affects the status of ions during dental mineralization, thereby resulting in different risks for the development of dental fluorosis upon exposure to the same level of fluoride. This finding indicates that the screening of the individuals is more susceptible to coal-burning type of endemic fluorosis, at least to some extent (Jiang et al. 2015). In addition, in the case of individuals exposed occupationally to fluoride, both a high fluoride load and carrying the TC or TT genotype were found to be risk factors for bone injury and, furthermore, there was a significant interaction between these two factors (Tu et al. 2011).

Collagen Type I

A variety of polymorphisms in the collagen type I α 2 (*COL1A2*) gene can lead to bone-related diseases, including achondroplasia and osteogenesis imperfecta. The frequency of the *COL1A2* PvuII PP genotype among children with dental fluorosis and living in areas where drinking-water fluorosis is endemic is significantly higher than that among those without this condition, but, however, with no appreciable difference in the case of the *COL1A2* RsaI RR genotype. Therefore, it was proposed that carrying the *COL1A2* PvuII PP genotype may increase the susceptibility of children to dental fluorosis. At the same time, the distribution frequency of the three *COL1A2* PvuII genotypes in adults was the same in the regions with and without high levels of fluoride area in the drinking water and the control area, indicating no apparent correlation between polymorphism at this site and skeletal fluorosis.

The mechanism by which the *COL1A2* PvuII polymorphism increases the susceptibility of children to dental fluorosis remains unclear and there may be linkage disequilibrium with other polymorphic sites within or adjacent to the *COL1A2* gene.

Sequence analysis revealed a C→A mutation at base 19,994 (in exon 25) of the gene that affected bone mineral density and bone metabolism (Huang et al. 2008). Among 6–12-year-old children in San Luis Potosí, Mexico, those exposed to high levels of fluoride and carrying this substitution had a higher risk of developing dental fluorosis (Escobar-García et al. 2016; Jarquín-Yñezá et al. 2018).

The collagen type I α3 (*COLIXA3*) gene encodes the α3 chain of type IX collagen and mutations in this gene are associated with lumbar disc diseases, multiple bone marrow dysplasia, and aberrant development of the peripheral nervous system (Paassilta et al. 2001; Asamura et al. 2005), with clinical symptoms similar to those characteristics of fluorosis. Several analyses of the frequency of the three polymorphisms in the exons of the *COLIXA3* in the peripheral blood of individuals with and without fluorosis revealed no relationship between the polymorphism in exon 5 and fluorosis. However, due to the small sample sizes involved, the possibility of a relationship between polymorphisms in this gene and fluorosis requires further evaluation.

Estrogen

Estrogen exerts pronounced effects on bone development, especially by stimulating the activity of osteoblasts and promoting the deposition of calcium and phosphorus. Thus, polymorphism in estrogen receptor (ESR) genes may have an impact on susceptibility to fluorosis. Although the distributions of the *RsaI*, *XbaI*, and *PvuII* variants of this gene show no significant association with dental fluorosis or abnormalities in bone mineral density, children with one normal (r) and one mutated (R) allele of the *RsaI* gene are at 1.83-fold higher risk of developing dental fluorosis, while those with the RR genotype also demonstrate an increased risk for dental fluorosis. The polymorphism in *Rsa I* is next to codon 328, which encodes valine, resulting in production of a synonymous mutation that may affect the synthesis, splicing, maturation, transport, translation, degradation, etc., of mRNA (Almeida et al. 2005).

Carrying allele X of XbaI can reduce the risk of dental fluorosis (Ba et al. 2011). The underlying mechanism may involve linkage between this locus and adjacent exons that result in changes in the structure and function of the ER and/or linkage to other introns and regulatory units that modulate expression of ER at the transcriptional level. In a recent study on male farmers with drinking-water type fluorosis in Henan Province, China, the polymorphisms *XbaI* and *rs3798577* in the *ESRα* gene were found to be correlated with the serum level of androgen binding protein (Ma et al. 2017; An et al. 2019).

Vitamin D Receptor (VDR)

VDR, located inside the nucleus, influences the transcription of genes whose products convert vitamin D into its biologically active form, 1,25-dihydroxyvitamin D3. The latter plays important roles in regulating calcium phosphate balance and bone metabolism, cell differentiation, and immunological reactions, as well as other important physiological processes. Our research team at Guizhou Medical University of China applied stratified sampling to the population of a region of coal-burning of endemic fluorosis in Guizhou, dividing the participants into an intervention group (villages and towns of endemic fluorosis areas with the improved stoves for at least 2 years), a non-intervention group (villages and towns of endemic fluorosis areas without improved stoves), and a control group (outside the area of endemic fluorosis). The polymerase chain reaction-restriction fragment length polymorphism procedure was then used to analyze the Bsm I and Fok I polymorphisms in the VDR gene in these three different groups.

We found that carriers of the Bsm I GG and Fok I CT/CC genotypes have a higher risk of developing fluorosis, in disagreement with certain previous reports. All of our participants were adults living in areas of coal-burning type of endemic fluorosis, where fluoride intake is higher than in regions with drinking-water type of endemic fluorosis. In addition, the dietary intake of calcium and protein by this local population was low. Altogether, these observations suggest that individual and regional variations in *VDR* gene polymorphisms exist and may be related to the type and severity of fluorosis (Yang et al. 2016).

Catechol-O-Methyltransferase (COMT)

COMT is responsible for eliminating approximately 60% of the dopamine in the prefrontal cortex, thereby influencing cognition and learning. One variant of the human *COMT* gene contains a G rather than an A in codon 158, causing a valine (Val) to be replaced by methionine (Met) and thereby reducing the activity of the enzyme. A report indicated an association between this polymorphism and children's cognition. Moreover, children living in area of drinking-water type of endemic fluorosis, Tianjin of China, who carry the val/val genotype, exhibited a higher average IQ than those with the met/val or met/met genotypes (Zhang et al. 2015).

Parathyroid Hormone (PTH)

PTH increases the level of calcium and decreases phosphorus in the blood. Moreover, in bone, PTH promotes the growth of osteoclasts and elevates the cytoplasmic concentration of calcium in these cells, as well as stabilizing their calcium status. In addition, the Bst BI polymorphism in the gene that encodes this hormone is associated with bone mineral density.

Although the Kazakh Russian and Han ethnic groups in Mongolia and Tibet differ with respect to the frequency of polymorphisms in this gene, the Bst BI genotype was not associated with any distinct different in the level of the hormone itself. Thus, these different ethnic groups differ with respect to bone metabolism, which may help explain differences in prevalence of brick-tea type fluorosis among them. However, among children living in regions with drinking-water type of endemic fluorosis, no association was observed between susceptibility to dental fluorosis and Bst BI polymorphism.

Although there is at present no way to alter genes that predispose to certain pathological conditions, it should be possible to identify individuals carrying these genes and suggest appropriate preventive measures, especially concerning risk factors in the environment. Future research in this area must encompass larger groups of people, as well as more candidate genes and environmental factors. Only a comprehensive understanding of the pathogenesis of endemic fluorosis can lead to evidence-based preventive measures.

The Epigenetics of Chronic Fluorosis

The Relationship between Epigenetics and Environmental Factors

The primary mechanisms underlying epigenetic regulation, including DNA methylation, histone modification, chromatin remodeling, genomic imprinting, chromosome inactivation, and non-coding RNA, are influenced profoundly by the environment in which we live. The food we eat, the water we drink, the air we breathe, and the mental states we experience can all have a tremendous impact, activating or shutting down the expressions of genes without altering the sequence of DNA.

Accumulating evidence indicates that environmental factors such as secondhand smoke, alcohol abuse, viral hepatitis, industrial pollution, and carbon emissions can play key roles in the initial stages of disease development and epigenetics may provide an explanation for such observations. Numerous studies have demonstrated that epigenetic alterations induced by exogenous compounds can lead to toxic effects or tumorigenesis. For instance, urinary and blood concentrations of arsenic were found to be positively correlated with genome-wide levels of DNA methylation in inhabitants of Bangladesh, where the level of maternal exposure to arsenic was also positively correlated with genome-wide DNA methylation in infant cord blood. Among individuals living in areas of China contaminated with arsenic, increased methylation in the promoter regions of the p16 and p53 anti-oncogenes lowers their levels of expression, which induces cancer. In addition, exposure to the heavy metal cadmium elevates the total genomic level of methylation, possibly by inhibiting the activation of methyltransferase. Altered methylation in the promoter regions of certain genes, histone modifications, and alterations in miRNA are also involved in the oncogenic process initiated by cadmium.

Some of the most important current issues in environmental epigenomic studies indicated that genes may increase susceptibility to disease when altered epigenetically, environmental factors can cause such alterations. Epigenome plays a role in connection with reproductive development, disease etiology, the potential for nutritional supplementation to attenuate the adverse effects of chemical and physical factors on the epigenome, and the identification of epigenetic markers of early-stage disease. The key to finding answers to these questions lies in the development of screening technologies that seems to enable rapid and accurate characterization of the entire epigenome. Indeed, during the past 30 years, many cellular and molecular biological techniques for the analysis of epigenetic phenomena such as DNA methylation and histone modifications have been developed. Application of these methodologies to both normal and abnormal cells has already and will continue to enhance our understanding of the involvement of epigenetic processes in both health and disease.

Correlation of Choric Fluorosis with Epigenetic Change

Recently, genome-wide bisulfite sequencing of the DNA of human osteosarcoma cell(HOS)exposed to fluoride found hypermethylation in the promoter regions of the bone morphogenetic protein 1, methionyl aminopeptidase 2, matrix metalloproteinase 11, and BTB domain and CNC homolog 1 genes that lowered the expression of mRNA. The products of these genes are involved in decomposition of the extracellular matrix, the organization and catabolism of collagen, angiogenesis, and osteoblast development, so such epigenetic changes could potentially be involved in the development of skeletal fluorosis (Daiwile et al. 2019).

Furthermore, in children with dental fluorosis the degree of methylation of the promoter region of the ERα gene correlates negatively with their serum level of calcium and positively with their urinary level of fluoride. However, this degree of methylation in adult male patients with skeletal fluorosis is lower than in those without this disease and is negatively correlated with the urinary concentration of fluoride (Zhang et al. 2017). It has been proposed that the level of methylation of the promoter region of the *ESRα* gene may modulate the bone damage caused by fluoride through its influence on the handling of calcium.

In addition, NaF reduced the levels of DNA methyltransferases (*DNMT*) 1, DNMT 3A, and DNMT 3B mRNAs in cultures of human embryonic hepatocyte L-02 cells in a dose-dependent manner. At the same time, the level of DNA methylation (determined by bisulfite sequencing) was lowered and epigenetic regulation of DNMTs altered. Furthermore, methylation of CpG islands in the c-Fos promoter was reduced and expression of this protein consequently upregulated, which might explain the apoptosis induced in L-02 cells by fluoride (Niu et al. 2012).

In patients with coal-burning fluorosis, the level of methylation of the genes encoding O6-methylguanine-DNA-methyltransferase and mismatch repair in blood cells was correlated with the severity of this condition. Moreover, the greater the severity, the lower the level of the corresponding mRNAs. Similarly, in fluorotic rats methylation of these same genes was elevated and the corresponding levels of mRNA and protein reduced, both in the blood and liver. These analogous findings in humans and animals indicate that lowering the expression of DNA repair genes may be related to the hepatic damage caused by fluoride (Wu et al. 2018).

Recently, attenuated expression of p16 due to hypermethylation of the promoter region of the corresponding gene, together with enhanced osteoblast proliferation with a prolonged S phase of the cell cycle has been observed both in osteoblasts exposed to fluoride and individuals with coal-burning fluorosis. Inhibition of methylation with 5-aza-2-deoxycytidine reversed these effects, clearly revealing the involvement of p16 hypermethylation in the osteoblast activation (Tu et al. 2011) associated with fluorogenic bone damage (Wu et al. 2019).

Fluoride and Histone Modifications

An unbalanced diet can lead to abnormal histone methylations that cause many diseases, including cardiovascular diseases, type II diabetes, and obesity, as well as tumors during early embryonic development (Rasmussen 2001). It was reported that female rats were exposed to different concentrations of fluoride in their drinking water (20, 60, or 120 mg per l) for 30 days and then subjected to superovulation (Fu et al. 2014). Thereafter, homogeneous cytoplasm-oocyte complex together with three dense layers of cumulus cells was subjected in vitro to maturation, fertilization, and embryo culture and all three of these processes found to be inhibited by fluoride in a dose-dependent manner. Flow cytometric analysis revealed that the impaired development of these embryos was related to apoptosis in response to DNA that had undergone abnormal epigenetic modification.

Subsequent immunofluorescent staining revealed that the extent of demethylation of H3K9 and H3K4 demonstrated a positive correlation to the dose of fluoride, whereas acetylation of H3K9 was not altered to any significant degree. At the same time, the levels of the mRNAs encoding the histone methyltransferases G9a, lysine methyltransferase 2A, and lysine demethylase 1A, as well as those of the corresponding proteins were elevated, while the levels of lysine demethylase 3A mRNA and protein were reduced. These observations indicate that fluoride can cause abnormal histone methylation in genes involved in early embryonic development.

In this same context, among the population in a region where coal-burning fluorosis is endemic, expression of the mRNA encoding histone deacetylase 1 (HDAC1) gradually decreased with an increase in the level of urinary fluoride, whereas, in contrast, the level of HDAC1 protein correlated positively with urinary fluoride. In addition, the level of this same protein in primary human osteoblasts exposed to fluoride for 72 hr. rose as the concentration of fluoride was increased. Moreover, no significant difference was found in the level of histone H3 acetylation in the peripheral blood mononuclear cells of individuals exposed to different levels of fluoride or with different levels of urinary fluoride or skeletal fluorosis among the Tibetan people in Qinghai and Kazakh people in Xinjiang, where brick-tea fluorosis is endemic.

P21 plays a key role in connection with the cell cycle checkpoint and exposure of human primary osteoblasts to fluoride reduced the level of histone acetylation in the gene that encodes this protein. As a consequence, histones bond less firmly to the promoter region of this gene, which, in turn, altered the appropriate binding of the required combination of transcription factors, thereby attenuating transcription. The resulting reduction in the level of the P21 protein promoted the transition of the human osteoblasts from the G1 to the S phase of the cell cycle, resulting in more rapid proliferation of these cells and acceleration of bone turnover. Such phenomena might be involved in the early development of fluorosis.

A recent study demonstrated that expression of the histone methyltransferases euchromatic histone lysine methyltransferase 1 and enhancer of zeste 2 polycomb repressive complex 2 subunit and the trimethylation levels of histone H3K9 and H3K27 in HOS were all elevated significantly upon exposure to a specific concentration of fluoride. Subsequent chromatin immunoprecipitation (CHIP) revealed that the degree of H3K9 trimethylation in the promoter regions of the genes encoding transforming growth factor beta receptor 2 (TGFBR2) and SMAD family member 3 (SMAD3) was higher, which decreased expression of these proteins. In addition, the expressions of COL1A1 and matrix metalloproteinases-13 (MMP-13), two other essential proteins related to the TGFBR2-SMAD signaling pathway, were also downregulated. STRING protein network analysis confirmed that the functions of TGFBR2, SMAD3, COL1A1, and MMP-13 are directly related to the formation of collagen, connective tissue, and bone. Accordingly, fluoride-induced hypertrimethylation of H3K9 with consequent inhibition of the expression of TGFBR2 and SMAD3 might contribute to the pathogenesis of fluorogenic bone damage (Daiwile et al. 2018).

Fluoride and miRNA

miRNAs, endogenous ncRNAs consisting of 22 bases, play key roles in connection with both normal physiological and pathological processes. Following exposure of human osteosarcoma cells to 8 or 20 mg NaF per liter, expression of 82 different miRNAs was found to be altered at the lower dose, 56 of which were upregulated; while the higher dose changed the expression of 62 miRNAs, with 21 upregulations. qPCR validation of the considerable changes in the expressions of miRNA-124 and miRNA-155 and target gene prediction indicated that these can regulate expression of Runx2 and receptor-activator of nuclear factor-kappa B ligand (RANKL), the osteoclast differentiation factor (Daiwile et al. 2015).

Chinese researchers also established an in vitro model based on exposure of MC3T3-E1 osteoblasts to 8 mg NaF per liter for 7 days, which resulted in upregulation of 45 and downregulation of 31 miRNAs. Functional analysis revealed that the miRNAs whose expression was changed by fluoride play a role in regulating signaling pathways, such as those involving Wnt, TGF-β, Hedgehog, vascular endothelial growth factor, and Notch (Wang et al. 2017). In another set of similar experiments, exposure of osteoclast-like cells to fluoride caused upregulation of 303 and downregulation of 61 miRNAs that might be involved in regulating pathways related to controlling osteoclasts, as well as in signaling pathways associated with cytokines and immunity (Yu et al. 2018).

In certain investigations by PCR CHIPs were employed to monitor changes in miRNAs related to apoptosis following exposure of human osteoblasts to high level of fluoride. Following treatment of fluoride for 24 hr., 48 miRNAs were upregulated and 4 downregulated, whereas after a 48-hr exposure of the ion, 21 miRNAs were upregulated and 2 downregulated, with the expression of 9 miRNAs being altered in a similar manner at both of these time-points. Bioinformatic analysis of the target genes of 10 of the miRNAs affected by fluoride indicated that (1) miRNA-451a can promote osteogenic differentiation by inhibiting expression of calcium-binding protein 39, thereby altering the activity of a membrane channel involved in the calcium ion transport; (2) miRNA-21 can promote RANKL-induced osteoclast differentiation via positive feedback through the c-Fos pathway/miRNA-21/ programmed cell death 4; (3) miRNA-125a, miRNA-155, miRNA-7b, and miRNA-34a, among others, can inhibit osteoclast formation through targeted regulation of macrophage-colony stimulating factor 1 and c-Fos and other key regulatory factors; (4) miRNA-9, miRNA-98, and miRNA-34a can inhibit IL-1-mediated downregulation of type II collagen and therefore inhibit chondrocyte differentiation; and (5) inhibitors of miR-200C-3p interfere with the bone morphogenetic protein 4/SMAD signaling pathway by targeting Noggin (Jiang et al. 2018).

In animal models exposure to fluoride can inhibit osteogenic activity along with an elevation in the expression of miRNA-23a in a manner that is dependent on the dose and time of exposure. At the same time, there was a tendency for this change in miRNA-23a expression to be related to the skeletal alterations in rats. Such observations have given rise to the proposal that miRNA-23a plays a role in the abnormal bone turnover caused in rats by fluoride poisoning.

There are numerous reports that fluoride can attenuate capacitation, hyperactivation, chemotaxis, acrosome reaction, and fertilization by sperm (Long et al. 2009; Feng et al. 2015). Following administration of 25 or 100 ppm fluoride to mice for 60 days, the concentration, survival, and motility, along with the mitochondrial membrane potential of their sperm were all reduced and, moreover, expression of 16 of 31 known miRNAs was altered. The 671 genes predicted to be targets for these altered miRNAs encode primarily proteins involved in protease inhibition, apoptosis, ubiquitin-mediated proteolysis, and signaling by Ca, Janus kinase-signal transducer and activator of transcription, mitogen activated protein kinase, P53, and

Wnt. Thus, it was proposed that changes in the expression of these miRNAs might serve as potential biomarkers for reproductive toxicity caused by fluoride (Raghunath et al. 2016; Sun et al. 2016).

Clearly, alterations in the levels of specific miRNAs upon exposure to fluoride can change the expressions of certain genes, including those encoding proteins involved in key signaling pathways, in a manner that can impair osteoclast differentiation and development. Such observations provide novel clues concerning the development of fluorosis.

Alteration of Imprinted Genes by Fluoride

A study (Zhu et al. 2014) showed that 100 mg fluoride was exposed to pregnant mice in their drinking water for 2 days and then BSP employed in combination with bisulfite restriction analysis to characterize the methylation of imprinted genes and long interspersed nuclear elements-1 (*LINE-1*) retrotransposon in their embryos during early development. Fluoride-induced aberrant DNA methylation of H19 associated with parentally imprinted genes, but not of this same histone associated with the maternally imprinted genes paternally expressed 3 (*Peg3*) and *LINE-1* retrotransposon. These findings show that fluoride can cause selective changes in the methylation of imprinted genes during early gestation.

In addition, DNA methylation during the development of embryos was investigated by incubating sperm exposed to fluoride with normal ova. As in the case of the embryos, fluoride reduced methylation of *H19*, although only to a small extent. In contrast, methylation of the *Peg3* gene was elevated greatly, which is difficult to understand since no accompanying alteration in DNA methylation or re-methylation was observed. Methylation of the *LINE-1* genes remained normal. These findings demonstrate that fluoride may indirectly change the DNA methylation of imprinted genes in embryos, but not in sperm, perhaps by disrupting a particular pathway in sperm that does not affect their DNA, whereas acts later during development of the embryo after fertilization.

After birth, inherited phenotypes can be influenced by environmental factors that result, e.g., in reprogramming of gamete development and the mechanism (s) underlying such apparent modification of phenotypes requires further elucidation (Zhu et al. 2014). In embryos lower DNA methylation of the parentally imprinted gene *H19* in Duchenne muscular dystrophy (DMD) induced by fluoride resulted in higher transcriptional activity of *H19*, whereas elevated methylation of the maternally imprinted gene insulin-like growth factors 2 in diffuse alveolar damage lowered its rate of transcription, and both of these effects were dose-dependent. Thus, fluorosis may induce general fluctuations in methylation in DMD, inducing the expression of imprinted genes (Fu et al. 2014).

On the whole, the observations of this nature indicate clearly that epigenetic modifications in response to environmental toxicants such as fluoride can influence gene expression in a manner that contributes to their toxicity. Importantly and in contrast to mutations, such epigenetic changes are reversible. Moreover, other environmental factors, such as the level of methyl donors in the diet, can interact with these effects of toxicants, a fact that might be exploited for prevention or treatment.

Currently, examination of the influence of epigenetics in connection with the development of chronic fluorosis is focused primarily on alternations in the methylation of specific genes, histone modifications, and changes in miRNA expression. In the future, this research must be extended to cover the potential effects of fluoride on genome-wide epigenetics. Such detailed information can aid in the prevention, early diagnosis, and treatment of chronic fluorosis.

Conclusions

Irreversible mutations and reversible epigenetic mechanisms both exert profound effects on gene expression and metabolic homeostasis. A database of genetic polymorphisms, in particular SNPs that influence the development of chronic fluorosis, is being established through epidemiological DNA sequencing. Such a database will be an invaluable aid in connection with prevention and treatment of this condition, as well as governmental policy-making.

Moreover, it is now clear that epigenetic modifications caused by exposure to environmental toxicants can alter gene expression in a manner that contributes to the development of their toxic effects, including environmental diseases. Since such effects are reversible, it might be possible to undo or perhaps even prevent them by manipulation of other environmental influences, such as the diet. Current studies on the epigenetics of chronic fluorosis focus mainly on the methylation of specific genes in cells and animal models and variations in epigenetic modifications caused by exposure to fluoride during embryonic development. The epigenetic status of the entire genome of individuals exposed to fluoride remains to be characterized and should provide invaluable insights for the prevention, early diagnosis, and treatment of chronic fluorosis.

References

Almeida S, Franken N, Zandoná MR, Osório-Wender MC, Hutz MH. Estrogen receptor 2 and progesterone receptor gene polymorphisms and lipid levels in women with different hormonal status. Pharmacogenomics J. 2005;5:30–4.

An N, Zhu J, Ren L, Liu X, Zhou T, Huang H, et al. Trends of SHBG and ABP levels in male farmers: influences of environmental fluoride exposure and ESR alpha gene polymorphisms. Ecotoxicol Environ Saf. 2019;172:40–4.

Asamura K, Abe S, Fukuoka H, Nakamura Y, Usami S. Mutation analysis of COL9A3, a gene highly expressed in the cochlea, in hearing loss patients. Auris Nasus Larynx. 2005;32:113–7.

Ba Y, Huang H, Yang Y, Cui L, Zhu J, Zhu C, et al. The association between osteocalcin gene polymorphism and dental fluorosis among children exposed to fluoride in People's Republic of China. Ecotoxicol Environ Saf. 2009;72:2158–61.
Ba Y, Zhang H, Wang G, Wen S, Yang Y, Zhu J, et al. Association of dental fluorosis with polymorphisms of estrogen receptor gene in Chinese children. Biol Trace Elem Res. 2011;143:87–96.
Daiwile AP, Sivanesan S, Izzotti A, Bafana A, Naoghare PK, Arrigo P, et al. Noncoding RNAs: possible players in the development of fluorosis. Biomed Res Int. 2015;2015:274852.
Daiwile AP, Sivanesan S, Tarale P, Naoghare PK, Bafana A, Parmar D, et al. Role of fluoride induced histone trimethylation in development of skeletal fluorosis. Environ Toxicol Pharmacol. 2018;57:159–65.
Daiwile AP, Tarale P, Sivanesan S, Naoghare PK, Bafana A, Parmar D, et al. Role of fluoride induced epigenetic alterations in the development of skeletal fluorosis. Ecotoxicol Environ Saf. 2019;169:410–7.
Escobar-García D, Mejía-Saavedra J, Jarquín-Yáñez L, Molina-Frechero N, Pozos-Guillén A. Collagenase 1A2 (COL1A2) gene A/C polymorphism in relation to severity of dental fluorosis. Commun Dent Oral Epidem. 2016;44:162–8.
Feng D, Huang H, Yang Y, Yan T, Jin Y, Cheng X, et al. Ameliorative effects of N-acetylcysteine on fluoride-induced oxidative stress and DNA damage in male rats' testis. Mutat Res Genet Toxicol Environ Mutagen. 2015;792:35–45.
Fu MZ, Wu XY, He J, Zhang Y, Hua S. Natrium fluoride influences methylation modifications and induces apoptosis in mouse early embryos. Environ Sci Technol. 2014;48:10398–405.
Huang H, Ba Y, Cui L, Cheng X, Zhu J, Zhang Y, et al. COL1A2 gene polymorphisms (Pvu II and Rsa I), serum calciotropic hormone levels, and dental fluorosis. Community Dent Oral Epidem. 2008;36:517–22.
Jarquín-Yñezá L, Alegría-Torres JA, Castillo CG, de Jesús Mejía-Saavedra J. Dental fluorosis and a polymorphism in the COL1A2 gene in Mexican children. Arch Oral Biol. 2018;96:21–5.
Jiang M, Mu L, Wang Y, Yan W, Jiao Y. The relationship between Alu I polymorphisms in the calcitonin receptor gene and fluorosis endemic to Chongqing, China. Med Princ Pract. 2015;24:80–3.
Jiang Y, Yang Y, Wang H, Darko GM, Sun D, Gao Y. Identification of miR-200c-3p as a major regulator of SaoS2 cells activation induced by fluoride. Chemosphere. 2018;199:694–701.
Long H, Jin Y, Lin M, Sun Y, Zhang L, Clinch C. Fluoride toxicity in the male reproductive system. Fluoride. 2009;42:260–76.
Ma Q, Huang H, Sun L, Zhou T, Zhu J, Cheng X, et al. Gene-environment interaction: does fluoride influence the reproductive hormones in male farmers modified by ERα gene polymorphisms? Chemosphere. 2017;188:525–31.
Niu Q, Liu H, Guan Z, Zeng Q, Guo S, He P, et al. The effect of c-Fos demethylation on sodium fluoride-induced apoptosis in L-02 cells. Biol Trace Elem Res. 2012;149:102–9.
Paassilta P, Lohiniva J, Göring HH, Perälä M, Räinä SS, Karppinen J, et al. Identification of a novel common genetic risk factor for lumbar disk disease. JAMA. 2001;285:1843–9.
Raghunath A, Jeyabaskar D, Sundarraj K, Panneerselvam L, Perumal E. In silico prediction of microRNAs on fluoride induced sperm toxicity in mice. Food Chem Toxicol. 2016;98:34–49.
Rasmussen KM. The "fetal origins" hypothesis: challenges and opportunities for maternal and child nutrition. Ann Rev Nut. 2001;21:73–95.
Sun Z, Zhang W, Li S, Xue X, Niu R, Shi L, et al. Altered miRNAs expression profiling in sperm of mice induced by fluoride. Chemosphere. 2016;155:109–14.
Tu J, Liu K, Song Y, Zhang Y, Cui C, Lu C. Interactive effect of fluoride burden with calcitonin receptor gene polymorphisms on the risk of F bone injury. Int Arch Occup Environ Health. 2011;84:533–8.
Wang Y, Zhang X, Zhao Z, Xu H. Preliminary analysis of micro-RNAs expression profiling in mc3t3-e1 cells exposed to fluoride. Biol Trace Elem Res. 2017;176:367–73.

Wu CX, Wang YH, Li Y, Guan ZZ, Qi XL. Changes of DNA repair gene methylation in blood of chronic fluorosis patients and rats. J Trace Elem Med Biol. 2018;50:223–8.
Wu S, Yan W, Qiu B, Liao Y, Gu J, Wei S, et al. Aberrant methylation-induced dysfunction of p16 is associated with osteoblast activation caused by fluoride. Environ Toxicol. 2019;34:37–47.
Yang D, Liu Y, Chu Y, Yang Q, Jiang W, Chen F, et al. Association between vitamin D receptor gene FokI polymorphism and skeletal fluorosis of the brick-tea type fluorosis: a cross sectional, case control study. BMJ Open. 2016;6:e011980.
Yu H, Jiang N, Yu X, Zhao Z, Zhang X, Xu H. The role of TGFβ receptor 1-smad3 signaling in regulating the osteoclastic mode affected by fluoride. Toxicology. 2018;393:73–82.
Zhang S, Zhang X, Liu H, Qu W, Guan Z, Zeng Q, et al. Modifying effect of COMT gene polymorphism and a predictive role for proteomics analysis in children's intelligence in endemic fluorosis area in Tianjin. China Toxicol Sci. 2015;144:238–45.
Zhang T, Shan KR, Tu X, He Y, Pei JJ, Guan ZZ. Myeloperoxidase activity and its corresponding mRNA expression as well as gene polymorphism in the population living in the coal-burning endemic fluorosis area in Guizhou of China. Biol Trace Elem Res. 2013;152:379–86.
Zhang Y, Huang H, Gong B, Duan L, Sun L, He T, et al. Do environmental fluoride exposure and ESRα genetic variation modulate methylation modification on bone changes in Chinese farmers? Chem Res Toxicol. 2017;30:1302–8.
Zhu JQ, Si YJ, Cheng LY, Xu BZ, Wang QW, Zhang X, et al. Sodium fluoride disrupts DNA methylation of H19 and Peg3 imprinted genes during the early development of mouse embryo. Arch Toxicol. 2014;88:241–8.

Part IV
Clinical Diagnosis and Treatments of Coal-Burning Type of Endemic Fluorosis

Chapter 21
Biological Indicators for Early Diagnosis of Endemic Fluorosis

Yan-Jie Liu and Zhi-Zhong Guan

Abstract Generally, clinical diagnosis of endemic fluorosis depends on tooth observation and bone imaging examination, in which dental and skeletal fluorosis can be determined. In addition, the fluoride contents in urine, blood, or other materials, such as nail and hair, may be valuable reference for the diagnosis of chronic fluorosis. Nevertheless, in the early stage of fluorosis, i.e., before obvious change of teeth and X-ray detectable findings of bones, some biological changes could be found since the disease influences almost every organ and system. Therefore, the detection of these biological changes could give an early diagnosis of chronic fluorosis, which would assist early intervention measures and prohibit the damages of fluorosis on multiple organs. Many biological parameters for the early diagnosis of chronic fluorosis have been investigated in clinic or basic laboratories. Serum enzymology and biological parameters in blood are often considered as an early diagnosis index of the disorder. Their detection methods are feasible and the results easily gotten since the assays are carried out by automatic biochemical detector. However, there are interference factors to the judgment of the development of the disease. The contents of calcium and collagen in the affected bone can reflect the damage of fluorosis, while the samples are difficult to be obtained. Some items in urine, including creatinine, urea nitrogen, calcium, phosphorus, and hydroxyproline, are considered as biochemical indicators for early diagnosis of endemic fluorosis, which can reflect the damage of chronic fluorosis on kidney and bone, respectively. In addition, the gene polymorphism or epigenetics can be used as index to predict susceptibility to the disease for the residents in area of endemic fluorosis. More importantly, the biological factors concerning oxidative stress in blood and urine,

Y.-J. Liu (✉)
Department of Pathology at the Affiliated Hospital of Guizhou Medical University, Guiyang, Guizhou, People's Republic of China
e-mail: gybllyj@gmc.edu.cn

Z.-Z. Guan
Department of Pathology at the Affiliated Hospital of Guizhou Medical University, Guiyang, Guizhou, People's Republic of China

The Key Laboratory of Endemic and Ethnic Diseases (Guizhou Medical University), Ministry of Education, Guiyang, Guizhou, People's Republic of China

Z.-Z. Guan (ed.), *Coal-burning Type of Endemic Fluorosis*,
https://doi.org/10.1007/978-981-16-1498-9_21

recognized by most researchers at present, are valuable for early diagnosis. Among them, the parameters relating to oxidative stress have high reference significance. For selecting the sensitive biomarkers to make an early diagnosis of chronic fluorosis, it is significantly valuable to detect these products of peroxidation or oxidation of lipid, protein or DNA, and the activities of antioxidative enzymes.

Keywords Biomarkers · Early diagnosis · Endemic fluorosis · Oxidative stress · Serum enzymology

Introduction

Endemic fluorosis causes systemic damage to the body. It affects bone, teeth, brain, liver, kidney, digestive tract, thyroid, and other vital organs of the body. In the world, there are over 35 counties with more than 200 million people live in the areas of endemic fluorosis with a potential risk, which may bring global concern on fluorosis research (Kabir et al. 2020). Generally, the clinical diagnosis of dental and skeletal fluorosis is mainly based on changes in teeth and bone tissues (Jiang et al. 2020). However, the early biochemical and metabolic changes (defined as biomarkers or indicators) of the body may occur before the typical changes in bone and teeth. Biomarkers are biochemical indicators that can mark the changes or possible changes in the structure or function of systems, organs, tissues, cells, and subcellular fractions (Mehta 2013) and can be used to diagnose a disease, determine the stage of a disease, or evaluate the safety and effectiveness of a new drug or therapy in a target population. The biomarker in endemic fluorosis could identify the specific biological alteration and be selected as one factor in the diagnosis or auxiliary diagnosis of fluorosis (Sampaio 2006). The susceptibility of biomarker to fluorosis is important for the individuals, which may indicate the related changes focusing on certain organ or system. Therefore, biomarker has become the much researched one for the early diagnosis of endemic fluorosis. It is of great clinical significance to establish the early diagnostic criteria for endemic fluorosis and avail patients to get effective treatment and rehabilitation. Although a large number of studies have addressed diagnostic criteria for endemic (or chronic) fluorosis, there is no clear consensus on early diagnostic indicators yet. Many studies have obtained valuable research results, which can be used for establishing early diagnosis of patients in future.

Clinical Indicators Closely Related to Diagnosing Endemic Fluorosis

Epidemiology

In the national classification standards for the areas of endemic fluorosis of China, if the prevalence rate of dental fluorosis among locally born and raised children aged 8–12 years old is more than 30%, and in drinking water the fluoride content is above 1 ppm, such area can be classified into the area with drinking water type of endemic fluorosis; if the total fluoride intake from burning coal is more than 3.5 mg/day, such area is labeled as the area with coal-burning type of endemic fluorosis.

In these areas, if prevalence of dental fluorosis is above 20% in 8–12 years old children, the patients have no skeletal fluorosis or only mild change of bone determined by imageological examination, and fluoride content is higher than 1 ppm in water or the total fluoride intake is more than 3.5 mg/day, such place may be considered as the mild endemic areas. If the prevalence of defective dental fluorosis is more than 30%, the patients have moderate skeletal fluorosis or the patients with a diagnosis as significant bone changes induced by fluoride are less than 2%, and the fluoride content is more than 2 ppm in water or total fluoride intake is more than 5.0 mg/day, the place is categorized as intermediate endemic area. If the prevalence of defective dental fluorosis is over 40%, the patients with severe skeletal changes of bone determined by imageological examination are more than 2%, the drinking water fluoride content is more than 4 ppm or the total fluoride intake is more than 7.0 mg/day, the area is classified as severe endemic area.

The people who live in the endemic fluorosis area for a long time have the potential danger of fluoride exposure. However, due to the difference of individuals exposed to fluoride varies, the response to the intake of excessive fluoride and susceptibility is different. Therefore, endemic fluorosis areas with different degrees of fluoride pollution are only chosen as auxiliary diagnostic indicators for fluoride damage after fluoride exposure. When the environmental fluoride content does not conform the individual condition, the diagnosis for endemic fluorosis should be subject to the actual condition of the resident. Epidemiological data are not very helpful in the early diagnosis of patients.

Many studies have shown that changed function and metabolism of multiple organs of the residents living in endemic fluorosis areas are closely related to fluorosis. An investigation attempted to define the correlation between fluoride concentrations in drinking water and liver function in populations with endemic fluorosis in India (Shashi and Bhardwaj 2011). It was found that serum fluoride concentration, cyclic adenosine phosphate (cAMP) content, and activities of alkaline phosphatase (ALP), acid phosphatase (ACP), aspartate aminotransferase (AST), and alanine aminotransferase (ALT) were significantly increased in patients with chronic fluorosis. The fluoride contents in water and serum were positively correlated with the biochemical indexes including AST, ALT, ALP, and ACP. The level of transaminases increased in patients with fluorosis, indicating a change in liver function.

Because ALP and ACP have specific catalytic activity on liver cells, the elevation of ALP and ACP levels caused by fluorosis can be considered as an early indicator of liver cell injury. Conversely, the changes in the activities of these enzymes also reflect the disorder of bone metabolism, which indicates the expression of increased activity of osteoblasts. Studies showed that chronic fluorosis enhanced the activity of liver functional enzymes. Chronic fluorosis has been shown to cause cardiovascular and renal dysfunction in the adult population. Fluoride exposure can lead to early vascular lesions in the population in the affected area, which increases the susceptibility of adults to cardiovascular diseases (Jiménez-Córdova et al. 2019).

Skeletal Fluorosis

Skeletal fluorosis was a toxic bone disease characterized by massive bone stiffness due to fluoride exposure (Sellami et al. 2020). Radiographic findings are primarily bony changes, including increased bone mass, ossification of the ligaments, and interosseous membrane, in which these changes may be confused with those seen in other rheumatic diseases. The major clinical manifestations of patients with skeletal fluorosis were stiffness of the joints, limb deformity, and pain (Majumdar and Sundarraj 2013). Patients with severe fluorosis may have spinal joint inactivity, often manifested as scoliosis, hunchback, limb stiffness, etc., which seriously affects the normal life of patients (Guan et al. 2017).

As for the early diagnosis of bone changes caused by endemic fluorosis, researchers studied bone mineral content, skeletal fluorosis degree, and cortical bone index, but still have not achieved satisfactory results. The X-ray manifestations of skeletal fluorosis include bone sclerosis, osteoporosis, osteomalacia, and ossification of parathyroid soft tissue. Imaging studies found that the patients had bone damage of different degrees. Some patients showed thickened bone trabeculae and gauze-shaped trabeculae or coarse network (Duan et al. 2012). In severe cases, the trabeculae fused and disappeared or replaced by osteosclerosis. It was also observed that patients with sparse and fuzzy trabeculae generally presented with the reduced bone density, and a double-concave wedge shape of the vertebral body.

In Guizhou of China, the area of coal-burning type of endemic fluorosis, four types of skeletal fluorosis can be identified among children and adolescents, i.e., dense (60.7%), osteoporotic (15.6%), mixed (16.4%), or softening (7.4%) skeletal fluorosis (Qin et al. 2009). Most diagnosed cases (91%) were mild or moderate in severity. For the fluorosis patients, the ossification of radial head annular ligament and the attachment of external and internal epicondylar extensor tendon of humerus were important characteristic signs of fluorosis, which could be used as the basis for X-ray diagnosis of fluorosis.

Micro-CT examination was a relatively new technique for clinical detection of bone phase change. It could be used to determine bone morphological parameters accurately and directly in micron level with high-resolution scanning, reconstruct trabecular three-dimensional imaging and unbiased structure. Micro-CT was used to observe bone damage in children, young people, and middle-aged patients with skeletal fluorosis (Li et al. 2014).

The use of radiographic data of skeletal fluorosis as a reference index for early diagnosis remains controversial, but it plays an important role as a basis to objectively reflect the damage of chronic fluorosis.

Dental Fluorosis

The first obvious clinical symptom of chronic fluorosis is dental fluorosis. The ameloblasts are very sensitive to toxicity of fluoride. The residents in endemic fluorosis areas can suffer from dental fluorosis exhibiting obvious spotted tooth when the total fluoride intake is more than 3.5 mg/day or content of fluoride in drinking water is above 1.0 ppm for a long term. The spot tooth rate of children in coal-burning type of fluorosis area can reach more than 50% and more than 20% in drinking water type of endemic fluorosis.

The use of dental fluorosis as a biomarker is limited because dental fluorosis is only formed by exposure to fluoride during the period of enamel formation (Den Besten 1994). Conversely, the severity of fluorosis is not only related to total fluoride intake, but also to the duration of fluorine exposure. Dental fluorosis does not determine whether they are still being afflicted by fluoride exposure or in a state of chronic fluorosis when they are grown up. Adolescents and adults who move into endemic fluorosis areas do not show dental fluorosis but may develop skeletal fluorosis and other systemic pathological lesions. Therefore, dental fluorosis can only be considered as an early indicator of chronic fluorosis in children, but not as a dependent indicator for the early diagnosis of endemic fluorosis in the whole population.

Fluoride Content in Different Organs or Tissues

Fluoride is thought to be an essential micronutrient that has some beneficial effects on the structural composition of teeth and bones (Styburski et al. 2017). Although fluoride had an anti-decay effect when it was applied partially to the teeth, more and more scientists considered that it was neither an essential trace element nor necessary for the development of healthy teeth and bones (Chouhan and Flora 2010; Shahab et al. 2017). Fluoride level in tissues and organs is often closely related to the degree of pathological damage induced by chronic fluorosis. With the upgraded detection methods and the increased sensitivity, fluoride content or its excretion of different tissues had certain role in the early diagnosis of chronic fluorosis (Kabir et al. 2020).

Fluoride Content in Blood

Fluoride exists in two forms after the absorption from blood: most of which binds to plasma albumin, and a few are free ions. Different forms of fluoride in blood can maintain a relative dynamic balance, which makes the level of fluoride in blood relatively constant. About 75% of the fluoride in blood is presented in plasma and about 25% in blood cells (mainly red blood cells). Fluoride content in plasma or serum is not affected by blood cell count, which is a more stable indicator. An animal study showed that the content of fluoride in serum or plasma was related to fluoride intake and excretion; the content of fluoride in fasting blood was parallel to that of urine and less affected by renal function. The fluoride level in blood may represent the content of fluoride exposure, in which fluoride content in serum from patients with chronic fluorosis was higher than that of the non-fluorosis population. So, fluoride content in blood is a useful index for early diagnosis of the chronic disorder with certain limits.

However, fluoride content in blood is also influenced by many factors, including inconsistent assay methods and sampling conditions. The correlation contents of fluoride with thyroid status in children from endemic fluorosis areas were evaluated by a research (Kumar et al. 2018). In the group with endemic fluorosis, the fluoride levels in serum and urine and thyroid stimulating hormone were significantly higher than those of controls. Furthermore, it was found that fluorosis was positively correlated with the status of thyroid function. But the thyroid hormone disorder was independent of the tooth eruption pattern.

Fluoride Content in Urine

Kidney is the most important organ of fluoride metabolism and about 75% of the fluoride in the body is excreted through urine. Urinary excretion is closely related to the deposition fluoride in the body. Under normal conditions, urinary fluoride content of non-fluorosis patients was less than 1 ppm, and patients with chronic fluorosis were often more than 2 ppm, or even 10 ppm (Elsair 1982). When urinary fluoride excretion increased, it often indicated that the body ingested excessive fluoride or accumulated more fluoride in the body. It was generally believed that the average level of urinary fluoride conformed to the incidence of dental fluorosis in a large number of residents (Ibiyemi et al. 2018). The fluoride level in urine is valuable index for the reliable diagnosis of fluorosis at present and a valuable monitoring index in the development of chronic fluorosis, which has certain reference value for the early diagnosis of individual chronic fluorosis.

The fluoride content in drinking water and urinary fluoride and indexes of bone metabolism of children living in the area of endemic fluorosis of southern China were studied after low-fluoride water was provided (Chen et al. 2013). The urinary fluoride and the levels of osteocalcin, calcitonin and ALP as well as bone mineral density were studied to understand the association of the most sensitive bone metabolic markers with fluoride exposure. The results showed that the fluoride concentration in drinking water in the intervention village was significantly decreased to 0.11 ppm after the fluoride content in the water source was reduced, which was close to the actual situation of the normal control village. Ten years after providing water with normal fluoride level, the urine fluoride contents of children in the intervention villages were close to those in the control villages. Children living in villages in the past endemic fluorosis area had similar skeletal indicators to those in the control village after the water fluorosis reduction intervention. The results indicated that the fluoride content of the drinking water in the original endemic fluorosis village in southern China had been reduced by reforming the water source, and the changes of various bone indexes of the local children were close to the normal level.

Eleven kinds of chemical elements in soil and pH value were investigated in two areas of coal-burning type of endemic fluorosis in Chongqing of China; at the same time, chemical elements in the blood from the local people were determined, which was used for analyzing the correlation between the relevant elements in soil and human blood and correlation with coal-burning fluorosis, as well as the relationship between these measurements and prevalence of dental fluorosis (Wang et al. 2014a). The results showed that Ni, I, F, Hg, and pH values in the soil were positively correlated with the prevalence of fluorosis in the local area, indicating that these soil parameters may be related to the occurrence and development of coal-burning type of fluorosis. The levels of copper, zinc, magnesium, and iron in the blood of children in the area of coal-burning type of endemic fluorosis were lower than those in control areas. The urinary fluoride levels of children with fluorosis were higher than those of health children in the same local and control areas. The levels of magnesium and iron in blood and urinary fluoride in adult patients with chronic fluorosis were higher than those of health persons in local and control areas. The area was deficient in elements that antagonize fluoride. However, due to differences in age, nutrition, individual functional status, and fluoride accumulation in the body, great impact on urinary fluoride excretion is presented. Thus, urine fluoride content as an early diagnostic index should exclude other factors.

Fluoride Levels in Bone, Nail, Hair, and Others

The fluoride contents in bone, nail, hair, and others have been considered as a diagnostic indicator of chronic fluorosis. Fluoride absorbed into the body mainly accumulates in bone tissue and it can represent the level of fluoride accumulation in the body. Fluoride concentrations are often elevated in volcanic eruptions due to the release of magma fluoride into the environment with the form of hydrogen fluoride. The study of active volcanic environment showed that the fluoride exposure in mountain environment was significantly correlated with the fluoride content in bone of Mus musculus, a wild breed, suggesting that the bone fluoride concentration could be used as a suitable biomarker for chronic environmental fluoride exposure. (Linhares et al. 2018). However, since bone tissue can only be obtained by biopsy clinically, which is not easily accepted by patients, it is difficult to carry out the examination.

The high fluoride concentrations of big toenail and hail of residents were found in the endemic fluorosis area as compared to those of the control region (Elekdag-Turk et al. 2019). Fingernails and hair have relatively stable levels of fluoride, making them an alternative biomarker for fluoride exposure. However, when the area under the curve (AUC) measurement was used to compare large toenail and hair samples, the sensitivity and specificity of the latter were lower than that of the former. Because individuals vary widely in the amount of fluoride ingested from drinking water, one study sought to identify the correlation of biomarkers of contemporary/recent fluoride exposure and children's total daily fluoride intake (TDFI) (Idowu et al. 2020). An investigation was carried out to assess some factors such as daily urinary fluoride excretion (UFE) and the levels of fluoride in plasma, fasting total saliva volume, hair and nails (toenails/fingernails) to explore their relationship with total fluoride exposure. The study found that TDFI, UFE, and fluoride concentrations in blood or different sources of organizations as biomarkers in high fluoride areas were significantly higher than those in low-fluoride areas. TDFI, UFE, plasma, and toenail fluoride concentrations were all positively correlated. Among them, TDFI was moderately positively correlated with fingernail fluoride concentration and fasting saliva total fluoride concentration and weakly with hair fluoride concentration. The study also confirmed the usefulness of 24-h urine samples for estimating fluoride exposure in children. Since children are a vulnerable population for the development of dental fluorosis, the strong association between TDFI and fluoride in plasma and toenails also suggests that these factors may serve as biomarkers to assess the health risk of fluorosis in children. It was established that TDFI was positively correlated with the amount of fluoride in nail clippings (Linhares et al. 2016). The fluoride level in nail clipping could be considered as a more reliable biomarker for chronic fluoride exposure than that in urine. In addition, the content of fluoride in nail clipping as a biomarker may be more appropriate for assessing long-term fluoride exposure in the areas where the fluoride content in drinking water is not too high.

Biological Parameters Correlated to Chronic Fluorosis

Biological parameters are biochemical indicators (the so-called biomarkers). They can mark possible changes in the structure or function at different levels, such as systems, organs, tissues, cells, or subcellular structures. Biomarkers can be used for disease diagnosis, disease staging, or evaluating the safety and efficacy of new drugs or therapies in the target population. Concerning environmental toxicology as the early diagnosis of toxic damage, the study of biomarkers has received increasing attention.

Serum Enzymology

Blood biochemical detection is important in the study of fluorosis-related biomarkers. Blood routine examination, specific enzyme activity, protein content, and blood electrolyte concentration are all necessary indicators for the diagnosis of endemic fluorosis.

Numerous studies observed significantly abnormal changes in oxidative stress-related enzymatic activity in multiple tissues or organs of chronic fluorosis patients, in which the activities of superoxide dismutase (SOD) and glutathione peroxidase (GSH-Px) were reduced (Guan 2014). Experimental results showed that excessive fluoride intake led to the decrease in total antioxidant capacity and the increase in malondialdehyde (MDA) as a metabolite product of lipid peroxidation, which was correlated negatively with the decreased SOD activity. The increased activity of myeloperoxidase (MPO) and the decreased glutathione-S-transferase (GSTs) in plasma were observed in the patients with coal-burning type of endemic fluorosis (Zhang et al. 2013).

The correlation of fluoride level in drinking water and hepatic function of the residents in endemic fluorosis areas of Punjab (India) was investigated (Shashi and Bhardwaj 2011). The results revealed the mean activities of cyclic adenosine monophosphate (AMP), ALP, ACP, AST, and ALT were significantly increased in fluorosis patients from all endemic fluorosis areas, suggesting the abnormal liver function in the patients with chronic fluorosis in these areas. ALP and ACP have specific and catalytic activity and can be considered as early markers of liver cell damage caused by chronic fluorosis. The elevated enzyme levels also reflect a bone disorder characterized by increased osteoblastic activity.

However, in an experiment of rats with chronic fluorosis, the activities of ALT and AST and the ratio of ALT/AST were obviously declined as compared to controls (He et al. 2014). At the same time, the level of activated transcription factor 4 (ATF4) decreased with the elevating content of fluoride, exhibiting a negative relationship between ATF4 content and fluoride exposure level. The results suggested that ALT and AST could not be considered as biomarkers of high fluoride exposure, while protein ATF4 might be an earlier sensitive indicator in epidemiology of chronic fluorosis.

Mineral Content in Blood

Some minerals in blood can indirectly reflect the condition of bone metabolism. Intake of high fluoride for a long period results in the formation of fluorapatite due to the replacement of hydroxyl ($OH^{-\bullet}$) in carbonate hydroxyapatite with fluoride ion (F^{-}). Due to the replacement of $OH^{-\bullet}$ by F^{-} in apatite, the metabolism of calcium and phosphorus is disturbed, the crystallization degree of calcium apatite is increased, the bone strength is reduced, and calcium and phosphorus are excreted from the bone. Such results lead to the decrease in bone inorganic salt content and the increase in organic matter ratio and result in bone structure transformation. Current studies showed that fluoride and calcium could form insoluble calcium fluoride precipitation, accelerate calcium salt deposition, which resulted in the decreased blood calcium concentration and increased blood phosphorus. Some studies showed that the changes of calcium and phosphorus in blood were not obvious in early stage of fluoride intake. However, changes in calcium and phosphorus in blood might occur in the early stages of chronic fluorosis under conditions of poor nutritional.

The following microelements, such as copper and zinc, and minerals, such as calcium, phosphorus, magnesium, sodium, and potassium, in serum of the fluorosis patients in area of endemic fluorosis were investigated (Ersoy et al. 2011). The results indicated that the level of copper, zinc, and magnesium in serum of patients with chronic fluorosis was significantly lower than those in controls, while the levels of sodium, potassium, calcium, and phosphorus did not exhibit significant changes. Studies concerning case-control were used as analyzing the etiology and pathology of bone deformities in children in areas of endemic fluorosis (Khandare et al. 2005). As compared with healthy children in non-fluorosis areas, the blood levels of some elements in children with chronic fluorosis were decreased, such as copper decreased by 14 times, calcium and zinc by 2.5 times, and selenium by 2 times (Tkachenko et al. 2021). On the contrary, as compared with the normal control group, the strontium and lead levels in blood of the children with chronic fluorosis significantly increased by 1.7 times and 1.4 times, respectively; the levels of sulfur, chlorine, and potassium were all lower than the reference values, and the levels of bromine and selenium were positively correlated with the levels of acyl hydrogen peroxide.

In children with skeletal deformity, the levels of serum 25-OH vitamin D3 and calcium level significantly declined, and ALP activity and intact parathyroid hormone (iPTH) level significantly higher as compared to controls. There were no significant differences in serum and urinary creatinine, copper and magnesium in patients with chronic fluorosis compared with the control group. The levels of ALP, calcium, and inorganic potassium in serum of children with endemic fluorosis were determined as the biochemical parameters (Ahmad et al. 2003). Higher serum ALP and calcium and lower phosphorus were detected in children with dental fluorosis, indicating that serum calcium and phosphorus may play a major role in the occurrence and development of skeletal fluorosis.

Whether the mineral content in blood can be used as early diagnosis of endemic fluorosis requires further study.

Changed Genetic Polymorphisms and Epigenetics

Polymorphism is the simultaneous and often discontinuous presence of two or more genotypes or allele in a biological population, also known as genetic polymorphism. Essentially, polymorphisms occur at the gene level, typically in regions of the gene sequence that do not encode proteins and do not have important regulatory functions. Studies on genetic epidemiology confirmed that genetic polymorphisms of candidate genes in individuals exposed to the same community and environment were associated with different types of fluorosis susceptibility (Pramanik and Saha 2017).

MPO, a member of the heme peroxidase superfamily, is secreted by activated neutrophils and active specie-producing enzymes that catalyze the production of hypochlorous acid to achieve host defense (Maruyama et al. 2004). Overexpression of MPO can cause hydrogen peroxide to produce more active molecules such as hypochlorous acid, which may stimulate lipid peroxidation. In previous study, the MPO activity, its mRNA expression and gene polymorphism in blood of residents in the area of coal-burning type of endemic fluorosis were investigated (Zhang et al. 2013). The results showed that as compared to non-fluorosis patients the blood MPO activity and mRNA level of the residents in the area of coal-burning type of endemic fluorosis were significantly higher. The results also showed that MPO genotypes with different frequencies (-1228g/A, -585t/C, -463g/A, -163c/T) and three haplotypes with higher frequencies (-163C-463G-585T-1228G-1276T, -163C-463G-585T-1228G-1276C, -163C-463G-585T 1228A 1276T) were obviously correlated to chronic fluorosis, which may be speculated that the increase of MPO activity caused by endemic fluorosis may be related to the stimulated MPO mRNA expression and the change of gene polymorphism. Glutathione S-transferase (GST) catalyzed the combinative reaction by nucleophilic glutathione with various electrophilic exogenous chemicals, which could prevent the exogenous chemicals from causing damage to the organism. A report concerning the correlation of GSTP1 rs1695 polymorphisms with the risk of brick tea type of endemic fluorosis indicated the difference of the prevalence rate in different ethnical groups (Wu et al. 2015). The numbers of homozygous wild type (GSTP1-AA) genotype of different ethnical participants (Tibetan, Mongolian, and Han subjects) were significantly higher than those of Kazakh and Russian participants. Participants carrying the G allele had a significantly reduced risk of skeletal fluorosis compared with Tibetan subjects carrying the homozygous GSTP1 Rs1695 allele. Therefore, it was suggested that G allele of GSTP1 Rs1695 might be a protective factor for brick tea type of skeletal fluorosis. FRZB1 rs2242070 AA genotype was only related to brick tea type of skeletal fluorosis in Kazakhs, instead of Tibetans (Yang et al. 2018).

Collagen was the main component in connective tissue and the most abundant and widely distributed functional protein in the body. The association of alleles and genotype frequencies of polymorphisms in the collagen type I α2 chain (COL1A2) gene rs412777 with the severity of child dental fluorosis from San Luis Potosi, Mexico was investigated (Jarquín-Yñezá et al. 2018). Fluoride concentrations in drinking water of the area ranged from 2.36 to 4.56 ppm, urine fluoride from 2.05 to 2.99 ppm, dental fluorosis prevalence was 100%, and COL1A2 allele was 67% wild type and 33% mutated. There was a significant correlation between the severity of dental fluorosis and the COL1A2 allele polymorphism in local residents. The conclusion of this study indicates that rs412777 polymorphism in the COL1A2 gene is associated with dental fluorosis as a related risk factor for the occurrence.

Val158Met polymorphism of the catechol O-methyltransferase (COMT) gene (rs4680) was found to be associated with cognitive performance. Since fluoride exposure is inversely associated with intelligence in children, it is suggested that the COMT gene VAL158MET polymorphism may increase the susceptibility to intelligence (IQ) deficits caused by fluoride exposure (Zhang et al. 2015). As compared to all participants, urine fluoride levels were more obviously associated with child IQ carrying dopamine receptor-2 Taq 1A TT genotype (Cui et al. 2018).

The effects of fluoride exposure and estrogen receptor α (ERα) gene polymorphism on reproductive hormone concentrations in women in endemic fluorosis areas have been highlighted (Zhou et al. 2016). An investigation found that serum gonadotropin-releasing hormone (GnRH) was lower in C/C genotypes than that in C/T and T/T genotypes in women with ERαrs3798577 genotype. In addition, serum GnRH of EFG women with ERα rs3798577 genotype C/C, C/T, or T/T was significantly lower than that of CG women. Serum GnRH of EFG women with PP, pp, XX and xx genotypes in PVUII and XBAI of ERα gene was lower than that of those with CG genotypes. The serum estradiol concentration of EFG was significantly lower than that of CG in women with PP, XX, and T/T genotypes in ERα gene. In the endemic fluorosis group, the serum GnRH of women with Pp genotype was lower than that of women with PP and pp in ERα PVUII genotype. The interaction between ERα gene and fluoride exposure may affect serum reproductive hormone concentrations in women. The association between ESR1, COL1A2, BGALAP, and SPARC gene polymorphisms and the risk of dental fluorosis was studied in 87 individuals from an endemic region of East India (Saha et al. 2021). Ten single nucleotide polymorphisms (SNPs) in ESR1 (rs2234693, rs2228480, rs3798577, rs2077647, and rs9340799), COL1A2 (rs42524 and rs412777), BGLAP (rs1800247), and SPARC (rs6579885 and rs4958278) were detected by PCR-RFLP. No significant association of the 10 SNPs was found with the risk of dental fluorosis, nor was there any change in the haplotype frequency of SNPs in ESR1, COL1A2, and SPARC genes. The haplotype frequency of SNPs in ESR1, COL1A2, and SPARC genes was not significantly different.

In biology, epigenetics refers to the "heritable" phenomenon of phenotypes altered by non-DNA mutations, which concerns DNA methylation, histone modification, chromatin remodeling, and regulation of non-coding RNA.

The level of DNA methylation was analyzed in the peripheral blood cells from the children with or without skeletal fluorosis in the area of coal-burning type of endemic fluorosis in Guizhou of China (Deng et al. 2019). The results showed that 15 differential hypermethylated and 22 hypomethylated CpG sites were determined in children with skeletal fluorosis, which involved 29 known genes. The majority of differentially methylated CpG loci was residing in gene body. Further GO and KEGG database for pathway analysis indicated that these differential genes were obviously related to synapse maturation, axon, neuron projection membrane, transcription coactivator activity, and notch signaling pathway. Pyrosequencing analysis showed changed hypomethylated gene MAML2 in children with skeletal fluorosis. A change concerning genome-wide DNA methylation in children with skeletal fluorosis was also found, suggesting that the changes of hypomethylation of

MAML2 may play an important role in the development procedure of skeletal fluorosis. The methylation rates of O^6-methylguanine-DNA-methyltransferase (MGMT), DNA repair gene, and DNA mismatch repair gene Mutl Homolog 1 (MLH1) were analyzed by specific PCR methylation method to explore the relationship between DNA repair gene methylation and chronic fluorosis in coal-burning type of endemic fluorosis area (Wu et al. 2018). It was observed that the methylation degree of MGMT and MLH1 genes in the blood of patients with endemic fluorosis increased, but their mRNA levels decreased, which were significantly correlated with the severity of fluorosis. The protein and mRNA levels of MGMT and MLH1 genes in blood and liver tissues of rats exposed to fluoride were also significantly reduced. These results suggest that the methylation of MGMT and MLH1 genes is altered by fluorosis, and the modified expression of these repair genes may be one of the mechanisms of liver injury induced by fluorosis.

It is important to note that an individual's genetic background may play an important role in influencing the risk of the development of chronic fluorosis when other factors such as specific environmental fluoride exposures, dietary patterns of fluoride, and intake of other nutrients remain unchanged (Pramanik and Saha 2017).

Oxidative Stress Indicators Related to Chronic Fluorosis

In early studies, the elevated level of lipid peroxide and reduced activities of antioxidant enzymes had been determined in serum and a variety of organs of animals with chronic fluorosis, and in serum of patients suffering from coal-burning type of endemic fluorosis (Guan et al. 1989, 1991, 1992). On the basis of such findings, one of such early studies first proposed that the increased level of free radicals might be a central link of different types of damage to the body associated with chronic fluorosis (Guan et al. 1989).

In order to understand the environmental characteristics of coal-burning type of endemic fluorosis in Bijie of China, the effects of oxidative stress on body injury were studied by monitoring fluoride content in air, water, food, coal, and clay binder samples (Wang et al. 2014b). Serum activities of copper/zinc SOD (Cu/Zn-SOD), catalase (CAT), glutathione peroxidase (GSH-Px), and lipid peroxidation (MDA) in endemic fluorosis patients were determined. In the results, the increased level of malondialdehyde (MDA) and the inhibited activity of antioxidant enzyme and the expression of Cu/Zn-SOD gene were found in the patients with endemic fluorosis. Chronic fluorosis decreased the activity of antioxidant enzymes in vivo, which was related to the level of fluoride exposure. The downregulation of Cu/Zn-SOD expression may play an aggravating role in the damage caused by oxidative stress of endemic fluorosis. The biomarkers of lipid peroxidation, such as acyl hydroperoxides and 2-thiobarbituric acid reaction substances (TBARS) were used as the investigation for the parameter in blood of children with chronic fluorosis in the endemic fluorosis areas (Tkachenko et al. 2021). The study showed that blood TBARS level was higher in children living in endemic fluorosis areas, while acyl hydroperoxide did not significantly change, as compared with healthy children living in non-fluorosis areas. Oxidative stress products in serum of residents living in coal-burning type of endemic fluorosis area were detected (Chen et al. 2017). The content of 8-hydroxy-2′-deoxyguanosine (8-OHdG), 8-iso-prostaglandin F2α (8-iso-PGF2α), and advanced oxidation protein products such as carbonyl group in the adults living in such area were significantly higher than that of controls. The children with minor dental fluorosis showed a significantly higher serum 8-iso-PGF2α than that of the children from non-endemic area. Since 8-iso-PGF2α was a stable and specific oxidation product produced by active radical attack on lipid peroxidation of arachidonic acid on biofilm, it could demonstrate early oxidative stress and might be used as a predictor for the early diagnosis of fluorosis.

Conclusion

A large number of studies have been conducted to screen biomarkers for the early diagnosis of chronic fluorosis. These biomarkers, including fluoride content, related enzymes, proteins, trace and major elements in blood and urine or fingernail have higher value for the early diagnosis of endemic fluorosis.

Serum enzymology and biological parameters in blood are often considered as an early diagnostic index of the disorder. Their detection methods are feasible and the results easily gotten since the assays are carried out by automatic biochemical detector. However, there are interference factors to the judgment of the development of the disease. The contents of calcium and collagen in the affected bone can reflect the damage of fluorosis, while the samples are difficult to be obtained. Some items in urine, including creatinine, urea nitrogen, calcium, phosphorus, and hydroxyproline, are considered as biochemical indicators for early diagnosis of endemic fluorosis, which can reflect the damage of chronic fluorosis on kidney and bone, respectively. In addition, the gene polymorphism or epigenetics can be used as index to predict susceptibility to the disease for the residents in the area of endemic fluorosis.

Among them, the parameters relating to oxidative stress have high reference significance. For selecting the sensitive biomarkers to make an early diagnosis of chronic fluorosis, it is significantly valuable to detect these products of peroxidation or oxidation, such as MDA and 8-iso-PGF2α (the metabolic products of lipid peroxidation), carbonyl group (a product of protein oxidation), and 8-OHdG (a primary marker after DNA oxidative damage) and the activities of antioxidative enzymes, such as SOD, CAT, and GSH-Px.

However, it is still necessary to make unremitting efforts to find sensitive biomarkers for the early diagnosis of endemic fluorosis, especially the signs of body damage before the occurrence of skeletal and dental fluorosis.

References

Ahmad M, Ahmad B, Nawazish-i-Hussain S, Mahmood S. Clinical investigations of skeletal fluorosis in children of Manga Mandi in Pakistan. Pak J Pharm Sci. 2003;16:9–11.

Chen S, Li B, Lin S, Huang Y, Zhao X, Zhang M, et al. Change of urinary fluoride and bone metabolism indicators in the endemic fluorosis areas of southern China after supplying low fluoride public water. BMC Public Health. 2013;13:156.

Chen XW, Wan CW, Zhang SY, Luo Y, Qi XL, Li Y. Serum oxidative stress products among residents living in endemic area of coal-burning fluorosis. Chin J Public Health. 2017;33:824–7.

Chouhan S, Flora SJ. Arsenic and fluoride: two major ground water pollutants. Indian J Exp Biol. 2010;48:666–78.

Cui Y, Zhang B, Ma J, Wang Y, Zhao L, Hou C, et al. Dopamine receptor D2 gene polymorphism, urine fluoride, and intelligence impairment of children in China: a school-based cross-sectional study. Ecotoxicol Environ Saf. 2018;165:270–7.

Den Besten PK. Dental fluorosis: its use as a biomarker. Adv Dent Res. 1994;8:105–10.

Deng J, Zeng XX, Liu XH, Chen D, Dong YT, Song H, et al. Screening and verification of differential DNA methylation genes in the peripheral blood cells of children with skeletal fluorosis in Guizhou province of China. Fluoride. 2019;52:135–45.

Duan QH, Chen XG, Wang XL, Jiao J, Hu J, Guan ZZ, et al. Diagnostic value of ossification at tendon attachment of ligament around elbow joint in fluorosis. J Pract Radiol. 2012;28:1738–40.

Elekdag-Turk S, Almuzian M, Turk T, Buzalaf MAR, Alnuaimi A, Dalci O, et al. Big toenail and hair samples as biomarkers for fluoride exposure - a pilot study. BMC Oral Health. 2019;19:82.

Elsair J. Fluoride content of urine, blood, nails and hair in endemic skeletal fluorosis. Fluoride. 1982;15:43–7.

Ersoy IH, Koroglu BK, Varol S, Ersoy S, Varol E, Aylak F, et al. Serum copper, zinc, and magnesium levels in patients with chronic fluorosis. Biol Trace Elem Res. 2011;143:619–24.

Guan ZZ. Research progress of oxidative stress in the pathogenesis of chronic fluorosis. Chin J Epidemiol. 2014;32:43–6.

Guan ZZ, Yang PS, Yu ND, Zhuang ZJ. An experimental study of blood biochemical diagnostic indices for chronic fluorosis. Fluoride. 1989;22:112–8.

Guan ZZ, Yang PS, Wang YN, Wei ZD, Zhou LY, Li FC. Study on antiaging index in plasma and erythrocytes of adult patients with fluorosis and aluminum poisoning. Guizhou Med. 1991;15:265–7.

Guan ZZ, Yang PS, Su YS, Wang YN. Changes of lipid peroxide content and anti-oxidant level in blood of the children living in the F-AI combined toxicosis area of Shuicheng of Guizhou. J Guiyang Med Coll. 1992;16:198–200.

Guan ZZ, Wang LH, Sun DJ. Endemic fluorosis. In: Sun DJ, editor. Endemic disease in China. Beijing: People's Medical Publishing House; 2017. p. 73–116.

He H, Wang H, Han M, Jiao Y, Ma C, Zhang H, et al. Study on changes of clinical indicators and key proteins from fluoride exposure. Biol Trace Elem Res. 2014;160:73–8.

Ibiyemi O, Zohoori FV, Valentine RA, Maguire A. Fluoride intake and urinary fluoride excretion in 4- and 8-year-old children living in urban and rural areas of Southwest Nigeria. Community Dent Oral Epidemiol. 2018;46:482–91.

Idowu OS, Duckworth RM, Valentine RA, Zohoori FV. Biomarkers for the assessment of fluoride exposure in children. Caries Res. 2020;54:134–43.

Jarquín-Yñezá L, Alegría-Torres JA, Castillo CG, de Jesús Mejía-Saavedra J. Dental fluorosis and a polymorphism in the COL1A2 gene in Mexican children. Arch Oral Biol. 2018;96:21–5.

Jiang N, Guo F, Sun B, Zhang X, Xu H. Different effects of fluoride exposure on the three major bone cell types. Biol Trace Elem Res. 2020;193:226–33.

Jiménez-Córdova MI, González-Horta C, Ayllón-Vergara JC, Arreola-Mendoza L, Aguilar-Madrid G, Villareal-Vega EE, et al. Evaluation of vascular and kidney injury biomarkers in Mexican children exposed to inorganic fluoride. Environ Res. 2019;169:220–8.

Kabir H, Gupta AK, Tripathy S. Fluoride and human health: systematic appraisal of sources, exposures, metabolism, and toxicity. Crit Rev Environ Sci Technol. 2020;50:1116–93.

Khandare AL, Harikumar R, Sivakumar B. Severe bone deformities in young children from vitamin D deficiency and fluorosis in Bihar-India. Calcif Tissue Int. 2005;76:412–8.

Kumar V, Chahar P, Kajjari S, Rahman F, Bansal DK, Kapadia JM. Fluoride, thyroid hormone derangements and its correlation with tooth eruption pattern among the pediatric population from endemic and non-endemic fluorosis areas. J Contemp Dent Pract. 2018;19:1512–6.

Li FC, Gao K, Cong XZ, Cong XZ, Yan SL. Analysis of bone micro-CT results of patients with bone deformation exposed to high aluminum high fluoride. Chin J Epidemiol. 2014;33:561–4.

Linhares DPS, Garcia PV, Amaral L, Ferreira T, Cury JA, Vieira W, et al. Sensitivity of two biomarkers for biomonitoring exposure to fluoride in children and women: a study in a volcanic area. Chemosphere. 2016;155:614–20.

Linhares D, Camarinho R, Garcia PV, Rodrigues ADS. Mus musculus bone fluoride concentration as a useful biomarker for risk assessment of skeletal fluorosis in volcanic areas. Chemosphere. 2018;205:540–4.

Majumdar KK, Sundarraj SN. Health impact of supplying safe drinking water on patients having various clinical manifestations of fluorosis in an endemic village of west bengal. J Family Med Prim Care. 2013;2:74–8.

Maruyama Y, Lindholm B, Stenvinkel P. Inflammation and oxidative stress in ESRD – The role of myeloperoxidase. J Nephrol. 2004;17:S72–6.

Mehta A. Biomarkers of fluoride exposure in human body. Ind J Dent. 2013;4:1–4.

Pramanik S, Saha D. The genetic influence in fluorosis. Environ Toxicol Pharmacol. 2017;56:157–62.

Qin X, Wang S, Yu M, Zhang L, Li X, Zuo Z, et al. Child skeletal fluorosis from indoor burning of coal in southwestern China. J Environ Public Health. 2009;2009:969764.

Saha D, Goswami R, Majumdar KK, Sikdar N, Pramanik S. Evaluating the association between dental fluorosis and polymorphisms in bone development and mineralization genes among population from a fluoride endemic region of eastern India. Biol Trace Elem Res. 2021;199:1–8.

Sampaio FC. Biomarkers of exposure to fluoride. J Appl Oral Sci. 2006;14:41–3.

Sellami M, Riahi H, Maatallah K, Ferjani H, Bouaziz MC, Ladeb MF. Skeletal fluorosis: don't miss the diagnosis! Skelet Radiol. 2020;49:345–57.

Shahab S, Mustafa G, Khan I, Zahid M, Yasinzai M, Ameer N, et al. Effects of fluoride ion toxicity on animals, plants and soil health: a review. Fluoride. 2017;50:393–408.

Shashi A, Bhardwaj M. Study on blood biochemical diagnostic indices for hepatic function biomarkers in endemic skeletal fluorosis. Biol Trace Elem Res. 2011;143:803–114.

Styburski D, Baranowska-Bosiacka I, Goschorska M, Chlubek D, Gutowska I. Beer as a rich source of fluoride delivered into the body. Biol Trace Elem Res. 2017;177:404–8.

Tkachenko H, Kurhaluk N, Skaletska N, Maksin V, Osadowski Z. Elemental status and lipid peroxidation in the blood of children with endemic fluorosis. Biol Trace Elem Res. 2021;199:1237–45.

Wang H, Mu L, Jiang M, Wang Y, Yan W, Jiao Y. The relationship between chemical elements in soil and whole blood, and fluorosis induced by coal-fired pollution. Environ Monit Assess. 2014a;186:2081–8.

Wang Q, Cui KP, Xu YY, Gao YL, Zhao J, Li DS, et al. Coal-burning endemic fluorosis is associated with reduced activity in antioxidative enzymes and Cu/Zn-SOD gene expression. Environ Geochem Health. 2014b;36:107–15.

Wu J, Wang W, Liu Y, Sun J, Ye Y, Li B, et al. Modifying role of gstp1 polymorphism on the association between tea fluoride exposure and the brick-tea type fluorosis. PLoS One. 2015;10: e0128280.

Wu CX, Wang YH, Li Y, Guan ZZ, Qi XL. Changes of DNA repair gene methylation in blood of chronic fluorosis patients and rats. J Trace Elem Med Biol. 2018;50:223–8.

Yang Y, Zhao Q, Liu Y, Liu X, Chu Y, Yan H, et al. FRZB1 rs2242070 polymorphisms is associated with brick tea type skeletal fluorosis in Kazakhs, but not in Tibetans, China. Arch Toxicol. 2018;92:2217–25.

Zhang T, Shan KR, Tu X, He Y, Pei JJ, Guan ZZ. Myeloperoxidase activity and its corresponding mRNA expression as well as gene polymorphism in the population living in the coal-burning endemic fluorosis area in Guizhou of China. Biol Trace Elem Res. 2013;152:379–86.

Zhang S, Zhang X, Liu H, Qu W, Guan Z, Zeng Q, et al. Modifying effect of COMT gene polymorphism and a predictive role for proteomics analysis in children's intelligence in endemic fluorosis area in Tianjin. Toxicol Sci. 2015;144:238–45.

Zhou GY, Ren LJ, Hou JX, Cui LX, Ding Z, Cheng XM, et al. Endemic fluorosis in Henan province, China: ERα gene polymorphisms and reproductive hormones among women. Asia Pac J Clin Nutr. 2016;25:911–9.

Chapter 22
Comprehensive Prevention and Control of Coal-burning Type of Endemic Fluorosis

Fu-Cheng Li and Chang-Xue Wu

Abstract Coal-burning type of endemic fluorosis is an endemic fluorosis unique to China. In addition to the environmental factors, its prevalence and control are closely related to the social and economic development, which features it as a poverty and underdevelopment caused disease. Concerning its prevention, control, and elimination, Chinese government draws experience and framework from other cases of endemic disease control. With integral effort between government departments and organizations, a network of disease control is established with collective effort. The fluorosis prevention and control are based on fluorosis prevention education, coupled with fluorosis prevention from heat source and stove improvement projects. With the assistance of *Chinese National Poverty Alleviation Project*, as well as the special fund support, fluorosis prevention publicity and education were promoted greatly in local society, and effects of these measures were paid off. This chapter focuses on implementation, evaluation, and acceptance of prevention and control measures of coal-burning type of endemic fluorosis.

Keywords Coal-burning contamination · Comprehensive measure · Education · Fluorosis · Prevention and control

---

F.-C. Li
The Key Laboratory of Endemic and Ethnic Diseases (Guizhou Medical University) of the Ministry of Education, Guiyang, People's Republic of China

Liupanshui Center for Disease Control and Prevention, Liupanshui, People's Republic of China

C.-X. Wu (✉)
The Key Laboratory of Endemic and Ethnic Diseases (Guizhou Medical University), Ministry of Education, Guiyang, Guizhou, People's Republic of China

The Provincial Key Laboratory of Medical Molecular Biology, Guiyang, Guizhou, People's Republic of China
e-mail: wuchangxue@gmc.edu.cn

Z.-Z. Guan (ed.), *Coal-burning Type of Endemic Fluorosis*,
https://doi.org/10.1007/978-981-16-1498-9_22

Introduction

Coal-burning type of endemic fluorosis is an endemic fluorosis unique to China. In addition to the environmental factors, its prevalence and control are closely related to the social and economic development, which features it as a poverty and underdevelopment caused disease (Li et al. 2007a, b; Xu et al. 2016). Prevention and control of coal-burning type of endemic fluorosis were executed by the related health departments in the Centers for Disease Control and Prevention (CDC), participated by research institutes and people from endemic area under the guidance of central government and health departments. For now, a comprehensive prevention and control system was established, featuring health education as basis, stove improvement project as the major measure, complemented by effective measures to block fluorosis contamination. As the government was working to promote the social-economic development, the government also adopted pertinent health education and interference measures: encouraging people from endemic areas to participate in the prevention measures, improving their awareness and methods of fluorosis prevention for solid healthy behaviors. These measures contributed to the sustainable control of this disease. For years, the central and local governments in China issued a series of preferential policies and fiscal fund for the prevention and control of coal-burning type of endemic fluorosis.

A National Measure Proposed by Government and Implemented by Medical Staff and the General Public

Since 1949, the central government has been assigning task teams to supervise the prevention and control of this endemic disease from national level. The task teams of all levels oversaw coordinating funds, personnel, prevention, medical treatment, scientific research, and supplies. Later, the prevention and control related work were reassigned to health departments.

Education on Prevention of Chronic Fluorosis

The health education publicity aims to help people correct poor health related activities and habits voluntarily through well designed, organized, and systematic societal or educational activities. With the elimination and alleviation of risk factors affecting health negatively, it is hopeful to form a living style which benefits health and improves quality of life. In coal-burning type of endemic fluorosis area, the following health education activities were often conducted.

Selected Tips of Fluorosis Prevention

Prevention and control tips on coal-burning type of endemic fluorosis concentrate on cause and prerequisite of fluorosis, main epidemiological factors, and requirements for fluorosis elimination in target population. Yellow teeth, arthralgia, skeletal deformity, and extension failure were all included as classic symptoms of endemic fluorosis. Local people should be aware that such disease is preventable yet not curable, open stove would contaminate the indoor air and cause fluorosis.

Preparation and Make of Fluorosis Prevention Education Materials

Fluorosis prevention education materials were prepared in accordance with target population's habit, concerning their specific conditions. Such materials were evaluated for its effect for further improvement and effectiveness.

Tips of Publicizing Fluorosis Prevention

Under the instruction of task team, local mass media, schools and villages carried out various publicizing activities. A catchy rhyme in Chinese language was composed for publicizing, which generally meant "Open stove dehydrating leads to fluorosis; solarized food is the best for prevention of fluorosis; raw feed for swine saves money and coal; exhaust pipe can get rid of soot and contamination; fluorosis prevention contributes to better life; damaged stove or its parts must be repaired in time to avoid soot contamination; foods must not hang over the stoves, air-tight storage can avoid soot contamination; rinsing before cooking." Such publicizing lend support to the fluorosis prevention project and change of living habit. It also laid a solid foundation for the intervention measures.

Publicity of Fluorosis Prevention via Media and Giveaways

Media resources in endemic areas (broadcast, TV program, newspaper, text-message, wall posters, outdoor printing, etc.) were fully utilized to present knowledge on prevention and control of coal-burning type of endemic fluorosis (Fig. 22.1). With the features of wide coverage and fast transmission, mass media became the major approach to publicize such knowledge in no time. With diversified publicity and activities, prevention tips and knowledge concerning the endemic aims to help

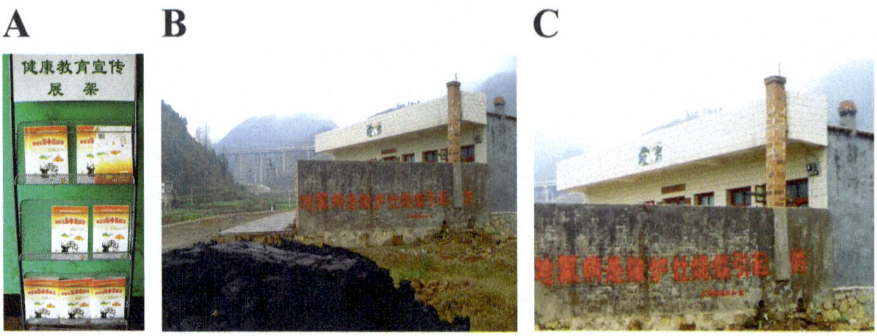

Fig. 22.1 Media resources (such as calendar, and brochure) utilized to present knowledge on prevention and control of endemic fluorosis. (**a**) Brochures showing the prevention and control of endemic fluorosis are available free of charge to local residents. (**b**) The propaganda on the wall beside the coal pile shows that endemic fluorosis is caused by burning coal in open stoves

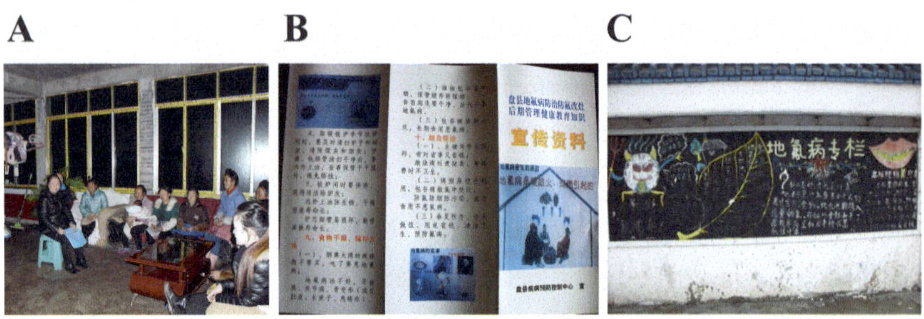

Fig. 22.2 The major tips on fluorosis prevention for housewives in Miao ethnic group. (**a**) A doctor with pamphlets teaches Miao housewives how to prevent fluorosis. (**b**) The pamphlets on the prevention of endemic fluorosis used by the doctor

people switch habits (heating and dehydrating corn/pepper on open stove) voluntarily.

Education on Fluorosis Prevention Taken by Housewives in the Endemic Areas

In the village, women are not only indispensable parts of farm work, but also the pillar of housework. Since they are more likely to stay at home, they are more prone to be the victim of soot contamination. The fluorosis prevention education for local women took the form of lecture and discussion in simple and straightforward dialect (Fig. 22.2). In addition, video-clips, brochures, and physical check were adopted to explain the danger of coal-burning type of endemic fluorosis. Such publicizing work was conducted by medical staff and officials in the village who had received fluorosis prevention training. With their introduction to the severity of endemic fluorosis, prevention and control measures, government subsidy policy on improved stoves project, it left a deep impression of local women, and they would serve as agents to pass the fluorosis prevention tips to each family member.

Education on Fluorosis Prevention Carried Out at School

Students in primary and secondary schools are fast learners. With the homogeneity in age group and social experience, together with the group living style and close family ties, it was ideal to organize lectures for students and then impart the knowledge to their family members (Fig. 22.3). Local education authorities organized fluorosis prevention courses for primary and secondary schools: local students from primary and secondary schools are required to have two hours fluorosis prevention course each semester with textbooks provided by local CDC. Multiple means were chosen to enhance fluorosis prevention tips for students: class meeting, wall posters, blackboard newspaper, themed essay, quiz, campus broadcast, and homework. Students were also encouraged to start "Youngster Initiative" to pass fluorosis prevention tips to family members, friends, and neighbors via family activity.

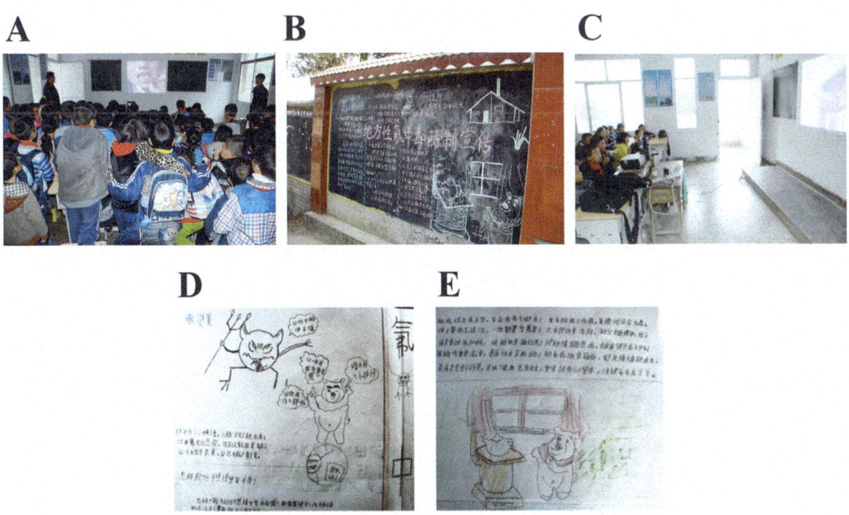

Fig. 22.3 Education concerning fluorosis prevention to primary school students. (**a**) To show primary school students the dangers of endemic fluorosis, and prevention and control of the disease with the propaganda video. (**b**) In a school wall newspaper, the harm and prevention of endemic fluorosis are painted with colored chalk

Education on Fluorosis Prevention Undertaken by Healthcare and Medical Institutions

Rural residents would first resort to township health centers and village clinics for medical help. Therefore, local people respect doctors dearly, and patients look up to health knowledge as the standard. They would easily accept doctors' suggestions on health. In such institutions, wall posters, leaflets, brochures, health tips, and fluorosis prevention consultation were all utilized to improve villagers' awareness (Fig. 22.4). On market-day, the following efforts were made to impart fluorosis prevention knowledge in fixed places: playing fluorosis prevention video, distributing paper materials, giveaways printed with fluorosis prevention core tips, responding to relevant consultation, and broadcasting fluorosis prevention information.

Improvement of Stoves

Coal-burning type of endemic fluorosis areas is often found in alpine areas (Fig. 22.5). The residents heavily rely on open stoves (without exhaust system) and coals for heating and cooking. Furthermore, the staple foods, corn and pepper, are often dehydrated on open stove soon after harvest to avoid dampening in storage.

Fig. 22.4 Education utilized to improve villagers' awareness in township health centers and village clinics. (**a**) A projection on the wall of a health center shows the integrated prevention and treatment of endemic fluorosis for patients. (**b**) Health centers display the comprehensive prevention and control of endemic fluorosis on a propaganda board

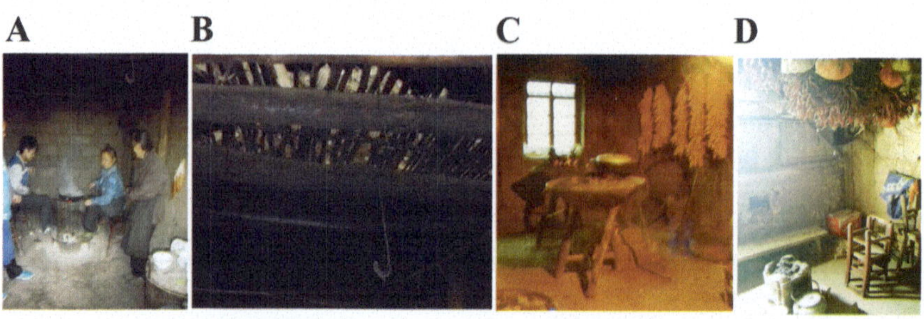

Fig. 22.5 The areas of coal-burning type of endemic fluorosis. Heating and cooking (corn and pepper are dehydrated soon after harvest to avoid dampening in storage) on open stove; high fluoride intake from contaminated air and food by soot which invariably leads to chronic fluorosis

Such practice results in high fluoride intake of contaminated air and food which invariably leads to chronic fluorosis. The prevention must focus on cooking fuel and food dehydration process/storage.

Improvement of Stoves for Fluorosis Prevention

The key of coal-burning type of fluorosis prevention lies in prevention. Raw coal is the primary source of heat in villages of endemic areas. The content of ash and coal tar in locally sold coal is relatively high. Furthermore, the habitual use of open stove together with its building material of fluoride containing coal-clay made the fluorosis prevention a dire situation. Local authorities successfully reduced fluoride content in soot and indoor air/food by improved stove project with exhaust system (Fig. 22.6). Such endeavor played a positive role in preventing and controlling the coal-burning type of fluorosis.

Fluorosis prevention stove is the major part of intervention measures. Manufacturers of stove, induction cooker, electrical cooking devices were invited to confirm possible parameters and design the proto-type for test according to the individual conditions of endemic areas. Only when the product meets the fluorosis prevention purposes, and the products are ideal in discharging the soot and reliable on cooking for both family and feed as well as heating, such products could be distributed to project families.

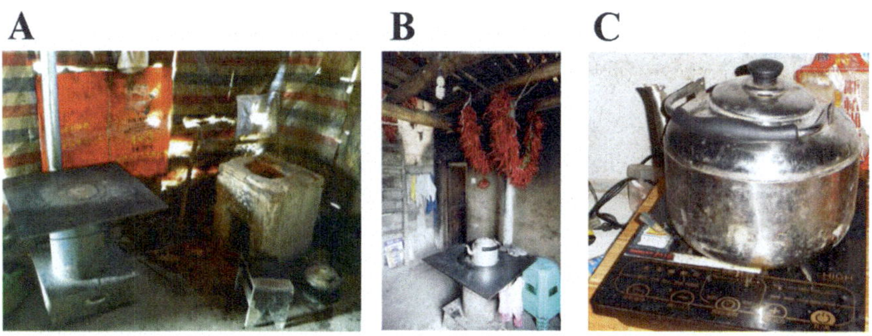

Fig. 22.6 Improved stoves. Government subsidized fluorosis prevention stove/oven and induction cooker to replace traditional open stove; fluorosis prevention stove is designed to dehydrate a small amount corns and pepper

Tailor-Made Fluorosis Prevention Equipment for Project Families

The project families were provided with thicker and rust-proof exhaust pipe when their current equipment is confirmed as low quality. Or families would be provided with fluorosis prevention stoves if there was absence of heating stove. Project families were subsidized with stoves, induction cooker, rice cooker, and biogas digester construction according to their practical need.

Standard of Fluorosis Prevention Stove

The fluorosis prevention stove was designed to prevent indoor contamination as an energy-saving device (heating, cooking food, and feed). Its measurements must fit for construction safety regulations and maintenance friendly. Vertical height of exhaust pipe must be no less than 1.8 m, that of traditional stove must be more than 2 m. The exhaust pipe must stretch out of the wall of more than 0.5 m with anti-downdraught and rain-proof features.

Instructing the Operation and Maintenance of Improved Stoves in Families of Endemic Areas

The long-term harmful habit of using coal in endemic areas could not be overturned overnight. Stove has limited operation life as a high temperature combustion apparatus. Its wearing parts, exhaust tube and rim, must be replaced regularly. In addition, the operation life of stove is determined by the routine maintenance. Hence, in endemic areas with improved stoves, village council appointed and trained supervisor who was in charge of instructing each and every villager to handle the improved stove in correct way. In this way, the villagers gradually switched from old open stove burning habit to correctly use the fluorosis prevention stove and were accustomed to seal the stove at all time. Villagers also learned to clean the tube and pipe for free flow of soot in time with correct maintenance of the wearing pars. Also, manufacturer of the stove set up after-sales shop in endemic areas for maintenance or replacement of malfunctions or damaged ones.

The Construction and Application of Clean Energy

Using raw coal for cooking fuel directly would discharge hazardous substance, like fluoride containing soot and sulfur dioxide. These substances contaminate both the indoor air and the atmosphere. With the assistance of central government, villages in coal-burning type of endemic fluorosis were aided with biogas digester construction and instruction. Meanwhile, electricity bill in the countryside was lowered significantly due to the power grid renovation, and diversified home appliances (rice cooker and electric heater) were available at cheap prices. All these contributes to the replacement of methane and electricity over raw coal. The application of clean energy as cooking and heating fuel blocked the coal-burning caused fluorosis from its source.

Behavioral Intervention

Promotion of Corn and Pepper Dehydration by Soot-Free Thermal Flux Technology

Corn is the only agricultural crop and staple food in coal-burning type of endemic fluorosis area in the alpine region. Every local household could harvest thousand kilograms of corn. In order to avoid raining season during harvest dampening the corns, residents use multiple open stove to dehydrate corns quickly. The global warming has forced endemic areas in west China (Guizhou, Shaanxi, Yunnan, Sichuan, Chongqing, and Hunan provinces) to forego open stoves but switch to solarization for corn and pepper dehydration and store the food in air-tight condition. Furthermore, the key economic crop in major coal-burning type of endemic fluorosis areas of China (Guizhou, Yunnan, Sichuan, Hubei, and Hunan provinces) is tobacco leaf, and most farmers own flue-cured tobacco chamber. Using such chamber for corn and pepper dehydration is equally effective to avoid fluoride contamination on food (Cong et al. 2009) (Fig. 22.7).

Fig. 22.7 Special flue-cured tobacco chamber. Villagers tried to dehydrate corn and pepper by solarization; dehydrating pepper in flue-cured tobacco chamber to avoid fluoride contamination

Adopting the Habit of Rinsing Before Cooking

For soot contaminated corn and pepper, certain amount of fluoride containing soot is attached on the surface of the food. If the food is rinsed before cooking, the fluoride content in corn will decrease over 15% (Li 1988), and that of pepper, 50%. Therefore, adopting the habit of rinsing before cooking will reduce the overall intake of fluoride and thus prevent the endemic fluorosis.

Free Lunch in Rural Schools to Reduce the Overall Intake of Fluoride for Students

According to relevant Chinese policies, students in rural primary and secondary schools are entitled with free lunch. After the implementation of such policy in endemic areas, the investigation found that urinary fluoride contents in students who enjoyed such policy decreased 35% compared with that of local adults. Meanwhile, the nutritional index of the students was increased and their physical condition improved.

National Ban on Small Scale and Illegal Coal Mines to Reduce the Use of Coal

In 2004, a strict national ban on small scale and illegal coal mines was practiced. Such practice decreased the coal supply in endemic fluorosis areas, hence, corn dehydration proportion by burning coal was also decreased substantially. The reduction of villagers eating fluoride containing corn also leads to the decreased overall fluoride content. Since the amount of edible corn with high fluoride by the villagers was decreased, the total fluoride intake in the body greatly decreased.

A Combined Practice of Poverty Alleviation Project to Promote the Prevention and Control of Coal-Burning Type of Fluorosis

Coal-burning type of fluorosis is a "poverty disease," in which patients are deprived with economic benefits by the disease or poverty stricken by the disease. For the poverty alleviation project, local authorities organized labor export or profit generating project to increase villagers' income. In this way, corn was no longer chosen as staple food or option for food, thus lowered the fluoride intake in local population. With such effort, staple food structure for population in endemic area was also improved. Publicity on fluorosis prevention was strengthened to update local people's understanding of its harm. Villagers gradually prefer to adopt air-tight stove for coal-burning or electricity powered device for cooking and heating. These efforts could eliminate the coal-burning contamination from its source.

Official Measures of Prevention and Control on Coal-Burning Type of Endemic Fluorosis

Prevention and Control Area Identification

Registered villages and families in coal-burning type of endemic fluorosis areas with no prevention and control measures were included as project target, and the whole village would receive prevention assistance for strengthening. The targets selection from endemic fluorosis area and non-fluorosis area was based on "Differentiation Standard on Endemic Fluorosis Areas" formulated by relevant Chinese authorities.

Budget, Funds, and Management for the Elimination of Fluorosis Hazards

The task team reported to superior team and peer entity for prevention and control funds according to the budget plan, concerning population, number of families, staff, facilities and equipment, transportation condition, prevention and control details, and potential funds from self-financing. The implementation of prevention measures strictly followed the prevention plan and central and local government accounting management regulations of special funds. The standardized finance management was under special accounting and dedicated for special use. The funds were guaranteed to be fully devoted for the prevention, control, and elimination of coal-burning type of fluorosis. The legitimate use of the funds was in full compliance with the regulations, which was free from illegal interception or embezzlement.

Prevention and Control Aim

In addition to the national aim of prevention and control, this project set a long-term aim tailored for individual areas, considering their compatibility between socio-economic development and fluorosis prevention work. With full dedication to the project, the general aim of fluorosis control or elimination was achieved ahead of scheduled time.

Specific Aim

There are quantified indexes in various phases to complete the general aim. (1) One hundred percent courses of fluorosis prevention were delivered to primary and secondary schools in towns and villages practicing the prevention measures. The coverage rate of families for fluorosis prevention publicizing reached 100%. (2) Fourth graders to ninth graders and housewives are familiar with >90% and >80% fluorosis prevention core tips, respectively. (3) Ownership rate of improved stoves in project households, qualification rate, and correct handling rate all exceeded 95%. (4) Satisfactory rate of fluorosis prevention stove >80%. (5) Serviceability rate of damaged stoves after repairmen >90%. (6) Correct dehydration rate of staple foods (corn and pepper) >70%, and proper storage rate >70%. (7) Rinsing rate of staple foods (corn and pepper) >70%.

Personnel Management

The management recruited or assigned staff for administrative management, service management, secretary, IT support, and archivist. The front-line staff are the cornerstone of prevention and control execution, including investigators, health instructors, purchasing staff, stove builders, technical instructor, administrative supervisors, acceptance staff, and general staff.

Mobilization

In initial stage of fully funded fluorosis prevention, the task team invited representatives from different communities and relevant offices to project convention. Such convention aims were to publicize the pressing situation of coal-burning type of endemic fluorosis, the necessity and feasibility of prevention and control, and its significance and purpose for the socio-economic development. This convention also served as prevention assigning tasks and mobilizing residents in endemic areas to finally achieve the goal of controlled situation and even fluorosis elimination.

Training

Relevant personnel must receive stratified and classified training in top-down fashion. Be it management, medical staff, teachers, local officials, or technicians of improved stoves, they should improve competence on fluorosis prevention and trouble shooting. In this way, the overall project quality and progress could be guaranteed.

Fluorosis Prevention Approach

Due to the differences in fluoride sources in endemic areas (endemic areas featured with cooking and heating caused indoor contamination and endemic areas featured with indoor coal-burning and locally produced corn dehydrating with open stove), there were differences in power grid, biogas digester, socio-economic development levels, and culture, the task team should identify individual approach for fluorosis prevention according to their specific conditions (choosing from improved exhaust pipe, stove, electrical devices, stove rim and taking into account the function of dehydrating corns).

Prevention and Control Progress Schedule

According to the general aim of prevention and control project, it was pivotal to pin-point progress of each stage and achieved the general aim in time. Such schedule was arranged according to the fiscal year appropriation and accounting. The following tasks were alternatively carried out in orderly manner to improve efficiency and timely execution: project initiation, training, background check, health education, stoves improvement, supervision, examination, and acceptance.

Procedures and Evaluation Standards of Prevention Effects

Procedures and Standards of Evaluation Application

The request for evaluating the prevention effects includes complete record of history of coal-burning type of endemic fluorosis, complete record of management and implementation of prevention measures, and self-evaluation meeting the standard of controlled situation or endemic elimination.

The contents of evaluation include self-evaluation with management on prevention and control work, implementation on prevention measures, endemic current status, and self-evaluation results; general statistics over the service condition improved stoves in project villages (including induction stove, liquified gas stove and methane gas stove); and current status of dental fluorosis in 8–12 years old children born in project villages.

The Manner of Acceptance and Evaluation

The manner concerns oral and written report on the history of coal-burning type of endemic fluorosis, status before prevention project, management of prevention measures, implementation of measures, and self-evaluation; reviewing prevention and control plans, implementation plans, record, summary report, endemic investigation files and data, endemic monitoring report; checking self-evaluation reports; on-site evaluation and relevant indexes that indicate the provincial and municipal task teams randomly selected and evaluated three project towns or villages from all levels of project. If there were less than three project towns or villages, all should be evaluated.

One village or residential area in project village will be selected and 10 in every other one household were evaluated. At the same time, dental fluorosis rate of 8–12 years old children born in project villages is checked. National task team randomly selected two project counties, towns, and villages for evaluation. If there are less than three project subjects, all should be evaluated. Meanwhile, dental fluorosis rate of children in primary schools of towns and villages should be evaluated.

Level by Level Evaluation and Acceptance Announcement

County level task team of endemic fluorosis assigned evaluation group for prevention or elimination of endemic fluorosis in project village. With the approval of this evaluation team, the results should be submitted to county level task team for evaluation and then to the higher level.

For the evaluation of provincial level task team over county, when over 95% project villages of the county reach standards of controlled situation or elimination of endemic fluorosis, county level task team will apply to provincial level task team for evaluation. Then, provincial level task team will assign experts team for evaluation. If the evaluation matched with the submitted reports, provincial level task team will record and submit the report to national level task team.

When over 95% project villages of the province reached standards of controlled situation or elimination of endemic fluorosis, provincial level task team applied to national level task team for evaluation. Then, national level task team would assign experts team for evaluation. If the evaluation is matched with the submitted reports, national level task team with record and publish report to the public.

Control Standard of Endemic Fluorosis

The control standard includes qualified rate of fluorosis prevention stoves $\geq$90%, correct dehydrating rate of staple food (corn and pepper) $\geq$90%; locally born and raised children (8–12 years old) with $\leq$30% dental fluorosis rate.

Elimination Standards of Endemic Fluorosis

Stoves, table cooker, exhaust pipe are all improved for fluorosis prevention purposes or switching to clean energy (methane, liquefied gas, electricity) in order to shield indoor coal-burning contamination. Qualified rate of fluorosis prevention rate is $\geq$95%. Residents in endemic area learn to use stoves in its air-tight condition, keep the exhaust pipe clear. When the stove is left unused, residents learn how to clean the stove and pipe, or to oil the stove parts; correct handling rate of fluorosis prevention rate is >95%. Over 95% projects households choose the dehydrated staple food (corn, pepper, or rice) >90% from soot-free sun-cured method or bake corn or chili in flue-cured tobacco chamber with complete exhaust pipe. Dental fluorosis rate of locally born and raised children (8–12 years old) is $\leq$15%.

Assessment of Prevention and Control Effect on Coal-Burning Endemic Fluorosis

Since 1987, the Three Gorges stove improvement for fluorosis prevention project (Sun and Lu 1995) kick off national level prevention and control of coal-burning type of endemic fluorosis. Since 2005, the prevention and control of coal-burning type of endemic fluorosis were backed by national finance. After years of diligent work, significant effects were achieved. General assessment of coal-burning type of endemic fluorosis condition in China, Guizhou province and Liu Pan-shui city was presented as follows:

Monitoring of Dental Fluorosis Status

In China, Guizhou province is worst stricken by coal-burning type of endemic fluorosis. It is estimated that 10 million patients suffer from dental fluorosis, 1 million patients suffer from skeletal fluorosis, which account for 30% and 50% of total estimates of China, respectively. In 2014, 11 mild endemic area counties reached the standard of endemic elimination, 1 county reached the standard of controlling the spread of endemic (Zhang et al. 2015a, b). Concerning the dental fluorosis rate in children aged 8–12 years old in 23 key endemic prevention counties, it was 73.29% in 2000, 45.76% in 2007, and 32.31% in 2014, presenting a gradually decreased manner. Fifteen counties out of 23 reached the standard of controlling the spread of endemic. Despite such number accounted for 65.22% of the total, the differences of the rest eight counties in reaching standards of controlling or elimination of the endemic areas showed great disparity (Zhang et al. 2017). The dental fluorosis prevalence rate in children aged 8–12 years old in Chongqing of China showed significant decrease from 70.00% (1985), 51.05% (2001), 27.09% (2009) to 11.28% (2015); the geometric mean values of urinary fluoride significantly decreased (Wei et al. 2017, 2018). In Sichuan province, 2015 monitoring results showed 16.29% dental fluorosis prevalence rate in children from five counties (Liu et al. 2018). In Yunnan province, 2017 random inspection showed 11.28% dental fluorosis prevalence rate in children from three counties, the fluoride level in urine obviously declined (Huang et al. 2019). In Shaanxi province, 1456 villages with coal-burning type of endemic fluorosis from eight counties were inspected from 2012 to 2015. It was found that 640 villages reached the standard of elimination (account for 43.96%), 419 villages reached the standard of controlling (account for 28.7%), and 397 villages were not up to the standard of control (account for 27.72%) (Fan et al. 2017). In Hubei province, dental fluorosis prevalence rate in children was 7.73% in the areas of coal-burning type of endemic fluorosis in 2014, in which five counties with endemic fluorosis reached the standard of controlling and elimination, and 1 failed to meet the standards of controlling (Zhang et al. 2015a, b). In Shanxi province, dental fluorosis prevalence rate in children was 6.41% in 2015. Among all the endemic counties detected, nine reached the standard of controlling and eleven reached the standard of elimination (Wu et al. 2017). In Jiangxi province, dental fluorosis prevalence rate in children of two counties was 19.92% in 2014 and results showed that the endemic was effectively controlled (Shangguan et al. 2017). In Henan province, inspection was carried on 13 counties with the areas of coal-burning type of endemic fluorosis. Among them, eight counties reached the standard of controlling, accounting for 61.54% (Yu et al. 2009). After decades of systemic prevention and control efforts on coal-burning type of endemic fluorosis, China has achieved significant results in controlling or eradicating the majority of coal-burning type of endemic fluorosis areas except certain counties/towns/villages in Guizhou and Shaanxi province.

Evaluation for the Typical Area of Coal-Burning Type of Endemic Fluorosis in Liupanshui City of Guizhou Province

Effective Control to the Disease

Liupanshui city is located in mountainous area of Wu-Meng of China, where featuring steep terrain in the Yangtze River-Zhujiang River watershed. In addition, the overdeveloped Karst landform leads to the scarcity of surface water, hence, this area is also nicknamed as the Sea of Coal in Southeast China. The 2001 survey revealed that 81% towns and villages in this area were the areas of coal-burning type of endemic fluorosis, presenting seven as severe endemic fluorosis areas and eight moderate endemic fluorosis areas (according to the "Differentiation Standard on Endemic Fluorosis Areas" formulated by relevant Chinese authorities (Li and Yan 1994)). In Shuicheng county of the city, dental fluorosis defect rate of primary school students aged 8–12 years old reached 23.63%. Such results of Shuicheng fell into the category of moderate endemic fluorosis areas/county, and its dental fluorosis defect rate was higher than that of Zhijin county (22.0%), the other significantly influenced area by this type of endemic fluorosis in Guizhou, at the same period. In the self-evaluation of 2015 on Liupanshui city, all counties, except Shuicheng, reached the standard of controlling or elimination in mild endemic area county. In Shuichen county, the inspection shifted its focus on four areas with severe endemic fluorosis with dental fluorosis defect rate ≥40%. After taking efficient works for the improved stoves and education for the prevention of endemic fluorosis, the previously mentioned towns and villages were categorized as none/mild endemic fluorosis areas instead of mild/moderate ones.

The Changes in Incidence Rate of Dental Fluorosis and the Urinary Fluoride Load

In 2001, dental fluorosis prevalence rate of the students in primary schools of Cichong village, Shuichen county was recorded by 83.6% and the defect rate of teeth by 28.4%; in 2015, these two items were 71.5% and 4.4% with significant improvement. However, in 2001, a minority group living in the area showed 76.8% dental fluorosis rate and 10.7% defect rate and in 2015, these two items were 81.8% and 10.6%, indicating no significant changes (Yang et al. 2015). At present, the contents of urinary fluoride from the children in the area of coal-burning type of endemic fluorosis are significantly lower than those of the endemic fluorosis level for children (1.40 ppm).

According to the mechanism of high fluoride damage, dental fluorosis generally forms when people are 2 years old. Even though dental fluorosis prevalence rate of primary school students fell into the category of endemic fluorosis area standard

(>30%), their urinary fluoride level might be lowered because of the improved nutrition by the free school lunch provided by government (Li et al. 2007a, b). Dental fluorosis of primary school students reflects the endemic fluorosis status in local areas a few years ago, while urinary fluoride level of population indicates real-time fluoride load status. Therefore, it was advisable to determine the endemic development and recent prevention/control effect from both the perspectives of primary school students' dental fluorosis status and urinary fluoride level of population (Li et al. 2009a, b).

Manifestation of Skeletal Fluorosis

Skeletal fluorosis manifests with over 90% fluoride deposited in hard tissues like teeth and bones. In 2014 survey conducted in Shaanxi province, there is no new patients with severe skeletal fluorosis after the implementation of prevention and control on coal-burning type of endemic fluorosis (Fan et al. 2017). In 2015 survey conducted in Sichuan province, the detection rate of dental fluorosis in children was 16.29%, while the rate of adults with skeletal fluorosis was 49.66%. Although the dental fluorosis in children was alleviated year by year, the detection rate in adults with skeletal fluorosis remained high (Liu et al. 2018). In Shuicheng county (moderate coal-burning type of endemic fluorosis area), two villages were randomly selected from the area of endemic fluorosis and one village from non-endemic area and the residents over 20 years old chosen for the survey. People suffered from limited multiple articular bending and stretch (excluded traumatic injury) in endemic village accounted for 21.0–45.4%, reached 8–17 times of those in non-endemic village (2.47%), which indicating that fluorosis damage on human skeleton is almost irreversible (Fig. 22.8). In addition, it was found that some residents were originated from endemic area when they were young and later developed into typical skeletal

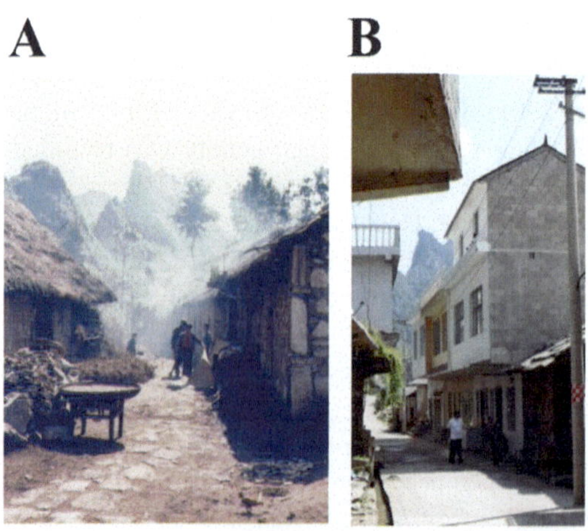

Fig. 22.8 Significant change of the resident situation. A comparison change of the living situation has been occurred before and after poverty eradication in village of the area of coal-burning type of endemic fluorosis. The left is a part of the village before poverty eradication; the right is the same part after poverty eradication

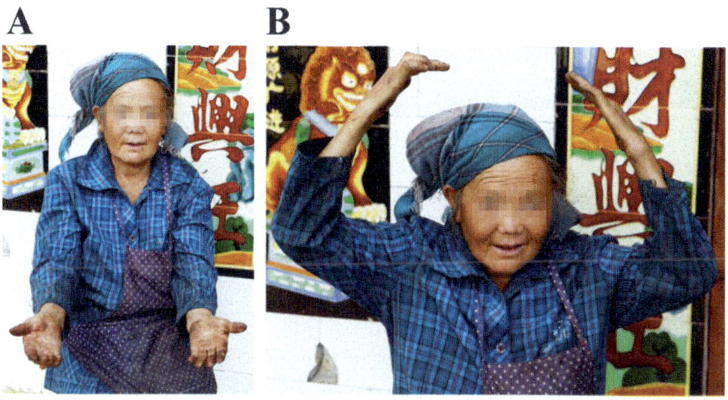

Fig. 22.9 A case still developed into typical skeletal fluorosis later even though moved to non-endemic area when she was young. A woman, moved to non-endemic fluorosis area before 20 years, was diagnosed with skeletal fluorosis when she was 68 years old with following symptoms: body aches, difficulty in walking, limited elbow, and shoulder joints function

fluorosis 10 years after they moved to non-endemic area because of marriage or migration (Fig. 22.9). Conversely, skeletal fluorosis had little effect on off-springs. For example, the residents in one village in the area of coal-burning type of endemic fluorosis used burning coal as fuel 15 years ago and then switched to eat rice or corn containing normal fluoride content as staple food. The residents over 45 years old with limited articular function were increased from 51.3% to 69.5% (132/190), there were also patients with skeletal fluorosis developed into semi-paralysis (Li et al. 2009a, b).

The above studies indicate that long-term bone damage with high fluoride intake in young children, even migration to non-endemic fluorosis areas, could still lead to skeletal fluorosis, which could be explained by the hypothesis, i.e., in the patients of skeletal fluorosis, the bone tissue can release fluoride sustainably, and the fluoride can be captured during bone reconstruction, thus causing secondary damage to the bone.

After the repair rate of bone and articular tissues falls behind fluorosis damage, potential fluorosis damage becomes more and more pronounced, while dental fluorosis does not change significantly over a lifetime. Since the rate of bone and joint tissue repair lags behind that of fluorosis injury, the potential for fluorosis injury becomes more and more apparent, while fluorosis does not change or worsen significantly over a lifetime.

To evaluate the effect of prevention and control interventions in coal-burning type of endemic fluorosis, a comprehensive judgment should be made based on sensitive urinary fluoride content, child-specific dental fluorosis and adults' characteristic signs of skeletal fluorosis and long-term bone damage, so as to avoid the prevention and control effect is covered up by long-term bone damage caused by high environmental fluoride.

Conclusion

Since 1980s, China has been actively carrying out the comprehensive prevention and control measures, including improved stoves, health education, and local economic improvement. In 2005, the Chinese government provided financial support for the comprehensive prevention and control of coal-burning type of endemic fluorosis. With the arduous effort from all levels of related authorities and communities, except for a few serious areas of coal-burning fluorosis in Guizhou and Shaanxi provinces, the remaining disease areas have reached the control standard of the prevalence rate of dental fluorosis in children aged 8–12 $\leq$30% or the elimination standard of $\leq$15%. Nevertheless, we should also pay attention to the treatment of existing patients with chronic fluorosis (including skeletal and dental fluorosis) and to the various situations that lead to the recurrence of endemic fluorosis. It is important to develop sustainable long-term prevention and control strategies to completely eliminate the harm of endemic fluorosis.

References

Cong XZ, Yang Y, Li FC, Yan SL, Luo Y. Comparison of fluoride content in corn roasted in Tobacco oven and open Coal fire. Chin J Endem. 2009;28:105.

Fan ZX, Li XQ, Zhou R, Cui CB, Li PA. Investigation on control effect of coal-burning endemic fluorosis areas in Shaanxi Province in 2012-2015. J Environ Health. 2017;34:529–32.

Huang KL, Wang AW, Li JG, Fan R, Xu SX, Dong X, et al. Investigation and analysis on the condition and prevention of endemic fluorosis caused by coal burning pollution in Yunnan Province. Chin J Endem. 2019;38:719–23.

Li FC. A report on a disease caused by corn baked by coal mixed with kaolin. Chin J Pre Med. 1988;22:225–9.

Li FC, Yan SL. Investigation report on fluorosis caused by coal burning pollution in Liupanshui, Guizhou. Chin J Endem. 1994;13:360–1.

Li FC, Wang Y, Yang Y, Zhao P. Preliminary observation of fluoride content in urine and its possible biological effect with fluoride toothpaste. Chin J Pre Med. 2007a;41:77.

Li FC, Yan SL, Qin AN, Liu GY, Xie WB, Gou JF. A preliminary study on the relationship between dental fluorosis and economic development in Guizhou coal baked corn fluorosis village. Chin J Pre Med. 2007b;41:119–22.

Li FC, Qin AN, Yan SL, Liu GY, Xie WB, Gou JF. Fluoride intake and effect of skeletal fluorosis in coal-baked corn fluorosis villages in Guizhou province. Mod Pre Med. 2009a;36:185–9.

Li FC, Yang Y, Yan SL, Luo Y, Qin AN. The correlation between dental fluorosis grade and urinary fluoride content in primary school students in coal-burning fluorosis area of Guizhou province and the monitoring value of urinary fluoride. J Mod Pre Med. 2009b;36:3865–7.

Liu P, Yang XJ, Chen J, Xu DY, Yuan P. Analysis of monitoring results of endemic fluorosis caused by coal burning pollution in Sichuan province in 2011-2015. J Pre Med Intell. 2018;34:70–4.

Shangguan J, Li ZH, Li A, Peng WF. Analysis of inspection and acceptance results of coal-burning fluorosis control in Shangli county and Yuanzhou district, Jiangxi province in 2014. Chin J Endem. 2017;36:265–8.

Sun YF, Lu QW. Investigation on the Improvement and consolidation of coal-burning fluorosis in the three gorges of the Yangtze River. Chin J Infect Control. 1995;1995:109–10.

Wei Y, Xie J, Li XS, Chen YL, Huang WL, Zhou S. Investigation and analysis on current situation of coal-burning fluorosis prevalence. Open J Epidem. 2017;7:115–23.

Wei Y, Xie J, Li XS, Zhou S. Analysis the results of control and elimination of coal-burning-borne endemic fluorosis in Chongqing city. Open J Epidem. 2018;8:145–52.

Wu ZM, Wang ZH, Li PF, Wu M, Jing YL, Wei H, et al. An analysis of the examination and evaluating results in controlling and eliminating coal-burning-borne endemic fluorosis in Shanxi Province. Chin J Endem. 2017;36:261–4.

Xu J, Hu TG, Li FC, Yang F, Deng J, Guan ZZ, et al. Analysis of the relationship between the family economic status of primary school students in Douqing Miao and Yi townships in Shuicheng County of Guizhou Province and the disease of dental fluorosis. Chin J Endem. 2016;35:834–7.

Yang F, Deng J, Li FC, Tang F, Sun ZG, Guan ZZ, et al. Observation on the prevention and control effect of endemic fluorosis caused by coal burning in Douqing Township, Shuicheng County, Guizhou Province. J Guiyang Med Coll. 2015;40:1020–3.

Yu B, Liu Y, Yuan CS, Kang JS, Huang HQ, Wei JJ, et al. Investigation on prevalence of endemic fluorosis caused by coal burning pollution in Henan Province. Chin J Endem. 2009;28:191–3.

Zhang BY, An D, Li DS, Yao DC, Ye HB, Gao J, et al. Evaluation of prevention and treatment effect of 11 counties in Guizhou Province with history mild disease of fluorosis caused by coal burning pollution in 2014. Chin J Endem. 2015a;34:511–3.

Zhang BY, Xiong PS, Zhu SZ, Zhou SH, Chen ZZ, Dai X, et al. Analysis of the investigation results of control and elimination of fluorosis caused by coal burning pollution in Hubei Province. Chin J Endem. 2015b;34:514–7.

Zhang BY, An D, Li DS, Yao DC, Zhang NH, Ye HB, et al. Analysis of dental fluorosis in children aged 8-12 years in the key areas of endemic fluorosis caused by coal burning in Guizhou Province. Chin J Endem. 2017;36:269–73.

Chapter 23
Surgical Treatment of Joint Damage in Skeletal Fluorosis

Ri-Guang Liu

Abstract Endemic skeletal fluorosis is caused by exposure to excessive amount of fluoride in water, air, or food. Guizhou province is a skeletal fluorosis severely afflicted area in China due to burning coal with high content of fluoride, especially in counties such as Zhijin, Dafang, and Shuicheng. The main clinical manifestation of such endemic is bone and joint with arthralgia, stiffness, and limb movement dysfunction. Imaging examination shows disorder structure of trabecular and cortical bone. There was more joint damage on spine and knee caused by skeletal fluorosis, but few on hip. Severe cases showed patients with disability and lifelong pain. Early diagnosis of chronic fluorosis and appropriate surgical methods can alleviate the sufferings and improve quality of life. The application of total hip arthroplasty and total knee arthroplasty for hip and knee injuries (induced by chronic fluorosis) can relieve arthralgia, improve joint function, and correct joint deformities. For spinal fluorosis, the osteosclerosis is the most. Symptoms were extensive joint hyperplasia of the spine and ossification of ligament hypertrophy, causing secondary spinal stenosis. Expansive laminoplasty can effectively relieve spinal nerve compression and restore spinal nerve function to variable degrees.

Keywords Endemic fluorosis · Hip arthroplasty · Joint damage · Knee arthroplasty · Skeletal fluorosis

Introduction

Fluoride is one of the essential micro-elements for animals and human, and its consumption at a reasonable amount protects teeth against microbial attack (especially in childhood). Fluoride is absorbed by human body through food, drinking water, toothpaste, mouthwash products, and air, among which drinking water is the

R.-G. Liu (✉)
Department of Orthopedics at the Affiliated Hospital of Guizhou Medical University, Guiyang, Guizhou, People's Republic of China
e-mail: riguangliu@gmc.edu.cn

Z.-Z. Guan (ed.), *Coal-burning Type of Endemic Fluorosis*,
https://doi.org/10.1007/978-981-16-1498-9_23

main source for fluoride absorption in most societies (Huang 2010). However, many epidemiological studies have shown that absorption of fluoride from drinking water in the long term leads to adverse effects on human skeletal tissue. Existing study demonstrates a significant relationship between fluoride concentration in the water and prevalence of skeletal fluorosis in an endemic fluorosis area (Srivastava and Flora 2020), such as Zhijin, Dafang, and Shuicheng in Guizhou province of China. Fluoride toxicity presents an extraordinary degree of consistency in its clinical manifestations, specifically joint arthralgia, stiffness, and limb movement dysfunction. Skeletal fluorosis shows increase in the bone density in adults, which is proven by X-ray and various radiological features. Such changes include osteosclerosis, osteomalacia, and osteoporosis, causing joint deformity and secondary spinal stenosis. Computer tomography (CT) is the best imaging measure for bone pathology visualization and provides more details than plain sciagrams. Magnetic resonance imaging (MRI) obviates the difficulties in myelography for being noninvasive and delineates soft tissue structures and changes in the spinal cord. Surgery has an important role in alleviating the joint arthralgia, disability, and improving the quality of life, which should be tailored to the individuals based on imaging findings. Surgery for joint reconstruction and spinal decompression is needed and is proven to be effective for severe cases: total hip arthroplasty (THA), total knee arthroplasty (TKA), and expansive laminoplasty.

THA Treatment for Severe Hip Osteoarthritis Induced by Fluorosis

Clinical Manifestation

Some patients are diagnosed with dental fluorosis, others with neck, waist, and large joints (more than three parts) persistent intermittent pain, which are free from seasonal and climate changes; some suffer from limb seizures, numbness, morning stiffness, waist-ache, motor dysfunction (in neck, waist, upper and lower limbs, especially in hip) (Huang 1999). It is so fast developing that patient cannot walk within 2 years. Tenderness is found in the femoral triangle, the end of adductor, as well as trochanteric pain. Movement range of the hip joint flexion and extension, adduction, abduction, internal rotation, and external rotation are all limited. “Four” shape test and Thomas motion test are positive.

Imaging Examination

According to the varied severity of disease, the X-ray shows differences in trabecular structure: grit-like or granular-like bone structure, bone plaque, or thin, sparse, fuzzy

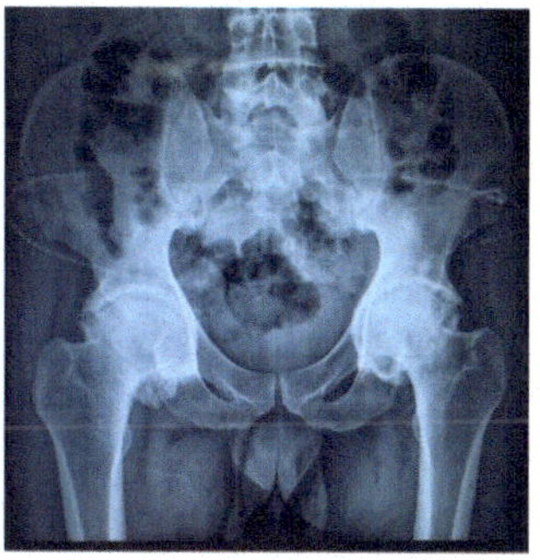

Fig. 23.1 X-rays manifestations of hip joint. X-rays manifestations of hip joint for severe fluoride hip osteoarthritis, which reveal periacetabular hyperplasia, osteophyte formation, and significantly narrowed joint space

bone trabeculae; or it is dense, coarse, cloth-shaped trabeculae, or trabecular bone fusion in most cases. The joints show degenerative changes. In severe cases of hip joints related disease, periacetabular hyperplasia and osteophyte formation can be seen. Osteosclerosis of the femoral head and acetabular subchondral bone are recorded; joint space is significantly narrowed, the femoral shaft cortex thickens, and the medullary cavity becomes narrowed (Fig. 23.1).

Laboratory Examination

Patients generally have mild to moderate anemia, which can be interpreted from following three aspects: (1) firstly, the high fluoride content affects metabolic and hematopoietic function of bone marrow; (2) secondly, high fluoride intake can cause nutritional absorption disorders in the digestive system, leading to the lack of hematopoietic materials and the possibility of anemia; (3) skeletal fluorosis causes pain, insomnia, affects basal metabolism, and high energy consumption will inevitably cause excessive nutrition consumption leading to anemia. Urinary fluoride in patients was elevated (Omid et al. 2017; Ibiyemi et al. 2018), but it did not have obvious clinical significance for diagnosis and judgment of disease severity. At the same time, there was no significant relationship between the level of urinary fluoride and the individual's condition (24-hour urinary fluoride levels in patients with severe skeletal fluorosis could fluctuate from 0.9 to 13.0 ppm). Immune-related indicators such as rheumatoid factor, anti-cyclic citrulline antibody, anti-nuclear antibody, and HLA-B27 were negative.

Diagnostic Criteria and Characteristics

Diagnostic criteria on skeletal fluorosis joint damage have been enumerated by *Grading Clinical Diagnosis of Endemic Skeletal Fluorosis* (GB 16396-1996) (Huang 2010): epidemiological history of living in high fluoride areas for long term; symptoms and signs of knee joint arthralgia and dental fluorosis; urinary fluoride levels exceed normal physiological values; X-ray results showing osteosclerosis, osteomalacia, and osteoporosis, or intermingle bone joint damage (such as bone plaque, structurally disordered bone trabeculae, medullary cavity stenosis, periacetabular osteophyte formation, and calcification of periarticular soft tissue).

The latest "*Diagnostic Criteria of Endemic Skeletal Fluorosis* (WS 192-2008)" in China presents following criteria: epidemiological history (born and lived in an endemic fluorosis area for more than 1 year); clinical manifestations: with neck, waist, and large joints (more than three parts) persistent intermittent pain, which is free from seasonal and climate changes or some suffer from limb seizures, numbness, morning stiffness, waist-ache, motor dysfunction in neck, waist, upper and lower limbs, and so on; X-ray findings in bone and joint. The entry on urinary fluoride in diagnosis was canceled.

Differential Diagnosis on Severe Hip Osteoarthritis Induced by Chronic Fluorosis

Rheumatic Arthritis

Rheumatic arthritis is a chronic and systemic autoimmune disease characterized by joint synovitis. With high morbidity and disability rate in young adults, such disease affects mostly on small joints and large joints in later stages. Positive rheumatoid factor and anti-cyclic citrullinated antibody have certain diagnostic significance. X-ray test shows trabecular bone decrease, atrophy, and bone thinning around the joint.

Pharyngeal pain, fever, and increased white blood cell count, as well as troubled large joints of four limbs are common in this disease. There are migratory joint swelling and pain, but no permanent damage is found after the joint symptoms disappear. Serum antistreptolysin "O," antistreptokinase, and antihyaluronidase are all positive, and salicylic acid preparations treatment is often rapid and significant.

Femoral Head Necrosis

Femoral head necrosis is quite common in patients received orthopedics. Except for the diseased hip joint, symptoms of joints arthralgia are generally not included. Early articular cartilage is not necrotic, and joint space narrowing is not obvious in femoral head necrosis. X-ray results show hip joint damage is caused by fluorosis with no obvious collapse of the femoral head and articular surface sclerosis. The gap between joints is narrow and fuzzy.

Primary Hip Osteoarthritis

The articular surface pairing is good and there is no abnormal line of force. Such disease develops slowly, with features affecting older patients and recurrence.

Secondary Hip Osteoarthritis

It is caused by overloaded articular cartilage and subchondral bone due to anatomical deformities and abnormal limb alignment, which can be identified by X-ray test.

Ankylosing Spondylitis

It is more common in younger males, initiating from the sacroiliac joints, facet joints of four limbs are not afflicted; the HLA-B27 test is generally positive.

Surgical Treatment

Operative Indication

For patients with severe hip osteoarthritis induced by fluorosis plan to receive THA, their clinical manifestations are joint activity dysfunction and X-ray diagnosed obvious hip damage, but patients with systemic or local infections and underlying diseases are absolutely prohibited to receive surgery (Su et al. 2012).

Preoperative Preparation and Evaluation

The evaluation can be divided into bone and soft tissue one. Pelvic radiographs and full-length femoral radiographs are used before operation to understand hip osteophyte hyperplasia, femoral medullary cavity stenosis, and primarily determining the type of prosthesis. Detailed physical examination should be conducted to understand the hip joint movement limitation, soft tissue contracture condition; especially after anesthesia (muscle cramp is relieved), further physical examination is necessary to understand the direction and degree of hip joint limitation. Such preliminary diagnosis on the direction and degree of loosening is vital for surgery.

Prosthesis Selection

Severe hip osteoarthritis induced by fluorosis is mostly found in young people. Existing prosthesis operates in limited time; therefore, selection of bioprosthesis (for the sake of convenient joint revision) with its characteristics is widely chosen: excellent bio-compatibility with blood and tissues and tested non-cytotoxic. Modular cemented prostheses is not excluded for patients with osteoporosis.

Anesthesia Methods

General anesthesia or continuous epidural anesthesia can be chosen considering patients' mental status and their cooperation.

Incision

The posterior lateral incision of the hip joint is selected mildly outside the midpoint of the connection between posterior superior iliac spine and the apex of the greater trochanter. Generally, the incision selected is slightly 1 cm beyond. This incision runs along the gluteus maximus fiber to reduce the transverse injury of the gluteus maximus, which can avoid damage to the gluteus medius.

Surgical Operation

Under general anesthesia or epidural anesthesia, the lateral position is chosen with the affected side upward. This side is strictly sterilized and disinfected. Iodine tincture is applied twice, and then de-iodized twice with alcohol. The non-bacterial sheets are prepared carefully. Choosing the posterolateral approach, the incision starts at the juncture of 1 cm slightly lower posterior superior iliac spine and 1/3 outside the central point of greater trochanter apex connection. Turning to the direction of the femoral shaft, broad fascia is incised and gluteus maximus is bluntly separated to expose the externally rotating small muscle group. Adducting the internally rotating hip joint, the stop of the externally rotating small muscle group is cut off to pull inwards to protect sciatic nerve. There is no need to separate sciatic nerve. Joint capsule is cut off and loosened firstly. The osteophytes in the posterior margin of the acetabulum is bitten off. Maintaining hip adduction and inward-rotation, calf is set vertically on operation table. Using Hohmann traction hooks on each side of the femoral neck to expose the femoral neck, osteotomy is performed 1.5 cm over the lesser trochanter to remove the femoral head. The acetabular marginal osteophytes and the granulation fibrous tissue in the acetabulum are cleaned. Synovial tissue and part of joint capsules are cleaned to expose transverse acetabular ligament and acetabular labrum. Soft tissue is loosened if needed. Suitable reamer is chosen to polish acetabulum, osteotome is necessary to notch on sclerotic articular surface. It is advisable to polish the acetabulum by using from small to big acetabular reamers until blood oozing.

Concerning the acetabular placement angle, the polishing of the acetabulum should be consistent with the opening angle of the acetabulum at later stage. Nevertheless, it is difficult to manage this angle because of the marginal osteophyte of the acetabulum and alteration of body position. Because of the acetabular lip, especially obvious osteophyte of posterior upper edge of the acetabulum, it is not advisable for installation angle. The anterior edge of the acetabular lip can be properly referenced as the installation mark: keeping the acetabular reamer edge approximately parallel to the anterior edge of the acetabular lip; the axis of the acetabulum passes right through the sacropromontory. Transverse acetabular ligament is a fixed mark and its fiber running can be used as a reliable reference for acetabular placement angle. If the osteophyte of transverse acetabular ligament obstructs the acetabular polishing and angle reference, it can be partially resected along the fiber path from the bone insertion for future reference. In no case should the incision be too deep to damage the branches of obturator artery. The acetabular placement angle is 45° abduction and 15° forward. A small amount of cancellous bone of removed femoral head can be implanted on the acetabular wall before installing the acetabular prosthesis. In this way, it can increase consistency of the compression and might accelerate the induction of bone growth. After implanting the acetabular prosthesis, one or two screws should be implanted in the upper posterosuperior area to ensure initial stability and facilitate early joint activity. Then, extra osteophytes around the acetabulum are bitten off once again.

Bending the knee and rotating the lower extremity inwardly to expose the osteotomy section of the femur. The medullary cavity is opened at about 10 °of the femur inclination angle. The specific direction can refer to the lesser trochanter. If there is no malformation of the femur, a finger width before lesser trochanter is exactly the inclination angle of the femoral neck. Also, the maximum diameter of the medullary cavity of the femoral neck osteotomy section can be used as a reference for the femoral neck inclination angle. The medullary cavity is enlarged by small to large reamers. When the standard test mold is installed, the depth of installing should generally exceed the standard line of enlarged medulla depth by 1–2 mm. After reset, move the limb in all directions within normal range: making sure the joints are stable, touching the knees and iliac bones to ensure equal limb length. Then withdrawing the test mold and medullary cavity reamer to install the same type of femoral stem bioprosthesis. Confirming the stability again after reset. Using normal saline to rinse scattered bone scraps may have certain effect on preventing heterotopic ossification. Whether or not placing drainage tube depends on the operation time: short operation time (within 1.5 h) and complete hemostasis, the incision is fine without drainage tube (since the role of drainage is to prevent infection and blood accumulation in the incision). Finally, each layer is sutured in turn.

There are four differences between THA induced by skeletal fluorosis and other causes. (1) Loosening joint capsule requires certain balance. Skeletal fluorosis causes calcification of the joint capsule and periarticular soft tissues and reduces elasticity. Patient's activities are also reduced due to arthralgia or joint contracture. When soft tissue is loosened, all directions of hip joint movement should be tested. If the soft tissue is loosened insufficient or imbalanced, the post-surgery joint function will be affected inevitably. (2) The synovium and hyperplastic soft tissue must be cleaned completely. Since patients have constant joint pain before surgery, it is highly possible that a large number of inflammatory factors in the joint synovial membrane and the periarticular osteophyte soft tissues involve in the stimulation of pain. (3) Acetabulum reaming must be notched with osteotome. In these cases, the acetabulum becomes sclerosis. After acetabular reaming, the acetabular surface is abnormally smooth without oozing of blood and resembles ivory bone with low friction; such surface is obviously detrimental to the placement of bioprosthesis and bone ingrowth. Carving notches with osteotome at the acetabular sclerosis site to increase friction, which could significantly reduce acetabular reaming time. (4) Existing bio-femoral stem prostheses are coated with 12 mm hydroxyapatite or other coatings to promote bone growth, while such coating is absent in femoral stem test model. Because of the osteosclerosis in skeletal fluorosis joint, small elastic range is found in medullary cavity, even 1–2 mm thickness can make the prostheses implantation difficult. Therefore, the reaming test mold must reach the standard or even exceed 1–2 mm. In this way, femoral stem prosthesis can be successfully placed to the right depth with one try.

Postoperative Management

Routine intake of antibiotics (first- or second-generation cephalosporins) for 3 days after surgery. On the first day, patients are instructed to do isometric contraction exercises with both lower limbs and take indomethacin orally 50 mg/time (twice a day for 3 weeks). On the third day, patients are instructed to exercise their lower limbs flexion and extension without load. Patients could get off bed by using walking aids one week after surgery, X-ray results are re-examined to monitor the status of the prosthesis. Then removing stitches on day 14. Skeletal fluorosis patients generally suffer from anemia, so they are instructed to change their dietary habits, and to stay away from the high fluoride content food made in coal-burning condition under intensive nutritional support treatment. Such measures guarantee long-term efficacy after surgery. Patients can walk on single crutch after 4–6 weeks and walk without crutch 8 weeks later.

Operative Complications

Skeletal fluorosis has no significant effect on incision healing, but ectopic ossification and calcification are more common in skeletal fluorosis; ectopic ossification may be more pronounced after surgery. It has been reported that a high proportion of ectopic ossification was observed after using indomethacin for prevention (Brooker grade I and II, no significant effect on joint function), but it shows that ectopic ossification is a common complication of such disease and requires special treatment. After the termination of high fluoride intake, the damage to the body can be reduced or even disappear, and of course it is also limited to certain symptoms of early stage and severe cases. In this process, the X-ray results changes are not obvious in the short term, but the function of avoiding high fluoride intake cannot be denied. No prosthesis loosening and subsidence in short term (Fig. 23.2). Although complications related to THA in other diseases have not been reported in the history of THA development, however, paying attention to complications of skeletal fluorosis after THA is necessary.

THA is effective in current systematic treatment of skeletal fluorosis induced severe hip osteoarthritis. It must be noticed that prevention of ectopic ossification after surgery is necessary. Due to limited cases and short observation period, the potential problems, complications, and long-term efficacy in the treatment require further observation and research.

TKA for the Treatment of Severe Knee Osteoarthritis Induced by Skeletal Fluorosis

Clinical Manifestation

With moderate and severe dental fluorosis, the early symptoms are systemic polyarticular pain, especially knee arthralgia, which gradually deteriorates with swelling, stiffness, deformity, and limited function (Kebede et al. 2016).

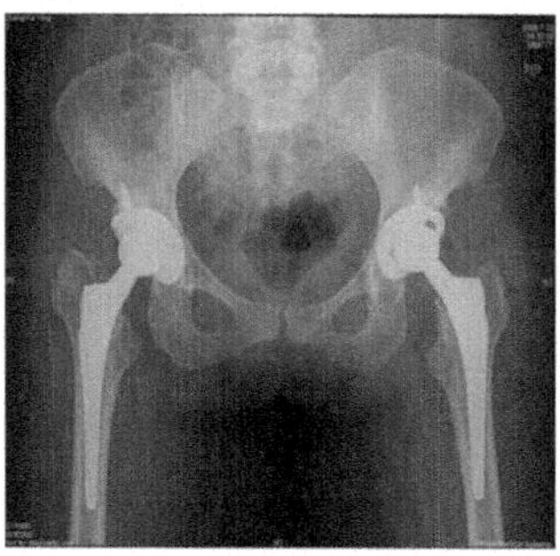

Fig. 23.2 X-rays image of bilateral hip arthroplasty. X-rays image of bilateral hip arthroplasty for severe fluoride hip osteoarthritis, the imaging results showed that the prosthesis aligned well

Medical Imaging Results

The anteroposterior (AP) and lateral (LAT) X-ray (knee joint AP and LAT X-ray) show that the bone mineral density is increased, the bone trabeculae transformed to be thickened, coarse, and fused; bone cortex thickened and medullar cavity narrowed or disappeared. Osteophyte formation is observed at the edge of articular facet and at the intercondylar eminence, knee joint space is narrowed; calcification and hyperplasia are observed in articular cartilage, together with swelling of knee joint. There are also different degrees of bone destruction, joint varus deformity, increased bone density, and uneven thickening around cartilage are observed (Fig. 23.3).

Laboratory examination

Even though urinary fluoride content in patients is increased, such value shows no obvious clinical significance for diagnosis of disease severity. Not only immunologic functions (such as rheumatoid factor, anti-cyclic citrulline antibody, antinuclear antibody, and HLA-B27) were negative, but also patients show symptoms of mild and moderate anemia (Saeed et al. 2020).

Diagnostic Criteria and Characteristics

It is hard to distinguish the symptoms, signs, and X-ray findings of arthropathy caused by degenerative/traumatic and rheumatism/rheumatoid diseases from that caused by skeletal fluorosis. *Diagnostic Criteria of Endemic Skeletal Fluorosis* and *Clinical Grading Diagnosis of Endemic Skeletal Fluorosis* by Chang-Qing Huang (Huang 2010) can be used for diagnostic reference. The body damage of endemic fluorosis is not only limited to bones and joints, but also on soft tissues. The changes

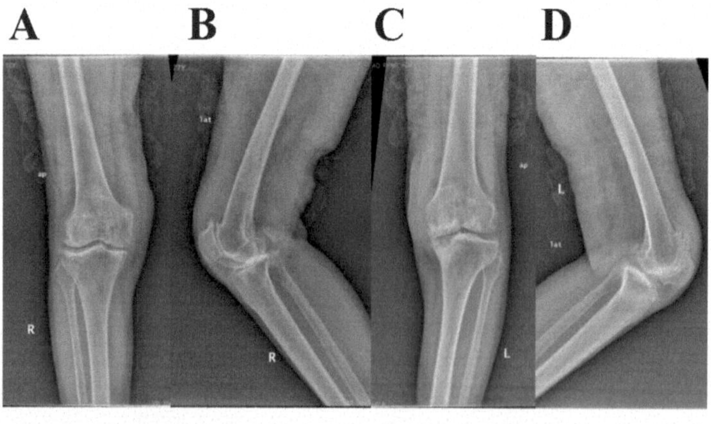

Fig. 23.3 Knee joint AP and LAT X-ray. The knee joint AP and LAT X-ray reveal bone destruction in medial tibial plateau of left knee and uneven thickening around cartilage of right knee

in imaging findings include bone, peripheral osseous tissues, and joint, which are the major basis for X-ray typing. There are eight most common manifestations: (1) trabecular bone abnormalities; (2) changes in bone density; (3) calcification of calf interosseous membrane; (4) interosseous membrane calcification of forearm together with changes in the attachment of the pronator muscle; (5) intercondylar eminence hyperplasia; (6) changes of elbow joint; (7) spinal changes with different degrees of vertebral ligament calcification; (8) calcification and degeneration of obturator membrane.

Skeletal fluorosis arthropathy is commonly seen in clinical practice, which patients are found with knee arthralgia, limited movement, joint dysfunction, ankylosis, morning stiffness, joint curve contracture deformity, and all found with dental fluorosis. The medical imaging test shows that the bone mineral density increased, the bone cortex thickened, the medullary cavity narrowed or disappeared, the knee joint space narrowed, intercondylar eminence with hyperplasia, bone articular surface eroded; also, there are arthrosclerosis, joint episome forming, periarticular edge osteophyte, calcified periarticular ligaments, calcified articular cartilage, and hyperplasia, together with different degrees of bone destruction.

Differential Diagnosis to Severe Knee Osteoarthritis Induced by Chronic Fluorosis

Severe knee osteoarthritis induced by fluorosis should be identified from the following diseases:

Rheumatoid Arthritis

Rheumatoid arthritis is a chronic systemic autoimmune disease characterized by joint synovitis. It has a high morbidity and disability rate in young adults, mostly affecting small joints and then large joints in later stages. Positive rheumatoid factor accounts for 3/4 in adult patients with rheumatoid arthritis, and anti-cyclocitrulline antibody has certain diagnostic significance. X-ray test showed trabecular bone decrease, atrophied, and thinning in periarticular bone.

Pharyngeal pain, fever, and increased white blood cells are common in sudden onset of such disease. It often involves with large joints of the extremities. Migratory joint swelling and pain show no permanent damage after symptoms disappear. Hyaluronidases are all positive, and salicylic acid preparations treatment is often rapid and significant.

Osteoarthritis

It is a comprehensive and multi-layered chronic inflammation caused by degenerative changes in articular cartilage, synovium, joint capsule, and other structures of the joints, affecting mainly the elderly. The most commonly affected parts are hand, knee, hip, and spinal joints. However, there are no thickened bone cortex, narrowed medullary cavity, or joint soft tissues sclerosis. Bone and arthralgia symptoms of skeletal fluorosis include consistent rest pain in neck, waist, extremities, and major joints, which are free from seasonal and climate changes. Symptoms might be accompanied with limb seizures, numbness, and morning stiffness.

Surgical Treatment of Knee Osteoarthritis Induced by Severe Fluorosis

Operation Indication

THA could serve as an option for patients suffering from severe arthralgia, instability, and deformity caused by various reasons affecting people's daily life seriously or for elderly patients received conservative treatment with no expected outcomes (Su et al. 2012). THA is also suitable for the young patient suffering from dysfunction because of hamarthritis. THA can alleviate pain obviously, correct severe flexion contracture deformity, and improve joint knee joint function as well as quality of life.

Preoperative Preparation and Evaluation

It is necessary to investigate patient's physical condition before surgery on the following items: history of diabetes, hypertension, coronary arteriosclerotic cardiopathy, anemia, hypoproteinemia, chronic infection disease, history of taking steroid hormone or non-steroid hormone drugs, and immunosuppressive agents. Examining deformity, muscle tone, muscle strength, and knee periarticular soft tissue to understand the condition of bone mass, knee joint, and treatment plan by knee joint weight-bearing X-ray test. Assessing the patient's desire and ability for rehabilitation, and evaluating the efficacy and risk of knee replacement objectively.

Prosthesis Selection

The knee joint damage caused by fluorosis often shows different degrees of calcification and cruciate ligament contracture. Therefore, the post-stable prosthesis is generally used. A restrictive prosthesis type can be chosen if knee deformity is severe.

Surgical Operation

After anesthesia, supine position is adopted on a sterilized drape. Affected lower limb is hemostasised with tourniquet. Median knee incision is performed in a genuflex position. By entering the joint from the patella medial side, osteophytes and synovium, meniscus, anterior/posterior cruciate ligament are removed. In this way, soft tissue can be released. Positioning the hole 7 mm before the hemostatic point of the posterior cruciate ligament in the intercondylar fossa. The distal femur osteotomy mold is installed after measuring external rotation eversion. Tibial plateau is prolapsed in genuflex and external rotation position and then tibial plateau femur osteotomy mold is installed. After confirming the line of force and determining the size, pendulum saw is used for the osteotomy. Osteophyte is removed from the lateral condyle of the tibia, femoral condyles, and lamina patella margin. Determining the size of femoral osteotomy mold, then the mold is installed and the external rotation of femur is confirmed. After that, four remaining facets of the distal femur are removed, and femur test model and tibial plateau test model are installed. The knee joint is stabilized when it is flexed at 90° and 45°, and range of motion (ROM) reaches 0° to 120°. The patella has a good motion trail, and the medial and lateral soft tissues are well balanced. After rinsing the osteotomy mold surface with saline, tibial prosthesis and femoral prosthesis are installed with antibiotic bone cement. Excessive bone cement is removed and liner is paved. Reset the joint and straightening the knee until the bone cement is solidified, and then suture the incision and place negative pressure drainage after rinsing and hemostasis (Luo et al. 2012).

The surgery for skeletal fluorosis joint is difficult, because the calcified tissue is sclerotic and thick. It is hard to reconstruct the correct line of force relationship of the lower limbs. Therefore, the following issues should be noted. (1) Correct osteotomy during surgery is important. (2) Soft tissue loosening is an important part of the operation, which directly affects joint stability and function after surgery. After resecting posterior cruciate ligament by conventional methods, the methods described in the literature should be applied: protecting medial posterior vascular and neural tissues of the knee, removing the capsule of both sides on the back of knee until the internal fat is exposed. After the complete loosening of the posterior articular capsule, the reflexed articular capsule at the back of the joint capsule (close to the femoral condyle and the tibial plateau) is bluntly separated and adhered, and the medial collateral ligament is fully loosened to the attachment point of the pes anserinus. Patients with skeletal fluorosis still have a high fluoride content in the body after surgery. There is possibility for calcified contracture of periarticular soft tissue after replacement, so flexion contracture deformity could be induced again. Therefore, it is important to maximize soft tissue loosening during the operation and keep knee prosthesis in extended position (overextended by 3° to 5° if necessary).

(3) All cases in this group are associated with flexion and varus deformity. If the traditional surgical approach is used to expose the knee joint and overturn the patella, it will lead to excessive tension on the patellar tendon insertion and even result in patellar tendon rupture. After the proximal quadriceps femoris is loosened and overturns the patella, it can avoid rupturing the patellar tendon. (4) The patellar edge should be cauterized for denervation and patellar articular surface should be drilled to reduce the pressure of patella. In this way, it can improve the degree of patella sclerosis and reducing articular pain caused by patellofemoral osteoarthritis after surgery. (5) The drilled holes on sclerotic articular surface of the tibial plateau can be used for prosthesis stabilizing by filling with bone cement.

Postoperative Management

Using antibiotics after surgery, anticoagulation treatment with low-molecular-weight heparin should be adopted for 1 week and conventional anesthesia treatment is performed. Drainage tube must be pulled out 24 h after operation. If the high fluoride state in the body of patients with skeletal fluorosis fails to be completely controlled after surgery, it may still cause the calcified contracture of the knee periarticular soft tissue after replacement, or the knee joint flexion contracture deformity may occur again due to improper postoperative treatment. To avoid such results, it is usually advisable to take the following measures. (1) Improving drinking water quality and changing coal-burning habits, and keeping zero-high fluoride intake, together with comprehensive treatment of skeletal fluorosis for long-term efficacy of surgery. (2) Knee flexion and extension functional exercises are necessary as soon as patient regains consciousness after anesthesia. The CPM functional exercise should start on the 3rd day after the operation (starting from 30° and increasing by 15° every day), and aiming for knee flexion over 93° when discharged. Start walking with a walker or crutches 5–7 days after surgery, practicing weight-bearing walk while strengthening gait and balance training, and aiming for flexion between 90° and 120° when discharged. When joint pain is relieved significantly, patients can practice weight-bearing walk. When knee joint flexion cannot flex 90° within 2 weeks, then the flexion deformity is very likely due to the poor postoperative efficacy. If so, massage correction under general anesthesia can be considered in the operating room, and patients should be advised to continue knee flexion and extension exercise after discharge.

Skeletal fluorosis is a complex chronic systemic disease and medication is of vital importance too (Yuan et al. 2020). Magnesium can reduce the absorption of fluoride in the body, adjust and repair the enzyme system dysfunction caused by fluoride, abnormal bone, and collagen metabolism, and also reduce the direct damage and toxicity of fluoride. Boron preparation can bond with fluoride ions to form insoluble fluoride to be excreted with feces, or excreted from the kidney in the form of BF^{4-}, hence, achieving the effect of preventing and treating skeletal fluorosis. Serum calcium of skeletal fluorosis is not constant. It might be caused by secondary

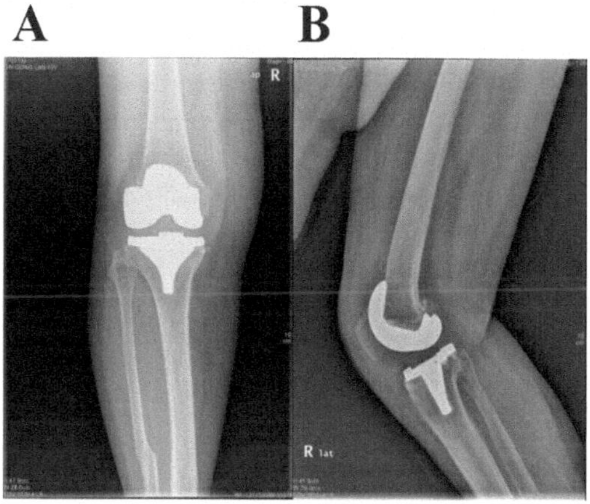

Fig. 23.4 X-rays image of total knee arthroplasty. Postoperational X-rays image of total knee arthroplasty, the imaging results showed that the prosthesis aligned well, the tibial surface was covered well, and the patella joint track was normal

hyperparathyroidism. Osteosclerosis is mainly the result of accelerated bone turnover and enhanced osteogenesis, thus, calcium supplement can inhibit hyperparathyroidism, stabilize blood calcium, and reduce bone turnover without aggravating sclerosis or ossification around the bone. At present, there is no agreed standard for fluorosis medication or systemic treatment and specific treatment standards.

This research chose 27 patients with total of 32 knees for treatment and research purpose. One week after the surgery, the patients were reported that the knee joint pain was significantly reduced, and the flexion deformity was significantly improved. After 2 weeks treatment, patients felt that the joint function, mobility, and muscle strength were improved, and the incision healed well without complications (such as wound infection). The knee HSS score before treatment was 38.4 points on average. After treatment, the score was 84.6 points, and the joint mobility was above 93°. After functional exercise, the knee joint flexion could reach above 100°, and patients were able to walk and use stairs normally, indicating an obviously increased mobility of patients. The imaging results showed that the prosthesis aligned well, the tibial surface was covered well, and the patella joint track was normal (Fig. 23.4).

Surgical Treatment of Spinal Stenosis Induced by Chronic Fluorosis

Clinical Manifestations

According to different segments of stenosis, the early symptoms are numbness and weakness in the lower limbs, cotton-like-ground feeling when walking, and bandaged feeling in chest and abdomen; both lower limbs have alternative spasms sometimes and are exacerbating progressively. Difficulty in walking in the advanced stage, walking assistance is required from others or crutches. There might be difficulty in urinating and bowel movement, and paraplegia is found in severe cases. Accompanied symptoms are bone and joint symptoms (such as back pain, limited mobility, and extremity joints pain). However, when the stenosis is severe, the pain will be relieved below the compressed segment. Most cervical and thoracic spinal stenosis show signs of damage to the upper motor neurons. Physical check reveals sensory disturbance below corresponding level, loss of muscle power in both lower limbs, increased muscle tone, weakened or disappeared abdominal reflexes, hyperactive knee reflexes, and Achilles tendon reflex. Hoffman sign and Babinski sign were positive. There are also signs of degenerative changes in dental fluorosis and large joints of the extremities (Zhu et al. 2017).

Medical Imaging Check

Imaging findings are important criteria for the diagnosis of skeletal fluorosis, spinal cord compression, and also serve to locate the position for surgical treatment. According to the classification of skeletal fluorosis, the X-ray results of the spine can also be divided into osteosclerosis, osteoporosis, osteomalacia, and mixed types. Osteosclerosis is the most common type, with different degrees of spinal degeneration, calcification/ossification of ligaments, and soft tissues of (surrounding) spinal canal, which lead to spinal stenosis. The conventional X-ray results of patients with osteosclerotic skeletal fluorosis show increased density of vertebral body and appendix, thickened and enlarged trabeculae, osteophytosis, and different degrees of bony bridge formation at vertebral edge. X-rays results of osteoporosis and osteomalacia type of skeletal fluorosis show reduced bone mass in spine and deformed spine. X-ray examination can reveal calcification or ossification of the anterior longitudinal ligament and the ligamentum flavum, hypertrophy of the intervertebral facet joints, and reduced spinal canal and intervertebral foramen. CT examination can determine the location, degree, and scope of spinal canal stenosis, which has important clinical value for treatment. However, it is worth noting that CT shows poor results for the degree of spinal cord compression and fluctuated ranges of lesions. MRI can clearly show the local and overall conditions of the spinal cord and compressed nerve, as well as deformation of the spinal cord, which is of great significance for the evaluation of treatment efficacy (Quadri et al. 2016).

Laboratory Examination

Fluoride level of blood and urine is checked for clinical reference. Patients generally have mild and medium anemia. Immune-related indicators such as rheumatoid factor, anti-cyclic citrulline antibody, anti-nuclear antibody, and HLA-B27 were negative.

Diagnosis and Differential Diagnosis

The diagnosis of spinal stenosis induced by fluorosis includes dual meanings: the diagnosis of skeletal fluorosis and the diagnosis of spinal stenosis induced by fluorosis. According to the *Diagnostic Standard for Endemic Fluorosis* (WS 192-2008), the diagnosis of fluorosis must meet the following standards of epidemiological history: patient was born and lived in the endemic fluorosis area or moved to such area after birth for more than 1 year; meeting the clinical manifestations and X-ray results of skeletal fluorosis. After being diagnosed as skeletal fluorosis, medical imaging examination showed calcification or ossification of ligamentum flavum, intervertebral facet joints, paravertebral and intercostal ligaments; hypertrophy of the intervertebral facet joints, reduced spinal canal, and intervertebral foramen. Spinal cord/nerve roots were compressed, there is even spinal cord degeneration. With these clinical symptoms, patients could be diagnosed as spinal stenosis induced by fluorosis (Sellami et al. 2020).

Surgical Indication

The surgical indications for spinal stenosis induced by fluorosis are the same as general spinal stenosis. Non-surgical treatment can be used for atypical cases, patients received non-surgical treatment without expected outcomes, patients with obvious symptoms and signs of nerve compression and confirmed cases by clinical imaging are absolute surgical indications.

Anesthesia

General anesthesia is appropriate.

Surgical Incision

The surgical approach generally chooses a spinal midline incision, and compression site determines the segments suggested by medical imaging (Xia et al. 2007).

Choice of Surgical Method

According to the results of epidemiological investigation, the main cause of stenosis compression is calcification or ossification of the ligamentum flavum, followed by calcification or ossification of the posterior longitudinal ligament, degeneration and hyperostosis of the vertebral body, degenerated and herniated disk, hypertrophy of small joints. The current surgical procedures for this disease mainly involve relieving the compression of the ossified and thickened ligamentum flavum as well as the compression of the intervertebral foramen due to hypertrophy of the facet joints (Zhu et al. 2017). Due to the obvious anatomical differences in spinal segments, choices in surgical methods vary too. Cervical spinal stenosis induced by fluorosis is generally treated with PCEML (Cheng et al. 2008), and thoracic and lumbar spinal stenosis induced by fluorosis are generally treated by posterior total laminectomy or partial facetectomy for decompression (Xia et al. 2007).

PCEML for Cervical Spinal Stenosis Induced by Chronic Fluorosis

Under general anesthesia condition, the patient lies down with prone position, neck flexion is exposed with skull traction, C2 to T1 spinous process median incision (about 12 cm in length) is performed. The skin, subcutaneous tissue, and neck ligaments are incised. Paravertebral muscles are dissected to reveal C2 to Tl bilateral lamina, spinous process and lateral mass of right side. C6 and C7 spinous process are bitten off. Bone surface is filled with bone wax. The spinal process connections of C3–C7 are drilled for suspension. The outer lamina cortex of C3 and C7 are bitten through at 0.5 cm from the inner edge of the right articular process with abrasive drill and rongeur. The inner lamina cortex is retained and trimmed into a V-shaped bone notch, which are sutured through the spinous process hole of the same side to the joint capsule of the same side. Generally, the door is opened from left to right. On the left side, the C3–C7 whole layer lamina cortex is bitten through with abrasive drill and rongeur at 0.5 cm away from the spinous process and the bottom, and the right intervertebral ligamentum flavum of C7, T1 and C2, C3 is removed. The door is opened from left to right. The left interlamina ligamentum flavum of C3 to C7 is disconnected and the open door lamina is gently unclenched. The adhesion between the dura mater and the ligamentum flavum is separated. The laminas are unclenched at an angle of 60° with width of 1.5 cm. When the dural sac swells and the pulse recovers, the incision should be tightened and knot tied. The wound should be completely rinsed and hemostasised and covered with gelatin sponge, or covered with subcutaneous fat tablets or exposing dura mater. For patients with C2 compression, the C2 spinous process is retained, but 1/3 below the C2 lamina and anterior 1/2 of the C2 lamina are abraded. The anterior cortex of residual lamina is bitten off with a narrow thin lamina rongeur to enlarge the C2 spinal canal into a dome shape. A negative pressure drainage tube is placed in the incision (Seichi et al. 2001).

Surgery for skeletal cervical spinal stenosis induced by fluorosis is difficult and easy to injure the spinal cord, due to the severe stenosis, sclerotic and thick ossified tissue, and tight adhesion or complete occlusion of epidural space. Therefore, the following measures should be taken. (1) A high-speed burr should be used together with rongeur to notch on lamina gently to increase the speed and reduce the frequency of spinal cord stimulation, which is better than by using rongeur only. (2) It is advisable to enter from the stenosis normal sides of lamina space rather than forcefully entering from the most severe stenosis side. (3) In order to prevent tearing the spinal meninges and causing cerebrospinal fluid leakage (CSFL) when opening the door, due to the heavy adhesion between the dura mater and the ligamentum flavum, the lamina should be reamed thinner first and then bitten off a thin lip lamina rongeur. Locating a gap to enter the vertebra tube, the adhesion connecting the dura mater and the ligamentum flavum is carefully separated. The nerve peeler is used to peel off the adhesion, and the ligamentum flavum is picked and cut with a sharp knife. After the ligamentum flavum is completely cut off and properly underminingly dissected, the door is opened again. The lamina is uncovered 1 or 2 segments at a time to facilitate adhesion separation and cut off under direct view, in this way, it could avoid tearing the dura mater without causing new damage. (4) In order to prevent the re-closing door phenomenon is made into a "V" shape. The angle of the lamina opening is maintained at 45° 60°. The spinous process hole is suture and suspended on the joint capsule of the same side, and the then knot tied tightly with silk thread to expand the spinal canal effectively. (5) For C2 spinal stenosis, the C2 spinous process should be kept during the operation. The drill bit is at an angle of 50° to the lamina of the spinal cord when 1/3 below the lamina and anterior 1/2 of the C2 lamina is removed. Using a small drill bit to ream the anterior cortex of the lamina thin. The remaining anterior cortex of the lamina is bitten off by using a narrow and thin-bladed lamina rongeur to make a dome shape to expand the C2 spinal canal and retain the previously mentioned muscle.

Total Laminectomy for Decompression and Partial Facetectomy for Decompression for Chest and Lumbar Spinal Stenosis Induced by Chronic Fluorosis

Under general anesthesia in prone position, soft pillows are placed under chest and iliac crest to prevent abdominal pressure. The posterior midline approach is adopted for the operation. The resection range includes the entire lamina and the medial facet joint process. The upper and lower range should be more than the lesion area by one lamina. The skin, subcutaneous tissue. and fascia are incised, and the paravertebral muscles are dissected to reveal the spinous processes, bilateral lamina, and articular process to the root of transverse process. Generally, an "uncovering" type of laminectomy is used to decompress. Locating the segments that need to be decompressed, the spinous process should be bitten off. A ball drill or rongeur is used for the "V"-shaped notch at the midline of the articular process on one side. It should be noticed that the depth of the bone notch must not penetrate the inner lamina, which might shake or destabilize the lamina when the contralateral side is notched and might injure the spinal cord. The midline of the contralateral articular process is notched and its depth is required to penetrate the ossified nest, and then return to the original side and being notched until the dura mater is exposed. Finally, using the rongeurs to carefully pick one end of the lamina and remove the entire lamina and ossified nest. For limited number of patients, it is difficult to remove the whole lamina due to severe ossification of the ligament at the joint capsule intruding into the spinal canal, or continuous ossification of the long segment ligament. The posterior wall of the spinal canal can be dissected by segmented "uncovering" method. For certain cases with local ossified blocks that are embedded in the dura mater, the "floating method" can be used to remove the surrounding ossification.

For thoracolumbar and lumbar spine surgery, nerve root canal stenosis should also be treated in surgery. For inferior nerve root compression, a narrow osteotome is adopted to remove the medial 1/2 of the inferior articular process, the inferior nerve root is pulled to the inside, and the neuroprosthesis is placed into the lateral recess to protect the nerve root. Then, osteotome is used to remove the medial 1/2 of the upper articular process. For patients with severe intervertebral foramen stenosis, the lower and inner edges of pedicle can also be removed. For the compression of the same nerve root, since the nerve root is in lower and outward edges of pedicle, it can be probed in the lower edge of the same side lamina to check whether the top of the superior articular process of the lower vertebra is compressed. If so, the top of the superior articular process is removed to expand the nerve root canal for decompression (Sodeyama et al. 1999; Juan et al. 2003).

The special pathological features of thoracic and lumbar spinal stenosis induced by fluorosis make surgery more difficult than usual. Especially for patients with thoracic spinal canal stenosis, due to the small and easy bleeding spinal canal, surgery for decompression is much more difficult. Therefore, the following points should be noted during surgery. (1) Entering lamina space of normal sides of stenosis rather than forcefully entering from the most severe stenosis side. It is not advisable to forcibly peel off when the adhesion is severe. (2) The operation should be stopped immediately if spinal cord irritation occurs, and switch to work on the contralateral lamina on both sides alternatively. (3) When the ossification of the ligamentum flavum is fused with the lamina without gaps, the lamina can be reamed thin with a micro-electric abrasive drill and then bitted off. Locating the gap to enter the spinal canal or allow it to float freely. Surrounding laminae should be eroded instead of forcefully entering with osteotome (avoid injuring the spinal cord). (4) The operation must be performed gently, since minor carelessness can lead to laceration of the spinal cord causing postoperative dysfunction. (5) The range of decompression should be appropriate, and the width and length of decompression should be sufficient. (6) Those who have dura mater rupture during surgery should be sutured or repaired as much as possible, and those cannot be sutured or repaired should be placed with drainage tube. (7) Patients with skeletal fluorosis bleed heavily during the operation, and more than 800–1200 ml blood should be prepared before operation. (8) A negative pressure drainage tube should be placed for 24–48 h after surgery to prevent postoperative hematoma compression. Careful preparations must be taken before surgery. Surgery must be performed carefully, hemostasia must be performed carefully, adhesions should be separated, and gradually expanded (Iwasaki et al. 2002; Cheng et al. 2008).

Postoperative Management

The drainage tube is usually removed 24–48 h after surgery. Antibiotics are routinely taken for 3 days. First- and second-generation cephalosporins are used as antibiotics. Dehydration drugs, dexamethasone, or methylprednisolone are generally used for 3–5 days. Patients with skeletal fluorosis generally suffer from anemia. Therefore, under intensive nutritional support treatment, patients are instructed to change their living habits, including changing traditional stoves and staying away from high fluoride intake of coal-burning food. Such measures are long-term guarantee after surgery. Cervical spine decompression surgery generally requires patients to have 3 months' immobilization of neck. Patients received thoracic spine surgery can usually get off bed after extubation and gradually increase the amount of activity. Patients received lumbar spine surgery must stay in bed for 3–5 days and then get off bed with waist support and do no excessive activity within 3 months.

Prevention and Treatment of Surgical Complications

Cerebrospinal Fluid Leakage

Gelatin sponge is applied to cover small laceration in dura mater which is difficult to suture or repair. Large lacerations should be sutured or repaired as much as possible. Broad fascia can also be used when the laceration is large. Each layer of the incision must be sutured tightly with local compression bandaging. The lateral prone position was taken after operation. Observe the volume of drainage and color at any time. The tube must be removed in time when the volume of drainage is large and the color was light, which imply cerebrospinal fluid leak. Then the drainage orifice should be sutured deeply through the wound and to be compressed with a salt bag.

Epidural Hematoma

Patients with skeletal fluorosis suffered from severe intraoperative bleeding. In order to prevent epidural hematoma caused by incomplete hemostasis during the operation or impeded drainage tube, sufficient blood (800–1200 ml) must be prepared before surgery. Hemostasis must be performed strictly during the operation, and a negative pressure drainage tube should be placed for 48 h to prevent the postoperative hematoma from compressing the spinal cord. For patients with highly suspected epidural hematomas with neurological symptoms, it is pressing to check the incision urgently, clean up the hematoma, and stop bleeding; then, the wound should be washed with iced saline, and drainage tubes should be placed. Dehydration agents and hormones should be applied after surgery.

Spinal Cord and Nerve Root Injuries

To prevent damage on spinal cord and nerve roots, it is necessary to properly master abrasive drilling techniques. For the cervical spine during surgery, the angle of uncovered lamina after the door is opened, the position of the portal axis, and the shape of notch on the portal axis must be maintained. It shall not be attempted to remove the anterior compression object from behind after expanding spinal canal sagittal diameter. During the operation, staying alert for hemostasis and normal blood volume must be maintained to prevent spinal nerve ischemic damage. The ossified ligamentum flavum is generally adhered with the dural sac, so it must be gentle and slow to uncover the posterior wall of the spinal canal to avoid spinal cord injury. For the cases of spinal cord injury, dehydration, hormones (conventional application of dexamethasone 20 mg, ivgtt for 2 weeks, gradually reduce the dose till stop) and neurotrophic drugs should be applied.

Infection

At all stages of the operation, strict aseptic procedures are key to eliminating potential infections, improving drug resistance, and preventing infection.

Conclusion

With the implementation of preventive measures, such as improved water and stove projects in the area stricken by fluorosis, new cases and incidence degree of endemic fluorosis will undoubtedly decrease year by year. However, there are still a large number of patients with endemic fluorosis, especially those who are disabled due to severe hip, knee, and spinal injuries. It is pivotal to explore effective treatment and improve their quality of life. The application of THA and total knee arthroplasty to treat hip and knee injuries induced by chronic fluorosis can relieve arthralgia, improve joint function, and correct joint deformities. Expansive laminoplasty can effectively relieve spinal nerve compression and restore its function to varying degrees. Appropriate surgical methods can not only relieve or alleviate the pain caused by bone and joint damage induced by fluorosis, but also greatly improve the quality of life and restore their self-care ability.

References

Cheng HY, Shang XW, Liu RG. Expansive open-door laminoplasty in the treatment of cervical spinal stenosis due to osteofluorosis. Chin J Spine Spinal Cord. 2008;18:37–40.

Huang CQ. Joint damage of endemic fluorosis and its significance. Chin J Endem Dis Ctrl. 1999;18:159–60.

Huang CQ. Clinical diagnosis of endemic fluorosis. Chin J Endem Dis Ctrl. 2010;29:231–3.

Ibiyemi O, Zohoori FV, Valentine RA, Maguire A. Fluoride intake and urinary fluoride excretion in 4- and 8-year-old children living in urban and rural areas of Southwest Nigeria. Community Dent Oral Epidem. 2018;46:482–91.

Iwasaki M, Kawaguchi Y, Kimura T, Yonenobu K. Long-term results of expansive laminoplasty for ossification of the posterior longitudinal ligament of the cervical spine: more than 10 years follow up. J Neurosurg. 2002;96:180–9.

Juan AM, Jose IM, Amarie BP, Laguia-Garzaran M. Expansive laminoplasty as a method for managing cervical multilevel spondylotic myelopathy. Spine. 2003;28:680–4.

Kebede A, Retta N, Abuye C, Whiting SJ, Kassaw M, Zeru T, et al. Dietary fluoride intake and associated skeletal and dental fluorosis in school age children in rural Ethiopian rift valley. Int J Environ Res Public Health. 2016;13:756.

Luo R, Liu RG, Ye C, Yu YN, Guan ZZ. Total knee arthroplasty for the treatment of knee osteoarthritis caused by endemic skeletal fluorosis. Chin J Tissue Engine Res. 2012;16:1555–650.

Omid N, Maguire A, O'Hare WT, Zohoori FV. Total daily fluoride intake and fractional urinary fluoride excretion in 4- to 6-year-old children living in a fluoridated area: weekly variation? Community Dent Oral Epidem. 2017;45:12–9.

Quadri JA, Alam MM, Sarwar S, Ghanai A, Shariff A, Das TK. Multiple myeloma-like spinal MRI findings in skeletal fluorosis: an unusual presentation of fluoride toxicity in human. Front Oncol. 2016;6:245.

Saeed M, Malik RN, Kamal A. Fluorosis and cognitive development among children (6-14 years of age) in the endemic areas of the world: a review and critical analysis. Environ Sci Pollut Res Int. 2020;27:2566–79.

Seichi A, Takeshita K, Ohishi I, Kawaguchi H, Akune T, Anamizu Y, et al. Long-term results of double-door laminoplasty for cervical stenotic myelopathy. Spine. 2001;26:479–87.

Sellami M, Riahi H, Maatallah K, Ferjani H, Bouaziz MC, Ladeb MF. Skeletal fluorosis: don't miss the diagnosis! Skelet Radiol. 2020;49:345–57.

Sodeyama T, Goto S, Mochizuki M, Takahashi J, Moriya H. Effect of decompression enlargement laminoplasty for posterior shifting of the spinal cord. Spine. 1999;24:1527–31.

Srivastava S, Flora S. Fluoride in drinking water and skeletal fluorosis: a review of the global impact. Curr Environ Health Rep. 2020;7:140–6.

Su WM, Liu RG, Ye C, Yu YN, Guan ZZ. Total hip arthroplasty for the treatment of severe hip osteoarthritis due to fluorosis. J Clin Rehabil Tissue Eng Res. 2012;16:1543–6.

Xia YP, Xu TT, Zhang LL, Zhu RS, Tian R, Jiang H. Relationship between the development of thoracic kyphosis and long-term outcomes of laminectomy for fluorosis thoracic stenosis. Chin J Spine Spinal Cord. 2007;17:426–9.

Yuan L, Fei W, Jia F, Jun-Ping L, Qi L, Fang-Ru N, et al. Health risk in children to fluoride exposure in a typical endemic fluorosis area on Loess Plateau, north China, in the last decade. Chemosphere. 2020;243:125451.

Zhu C, Zhou S, Liu L, Song Y. Multiple levels of spinal canal stenosis in endemic skeletal fluorosis. Joint Bone Spine. 2017;84:737.

Chapter 24
Clinical Prevention and Treatment of Coal-burning Type of Endemic Fluorosis

Ling Mao and Yong Wang

Abstract Dental fluorosis is the result of excessive fluoride intake during tooth formation. Exposure to a high level of fluorides can cause enamel fluorosis and primary dentin fluorosis during childhood. Coal-burning type of endemic fluorosis is one type of chronic fluorosis. This fluorosis has a high prevalence rate in China. The clinical manifestations of the patients with dental fluorosis are mainly the changes in the color, texture, and morphology of enamel and dentin. According to the degrees of tooth damage, dental fluorosis is divided into three types: mild one shows slight accentuation of the perikymata with white opaque plaque (chalk type), moderate one shows white opaque enamel with some discoloring and pitting (colored type), and severe one with obvious plaque (defective type). Dental fluorosis affects not only the normal masticatory function, but also the physical appearance of the patients, which brings adverse effects of social, psychological, and other aspects to the patients. Effective prevention and treatment for the coloring and morphology change of dental fluorosis have long been the focus of dentists.

Keywords Clinic treatment · Dental fluorosis · Personalized staining · Prevention

Introduction

There are three types of endemic fluorosis, namely drinking water, coal-burning, and brick-tea type of endemic fluorosis. The most typical clinical manifestations are dental and skeletal fluorosis. Dental fluorosis is often found with age related characteristics. Fluoride mainly damages the ameloblast of the tooth germ during the period of enamel development. Excessive fluorine enters the body during the period of tooth development and mineralization and results in permanent dental fluorosis. Deciduous dental fluorosis indicates that the calcification of deciduous

L. Mao (✉) · Y. Wang
The Key Laboratory of Head and Neck Cancer Research, Hospital and School of Stomatology, Guizhou Medical University, Guiyang, Guizhou, People's Republic of China
e-mail: maoling@gmc.edu.cn

Z.-Z. Guan (ed.), *Coal-burning Type of Endemic Fluorosis*,
https://doi.org/10.1007/978-981-16-1498-9_24

teeth begins in the embryo and is fully developed within 11 months after birth. Therefore, the deciduous dental fluorosis is found in 11 months old infants. Permanent dental fluorosis means that children, who were born in the endemic fluorosis area and raised in such area before the age of 7–8 years old were affected by excessive intake of fluoride. Permanent dental fluorosis cannot be changed once it is formed. However, children who move into the high fluoride content area after permanent tooth eruption will not suffer from fluorosis plaque. In addition to damages of the teeth, dental fluorosis deforms physical appearance of the teeth, so that patients also suffer from psychological pressure in everyday contact. Therefore, the clinical treatment of dental fluorosis has important medical and social significance. This essay focuses on how to classify the severity of dental fluorosis and how to provide a suitable treatment plan for the patients.

Index, Clinical and Differential Diagnosis of Dental Fluorosis

Index of Dental Fluorosis

Dean index (DI) is known as the "gold standard" in classifications of the dental fluorosis index. In 1934, Dr. Dean developed an index to describe and diagnose enamel fluorosis based on his interpretation of clinical manifestations (Dean 1934). This classification is still widely used at present after being modified in 1942 (Dean 1956). DI classifies teeth with fluorosis into six categories based on clinical manifestations.

The *Community Fluorosis Index* (CFI) was developed by researchers in 1946 to measure the burden of fluorosis in a population. Instead of merely measuring the overall prevalence of dental fluorosis (i.e., proportion of people suffer from fluorosis), it takes the severity of occurring fluorosis into account. The CFI is thus a measurement of both prevalence and severity.

Although the DI is still widely used in the investigation of fluorosis, more sophisticated diagnostic scales have been developed. These include the *Thylstrup-Fejerskov Index* (TFI) and the *Tooth Surface Index of Fluorosis* (TSIF) (Thylstrup et al. 1978) which can characterize the macroscopic appearance of teeth in relation to the underlying histological condition of enamel. It expands detection on the range of more severe forms of dental fluorosis. The DI identifies five types of dental fluorosis (suspected, very mild, mild, moderate, and severe), while the TFI identified nine.

In 2011, the *Clinical Diagnostic Criteria for Dental Fluorosis* (WS/T 208-2011) was formulated by the Professional Committee of Endemic Disease Standards in China, which was later approved by the Ministry of Health of China. In the clinic diagnostic criteria, dental fluorosis was classified into five types according to the gloss of enamel surface, the degrees of color change, and defects (Table 24.1), which is issued by the Ministry of Health of the People's Republic of China.

Table 24.1 Clinical diagnostic criteria for dental fluorosis (WS/T 208-2011)[a]

Classification	Criteria
Normal	Enamel is translucent, pale, and creamy white, the surface is smooth and glossy
Suspected	Enamel is found with slight aberrations from the translucency of normal enamel, ranging from a few white flecks to occasional white spots. This classification is utilized in those cases when a definite diagnosis is pending and "normal" classification cannot be justified
Very mild	Small, opaque, paper white areas scatter irregularly over the tooth, approximately 25% of the tooth surface. In general, tooth show no more than 1–2 mm of white opacity at the tip of the cusps, tip of the bicuspids, or second molars
Mild	The white opaque areas on enamel affect 1/4 of the tooth surface, or even the entire tooth surface, and the teeth are dull. Some areas of the tooth surface show abrasion, and the anterior teeth are sometimes occasionally blurry stained
Moderate	The white opaque area spreads over the entire tooth surface, and there are tiny independent pit-shaped defects on the labial side or buccal side of the tooth. Tooth may show significant abrasion, but tooth morphology does not change significantly, and brown stain is frequently a disfiguring feature
Severe	All enamel surfaces are affected. Hypoplasia is so significant that the general form of the tooth may be affected. The enamel defect appears to be fused, band-like, or flaky, teeth color can range from brown to near black; teeth present a corroded-like appearance

[a]Diagnosis of dental fluorosis, cited from Health Care Standard of the People's Republic of China (WS/T 208-2011), which is issued by the Ministry of Health of the People's Republic of China

Clinical Diagnosis

The diagnosis of dental fluorosis takes into account clinical examination, medical history, and history of long-term excessive fluoride exposure during tooth development. Concerning the clinical examination, it is pivotal to clean the labial and buccal surface of teeth, so that the dental surfaces remain clean and dry. After examining the enamel damage of each tooth, two teeth with the most severe damage are chosen for fluorosis classification diagnosis based on the degree of enamel damage. If the selected teeth fall into different classifications, the diagnostic classification of patients with dental fluorosis shall be based on the less damaged fluorosis tooth. Primary teeth and permanent teeth with fluorosis should be recorded separately. Only permanent teeth with fluorosis should be checked when both primary and permanent teeth are presented in the dentition. Combined with clinical examination, there are individual variations in clinical fluorosis manifestations which are highly dependent on the duration, timing, and amount of fluoride exposure. Enamel shows slight aberrations related to the translucency of normal enamel. The sporadic white spots on the teeth were combined with a hard and smooth enamel surface (Abanto Alvarez et al. 2009) with increasing severity; the subsurface enamel became porous. Enamel might show yellow or brown discoloring or pitted white-brown lesions. However, fluorosis is not the cause of discoloring to the enamel directly. Fluorosis caused subsurface porosity or hypo-mineralization, which infested the dentinal-enamel junction as severity increased (Ritter 2005). Hence, affected teeth were more susceptible to coloring. Due to diffusion of exogenous ions, these colorings could develop into an abnormally porous enamel which then would exacerbate severe dental fluorosis.

Differential Diagnosis

Exogenous stain colored teeth should be distinguished from dental fluorosis. Exogenous stain colored teeth show manifestations of plaque, tartar, soft tartar, and pigment (smoke, tea) stains deposited on the tooth surface, etc. Coloring is often heavier on the lingual side than on the labial side and buccal side, and the mandibular teeth coloring is heavier than the upper teeth. Such exogenous stain on tooth surface can be removed.

Enamel defects: undiagnosed and untreated celiac disease can cause enamel defects.

Enamel hypoplasia: during tooth development and mineralization, enamel developmental disorders are caused by nutritional deficiencies, endocrine disorders, or hyperthermia in infants and mothers. The formation of banded or dimpled depressions on the enamel surface is the main feature of the disease. The depressions are often brown colored. The main proof of diagnosis is the disease occurs in the same period when the teeth are formed and developed. During the physical check, if the surface of the suspected tooth is smooth and hard, and the color and transparency of enamel are not affected, the enamel is diagnosed as normal.

Tetracycline teeth: taking tetracyclines during tooth development and mineralization leads to tetracycline calcium orthophosphate complex and diffuses the teeth colors ranging from light gray, light yellow, to dark gray, dark yellow, or dark brown.

Dental trauma: trauma to deciduous teeth may affect mature stage of enamel formation, which may lead to opaque enamel in permanent teeth (Skaare et al. 2013).

Dental caries: fluorosis-resembling enamel defects are often misdiagnosed as dental caries (Sabokseir et al. 2016). Such disease shows features of progressive damage on enamel, dentin, or cementum under the influence of external factors. Dental caries mostly occur in the gaps and adjacent spaces of the teeth. The lighter ones can be stained to brown-black spots, and the surface is tarnished. Severer ones can be seen with caries pit, and the enamel area at the caries lesion is rough and soft. The subjects are more sensitive to cold, hot, sour, and sweetness.

Treatments of Dental Fluorosis

At present, there is no standard for clinical treatment of dental fluorosis, but there are some mature treatment experiences. In the case of mild fluorosis, no treatment is required. Bleaching can be used to treat stained fluorosis teeth without substantial defects; more repair methods are required for those with substantial defects. The main types of treatment are described below.

Tooth Whitening

The tooth color change of fluorosis plaque teeth seriously affects the esthetic appearance of patients. For mild dental fluorosis, tooth whitening is an effective method. Such method aims either changing the intrinsic color or removing and controlling extrinsic stains (Joiner and Luo 2017). The chemical degradation of the chromogens within or on the tooth is reversed by bleaching (Fig. 24.1).

Night whitening treatment at home: home whitening is a patient self-performed therapy that is prescribed and prepared by a dentist. Patients need to actively participate in their treatment and follow the guidelines accurately. The method is to conduct a conventional check first. A plaster model is prepared for the occlusal tray; the decoloring reagent is then placed on the occlusal tray; after wearing the occlusal tray, the decoloring reagent is applied to cover the entire lip surface, and each decoloring time should be no less than 10 h. The 11% urea peroxide gel preparation is also used to make the support film. The decoloring reagent is placed in the support film to guide the patient to wear on the stained teeth after cleaning the teeth every night and should be removed the next morning (each time should be no less than 2–3 h). 20 d is a course of treatment, each course requires 10–15 ml of 11% urea peroxide gel. Compared with the past bleaching technology, this treatment can release peroxide slowly for a longer time in low concentration. It has the advantages of safety, easy to handle, and effectiveness (Carey 2014). Home-based bleaching would result in less tooth sensitivity than in-clinic treatment. Mishandling or inaccurate use of the bleaching trays could cause harm to the patient, such as blistering or sensitivity of the teeth and the surrounding soft tissue. Inconsistent use of the bleaching trays can lead to slowing-down and irregular of whitening process (Llena et al. 2020). Some patients with serious gag reflex may not tolerate the trays and would need to consider other methods of teeth whitening.

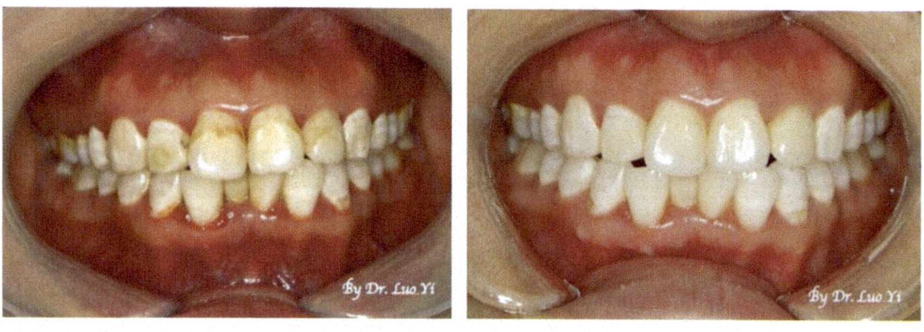

Fig. 24.1 Comparison of before and after grinding and bleaching against fluorosis

Whitening treatment in-clinic: a cotton swab is used to dip into bleach solution (36% hydrochloric acid, 30% hydrogen peroxide, and ether) and then it is applied on the targeting tooth surface. Such process is repeated until reaching satisfactory results. The bleach solution containing hydrochloric acid will cause the surface of enamel to be rough and easy to be stained again easily. In order to overcome this shortcoming, crack-blocking reagent is applied on the tooth surface after decolorization, and the teeth are irradiated for 40 s. Then, the teeth are observed by an electron microscope to check if the effect. The essence of chemical decolorization is enamel demineralization, and the staining substance is removed as demineralizing. Hydrochloric acid has a strong demineralization capacity, and the penetration of hydrochloric acid demineralization shares a very close relationship with the density of dental fluorosis. This layer is gently polished for loosening before decoloring to make the enamel surface rough. In this way, the wettability of the hydrochloric acid solution is improved on the enamel surface, which can promote the penetration of the decolorizing reagent and superficial demineralization. Mechanism of decoloring may be the hydrogen peroxide caused demineralization of the surface layer, which then changes the enamel structure and its transparency; at the same time, hydrogen peroxide can rapidly decompose at tissues and trace heavy metal ions, then release active nascent oxygen to bleach colored substances. When 30% hydrogen peroxide is used in treatment, not only can the colored area be decolorized, but the non-colored area can also be whitened. When hydrochloric acid is used for decolorization, different degrees of chalk-like changes can be seen after the enamel surface is dried, but the gloss disappears. It takes 3 months to restore the gloss. Successful bleaching largely depends on the degree of tooth decoloring, the cause of staining, and the length of time the pigment penetrates the tooth tissue. Hydrogen peroxide works not only on enamel, but also on dentin, cementum, and even dental pulp tissue. Sixty-seven percent to seventy-eight percent of patients have symptoms of tooth sensitivity. The entire treatment process can be completed in about 40 min, indicating great convenience. This technology received special optical improvement to reduce the irradiation temperature greatly, hence the damage to dental pulp tissue is reduced during operation, and occurrence of tooth sensitivity symptoms is also reduced greatly. Such technology shows ideal bleaching effect and short treatment time for most patients with mild fluorosis. Soft and hard tissue damage in the mouth is not obvious.

Light-accelerated bleaching: light-accelerated whitening technology is a new type of whitening treatment compared to home bleaching. Apart from shorter treatment time, it also reduces the toxic and side effects from bleaching. Therefore, patients show better co-operation in treatment. The combination of light sources allows the whitening reagent to quickly penetrate the surface and deep layers of the teeth to exert better bleaching effect. The principle of light-accelerated bleaching is to transmit high-intensity blue light (with a wavelength between 480 and 520 nm) through an optical fiber and then through two optical lenses which are coated more than 30 times to remove harmful ultraviolet and infrared. The whitening reagent (stored in a carrier with a diameter of 20 nm) causes redox through the enamel, dentin tubules, pigments deposited on the surface and deep layers of the teeth in the shortest time to achieve the effect of whitening. The whitening reagent has excellent hydrophilicity and whitening effect with no need of touching the gums at all and will not cause damage to teeth and periodontal tissues. It has the advantages of less irritation to the pulp, short treatment time, and obvious efficacy. Under the same condition (high concentrations of hydrogen peroxide are used), use of light during bleaching increases the risk of tooth sensitivity and may not be more effective than without light (He et al. 2012). A 2015 study showed that the use of light activator did not improve bleaching with measurable effects and was likely to increase the temperature of the associated tissues resulting in damaging gums and dental tissue (Baroudi and Hassan 2014). However, the long-term efficacy of cold light whitening technique requires further observation, since it showed poor efficacy to moderate/severe fluorosis.

Composite Resin

The method of enamel abrasive erosion includes: (1) patient is over 18 years old; (2) the tooth is evaluated as mild or low in Dean's fluorosis index, without dental tissue defects (mild, very light, and suspected); (3) the tooth is evaluated as moderate in the Dean's fluorosis index but patients with relatively lower requirements on decolorization.

It is necessary to communicate with patients before treatment to clarify the goals of treatment, routinely take preoperative photos, and prepare research models if necessary. First, wet blasting is used to remove the pigment on the surface; the degree of enamel deformity of the anterior teeth plaque is observed to determine the amount of grinding and cutting is required in the enamel layer for shaping. A diamond bur is chosen for later use. Mesial and gingival enamel layer are cut. The tooth cusp surface should be kept evenly and stained layer should be removed by about 0.1–0.2 mm in wet condition. The enamel is ground and cut gradually. In this process, observing the effect and leaving space for polishing. Observing the effect of grinding from different angles, and the shape and radian of teeth and their counterparts from opposite side should be as symmetrical as possible. Shape of the teeth, from the mesial surface to the point of abutment, should be noted. The abutment relationship should not be damaged and it is not suitable to grind the stained plaque area too deep and leave dents. Sufficient cooling water should be prepared and used during grinding to prevent heat from damaging the pulp tissue. The tooth surface is cleaned with 95% alcohol to remove the organic matter on the tooth surface. The tooth surface is etched with 35–50% phosphoric acid for 2 min, then is rinsed with clean water and blown dry. Adhesive is evenly applied on teeth and is treated with a light-curing machine for 40 s. Teeth is polished with polishing paste. The treatment of abrasive erosion can not only meet the esthetic requirements of the patient, but also avoid damaging the dental tissues or affecting the tooth function. In addition, it has preventive values of anti-caries, desensitization, and polishing compared with other methods. For deeper pigmented parts, it can be combined with the composite resin repair technique.

Light-curing composite resin veneer: this repairing method features with being simple, flexible, less grinding, low cost and can be completed with one try. However, since intra-oral operation is affected by multiple factors, it is also difficult to meet the ideal requirements of edge contour and surface texture of the veneer. The long operation time next to the chair also limits its clinical application. Photosensitive resins have defects such as poor density, incomplete surface curing, discoloration of edges and bodies, and poor surface quality. This method refers to the use of light-curing composite resin to directly shape, cure, and polish the dental cavity and complete dental restorations in the end. Therefore, light-curing veneers are now mostly used for small-scale, individual tooth restorations or temporary veneer restorations. For cases with severe stained fluorosis, in order to achieve satisfactory color, the thickness of the resin needs to be increased. As the thickness of the resin increases, comfort of the patient decreases, the esthetic appearance of the teeth becomes worse. In addition, the resin veneer is prone to wear off and age, and plaque accumulation on the edges would affect periodontal health. However, as long as veneer is tightly insulated, thoroughly cleaned, fully acid-etched, and the edges of the prosthesis are tight, the thickness is suitable after grinding and polishing, and oral hygiene and proper diet are maintained, the repair success rate can be greatly improved. There are two kinds of direct and indirect methods.

Specific steps of direct method are as follows. (1) Sharp, cylindrical, flame type diamond burs are chosen to grind the colored or loose enamel surface (the thickness is generally 0.3–0.5 mm). It can be slightly deeper as long as it does not damage dentin. (2) Color comparison: according to the patient's age, face color, and color of adjacent teeth, the color is compared under natural light. There are many types of composite resins on the market, and their performance and usage are different. Doctors should carefully read each factory's instruction manual and then practice in line with instructions to achieve good results. (3) Erosion of affected teeth: a rubber barrier is used to fully isolate moisture to prevent saliva contamination. A special small hairbrush is dipped in 35–55% phosphoric acid to rub the tooth surface for 30–60 s, and then the teeth are repeatedly rinsed with distilled or running water. Patients should not gargle during the procedure. Finally, the tooth surface is blown dry, and the tooth surface is chalky or beige. (4) Coating adhesive: a small brush is used to apply the adhesive on tooth surface after acid etching, then it is blown gently with an air gun to make it evenly cured. After 20 s of visible light irradiation, the adhesive is primarily cured. Then the color-blocking reagent is applied 1–2 times according to the esthetic needs. (5) Repair form: stainless steel sculpting knife is chosen to push appropriate amount of composite resin to the required part. The resin must not be pushed beyond the edge or deep into the gingival sulcus and compress the gums; in addition, the pressed composite resin must have no air bubbles. Doctors should try to apply restoration to reflect the structure of the tooth surface and complete curing. After satisfactory shaping, the resin can be irradiated with visible light for 40 to 60 seconds according to the thickness and color of the material. (6) Trimming and polishing: the resin surface is removed and polished. Articulating

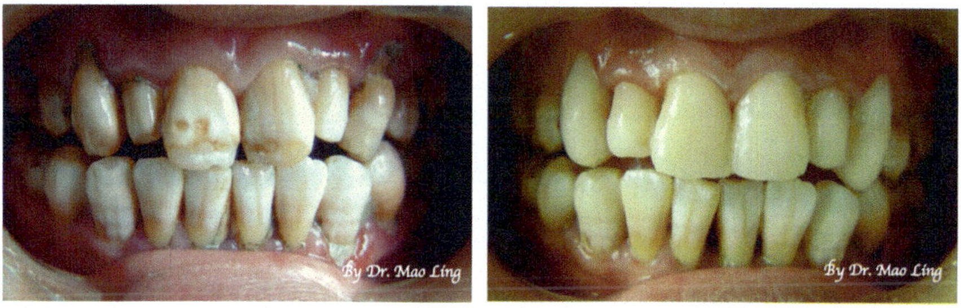

Fig. 24.2 Comparison of before and after light-curing resin treatment

paper is used for centric, protrusive, and lateral articulation to detect early contact, grinding and polishing requires high precision, and operating principle is from coarse to precise. Such process is carried out with dripping water with grinding speed about 1000 r/min. Finally, resin is polished with a rubber grinding cup or cloth wheel dipped in extremely fine abrasive. After polishing, the resin has the same refractive index. The surface is extremely smooth, and it is difficult for plaque depositing.

The composite resin prosthesis should be carefully taken care of and be maintained to extend its operating life. Practical way is to keep the mouth clean. It should be noted to choose soft toothbrush when brushing teeth to reduce resin wearing and prevent gingivitis. Drinks and foods which might stain the teeth should be consumed less, such as strong tea, coffee, and vinegar. Patients should bite hard objects under no circumstance, because the cut edge is the easiest to break off. In addition, repaired part should receive physical check regularly to detect luster, abrasion, oral cleanliness, etc. Further treatment could be necessary according to the teeth condition (Fig. 24.2).

In the oral environment, aging and discoloring of the composite resin are inevitable. The reasons that caused inconsistent color are: rough and unsmooth surface of composite resin, the infiltration and deposition of pigments after eating, the untreated original gum inflammation, and the resin aging. In order to avoid discoloring of the composite resin as much as possible, it is best to use silicon chips for polishing. After polishing the tooth surface, a layer of adhesive is applied and irradiated for 20 s to increase the gloss for better results.

The indirect method is to mod for impression and complete the indirect veneering on the model after the preparation of prosthesis. Then it is bonded by adhesive. It is particularly suitable for cosmetic repair of large-area multiple fluorosis teeth, which can greatly save the operation time. Also, such method is well designed and boasts good results (Peumans et al. 2004).

Porcelain Veneer

The selection of an appropriate treatment plan depends on the severity of fluorosis (Akpata 2001). Bleaching and micro-abrasion have been recommended for treating mild cases of fluorosis; however, in moderate to severe cases, these two are either ineffective or may lead to only transient improvement (DenBesten and Li 2011). Composite resin restorations are prone to discoloring, chipping, and debonding. Porcelain veneer is considered as ideal treatment for moderate to severe cases of fluorosis with optimum esthetics results, wear resistance, biocompatibility, and long operating life of veneers (Rotoli et al. 2013) (Fig. 24.3).

The use of ceramic laminate veneers as an alternative over full dental crown-coverage is clinically prevalent because of its minimum invasion and esthetic superiority (Taşın et al. 2020). Modern dentistry should keep pace with biomimetics or bio-emulation concepts (Bazos and Magne 2011, 2014): restorations should yield as far physiologic function of the natural tooth as possible, integrated with biologic, bio-mechanic, functional, and esthetic functions (Tirlet et al. 2014). A number of clinical studies have found that laminate veneers bonded upon enamel substrates yielded positive clinical outcomes over a period of 10 years or more (Menini et al. 2018; Swain et al. 2018). Different treatment options should be considered when adhesive restorations in the anterior region are required. The selection of restorations is based on the following criteria: tooth vitality preservation, minimum loss of dental tissue, minimally invasive approach on gingival health, esthetic demands, patient's age, total cost, and total treatment time. Further parameters that should be considered are as follows: the number and extent of involved teeth, type of function, feasibility of functional and anatomical recovery of the restored tooth, and biomechanical resistance of the restored tooth. The treatment order is as follows: intra-oral and extra-oral esthetic analysis of patients through static photo documents and dynamic video coupled with digital previsualization by means of DSD. Clinical previsualization adopts mock (Magne and Belser 2004) anesthetic pre-evaluative temporaries (APTs) (Gurel et al. 2012) based on a diagnostic wax-up. Evaluation and treatment of endodontic, mucogingival, orthodontic problems should be performed if necessary. Minimum invasive preparation is performed based on both the mock-up and the silicon keys. Feldspathic or lithium disilicate veneers are

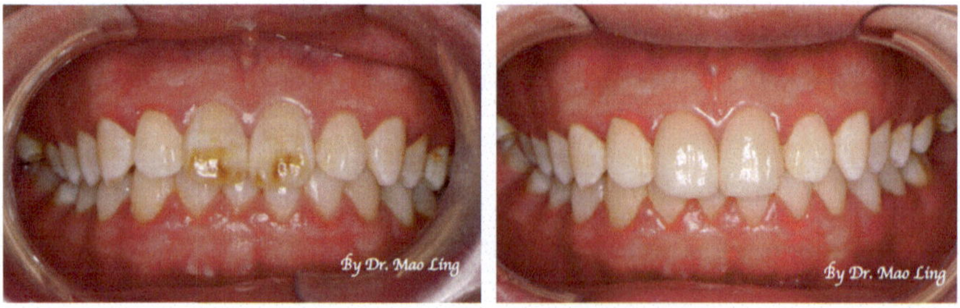

Fig. 24.3 Comparison of before and after porcelain veneer treatment for fluorosis

produced according to design requirements. Light-curing resin cements are chosen to match natural teeth color. Try-in and adhesive is applied under rubber isolation (Chankanka et al. 2010).

Compared with resin veneer, the advantages of porcelain veneer include as follows. (1) Porcelain veneer shows good transparency and gloss, especially the transparent layer at the cut end, which can simulate natural teeth well. (2) Porcelain veneer has better biocompatibility and avoids irritation in gingival margin. (3) The color of porcelain veneer is relatively wear-resistant over a long period of time, and it is not easy to age and change color. (4) The density and smoothness of porcelain veneer are not easy to be affected by bacteria. Compared with metal porcelain crowns, porcelain veneers require less dental tissue to be removed, which can save dental tissues as much as possible and avoid damaging dental pulp tissues. At the same time, due to the use of the upper gingival edge or the flat gingival edge for the tooth preparation form, the potential harm of the porcelain attached to the periodontal area is also less than the metal porcelain crown and all-ceramic crowns. In order to reach a natural-look/esthetic restoration, the TP value of veneer materials should be close to that of natural enamel (Arif et al. 2019). Porcelain veneering has become a relatively mature technique for minimum invasive dental restorations.

Full Crown

For severe dental fluorosis, especially when the tooth changes into dark color, it cannot be repaired by other methods but full crown restoration (Fig. 24.4). The full crown restoration includes porcelain metal full crowns and porcelain full crowns. Porcelain metal full crowns have the strength of metal and the beauty of porcelain, boasting features of low cost, strong opacity, realistic texture, smooth surface, good histocompatibility, color matching, close to natural teeth, compression/corrosion/wear/discoloring resistance. At the same time, it can be carried out in hospitals with primary facilities. Since the application of porcelain metal full crowns in clinical practice, this technique has been widely used in the treatment of dental fluorosis. However, the metal base crown directly affects the light transmission of the

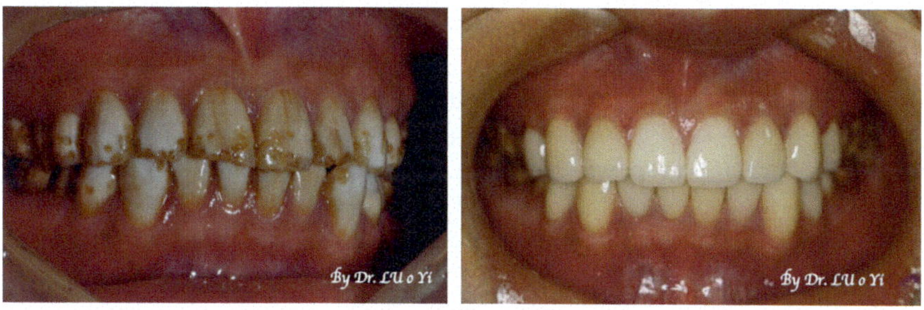

Fig. 24.4 Comparison of before and after full crown treatment for dental fluorosis

prosthesis, leading to the lack of layered sense of natural teeth. At the same time, the metal ions in the alloy not only discolor the porcelain, but also it was reported that some patients are allergic to the metal. The above-mentioned shortcomings all contribute to the development of all-ceramic teeth to replace the metal base structures. All-ceramic materials have the features of dense texture, wear resistance, smooth surface, and excellent biocompatibility. Such material boasts with transparency and refractive index close to that of enamel. Its appearance color is almost the same as that of natural teeth, which can achieve the best esthetic effect. Its unique esthetic properties are superior to metal materials and other polymer materials, and its retention effect is better than that of veneer repair.

However, full crown restoration requires abrasion of hard tooth tissues, resulting in 60–70% loss of healthy enamel. When the abutment tooth is a living pulp tooth, full crown restoration will easily have an adverse effect on the pulp or even affect the pulp negatively. Exposure causes tooth pain and inflammation. As for esthetic reasons, the edges of full crowns often require stretching below the gingival margin for esthetic purpose (or concealment), thus it increases risks of periodontal tissue disease. In addition, due to the limitations of naked-eye colorimetric and colorimetric technique, porcelain restoration often fails to achieve satisfactory treatment results for patients with fluorosis, and it cannot meet the requirements of individual patients to restore the original teeth color which was discolored by fluorosis.

Personalized need:personal customized colorimetric guide for dental fluorosis is required for crown restoration. Metal crown restoration is a cap-shaped restoration made of metal to cover the entire crown surface. The advantage of this method is the large contact surface with the teeth, strong retention effect, and good protection of teeth. It is often used for repairing large tooth defects or retaining the fixed dentures. The disadvantage is that the color of the metal affects the esthetic appearance. It is often used to repair posterior teeth.

The clinical treatment and preparation procedures for crown restoration are as follows: (1) Stipulating treatment plan and completing the pre-repair preparations. The doctor will stipulate comprehensive prosthetic treatment plan and choose the appropriate prosthesis type according to the patient's dental defect etiology, defect size, position of the missing tooth, occlusion, remaining teeth in the mouth, eating habits, and patient's requirements. Various preparations should be made before restoration, including dental pulp and root canal treatment for apical disease, periodontal treatment, orthodontic treatment before restoration, etc. Following procedures/data should be prepared: preoperative complaints, current medical history, clinical examination, X-ray reports, prognosis, informed consent (concerning repair method, process, time, cost, possible complications, choosing root canal treatment or not, etc.; retaining preoperative photos, X-rays reports and observation models), and patient signature if necessary. (2) Dental preparation: doctor should prepare the dental prosthesis according to the selected type of restoration, and check different aspects (the condition of the remaining teeth, the position of the missing tooth, and the occlusion, etc.) for restoration space. Treatment procedures are: local anesthesia → colorimetry → tooth preparation → gingival resection→ shoulder preparation→ finishing → inspection. (3) Preparation of models and temporary crown restoration:

after preparation of the teeth, the doctor performs gingival retraction through the gingival retraction line; gingival retraction fluid/gingival paste is prepared, and suitable mod impression materials are selected to make impressions and casts plaster model later. (4) Prosthetic colorimetry: (1) the doctor observes and discriminates the color of the remaining teeth; (2) colorimetric choices (conventional, special, custom colorimeter) are considered and colorimetric photos are recorded for the color characteristics of affected teeth; (3) the tooth color information is sent to the porcelain technician accurately. (4) Production of restoration parts: individualized restorations for severe fluorosis require internal and external staining to adjust the color. (5) Clinical trials: contact relationship between crowns and adjacent teeth is checked. It is necessary to check whether the edge is tight or not according to the bite test. Sometimes it is necessary to modify and adjust the shape of the full crown, especially the color, polishing, glazing, and cementing. The restored teeth need to be evaluated during regular oral examinations. The assessment includes the edge integrity of the restoration, staining, abrasion, secondary caries, and gum condition. It is necessary to deal with secondary caries and gum inflammation in time to ensure that oral health care is implemented correctly and effectively. (6) Precautions after treatment: (1) good oral hygiene habits should be cultivated and maintained (like brushing teeth and using dental floss regularly); patients with periodontitis and gingival recession should use cleaning tools such as interdental brushes and oral irrigator; (2) rock hard food should be avoided; (3) patients should have regular return visit with doctors.

Glass Ionomer Cement

Glass ionomer cement is a compound mainly composed of aluminum glass and polyacrylic acid. It is mainly applied to repair worn teeth. Due to its low abrasion resistance, it is mostly used as repairing material. It bonds with the teeth without traditional acid etching, which can preserve healthy dental tissue to the greatest extent. Even if the dental issue is cured, the material will not polymerize and shrink. It has the closest thermal expansion coefficient to dental tissue, together with features of good edge sealing and good X-ray block resistance, which is convenient for postoperative follow-up diagnosis. In recent years, the low solubility of the new glass ionomer cement guarantees excellent durability, low cost, feasibility and fast use. It is also suitable for hospitals with only primary facilities or patients with severe dental fluorosis but low color requirements.

Notices During Treatment

Dental Fluorosis Color

The clinical manifestations of fluorosis teeth are tarnished tooth surface, opaque white plaques in the form of narrow strips or spots on the enamel, and yellow or brownish black staining in the opaque area. These manifestations must be taken into consideration of colorimetric comparison of fluorosis teeth during porcelain repair. The problem becomes extremely difficult due to following reasons: on the one hand, it is impossible to use specific and accurate chromatics language to express the required fluorosis tooth color; on the other hand, there is no wide range of color palette to compare and dedicate to fluorosis colorimetry. The colorimetric palette commonly used in clinics cannot completely cover the wide range formed in natural tooth colors, and even less in targeting color of fluorosis plaque teeth. Some scholars have performed some measurements on fluorosis in some areas by spectrophotometers, color measuring instruments. However, due to factors such as number of samples, color distribution, and measurement standards, such measurements cannot fully reflect the fluorosis tooth color in China (Mao et al. 2012).

Shade Selection of Dental Fluorosis

There are many involving factors to obtain a record that matches the tooth color. Research on matching tooth color requires multidisciplinary knowledge such as physiology, physics, optics, materials science, and understanding characteristics between normal natural teeth and abnormal teeth such as fluorosis (its structural analysis and measurement). At the same time, the matching is also closely related to materials and manufacturing processes. Colorimetry is the record and transmission of color through some technical methods. The following guide and measurement can be comprehensively applied in together with specific conditions.

Conventional Shade Guide

There are various shade guides from manufacturers such as Vita and Shofu as well as many color systems, but the material of such guides is different from the materials used in clinical and technical production, so does the thickness. The shade guide does not include all-natural tooth colors, and it is impossible to compare the color of fluorosis plaque teeth. For example, when performing porcelain restoration, the color of the shade guide does not take the metal inner layer into account. There are many interference factors due to visual inspection. Therefore, it is more difficult to obtain accurate values by using a conventional shade guide.

Instrument Measurement

There are mainly spectrophotometers and color measuring instruments which have the advantages of being fast, stable, repeatable, digitalized, and highly credible results. Of all the choices, Olympus colorimeter is currently the most advanced and convenient dental colorimetric system in the world due to its advanced optical research. Its colorimeter is developed and produced with high accuracy and color reduction. Compared with the colorimetric products in the market before and at present, this particular system has the following characteristics: spectrophotometric colorimetry, the use of a hood, the colorimetric results are free from external factors (such as lights, sunlight), accurate values, and low error rate. Its software is used for data analysis after taking pictures to avoid human error caused by point colorimetry. The software is powerful and can be installed on multiple computers. As long as the photos taken with the colorimeter are shared, they can be used for analysis to facilitate communication and statistics study between doctors. In addition to the teeth, pictures of the dentition and face should be taken to judge the shape and color to coordinate the whole process. This system can analyze and determine tooth color, pigmentation point, and various data, which is convenient for choosing tooth color and statistics analysis.

Development and Application of Customized Fluorosis Teeth Shade Guide

Due to economic conditions, dental fear, and natural tendency, many patients with fluorosis plaques cannot accept the porcelain restoration for all the anterior teeth, but they are more willing to accept personalized porcelain restoration for consistent color and morphological features of adjacent teeth and counterparts from opposite side. The surface morphology of fluorosis teeth is complex and diverse with irregular flat surface. The diversity of color and transparency, combined with different levels of fluorosis, makes it difficult to customize restoration of fluorosis teeth. The color of the restoration is determined by the inherent color of the ceramic. The body porcelain is the main part of restoration, and it is also the main factor that forms the inherent color of restoration. The esthetic and restorative dentistry has long been hampered by the shade matching problem of ceramic materials (Bai et al. 2020). Therefore, it is necessary to start from the body porcelain layer of the metal porcelain restoration to study the basic shade of fluorosis. The color modification methods of porcelain restorations can be divided into external staining method and internal staining method. The internal staining method is to add dentin modified porcelain to the body porcelain layer to change its color. For most economically underdeveloped regions, the absence of computer colorimeter forces most doctors or technicians to use external staining when repairing fluorosis. The commonly chosen reddish-yellow shade in the colorimetric plates is used as the basic color, and it is externally

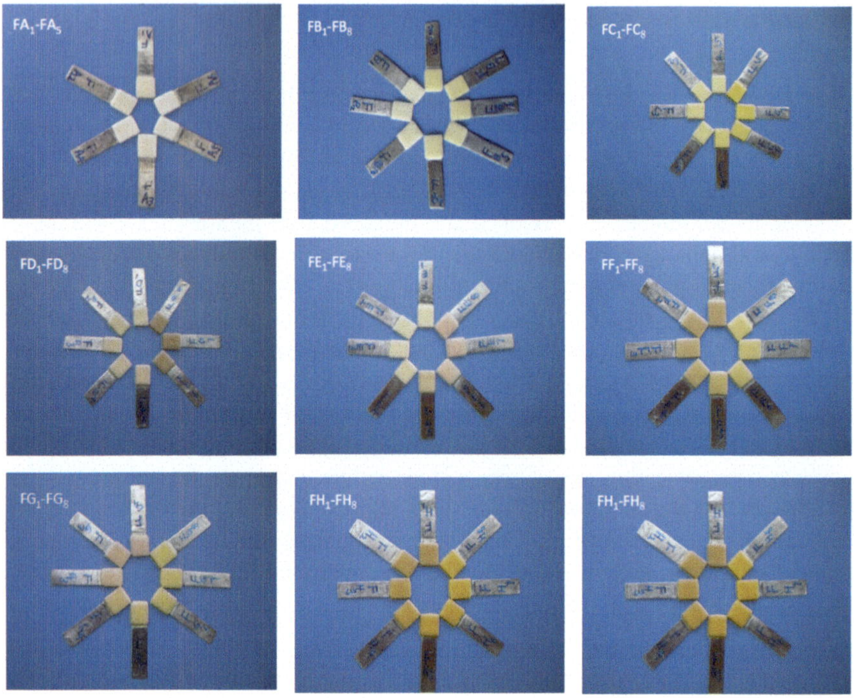

Fig. 24.5 Custom-made fluorosis shade guide FA to FI in nine groups

dyed when the patient revisits. This is not feasible for dental clinics that rely on external processing. The external dyeing method works on only superficial level. For long-term effect, the prosthesis can lose or partially lose its surface colorant due to long-term brushing abrasion and biting contact, hence prosthesis color is changed inevitably. To tackle the issue of reproducing the color of fluorosis teeth and custom-made colorimetric restoration, this research group developed a custom-made porcelain for fluorosis tooth based on the "secondary colorimetric method" by studying chromaticity distribution range of fluorosis teeth. The customized fluorosis dental shade guide is a custom-made multi-shade guide. This type of shade guide is composed of dentin modified porcelain and dentin porcelain. It is different from the hue and chroma colorimeter of opaque porcelain. The customized fluorosis shade guide is divided into 9 groups ranging from FA to FI: the FA group has five pieces to simulate the chalk color of mild fluorosis tooth. The remaining eight groups (FB to FI) have their own colorimetric palette, respectively, each simulates the color of moderate and severe fluorosis tooth (Fig. 24.5). When the intentional shade range exceeds the commonly used shade guide, the custom-made multi-shade guide can complete color comparison as an independent system, so the crown restoration could be closer to the color of fluorosis tooth compound.

Color Modification of Restoration

The color modification methods of porcelain restorations can be divided into two methods: external and internal dyeing methods.

Internal dyeing color modification: colorimetric palettes are used with dentin as the main color shade. After tooth preparation, the dentin color of the remaining tooth tissue after tooth preparation was first referred, and then the adjacent tooth color and surface color. The special structure requires colorimetric comparisons and records them in details. If the color of tooth body is not a choice in the color palette, doctors can mix the colors of tooth body and adjacent teeth for a special porcelain block. Comparison between the color of newly generated porcelain block and the abutment/adjacent teeth color can increase the choice of color. The color of porcelain-fused metal restoration includes its inherent color, light source color, and ambient color. Concerning light source and ambient color, restoration color is determined by the inherent color of the porcelain, which is the main part of porcelain-fused metal restoration, and it is also the main factor that forms the inherent color of the restoration.

Special Dyeing

Special dyeing is to mix the coloring reagent in the porcelain powder at a certain ratio or use the coloring reagent alone, so that the porcelain crown has a dyeing effect and is consistent with the natural tooth color. Generally, the coloring reagent is a combination of internal and external dyeing. After the prosthesis is initially formed, it can be modified repeatedly to verify the dyeing effect without affecting its shape and quality. However, due to deep coloring of severe fluorosis, coloring such as brown or gray should be selected for coloring modification. In addition, severe dental fluorosis is often accompanied with enamel deficiency, and its morphology can be modified after the prosthesis is finished. In order to simulate the shape and depth of the enamel defect of the tooth with the same name or the adjacent tooth, after the porcelain sintering is completed, the shape of enamel is trimmed, a proper amount of porcelain is removed from the surface, and the color is added. For the chalk color of dental fluorosis, white is used in the concentrated color to mix with the incisor porcelain powder. The more white colorant is used, the brighter the porcelain changes. The proportion of white porcelain powder should be increased accordingly. The mixing ratio of the concentrated color porcelain powder and the incisor porcelain powder should be chroma adjusted for coloring. The greater the chroma, the higher the proportion of concentrated color porcelain powder, and vice versa. After trimming the enamel, colorant is applied on the surface of the porcelain to make the surface of the restoration closer to the natural tooth color. However, since the color of colorant will be different before and after sintering when using the same colorant such as brown, yellow, etc., the chroma should be adjusted to be higher in order to compensate the chroma after sintering. For individual cases of dental fluorosis with tetracycline teeth, a certain proportion of colored porcelain powder such as gray or brown should be mixed in the body porcelain. For cases with adjacent discoloring, enamel cracks, pigmentation, abrasion of the incision, it should be matched and externally stained to simulate the color of natural teeth from every detail.

Tooth Preparation is the Foundation of Restoration

Tooth preparation is one of the key links to reproduce the color characteristics of fluorosis teeth. If the grinding amount is insufficient, the porcelain body and enamel porcelain cannot be made as thick as possible, so that the color of the opaque layer could be revealed, the brightness could be increased, and the chroma is reduced. It is difficult to achieve the special staining of fluorosis teeth. Therefore, it is effective to appropriately increase the space for restorative materials during the preparation of the abutment for esthetic effect of porcelain-fused metal restoration. In addition, the teeth are generally divided into several parts during general colorimetric comparison, and the colors of these parts are recorded with charts. However, the staining of fluorosis teeth is special, and lightness or color should be added in the stained and non-stained areas. The low ratio of mixed porcelain powder makes the transition between the dyed area and the non-stained area more natural. It is best to invite the producer to the clinic to observe the patient's coloring process carefully and take pictures or draw the stained part. Thus, the details in shape and color could provide a basis for making satisfactory restorations (Niu 2014).

Prevention of Dental Fluorosis

Fluoride exists in an ionized form in ground waters. In endemic fluorosis areas, such as those in China, India, the Middle East, Africa, etc., groundwater containes high fluoride, or food and air are contaminated by fluoride from burning coal. Dental fluorosis can occur if the residents in these areas get excessive amounts of fluoride for a long-term.

There are several ways to prevent fluorosis. (1) Avoid drinking high-fluorine water and strictly control the concentration of fluoride in the water. (2) the following measures can be taken against coal-burning type of fluorosis: improving stoves and reducing indoor air fluoride contamination are reliable and preventive measures at this stage (Zhang et al. 2013). Relying solely on improved stoves cannot satisfactorily solve the fluoride contamination of food. Improved dehydrating methods for crops avoid direct contact of soot. Reducing fluoride contamination of crops is a key measure to reduce fluoride intake in coal-burning type of fluorosis prevention. (3) Avoid drinking/eating tea, vegetables, and fruits contaminated by fluoride. The prevention and control of fluorosis require a comprehensive measure, residents need to change their living habits, eating habits, and improve nutrition conditions.

Conclusion

There are various treatments for dental fluorosis, and each method has its own merits. Bleaching treatment is mainly used for patients with mild to moderate dental fluorosis. Compared with crown restoration treatment, resin and porcelain veneer repair do not require a lot of grinding and modification of teeth, which preserve the integrity of the tooth tissue to the greatest extent. They are safe and effective since there will be no irreversible damage to the pulp tissue. For patients with severe dental fluorosis and enamel defects or deep staining, bleaching treatment alone cannot achieve the goals of improving esthetics requirement and restoring functions. Therefore, it must be repaired by photosensitive curing restoration, porcelain veneer restoration, full crown restoration, or other methods. Once fluorosis is formed, it remains for a lifetime. However, as long as appropriate treatment is chosen according to the real situation and actual needs of the patient, the symptoms and appearance could be improved, and the dental function can be restored. It should be noted that the treatment of fluorosis should focus on its prevention. The implementation of water improvement and stove improvement projects is the key to the eradication of fluorosis.

References

Abanto Alvarez J, Rezende KM, Marocho SM, Alves FB, Celiberti P, Ciamponi AL. Dental fluorosis: exposure, prevention and management. Med Oral Patol Oral Cir Bucal. 2009;14: E103–7.

Akpata ES. Occurrence and management of dental fluorosis. Int Dent J. 2001;51:325–33.

Arif R, Yilmaz B, Johnston WM. In vitro color stainability and relative translucency of cad-cam restorative materials used for laminate veneers and complete crowns. J Prosthet Dent. 2019;122:160–6.

Bai XD, Yan YZ, Li Q. Spectrophotometric evaluation of color errors generated in the visual color duplication procedure for current ceramic veneers. J Dent Sci. 2020;6:1–9.

Baroudi K, Hassan NA. The effect of light-activation sources on tooth bleaching. Niger Med J. 2014;55:363–8.

Bazos P, Magne P. Bio-emulation: biomimetically emulating nature utilizing a histo-anatomic approach; structural analysis. Eur J Esthet Dent. 2011;6:8–19.

Bazos P, Magne P. Bio-emulation: biomimetically emulating nature utilizing a histoanatomic approach; visual synthesis. Int J Esthet Dent. 2014;9:330–52.

Carey CM. Tooth whitening: what we now know. J Evid Based Dent Pract. 2014;14:70–6.

Chankanka O, Levy SM, Warren JJ, Chalmers JM. A literature review of aesthetic perceptions of dental fluorosis and relationships with psychosocial aspects/oral health-related quality of life. Community Dent Oral Epidemiol. 2010;38:97–109.

Dean HT. Classification of mottled enamel diagnosis. J Am Dent Assoc. 1934;21:1421–6.

Dean HT. Fluorine in the control of dental caries. J Am Dent Assoc. 1956;52:1–8.
DenBesten P, Li W. Chronic fluoride toxicity: dental fluorosis. Monogr Oral Sci. 2011;22:81–96.
Gurel G, Morimoto S, Calamita MA, Coachman C, Sesma N. Clinical performance of porcelain laminate veneers: outcomes of the aesthetic pre-evaluative temporary (apt) technique. Int J Periodontics Restorative Dent. 2012;32:625–35.
He LB, Shao MY, Tan K, Xu X, Li JY. The effects of light on bleaching and tooth sensitivity during in-office vital bleaching: a systematic review and meta-analysis. J Dent. 2012;40:644–53.
Joiner A, Luo W. Tooth colour and whiteness: a review. J Dent. 2017;67S:S3–S10.
Llena C, Villanueva A, Mejias E, Forner L. Bleaching efficacy of at home 16% carbamide peroxide. A long-term clinical follow-up study. J Esthet Restor Dent. 2020;32:12–8.
Magne P, Belser UC. Novel porcelain laminate preparation approach driven by a diagnostic mock-up. J Esthet Restor Dent. 2004;16:7–18.
Mao L, Liu W, Liao J, Wu L, Wang Y. Initial development and chromatic value analysis of customized dental fluorosis shade guide. Chongqing Med. 2012;41:227–9.
Menini M, Pera F, Barberis F, Rosenberg G, Bagnascoa F, Pesce P. Evaluation of adhesion between carbon fiber frameworks and esthetic veneering materials. Int J Prosthodont. 2018;31:453–5.
Niu JJ. Clinical application of anterior teeth porcelain veneer restoration. Chin Pract Med. 2014;9:98–9.
Peumans M, De Munck J, Fieuws S, Lambrechts P, Vanherle G, Van Meerbeek B. A prospective ten-year clinical trial of porcelain veneers. J Adhes Dent. 2004;6:65–76.
Ritter AV. Dental fluorosis. J Esthet Restor Dent. 2005;17:326–7.
Rotoli BT, Lima DA, Pini NP, Aguiar FH, Pereira GD, Paulillo LA. Porcelain veneers as an alternative for esthetic treatment: clinical report. Oper Dent. 2013;38:459–66.
Sabokseir A, Golkari A, Sheiham A. Distinguishing between enamel fluorosis and other enamel defects in permanent teeth of children. PeerJ. 2016;4:e1745.
Skaare AB, Maseng Aas AL, Wang NJ. Enamel defects in permanent incisors after trauma to primary predecessors: inter-observer agreement based on photographs. Dent Traumatol. 2013;29:79–83.
Swain MV, Gee C, Li KC. Influence of ageing on glass and resin bonding of dental glass-ceramic veneer adhesion to zirconia: a fracture mechanics analysis and interpretation. Acta Biomater. 2018;74:454–63.
Taşın S, Celik G, İsmatullaev A, Usumez A. The effect of artificial accelerated aging on the color stability, microhardness, and surface roughness of different dental laminate veneer materials. J Esthet Restor Dent. 2020;23:1–7.
Thylstrup A, Fejerskov O, Mosha HJ. A polarized light and microradiographic study of enamel in human primary teeth from a high fluoride area. Arch Oral Biol. 1978;23:373–80.
Tirlet G, Crescenzo H, Crescenzo D, Bazos P. Ceramic adhesive restorations and biomimetic dentistry: tissue preservation and adhesion. Int J Esthet Dent. 2014;9:354–69.
Zhang BY, Li DS, Zhang RZ, Gao J, Hu XQ, Wang XM, et al. Health education impact evaluation of later-stage management of control of endemic fluorosis by coal-burning pollution. Studies Trace Elem Health. 2013;30:48–9.

Chapter 25
Drug Development

Zhi-Zhong Guan, Yi Zhao, and Chang-Xue Wu

Abstract Endemic fluorosis has a great impact on human body and so far, no effective drug has been successfully developed. More commonly used drugs to treat chronic fluorosis contain elements such as boron, magnesium, calcium and trace elements such as selenium and zinc preparation, and also antioxidant materials to combat the damage of free radicals and improve the activity of antioxidant enzymes to treat chronic fluorosis. These drugs and some natural plant compositions have been confirmed to achieve a certain therapeutic effect, but still do not solve the underlying problem to the patients with chronic fluorosis. Therefore, the research and development of drugs to treat endemic fluorosis is still a hot research direction at present.

Keywords Antioxidants · Endemic fluorosis · Free radicals · General and trace elements · Therapeutic drugs

Z.-Z. Guan (✉) · C.-X. Wu
The Key Laboratory of Endemic and Ethnic Diseases (Guizhou Medical University), Ministry of Education, Guiyang, Guizhou, People's Republic of China

Department of Pathology at the Affiliated Hospital of Guizhou Medical University, Guiyang, Guizhou, People's Republic of China

The Provincial Key Laboratory of Medical Molecular Biology, Guiyang, Guizhou, People's Republic of China
e-mail: zzg@gmc.edu.cn

Y. Zhao
College of Foreign Languages, Guizhou Medical University, Guiyang, Guizhou, People's Republic of China

Z.-Z. Guan (ed.), *Coal-burning Type of Endemic Fluorosis*,
https://doi.org/10.1007/978-981-16-1498-9_25

Introduction

Endemic fluorosis is widely distributed and seriously endangers human health, especially in developing countries such as China and India. Because of the complicated pathological damage mechanism of endemic fluorosis, it is difficult to develop drugs to treat fluorosis (Guan 2014). Up to now, there is no drug that has been approved by the Food and Drug Administration to treat the disease. For many years, the prevention and control of endemic fluorosis have been taken at all levels and a large amount of human and material resources given to the endemic areas of the disease in the long-term strategy. Since the 1980s, pilot studies on the prevention and treatment of coal-burning type of endemic fluorosis have been carried out first in Guizhou province of China, forming an intervention model, i.e., modifying cooking stoves, improving health education, and raising prevention consciousness in the areas, which has made remarkable achievements in eliminating fluorine sources, reducing the generation of new patients, and preventing the harm of chronic fluorosis. However, because there is no effective drug treatment, a large number of patients with chronic fluorosis are still suffering. Therefore, it is an urgent task to develop effective drugs for the treatment of endemic fluorosis in order to thoroughly relieve chronic fluorosis (Jiang et al. 2020). Currently, two main types of drugs for the treatment of endemic fluorosis: chemical elements and Chinese herbs (including extracts and compound preparations) have been attracted high attention.

Chemical Elements Commonly Used to Treat Endemic Fluorosis

Fluorine is a bone-loving element. About 96% of fluorine in endemic fluorosis patients is deposited in bone tissue, affecting bone formation, mineralization, and absorption, leading to skeletal fluorosis and dental fluorosis. For a long time, many macro elements with antagonistic effect of fluorine have been selected to treat endemic fluorosis. Among them, these elements such as boron, magnesium, aluminum, and calcium can combine fluoride to form the compounds that are not easily absorbed in the gastrointestinal tract, therefore reducing the absorption of fluorine. They can form a complex with fluoride in the blood, excrete through urine or stool, and finally reduce the accumulation of fluoride in the body and toxicity. These chemical elements play a certain role in preventing the occurrence and development of chronic fluorosis, but they cannot really relieve the damages caused by the disease. Conversely, some trace elements, such as selenium, zinc, molybdenum, and copper, are very important to treat chronic fluorosis by affecting the activity of certain antioxidant enzymes and reducing the toxicity of free radicals in the body, thus achieving the therapeutic purpose for fluorosis to a certain degree.

Boron

For a long time, boron has been proved to be an effective fluoride antagonist and widely used in the treatment of chronic fluorosis. As early as the 1970s, many studies through animal experiments indicated that the bone and tooth damages of mice exposed to fluoride were significantly reduced after giving boron, so it was considered that boron was a fluoride antagonist. A series of experiments were carried out in rabbits, focusing on the effect of boron on calcium and phosphorus metabolism and bone structure after antagonizing fluoride. The experimental results showed that in the acute fluorosis group the rabbits exhibited obvious hypocalcemia and mild hyperphosphatemia, while in the fluoride plus boron group the calcium and phosphorus metabolism was normal and maintained balance (Elsair et al. 1980). The results of the subacute experiment indicated that the femoral and tibial cortex of rabbits in the fluorosis group was thickened, but there was not any significant change in the bone of rabbits in the fluoride plus boron group. At the middle period of the subacute trial or after ending the exposure of fluoride in the experiment, boron administration improved bone X-ray changes induced by fluorosis. The exposure of fluoride with low dose fluoride (15–30 mg/kg body weight/day) to animals for 8 months and then the treatment of boron could correct the disorder of calcium and phosphorus metabolism in the animals. The above experimental results confirm the antagonistic effect of boron on fluorosis, the mechanism of which may be related to that fluoroborate (BF_4^-) ions formed by boron and fluoride in vivo were excreted in urine and therefore the accumulation of fluoride in the body was reduced, thus weakening the toxic effect of fluoride. The results of a year-long experiment in pigs showed that excessive sodium fluoride caused the thickening of the long bone cortex in pigs, while the toxic effect was alleviated by the administration of high doses of borax (Seeffner and Teubene 1983). By employing rat experiment, it was proved that boron had the effect of promoting the excretion of fluoride in urine. A research group in Guizhou Medical University conducted experiments on reducing the toxicity of sodium fluoride with borax in rabbits. In the study, the animals were exposed to 2% sodium fluoride solution every Monday, Wednesday, and Friday and treated with 2% borax solution every Tuesday, Thursday, and Saturday for an experimental period of 6 months, as compared with the rabbits given sodium fluoride alone. The experimental results showed that the rabbits fed with fluoride and boron had higher urinary fluoride excretion and lower fluoride content in bone, suggesting that boron had a role in reducing fluoride toxicity. When rabbits were given sodium fluoride (20 mg/kg body weight/day) for the experiments with subacute and chronic fluorosis, the activity of glutamyl transpeptidase in urine was increased, the activities of alkaline phosphatase (ALP), mitochondrial succinic acid dehydrogenase, and lysosomal acid phosphatase in renal tissue were decreased, and various organelles in renal tubular cells were damaged (Li et al. 1989). At the same time, animals treated with sodium tetraborate at 30 mg/kg body weight/day were found to increase urinary fluoride excretion and decrease renal fluoride content. Boron can not only promote the excretion of fluoride from the kidney, but also change the adverse effects of fluoride on glutamyl transpeptidase, ALP, mitochondrial succinic acid dehydrogenase, lysosomal acid phosphatase, and so on and antagonize the damage effect of fluoride on the ultrastructure of renal tissues. In addition, studies on the isolated cells showed that the anabolisms of protein, RNA, and DNA in rat kidney cells were affected to different degrees by the culture medium containing fluoride, while sodium tetraborate had a comprehensive antagonistic effect on the above toxic effects of fluoride. It was also found that sodium tetraborate reduced the fluoride content in human amniotic cells and rat kidney cells exposed to sodium fluoride. According to the determination of fluoride content in cultured cells and medium under different conditions, it is speculated that boron can directly reduce the permeability of cell membrane to fluoride ion, thus increasing the body's resistance to fluorosis. The mechanism of boron antagonism against fluorosis is multifaceted, as follows: (1) Boron directly interacts with fluorine in the body to form an insoluble complex BF_4^-. On the one hand, it promotes fluoride to be discharged with urine and reduces fluoride load in the body; on the other hand, it changes the existing form of fluoride to reduce fluoride toxicity, so as to achieve the purpose of detoxification. (2) Boron reduces the permeability of cell membrane to fluoride ions and therefore declines the fluoride content in cells. (3) Boron can antagonize the abnormalities of some necessary trace elements caused by fluoride and correct the balance disorders of trace elements.

In terms of clinical research, there have also been a large number of reports on the use of boron in the treatment of endemic fluorosis. Under the condition of not continuing to be exposed to fluoride, taking borax can improve the adverse reactions caused by fluoride, such as liver and blood coagulation injury, abnormal bone imaging, calcium and phosphorus balance disorder, etc., and the symptoms and signs of patients have different degrees of relief (Frank et al. 1985). It has been reported that after 400 days of borate administration, the clinical symptoms of patients with chronic fluorosis were significantly improved, urine fluoride output increased, hydroxyproline excretion decreased, and hair fluoride content decreased. A trial of borax treatment in 31 patients with endemic skeletal fluorosis was conducted (Zhou et al. 1987). Among the 10 patients, the dose was 90–135 mg/day for 95 days, and the effective rate of clinical symptom improvement was 50% after taking the drug, and the urine fluoride excretion was significantly increased compared with that before taking the drug. The remaining 21 patients were divided into two groups. In the experimental group, the patients were given borax tablets with gradually increasing doses from 300 mg/day to 1100 mg/day for 3 months. On the other hand, the control group was used with yeast tablets instead of borax. The results showed that headache, dizziness, numbness, limb pain, shoulder joint activity obstruction, squatting difficulty, and other symptoms and signs in the patients with borax tablets were significantly improved, the effective rate was up to 80% and the urine fluoride output was significantly increased, which was positively correlated with the dosage of borax. Some authors treated 105 patients with endemic fluorosis with borax, and the total clinical effective rate was 94.6% after 6 months of treatment, of which 30.4% were cured and 57.1% were effective. Biochemical examination showed that the calcium content in blood and urine increased significantly, while the phosphorus content did not change significantly, and the activity of ALP in blood, total hydroxyproline in urine, and cAMP content decreased significantly. Therefore, it was believed that boron might play a good role in improving the abnormal bone metabolism caused by fluorosis. A research used borax and traditional Chinese medicine compound preparations containing borax to treat the patients with endemic skeletal fluorosis for 2 years found that more than 90% of the subjects presented the improvements of clinical symptoms, signs, joint function, biochemical parameters, and bone mineral density, including two cases with the improvement of bone X-ray.

Based on the above data, the antagonistic effect of boron on fluorosis is obvious, and the therapeutic effect is definite. However, the mechanism of action of boron on fluoride toxicity in bone or other organs remains to be further studied. Although boron is a low-toxic substance, it still has certain reproductive toxicity and cytotoxicity. Boron by long-term use can be accumulated in the body. The side effects of treating endemic fluorosis still need to be further investigated.

Magnesium

As early as the 1950s, magnesium was found to have an antagonistic effect on fluoride. Magnesium and fluoride can form magnesium fluoride (MgF_2), which is an insoluble substance and can reduce the absorption of fluoride in the intestinal tract. Magnesium is also an activator of many enzymes in the human body. Increasing the content of magnesium in the body can antagonize the adverse effects of fluoride on some enzymes. Accordingly, there have been reports of the use of magnesium in the treatment of endemic fluorosis. Earlier research reports showed that the treatment of 80 cases of endemic fluorosis with haloalkaloid (mainly containing magnesium chloride) achieved an effect of 97.5%. After treatment, the levels of blood calcium and phosphorus decreased, urinary calcium excreted decreased, and urinary phosphorus increased, which was considered as evidence that magnesium could improve osteolysis caused by skeletal fluorosis.

In the early 1970s, Indian researchers found that natural ores, serpentine (magnesium silicate), adsorbed large amounts of fluoride in aqueous solutions with a certain pH range. The tests in vitro confirmed that serpentine promotes the release of fluoride in bone. Valuable results were obtained by using serpentine to treat 10 patients with skeletal fluorosis and 16 patients with nerve root symptoms of fluorosis (Rao and Marty 1975). Chinese researchers have conducted a large number of studies on serpentine. Using serpentine containing preparations, such as flurping, self-made serpentine tablets, and compound mixtures combined with Chinese herbal medicines, the clinical treatment efficiency is up to 62.5–100%. Most of the clinical studies on serpentine in the treatment of skeletal fluorosis have achieved positive results, and the symptoms, X-ray and biochemical tests, and other indicators of patients before and after treatment have obvious changes, especially the improvement of bone X-ray pictures. The results from 10 cases of fluorosis treated with flurping (serpentine) showed that the symptom of the patients was significantly alleviated, joint movement improved, muscle strength enhanced, urine fluorine excretion increased, and fluoride content in blood and hair decreased. In addition, an effective rate of 96.8% is observed by using serpentine to treat endemic fluorosis, in which the effect of reducing the content of acetylcholine and increasing the activity of cholinesterase was found (Xi et al. 2008). Nevertheless, the mechanism of serpentine in the treatment of skeletal fluorosis is not very clear. Some scholars believed that the hydroxyl group in the crystal structure of serpentine could replace fluorine ion since fluoride excretion in urine and stool of patients was increased after taking serpentine. It indicates that serpentine can free a large number of calcium and magnesium ions under a certain pH value environment, so as to promote fluoride discharge; serpentine as a complex of mineral mixture plays an integral role; serpentine is rich in a variety of inorganic elements, especially trace elements. The effect of serpentine on urinary fluoride excretion has been regarded as an important basis for the treatment of skeletal fluorosis. However, some reports have found no effect of serpentine on urinary fluoride excretion.

It should be noted that some types of serpentine contain asbestos fibers, which has been classified as a carcinogen by the International Center for Research on Cancer and has the risk of cancer, so it has not been widely used. Therefore, the application of serpentine in the prevention and treatment of endemic fluorosis should pay special attention to strict selection and laboratory identification. In addition, further toxicological studies are necessary.

Calcium

Many scholars have used animals to conduct experiments and prove whether calcium has an effect of anti-fluorosis. When 0.1% fluoride was added to the diet of weaning mice, these animals died within 4–6 weeks, but after 4 weeks, adding 3% calcium citrate to the diet containing fluoride kept the mice alive and gradually alleviated their toxic manifestations. Therefore, calcium is considered to have an anti-fluorosis effect. In a prevalence study of 260 schoolchildren living in an endemic fluorosis area in South Africa (water fluoride content 8–12 ppm), hypocalcemia was documented in 23% (Pettifor et al. 1989). Furthermore, in a separate study of nine children with skeletal symptoms due to endemic fluorosis, hypocalcemia was found in six. Serum calcium values also correlated inversely with the degree of osteomalacia on iliac crest bone histomorphometry. It is suggested that low dietary calcium intakes might exacerbate the severity of the bone lesions in children living in areas of endemic fluorosis. In a study, 13 patients with endemic fluorosis were treated with calcium and vitamin D for 6 months. Among them four cases showed significantly better effect and nine cases improvement, representing the decreased level of fluoride and the increased calcium in blood. Calcium plus vitamin D has also been used in the treatment of endemic fluorosis with a reported efficacy of 86%. However, studies have pointed out that calcium cannot reduce the absorption of fluoride in the intestinal tract, and intake of different doses of calcium has little effect on the excretion and balance of fluoride. Experiments in rats have shown that calcium supplements do not effectively increase the excretion of fluoride. It is generally believed that long-term use of calcium, especially combined with vitamin D, will absorb a large amount of calcium into the blood, but will promote the occurrence and development of skeletal fluorosis due to calcification and osteosclerosis. It has also been suggested that calcium supplements may be used in such skeletal fluorosis patients with osteoporotic or osteomalacia. For these patients, daily oral administration of 8 g calcium lactate or 11 g calcium gluconate is beneficial to the improvement of bone lesions.

Recently, a study investigated the effects of calcium on fluorosis-induced impairment in learning and memory of offspring rats (Zhang et al. 2017). The results showed that high fluorosis significantly reduced synapse density, length of synaptic active zone, and thickness of postsynaptic density, and led to abnormal changes in the structural parameter of synaptic gap width, which was significantly reduced or increased. However, high dietary calcium significantly reversed the abnormal changes in structural parameters, and low calcium aggravated these variations. The results suggested that dietary calcium significantly affected hippocampal synaptic plasticity of offspring of mothers exposed to water fluorosis. Furthermore, a study was carried out to evaluate the prevalence of dental and skeletal fluorosis in adolescents and association of fluorosis with calcium intake and vitamin D status (Patel et al. 2017). The results indicated that fluorosis was more common in lower socioeconomic strata adolescents, and relatively adequate calcium intake and serum 25 hydroxyvitamin D3 concentrations may increase the efficiency of dietary calcium absorption, thus preventing severe fluorosis.

Aluminum

It was found that increased level of aluminum, a complexing agent of fluoride, in water and food can interfere with the absorption of fluoride in the gut. Aluminum was confirmed to have the anti-fluoride effect and 40–60% of fluoride in the intestinal tract can be bound by aluminum (Spencer et al. 1981). Taking a small dose of aluminum hydroxide can increase the excretion of fluoride in stools and also reduce the content of bone fluoride. In China, the patients with endemic skeletal fluorosis were treated with aluminum hydroxide gel or tablet and the symptoms of some cases were significantly improved reaching to 78% after taking medicine for 3 months, during which the urinary fluoride was reduced. A study on the combined effect on metabolism of fluoride and aluminum showed that the absorption of fluoride could be significantly reduced when the chickens were treated with relatively small dose of fluoride (12 mg/kg body weight/day) and aluminum (150 mg/kg/day, but the effect on the larger dose (300 mg/kg/day) of fluoride was not obvious (Xu et al. 1991). However, higher dose of fluoride can significantly increase the absorption, positive balance, and bone accumulation of aluminum. In each link of metabolism, the combined action of fluoride and aluminum is that aluminum is the antagonist of fluoride, but its antagonistic ability is limited. Fluoride is a promoter to the accumulation of aluminum, leading to toxic effect. An experimental result combined the action of fluoride and aluminum in rabbits showed that fluoride could promote the accumulation of aluminum in bone, and aluminum was believed to play a promoting role in fluorotoxic bone lesions. The morphometric changes of the bones of rats poisoned with aluminum and fluoride were investigated by a research group (Yu et al. 1993). They mixed the coal with kaolin in the area of coal-burning endemic fluorosis and roasted the corn and then fed the food to animals with low or high contents of aluminum plus fluoride diets for 14–24 weeks. Compared with the control group, the animals in the two groups presented osteomalacia lesions in the bones. The fluoride/aluminum ratio in diet has a great influence on the aluminum plus fluoride combination properties, i.e., when the ratio of aluminum to fluoride in diet is 2:1, the increased incidence of dental fluorosis and a large number of accumulated aluminum and fluoride in bone occurred, leading to serious damage of bone tissues with a synergy effect of these two elements (Song et al. 1991). Based on the above data, taking aluminum agent can reduce the absorption of fluoride and achieve the antagonistic effect of fluoride. However, under certain conditions, aluminum may have synergistic effect with fluoride and aggravate the bone damage. There are synergistic toxic hazards of high fluoride and high aluminum to human body in the areas of coal-burning type of endemic fluorosis of Liuanshui, China, in which the patients manifested the increased activity of ALP and lactate dehydrogenase, the decreased creatine phosphokinase and obvious fluorosis, especially severe fluorotoxic osteomalacia in children (Li and Guan 2014). A high autism spectrum disorder (ASD) prevalence has been reported from countries with water fluoridation as well as from endemic fluorosis areas, in which fluoride in synergistic interactions with aluminum's free metal cation (Al_3^+) can reinforce the pathological symptoms of ASD and this reinforcement takes place at concentrations several times lower than when acting alone (Strunecka and Strunecky 2019).

Selenium

A large number of data show that when endemic fluorosis occurs, the content of many trace elements in the body changes. In recent years, great progress has been made in the study of the relationship between fluoride and trace elements, especially the use of some trace elements to antagonize fluorosis.

Selenium is an essential trace element with various functions. As an important component of glutathione peroxidase (GSH-px), an antioxidant enzyme, selenium can improve the activity of GSH-px and antagonize the toxic effect of free radicals in oxidative stress reaction. In the late 1980s, people began to pay attention to the relationship between selenium and fluoride. A large number of data have shown that the mechanism of chronic fluorosis may involve that excessive fluoride leads to increased oxidative stress response, causing extensive damage to the body. Selenium preparation can improve the activity of GSH-px and reduce the damage of free radical and lipid peroxidation product malondialdehyde to cell membrane structure. Animal experiments showed that the content of collagen in animal cartilage increased significantly after selenium preparation was given to rats with chronic fluorosis, suggesting that selenium could antagonize the inhibition effect of fluoride on hydroxylase and enhance the synthesis of collagen in cartilage tissue (Mo et al. 1991). Selenium supplementation has a certain effect on rats with fluorosis. The selenium content in blood after selenium supplementation is positively correlated with urinary fluoride excretion, indicating that selenium can promote urinary fluoride excretion (Li et al. 1995). Rats with chronic fluorosis caused by fluoride supplementation in drinking water were given sodium selenite feed at the same time. After 1 month, with the increase of selenium content in urine and stool, the excretion of fluoride in urine also increased significantly. After giving selenium, the activities of ALP, aspartate aminotransferase, alanine aminotransferase, and lactate dehydrogenase in serum as well as the level of lipid peroxide in brain and kidney tissues were significantly decreased and while GSH-px in serum had a rising trend. Treatment with selenium to the rats with chronic fluorosis can improve their eating desire and weight gain, promote urinary fluoride excretion, delay skeletal fluorosis, inhibit bone mineralization disorder, enhance antioxidant enzyme activity, adjust the thyroid hormone levels, release the change of electrocardiogram, and confront myocardial injury (Bian et al. 2004). Selenium in vitro also antagonized the cytotoxic effects of fluoride. The mitochondrial cytochrome C/caspase-9/caspase-3 apoptosis pathway in the hippocampus was one of the mechanisms leading to fluorosis-induced brain damage, and whereas selenium could alleviate fluorosis induced brain injury by influencing this pathway (Zheng et al. 2016). Selenium can alleviate apoptosis induced by fluorosis, the mechanism of which may be through improving the expression of heat shock protein 70 and reduce oxidative stress by regulating levels of NAD-dependent deacetylase sirtuin-1 and antioxidant enzymes and the secretion of certain reproductive hormones (Li et al. 2020).

It has been reported that selenium has an obvious protective effect on the ultrastructure damage of neuronal cells in human brain influenced by fluorosis (Pang et al. 1994). Appropriate amount of selenium can antagonize DNA damage and cell apoptosis caused by fluorine. In addition, it was also found that selenium had an antagonistic effect on the reproductive toxicity of fluoride. As compared with the rats in the group given fluoride alone, sperm with fertilization activity increased significantly and sperm deformity rate decreased in the animals by giving selenium and fluorine. It is generally believed that selenium, as an important component of GSH-px, can improve the activity of GSH-px, reduce lipid peroxidation, and stabilize cell membrane, thus antagonizing the oxidative stress damage caused by chronic fluorosis.

Although the existing data confirm that selenium has the effect of combating chronic fluorosis, most of them are confined to animal experiments and involve little clinical application. In addition, it has been reported that selenium may aggravate the incidence of fluorosis. For example, the selenium content in drinking water in the area of endemic fluorosis is closely related to the prevalence of chronic fluorosis, and the increase of selenium content may aggravate the harmful effect of fluoride on human body. The combined effects of different doses of selenium and fluoride on rats showed that different dosage ratios of the two elements could vary (Deng et al. 1996). The adverse effect of high dose selenium on fluoride was no longer antagonistic but synergistic. Therefore, it is reasonable to believe that selenium may play a dual role. An appropriate concentration of selenium may antagonize the adverse reactions of chronic fluorosis, but beyond a certain limit, it may lead to imbalance of trace elements and has a synergistic effect on the occurrence of fluorosis.

Zinc

Zinc is an essential trace element in human body and a component of various enzymes. It is very important for bone growth and development. Zinc can also maintain the normal structure and function of cell biofilms. Fluoride is known to inhibit ATPase activity on the biofilm and affect the stability of the cell membrane, while zinc may be an important anti-fluoride mechanism. Epidemiological investigation showed that the content of zinc in serum of patients with skeletal fluorosis in endemic fluorosis area was significantly lower than that of healthy control group (Ersoy et al. 2011). A study in trace elements in the hair of children aged 7–13 in the area of coal-burning type of endemic fluorosis found that the hair zinc of children in the disease area was significantly lower than that of children in the non-disease area, indicating that there was zinc deficiency in children in the area of endemic fluorosis.

The results of animal experiments showed that zinc blocked the accumulation of fluoride in bone, antagonized the renal toxicity of fluoride, and reduced the effect of fluoride on cholinesterase activity (Bai et al. 1993). The rats were fed with drinking water containing 50 ppm fluoride and at same time given different doses of selenium and zinc preparations by gavage (Xia et al. 2001). After 2 months of experiment, fluoride content in urine, activities of GSH-px, L-glutathione transferase, and superoxide dismutase (SOD), level of lipid peroxide and glutathione in blood or tissues of the rats were measured. The results showed that selenium–zinc agents attenuated the increased lipid peroxidation and the activity of L-glutamyl transferase, and the decreased activities of GHS-px and SOD as well as glutathione induced by chronic fluorosis. These results suggest that a certain dose of zinc has an obvious antagonistic effect on chronic fluorosis, the mechanism of which may be related to promoting urinary fluoride excretion and anti-oxidative stress. However, there is a few of clinical evidence of zinc in the treatment to endemic fluorosis, which needs to be further investigation.

Molybdenum

Molybdenum is an essential trace element in human body and a constituent of many enzymes, which interacts with some trace elements in body. More and more attention has been paid to the relationship between molybdenum and endemic fluorosis. The studies focusing endemic fluorosis revealed that the suffering degree of the disease in those areas of endemic fluorosis with the closed content of fluoride in drinking water was negatively depended on the level of molybdenum content and/or the ratio of molybdenum to fluorine ratio in water (Yao 1992). Interestingly, the study on the relationship between the content of molybdenum in drinking water and the incidence rate of dental fluorosis of children found that when the content of molybdenum in water was more than 2.0 mg/l, the incidence of dental fluorosis was less severe and vice versa (Yao and Zhang 1992). Some researchers measured the contents of molybdenum in water, main food, and human hair in the areas of drinking water type of endemic fluorosis and found that content of both fluorine and molybdenum in water in these areas was significantly higher than that in non-fluorosis areas, while there was no significant difference in the content of molybdenum. Therefore, the authors believe that the high levels of fluoride and molybdenum in the fluorosis areas may be just a kind of coexistence relationship, which is the result of element enrichment, instead of aggravating fluorosis. However, the treatment of molybdenum could have an interfering effect on fluorosis.

The antagonistic mechanism of molybdenum against chronic fluorosis may be related that molybdenum and fluoride enter the digestive tract at the same time and form a complex, which affects the absorption of fluoride in the digestive system and thus reduces the toxic effect of fluoride. Another possible mechanism might be that after the formation of the complex between molybdenum and fluorine ions in the body, the conversion of fluorine ions is restricted, so as to promote the excretion of fluoride and reduce the accumulation of the ion in the body and therefore reduce the toxic effect of fluoride.

Copper

Copper is one of the essential trace elements in human body and a catalyst in the redox system, which relates to the activity of many enzymes. The effect of copper on fluorosis has long been studied. Singh et al. reported that copper deficiency in animals with fluorosis caused poor utilization of iron and abnormal contents of zinc and manganese in tissues, suggesting that abnormal changes in trace element content were related to the incidence of fluorosis (Singh 1984). Indian scholars reported that in fluorosis patients with genu valgus syndrome, characterized by osteomalacia, osteoporosis, and bone malformation, copper and manganese were deficient in the body and copper content in the bone is obviously low. The effect of chronic fluorosis may be related to low level of copper in the environment and insufficient copper intake in the human body. A study with experiment in rats by Zhang Hua confirmed that the fluorosis could interfere with the levels of elements, such as copper, zinc, iron, and manganese, destroy the normal balance of trace elements in the body. Whereas supplement of copper to attenuate the toxic effect of excessive fluoride, weak the influence of fluoride on bone shown by X-ray and increase urinary fluoride excretion. It reported that the content of copper in bone and other tissues of animals with fluorosis was relatively low, and the collagen breakdown products increased (Susheela and Sharma 1982). The authors believed that the mechanism of accelerated collagen decomposition caused by high fluoride might be related to the effect of fluoride on the content of copper in the body and the activity of lysyl oxidase containing copper, leading to the insufficient collagen cross-linking and the decreased stability. A study on the relationship between the content of 20 elements in drinking water and the prevalence of fluorosis showed that the prevalence of dental fluorosis decreased significantly with the increase of copper, zinc, and calcium and skeletal fluorosis with copper, chromium, and calcium in drinking water.

Although most studies have shown that copper has an antagonistic effect on fluorosis, contrary data indicated that copper and fluorosis have a certain synergistic effect. The contents of fluoride in urine, copper, calcium, magnesium, and zinc in serum from the patients with delayed bone age in endemic fluorosis areas were measured and the results showed that urinary fluoride and serum copper were the risk factors for delayed bone age in the children and adolescent with high fluoride content in urine (Qian et al. 1993). The risk of delayed bone age was 4.9 times higher in children with urine fluoride content of 316–1602 mol/l than in children with urine fluoride content of 35–79 mol/l. The risk of delayed bone age of serum copper at 14–23 mol/l was four times higher than that of serum copper at 7–14 mol/l, indicating that copper and fluoride had synergic effect on delayed bone age. Nevertheless, the relationship between copper and fluorosis is limited to epidemiological investigation and animal experiments, and the clinical application of copper agent in the treatment of fluorosis needs further in-depth study.

Traditional Chinese Medicine (TCM) Prescription for the Treatment of Endemic Fluorosis

In recent years, Chinese herbal medicine for the treatment of endemic fluorosis has been advocated. In literature, there are many reports of animal experiments and clinical studies on the treatment of chronic fluorosis with TCM prescriptions. Most of these formulations understood by researchers are based on the theory of by using TCM to treat endemic fluorosis and the composition of these Chinese medicine herbs is strictly and individually targeted selected. It is reported that all of them have certain efficacy, but almost all of them are in the experimental stage and have not put on the market to the treatment of endemic fluorosis.

In TCM books, there are no exact records of endemic fluorosis. Some doctors in TCM think endemic fluorosis belongs to category of "Bi disease" (related to the arthralgia due to spinal and joint disorders); some believe it to be classified into "kidney Bi disease" or "bone Bi disease"; some ascribe the performance of this disease such as stiffness, numbness, hunchback, and atrophy as "bone wilt." It is generally believed that fluorine poisoning leads to "Qi" (according to TCM, "Qi" is the basis of most materials of the human body) weakness and/or blood weakness, dysfunction of the "Zang-Fu organs" (meaning anatomical patterns and physiological features of organs in TCM), especially presenting kidney deficiency without nourishing the bones and causing the disease. Therefore, it is more to use the medicine formula of tonifying liver and kidney to regulate "Qi" and function of blood. In China, several Chinese medicine formulations from famous doctors in TCM have been investigated and the beneficial effects against chronic fluorosis obtained. Here, two Chinese medicine formulations are briefly described as examples.

"Dan-Lan-Xian Boron" Therapy Fluoride Capsule

"Dan-Lan-Xian Boron" therapy fluoride capsule is a TCM compound developed by combining the research group of endemic fluorosis of Guizhou Medical University and Guiyang Run-Feng Pharmaceutical Co., Ltd., China and the pre-clinical trial study has been completed (He et al. 1999). The results showed that this preparation had good efficacy and high safety, which has passed the review of State Food and Drug Administration, obtained the drug clinical trial approval, and is undergoing clinical trial.

"Dan-Lan-Xian Boron" therapy fluoride capsule is based on the theory of combination of Chinese and western medicine, aiming at several main links of endemic fluorosis, using seven medicine composition. The drug has achieved the following several aspects of the effect. (1) Tonifying the kidney and strengthening the bones. Traditional medicine on fluorosis is described as: "Fluoride invades the body, resulting consumption and loss of Qi and blood, no nourishment to the Zang organs and filling the kidney essence, and leading to kidney fatigue. Since kidney is responsible to bone and generates marrow, kidney fatigue appears a series of bone lesions and systemic symptoms. This formula uses traditional Chinese medicine such as Salvia miltiorrhiza, Epimedium, and Yanhusao, aiming to enrich Qi and tonify kidney, strengthen tendon and bone, dehumidify and activate collateral channels, eliminate toxin and set pain, produce saliva and nourish blood in order to alleviate bone lesions and deformities, bone and joint pain and other symptoms, and relieve the symptoms of patients. (2) Anti-oxidation damage. Modern medical research has proved that high amount of fluoride results in higher level of oxidative stress, i.e., too much produce of free radicals and too weakness of the activity of antioxidant enzymes or antioxidant materials. This prescription employs Salvia miltiorrhiza, Epimedium, Gynostemma pentaphylla, and other herbs, the effective components of which are all strong antioxidants. These substances synergy can enhance the ability of antioxidant enzyme system, remove excess free radicals, and prevent the damage to the organization, so help to block the main steps of fluoride poisoning and prevent the development of fluorosis. (3) Elimination of excess fluoride stored in the body. In the treatment of fluorosis, fluoride accumulates in the body and its etiology should be eliminated in addition to against its symptoms, pathological basis, and pathogenesis. Borax is commonly used to treat chronic fluorosis, in which its mechanism is to expel fluoride from the body. In the recipe borax is taken as medicine to form a complex with fluoride and thereafter to make the ion to be eliminated from the body. In addition to borax direct discharge of fluoride, Salvia miltiorrhiza and other medicinal materials can strengthen the body viscera function, but also promote the elimination of fluoride from the body and decrease its accumulation. In the therapy of TCM, borax is generally believed to be used outside to give its priority, while by internal use it should be appropriate cautious since borax has been reported as cytotoxic and reproductive toxin in certain situation. To solve this problem, the research group conducted long-term toxicity tests on "Dan-Lan-Xian Boron" therapy fluoride capsule in rats for 25 weeks and beagles for 40 weeks. According to the test results, it was concluded that the safe dose range was less than 20 times of the clinical dose.

"Guling-Tongbi" Pill

"Guling-Tongbi"pill is a pure Chinese medicine preparation developed by a research group in Luzhou Medical College (it is currently being renamed as Southwest Medical University) according to a prescription from a senior doctor in TCM for treating bone disorders (Liu and An 2006). Its main ingredients include 12 kinds of herbs. The functions and effects of the prescription include tonifying the kidney, eliminating phlegm, and clearing damp and nourishing the liver. It is used for "Bi" diseases (such as rheumatism and rheumatoid arthritis) caused by obstruction of collaterals and deficiency of liver and kidney, including joint pain, swelling, stiffness, adverse flexion and extension, or even enlarged deformity, accompanied the pain of waist and knee pain, chilly of body and cold limbs. Due to the similar symptoms of skeletal fluorosis with these disorders, the prescription is selected as a prescription of TCM to treat chronic fluorosis. "Guling-Tongbi" pill can reduce the absorption of fluoride through digestive tract, increase excretion of fluoride in feces, slow down the increase of fluoride in serum, liver, kidney, bone, and urine, reduce pathological damage of liver, kidney, and bone, and release the clinical symptoms of fluorosis.

Development Trend of Antioxidant Drugs

Since the theory of oxidative stress has been widely recognized in the pathogenesis of systemic damages caused by endemic fluorosis, the development of antioxidant drugs has become a very important development direction in recent years (Tian et al. 2018). The oxidative stress injury caused by excessive fluoride in rats treated with vitamin E and vitamin C can be significantly reduced in vivo studies (Basha and Sujitha 2011). The high toxic lipid peroxidation end product, 4-hydroxiynonenal (4-HNE), is a nonprotein mediator of apoptosis and involves the pathogenesis of chronic fluorosis. The abundant presence of HNE conjugated proteins in the fluoride-exposed cells was confirmed (Singh and Das 2019). The co-treatment with the combination of the dietary antioxidants vitamins C and E was effective in protecting the cells and reducing the fluoride-induced burden of HNE modified proteins in the cells. The significantly protective and therapeutic results of vitamin E and C treatment were obtained, in which the damage of DNA oxidation induced by experiment chronic fluorosis was attenuated by the antioxidants (Oner et al. 2020).

Epigallocatechin gallate in green tea has a strong inhibitory effect on oxidative stress and DNA damage in brain, liver and kidney caused by fluorosis and can reduce the neurotoxic damage caused by fluorosis (Thangapandiyan and Miltonprabu 2013). The level of oxidative stress in the brain tissue of the rats with chronic fluorosis was increased and the behavior of the rats was changed. After treatment with Spirulina platensis, the level of oxidative stress was reduced and abnormal behavior improved. A study showed that high fluoride increased malondialdehyde level, decreased the content of glutathione and the activities of catalase and SOD in myocardial tissues and erythrocytes of rats. While giving Quercetin and Silymarin combated the toxicity of fluoride and declined the elevated levels of oxidative stress by fluorosis (Nabavi et al. 2012). Recent studies have found that melatonin has an inhibitory effect on the elevated level of oxidative stress induced by chronic fluorosis (Chawla and Rao 2012).

In our previous studies, we found that resveratrol, a polyphenol to regulate cellular protection against oxidative stress, attenuated the raised level of oxidative stress and the lowered ability for learning and memory of rats resulted from the chronic fluorosis, indicating that resveratrol may have a neuroprotective effect on fluoride toxicity by stimulating the express of neuronal nicotinic acetylcholine receptors (Zeng et al. 2019, 2020). In addition, resveratrol also provides a protective effect against the increased blood pressure and the potential endothelial damage caused by fluorosis, in which one of the mechanism may be related to the capability of the compound to reduce fluoride-induced oxidative stress and endothelial tissue damage (Bulduk et al. 2020). The exposure of high fluoride reduced the protein expressions of both M1 and M3 muscarinic acetylcholine receptors (mAChRs) and elevated the levels of free radical and MDA in rat brains (Dong et al. 2017). The treatment of vitamin E prevented the increased production of free radicals in the brains of rats fed with high fluoride, in which the mechanism for this may involve the antioxidant allowing the expression of mAChRs to be recovered.

Current Situation and Problems in the Study of Endemic Fluorosis Drugs

Since 1970s, more than 50 kinds of drugs have been reported to treat endemic fluorosis. Chemical elements such as boron, magnesium, and calcium were used for the treatment of the disease. However, there is no obvious progress in drug research and development to the efficient treatment of endemic or chronic fluorosis.

In China, TMC is advocated in the treatment of endemic fluorosis and the advantages of Chinese medicine have been exhibited. However, there still present many shortcomings, such as the lack of a unified method to evaluate the effect of clinical treatments and no-confirmation to the efficacy of various drugs. Although researchers have formulated various formulations, conducted pharmacodynamic animal experiments and clinical treatment studies, most of them failed to meet the requirements of national requirement of new drug.

Over the years, large amounts of funds and manpower in the improvement of water supply, cooking stoves, elimination of fluoride sources, and other aspects have been obviously invested in the world and the incidence of endemic fluorosis greatly reduced, but tens of millions of patients with endemic fluorosis, especially those with typical skeletal and dental fluorosis, are still suffering from endemic fluorosis. Therefore, it is still an urgent task to develop ideal therapeutic drugs to completely eliminate the hazard of endemic fluorosis.

Conclusion

Over the years, the researchers and medical staff in the prevention and treatment of endemic fluorosis have done a lot of work in the development of drugs to treat the disease by using boron, magnesium, calcium, selenium, zinc, molybdenum, copper, and other major and trace elements as well as antioxidant materials and certain positive results have been achieved. On the other hand, Chinese researchers and doctors have also employed Chinese medicine prescription and Chinese herb extracts to treat the patients with endemic fluorosis, which opens up a new way to the therapy of the disease. Although there is no consistent program and recognized therapeutic effect for the drug treatment of endemic fluorosis yet, the researchers on the prevention and treatment of endemic fluorosis are still making unremitting efforts to carry out the investigations of drug therapy in order to relieve the harm of the disease on the related population and make contributions to alleviating the pain of patients with endemic fluorosis.

References

Bai XT, Bao KG, Liu AP. Blocking effect of zinc, manganese and molybdenum on fluoride bone deposition in rat. Chin J Ctrl Endem Dis. 1993;8:267–8.

Basha MP, Sujitha NS. Chronic fluoride toxicity and myocardial damage: antioxidant offered protection in second generation rats. Toxicol Int. 2011;18:99–104.

Bian JC, Xian SM, Ye P, Yang XX, Liu YP, Lu CH, et al. Experimental study on the protective effect of selenium on body damage caused by fluoride. Chin J Ctrl Endem Dis. 2004;19:257–9.

Bulduk M, Oto G, Ozdemir H, Demirel-Yilmaz E. The effect of resveratrol therapy on the vascular responses caused by chronic fluorosis. Fluoride. 2020;53:23–39.

Chawla SL, Rao MV. Protective effect of melatonin against fluoride induced oxidative stress in the mouse ovary. Fluoride. 2012;45:125–32.

Deng H, Liang YX, Cao JX, Xu ZJ, Fang X, Sun SZ. Study on the combined effects of selenium and fluorine in rats at different doses. Health Res. 1996;25:106–8.

Dong YT, Wei N, Qi XL, Liu XH, Chen D, Zeng XX, et al. Attenuating effect of vitamin E on the deficit of learning and memory of rats with chronic fluorosis - the mechanism may involve muscarinic acetylcholine receptors. Fluoride. 2017;50:354–64.

Elsair J, Merad R, Denine R, Reggabi M, Alamir B, Benali S, et al. Boron as a preventive antidote in acute and subacute fluoride intoxication in rabbits, its action on fluoride and calcium-phosphorus metabolism. Fluoride. 1980;13:129–36.

Ersoy IH, Koroglu BK, Varol S, Ersoy S, Varol E, Aylak F, et al. Serum copper, zinc, and magnesium levels in patients with chronic fluorosis. Biol Trace Elem Res. 2011;143:619–24.

Frank J, Runge H, Bech R, Wiedner W, Kramer W, Kochmann W, et al. Boron as an antidote to fluorosis. Part 1: studies on the skeletal system. Fluoride. 1985;18:187–95.

Guan ZZ. The importance of studying the pathogenesis of endemic diseases. Chin J Endem. 2014;33:119–20.

He XY, Zhang XL, Zhao MX, Liu JL, Wang W. Therapeutic effect of Gynosaponin-Salvia miltiorrhiza compound on industrial fluorosis. Guizhou Med. 1999;3:165–7.

Jiang S, Chen QG, Zhang HT. Overview of the treatment of endemic fluorosis in recent 10 years. Chin J Ctrl Endem Dis. 2020;35:212–3.

Li FC, Guan ZZ. Synergistic intoxication with aluminum and fluoride in patients in area of coal burning endemic fluorosis. Fluoride. 2014;47:283–6.

Li JX, Chen RA, Zhang MB, Yang DJ, Yu LL. Experimental study on boron resistance to fluoride bone injury. Chin J Public Health. 1989;8:161–3.

Li SC, Lv SM, Xiong YM, Zhang YY. Selenium supplementation promoted urinary fluoride excretion in rats with fluorosis. Chin J Ctrl Endem Dis. 1995;10:11–4.

Li W, Dong S, Chen Q, Chen C, Dong Z. Selenium may suppress peripheral blood mononuclear cell apoptosis by modulating HSP70 and regulate levels of SIRT1 through reproductive hormone secretion and oxidant stress in women suffering fluorosis. Eur J Pharmacol. 2020;878:173098.

Liu XT, An SK. Observation on effect of Gulingtongbi pill on 120 cases of skeletal fluorosis. Chin J Endem. 2006;21:724–6.

Mo XY, Zhang YY, Li SC, Fan WZ. Study on collagen metabolism in rats with experimental fluorosis - on the anti-fluorosis effects of drugs such as boron, selenium, fluning and cistanche pills. Chin J Ctrl Endem Dis. 1991;6:321–3.

Nabavi SF, Nabavi SM, Latifi AM, Mirzaei M, Habtemariam S, Moghaddam AH. Mitigating role of quercetin against sodium fluoride-induced oxidative stress in the rat brain. Pharm Biol. 2012;50:1380–3.

Oner AC, Dede S, Yur F, Oner A. The effect of vitamin C and vitamin E on DNA damage oxidative status, and some biochemical parameters in rats with experimental fluorosis. Fluoride. 2020;53:154–63.

Pang YX, Zhu P, Zhang SC, Fu KW, Sun YF, Tang RQ. Protective effects of magnesium and selenium on brain cell damage induced by fluoride. Chin J Endem. 1994;9:329–30.

Patel PP, Patel PA, Zulf MM, Yagnik B, Kajale N, Mandlik R, et al. Association of dental and skeletal fluorosis with calcium intake and serum vitamin D concentration in adolescents from a region endemic for fluorosis. Indian J Endocrinol Metab. 2017;21:190–5.

Pettifor JM, Schnitzler CM, Ross FP, Moodley GP. Endemic skeletal fluorosis in children: hypocalcemia and the presence of renal resistance to parathyroid hormone. Bone Miner. 1989;7:275–88.

Qian C, Zhang ZY, Dai GJ, Li J, Zhai C, He W. Study on the interaction of fluoride, copper, calcium, magnesium and zinc on bone age delay in high fluoride area. J Chin Med Univ. 1993;2:37–9.

Rao SR, Marty KLR. Treatment of fluorosis in human beings. Fluoride. 1975;8:12–6.

Seeffner W, Teubene M. Antidotes in experimental fluorosis on pigs, morphological studies. Fluoride. 1983;16:33–7.

Singh M. Biochemical and cytochemical alterations in liver and kidney following experimental fluorosis. Fluoride. 1984;17:81–93.

Singh P, Das TK. Ultrastructural localization of 4-hydroxiynonenal adducts in fluoride -exposed cells: protective role of dietary antioxidants. Fluoride. 2019;52:49–58.

Song SC, Wei ZD, Zhou LY, Zhang H, Fang SJ, Cai QH. An experimental study on the role of aluminum in the etiology of fluorosis. J Guiyang Med Coll. 1991;16:217–25.

Spencer H, Kramer L, Norris C, Wiatrowski E. Effect of aluminum hydroxide on plasma fluoride and fluoride excretion during a high fluoride intake in man. Toxicol Appl Pharmacol. 1981;58:140–4.

Strunecka A, Strunecky O. Chronic fluoride exposure and the risk of autism spectrum disorder. Int J Environ Res Public Health. 2019;16:3431.

Susheela AK, Sharma YD. Certain facets of F- action on collagen protein in osseous and non-osseous tissues. Fluoride. 1982;15:177–90.

Thangapandiyan S, Miltonprabu S. Epigallocatechin gallate effectively ameliorates fluoride-induced oxidative stress and DNA damage in the liver of rats. Can J Physiol Pharmacol. 2013;91:528–37.

Tian Y, Xiao Y, Wang B, Sun C, Tang K, Sun F. Vitamin E and lycopene reduce coal burning fluorosis-induced spermatogenic cell apoptosis via oxidative stress-mediated JNK and ERK signaling pathways. Biosci Rep. 2018;38:1003.

Xi Q, Yang LL, Han JH, Liu YY. Current situation and comment on drug treatment of fluorosis. Chin J Control Endemic Dis. 2008;23:182–3.

Xia T, Wang AG, Yu RA, Tan LP, Chen JX, Chen XM. Experimental study on antagonism of selenium and zinc preparation against chronic fluorosis. Chin J Endem. 2001;16:180–2.

Xu EL, Wei ZD, Zhou LY, Zhang H, Wang SX, Wu FY, et al. Difference of the interaction between fluoride and different kinds of aluminum compounds in metabolism during the combined action of fluoride and aluminum. J Guiyang Med Coll. 1991;16:207–11.

Yao ZM. Investigation on the relationship between trace elements and endemic fluorosis. Chin J Endem. 1992;7:100–2.

Yao ZM, Zhang LH. Treatment and experimental study of trace element molybdenum and dental fluorosis in children. Chin J Pre Med. 1992;26:28–30.

Yu YN, Xiao KQ, Cao XM, Deng CJ, Liu QL, Zhou LY, et al. Bone morphometrology of rats with experimental combined toxicity of aluminum and fluoride. Chin J Endem. 1993;8:332–4.

Zeng XX, Deng J, Xiang J, Dong YT, Cao K, Liu XH, et al. Resveratrol attenuated the increased level of oxidative stress in brains and the deficit of learning and memory of rats with chronic fluorosis. Fluoride. 2019;52:149–60.

Zeng XX, Deng J, Xiang J, Dong YT, Cao K, Liu XH, et al. Protections against toxicity in the brains of rat with chronic fluorosis and primary neurons exposed to fluoride by resveratrol involves nicotinic acetylcholine receptors. J Trace Elem Med Biol. 2020;60:126475.

Zhang Z, Wang X, Nian W, Liao Q, Zhang R, Ouyang W. Effects of calcium on drinking fluorosis-induced hippocampal synaptic plasticity impairment in the offspring of rats. Transl Neurosci. 2017;8:191–200.

Zheng X, Sun Y, Ke L, Ouyang W, Zhang Z. Molecular mechanism of brain impairment caused by drinking-acquired fluorosis and selenium intervention. Environ Toxicol Pharmacol. 2016;43:134–9.

Zhou LY, Wei ZD, Liu SZ. Effect of borax in treatment of skeletal fluorosis. Fluoride. 1987;20:24–7.

Chapter 26
Improvement of the Health Condition of Local Residents After Comprehensive Treatments of Coal-Burning Type of Endemic Fluorosis

Zhi-Zhong Guan, Yan-Long Yu, and Yi Zhao

Abstract It is an important strategy to give a sustainable prevention and control in order to obtain the achievement with elimination of endemic fluorosis. In reviewing the results of comprehensive measures (improved stoves and health education) on coal-burning type of endemic fluorosis in China, this paper is devoted to the survey and study on medical and health effects of the endemic areas received such treatment. The findings in the areas of coal-burning type of endemic fluorosis after taking comprehensive measures revealed that the prevalence rate and the severity of dental fluorosis in primary school students were obviously reduced; the situation in younger children with mental retardation was significantly improved; the symptom of patients with skeletal fluorosis was alleviated with a moderate degree. The results by the examination of blood biochemistry indicated the following improvements in patients with chronic fluorosis in the areas of the fluoride pollution by burning coal: the improved condition for anemia, the raised level of serum high-density lipoprotein, the reduced activities of serum glutamic-pyruvic transaminase and

Z.-Z. Guan (✉)
The Key Laboratory of Endemic and Ethnic Diseases (Guizhou Medical University), Ministry of Education, Guiyang, Guizhou, People's Republic of China

Department of Pathology at the Affiliated Hospital of Guizhou Medical University, Guiyang, Guizhou, People's Republic of China

The Provincial Key Laboratory of Medical Molecular Biology, Guiyang, Guizhou, People's Republic of China
e-mail: zzg@gmc.edu.cn

Y.-L. Yu
The Key Laboratory of Endemic and Ethnic Diseases (Guizhou Medical University), Ministry of Education, Guiyang, Guizhou, People's Republic of China

The Provincial Key Laboratory of Medical Molecular Biology, Guiyang, Guizhou, People's Republic of China

Y. Zhao
College of Foreign Languages, Guizhou Medical University, Guiyang, Guizhou, People's Republic of China

Z.-Z. Guan (ed.), *Coal-burning Type of Endemic Fluorosis*,
https://doi.org/10.1007/978-981-16-1498-9_26

glutamic-oxalacetic transaminase, the increased activities of catalase and glutathione-S-transferase, the lowered activity of myeloperoxidase, and the weakened effects of fluorosis on the inhibited mRNA expression of glutathione-S-transferase and the stimulated myeloperoxidase. In addition, it was also found that the quality of life for the residents in such areas was significantly improved. Multiple parameters on health improvement could be considered as referential index in evaluating the health condition of patients from such areas. Yet, of all the tested indexes, certain parameters failed to reach the health standard. Namely, recovery of senior patients was far from satisfaction. Sustainable and comprehensive intervention measures as well as long-term management mechanism must be adopted to ultimately eradicate the harm of coal-burning type of endemic fluorosis.

Keywords Coal-burning type of endemic fluorosis · Health evaluation index · Quality of life · Parameters in blood chemistry · Strategy of prevention and control

Introduction

Coal-burning type of endemic fluorosis is the driving force to impoverish or return to poverty by illness in the endemic areas of the disease. Traditionally, the Chinese government attached great importance to the health condition of people living in the areas of endemic fluorosis and listed the eradication for coal-burning type of endemic fluorosis as key public health service of national health care reform (Guan et al. 2017). Chinese scientists in Guizhou province broke the ice of fluorosis prevention and control research on coal-burning type of endemic fluorosis since the 1980s. Since then, a model with comprehensive prevention and control took shape with improving stoves, carrying out health education, boosting economy, and raising health care ability. The measure with large-scale prevention and control on the areas of coal-burning type of endemic fluorosis had effectively reduced the number of new patients and alleviated the severity of such disease to a large extent. To better understand the health condition of people living in endemic fluorosis areas received comprehensive measures, the patients and relevant residents from the rural endemic areas of Guizhou province of China were selected for the study. Study subjects were informed with relevant information and signed the Informed Consent Agreement, then they went through the following check-ups: clinical manifestation, related symptoms, relevant biochemical indexes on blood and urine, bone imaging, check-ups on multiple systems (neurotic, cardio-vascular, digestive, and respiratory systems). Meanwhile, healthy people from the non-endemic areas or the residents from the endemic areas but not receiving effectively comprehensive treatment were chosen as controls. A medical evaluation for health condition of patients with chronic fluorosis who had received comprehensive treatments was made.

Research Subjects

The areas of coal-burning type of endemic fluorosis in Guizhou of China were chosen to conduct this study. The chosen areas included: the endemic fluorosis areas with the treatment of long-term comprehensive measures (>3 years), the endemic areas with the treatment of short-term comprehensive measures (around 1 year), and the endemic areas without obvious treatment; the critical and mild endemic fluorosis areas were also divided. Meanwhile, the areas of non-coal-burning type of endemic fluorosis in Guizhou was chosen as the control group. The research was further divided into the endemic fluorosis areas inhabited with minority group and non-minority group. The total of over 5000 research subjects were divided into child and adult groups, in which the latter was divided into different age groups (39 years old and below, 40 to 59 years old, and 60 years old and older).

Effect of Comprehensive Treatments on Children with Dental Fluorosis

The prevalence rate and classification of dental fluorosis among primary school students aged 8–12 years in three schools in Guizhou province with severe endemic fluorosis caused by coal burning were continuously monitored over the past 30 years to observe the prevention and control effects of the disease. Dean division test was adopted to check the condition of dental fluorosis, together with investigation on staple food, schooling census, minority groups, and so on. This 30 year long prevention and control measures had significantly lowered the prevalence rate and severity of dental fluorosis in primary students of these areas. In particular, the full coverage of the prevention and control to the coal-based endemic fluorosis since 2006 has achieved remarkable results. The study also found that children from endemic areas with corn as staple food showed worse dental fluorosis conditions than those from the areas with rice as staple food. Prevalence rate of children with dental fluorosis was higher in minority group living in the endemic fluorosis areas than that of Han group. The overall results suggest that to reduce the incidence of dental fluorosis in children to a great extent and consolidate the achievements in prevention and control of fluorosis, comprehensive treatments to the disease should be continued, and special attention should be paid to the areas with ethnic minority group.

Effect of Comprehensive Treatments on Intelligence of Children

It was proven that excessive fluoride intake could be deposited in brain tissues through blood–brain barrier (Cao et al. 2019). For children, their brains are undergoing a key period of growing and differentiation, hence, the change and stimulation from both internal and external environment can pose a negative effect on brain development and ultimately cause mental retardation. Researches had confirmed that chronic fluorosis caused child mental retardation in the areas of endemic fluorosis (Li et al. 1995). In our previous study, primary school students (8–12 years old) from the areas of coal-burning type of endemic fluorosis in Guizhou were selected as the examined subjects and the age-mated subjects from the primary schools of the non-endemic areas as controls to compare the effects of different comprehensive treatment time on the intervention of abnormal intelligence of children (Wei et al. 2014). This research checked relevant conditions (dental fluorosis condition, urine fluoride level, and intellectual ability) of school age children to analyze the effect of comprehensive measures on intelligence development of children from the areas of endemic fluorosis. Dean assay was adopted to test dental fluorosis, fluorine ion selective electrode was adopted to detect urinary fluoride, and *Raven's Standard Progressive Matrices* was adopted to check the intelligence level of children. The results showed that the positive rate of dental fluorosis was 72.5% in long-term comprehensive measures group, 85.3% in short-term comprehensive measures group, and no case in control group. The differences of dental fluorosis positive rates between the groups with long-term and short-term comprehensive measures were statistically significant. Urinary fluoride contents were significantly higher from the children in the areas of endemic fluorosis than that of the control group and lower from the long-term treatment group than that of the short-term treatment. After long-term comprehensive treatment, the positive rate of mental retardation in the younger group (8–10 years old) was significantly lower than that in the short-term group. Urinary fluoride content was negatively correlated with intelligence level of these subjects. The research finding indicated that intelligence development of the children in the areas of endemic fluorosis was obviously slower than that of the control group. Interestingly, such condition could be improved in younger age children group after a certain period for comprehensive treatment.

Effect of Comprehensive Measures on Skeletal Fluorosis

A study conducted substantial surveys and researches on the conditions of patients with skeletal fluorosis from the areas of coal-burning type of endemic fluorosis receiving comprehensive treatments and evaluated the effect of such measures (Duan et al. 2013). Digital radiography was adopted for X-ray image of the body positions, such as pelvis, right forearm, and right shin of the subjects. Then, the results were evaluated for skeletal fluorosis according to *Diagnostic Criteria for Skeletal Fluorosis* (WS192-2008) issued by China. The results showed that the total X-ray positive rate of skeletal fluorosis was 28.3%, among them 22.9% with mild and 5.4% with moderate skeletal fluorosis positive rate. No patients with severe skeletal fluorosis were found in this study. Despite the lowered positive rates compared with the past, the total positive rate was still higher than the national average level for skeletal fluorosis in the endemic areas. Such results indicated that the areas of coal-burning type of endemic fluorosis in Guizhou were still the worst affected areas in China. Furthermore, the past population survey on coal-burning type of endemic fluorosis in Guizhou revealed certain proportion of patients with moderate and severe skeletal fluorosis, while this study discovered no patients with severe skeletal fluorosis but mild ones. With the differences in the present and past studies, it implied that the comprehensive measures targeting coal-burning type of endemic fluorosis reduced the number of patients with skeletal fluorosis and made significant achievement in prevention and control. Comparing the duration of comprehensive measure, the endemic fluorosis area with longer period of treatment showed 20.4% of positive rate of skeletal fluorosis, while the areas with shorter time kept 32.4% with a significant difference. The results suggest that after a long period of comprehensive prevention and treatments in the areas of coal-burning type of endemic fluorosis, skeletal fluorosis has been effectively controlled.

Changes in Common Blood Biochemistry Parameters

Blood biochemistry tested parameters refer to the indexes which are easily to evaluate the improvement of health condition of patients with endemic fluorosis in the areas receiving comprehensive measure (improved stoves and health education). Positive results were achieved by comparison with the tested biochemical parameters in blood obtained from the residents living in the areas of endemic fluorosis with or without comprehensive treatments and non-endemic fluorosis area, as well as the fluorosis areas with the different treatment time (Wu et al. 2010).

Routine Measurement of Blood Biochemistry

Routine blood test results were compared between two groups received comprehensive measures for years and those received no improved stoves and other measures in the areas of coal-burning type of endemic fluorosis. Tested parameters included red blood cells (RBC), white blood cells (WBC), hemoglobin (Hb), hematocrit (HCT), mean corpuscular volume (MCV), mean corpuscular hemoglobin concentration (MCHC), coefficient of variation in red blood cell distribution width (RDW-CV), and platelets (PLT). Investigation had confirmed that the proportion of patients with anemia rose in the area of endemic fluorosis with the aggravated condition. Anemia involved with multiple factors, but RBC count and Hb test would aid the diagnosis of anemia in patients with endemic fluorosis, which is correlated with the situation of fluoride intake, nutrition status, and generic background. This study showed that certain parameters (RBC, Hb, HCT, MCHC, and PLT) of population from endemic fluorosis areas received comprehensive treatments obviously tended to get better than those received no treatment; in particular, the number of people with lower RBC and Hb level than normal value were significantly reduced. These results indicated that anemia condition in the residents from endemic fluorosis areas was improving after years of comprehensive treatments.

Activity of GPT and GOT

Many researches proved that chronic fluorosis would cause severe damage of liver and heart. The activity of GPT or GOT is the index of detecting hepatocellular or cardiomyocyte injury. The study on population from the areas of coal-burning type of endemic fluorosis revealed that the values of GPT or GOT activities in the residents from mild endemic fluorosis areas with long-term comprehensive treatments were close to those of normal values as compared with healthy population from non-endemic areas; while other groups (people from the critical or mild endemic areas with short-term treatment) showed obviously higher activities of GPT or GOT. In addition, the activity and the abnormal increase rate of GPT or GOT in the residents from endemic fluorosis areas with long-term comprehensive treatment were lower than those from endemic areas with short-term treatment. The research proved that activity of A GPT or GOT in population from endemic fluorosis areas was declined after a certain period of comprehensive treatments, in which its mechanism might be related to less damage of fluoride on liver and heart after taking comprehensive measure.

Level of Blood Lipids

Total cholesterol (TC), triglyceride (TG), and high-density lipoprotein (HDL) are the factors closely correlated with arteriosclerosis. Study had shown that high fluoride intake triggered the increases of TC and TG, and the decrease of HDL, which then caused arteriosclerosis. The level of blood lipids of the residents living in the areas of coal-burning type of endemic fluorosis received comprehensive treatments and non-endemic areas had been observed (Liao et al. 2013a). These subjects were divided into five groups according to the severity of fluorosis and the differences in periods of comprehensive treatments, i.e. (1) mild endemic area with short-term treatment group: the residents with mild symptom of chronic fluorosis and the time of comprehensive treatments for less than 1 year; (2) mild endemic area with long-term treatment group: the residents with mild symptom of chronic fluorosis and the time of comprehensive treatments for more than 3 year; (3) critical endemic area with short-term treatment group: the residents with severe symptom of chronic fluorosis and the time of comprehensive treatments for less than 1 year; (4) critical endemic area with long-term treatment group: the residents with severe symptoms of chronic fluorosis and the time of comprehensive treatments for more than 3 year; (5) control group: subjects from non-endemic fluorosis areas.

Fasting blood was extracted from the above people in the morning and TC, TG, and HDL in serum were measured. The test results showed that levels of TC and TG of subjects from endemic fluorosis areas increased, while HDL was lowered as compared with the control group (Liao et al. 2013a). The differences between TC and TG in groups with same symptoms but different treatment time were not statistically significant. However, the HDL level of the residents from endemic fluorosis areas received long-term comprehensive treatments was significantly higher than that of the residents with short-term treatment areas. The levels of TC and TG were increased, while HDL was lowered in the residents from the fluorosis areas with comprehensive treatments as compared those of control subjects, indicating that the recovery of blood lipid level was not unsatisfactory. Whereas HDL level of the residents from long-term comprehensive treatment areas significantly increased as compared with short-term comprehensive treatment areas, indicating a certain recovery of the parameter induced by comprehensive treatments.

Level of Blood Glucose

The level of blood glucose was detected in the residents living in the areas of coal-burning type of endemic fluorosis to evaluate whether the parameter could be used as one of the evaluating indexes for the health improvement after comprehensive treatments to the disease (OuYang et al. 2013). The results indicated that the level of blood glucose of the residents from endemic fluorosis areas was obviously lower than that of control group from non-endemic areas, but the values still felled into the normal ranges. Significant differences in blood glucose level were not detected in these groups from endemic fluorosis areas received different treatment time. Furthermore, hyperglycemia positive rate (blood glucose level higher than normal range) of population from endemic fluorosis areas was no high than that of the residents from non-endemic fluorosis areas. The influence factors on the level of blood glucose could be associated with diet structure, dietary intake, nutrition condition, and fluoride effect on metabolic process of blood glucose.

Changes of Blood Biochemistry Parameters in Elder Population

The health condition of elder population living in the areas of coal-burning type of endemic fluorosis received comprehensive treatments was investigated (Wang et al. 2020). The following parameters were surveyed: RBC, WBC, PLT, Hb, GPT, GOT, urine albumin (UA), blood urea nitrogen (BUN), creatinine (CREA) as well as education background and family economic status. The results showed that the comprehensive treatments in endemic fluorosis areas obviously helped these blood biochemistry parameters of population below 60 years old to be close to the those of the control group, while levels of RBC, Hb, PLT, and BUN of population in the endemic fluorosis areas over 60 years old were still lower than those from non-endemic areas, yet the values of ALT, AST, and CREA increased. There were no significant correlations of such changes with education background and family economic status. Despite the health condition improved obviously in population below 60 years old in endemic fluorosis areas by means of comprehensive treatments, population over 60 years old in the areas showed no significant improvement, which may be related to a more severe damage due to suffering ca longer course of the disorder and difficulty in recovery for elder patients with chronic fluorosis.

Changes of Cardiac Function

It was reported that endemic fluorosis triggered abnormal changes in electrocardiograph (ECG) of population from the areas of endemic fluorosis. The staff working in the fluoride related company showed relatively higher abnormal detection rate of ECG and manifested arrhythmia, incomplete left bundle branch block (LBBB), left ventricular hypertrophy (LVH), left ventricular high voltage, and limb lead low voltage. To better understand whether comprehensive treatments alleviated the damage of cardiac function induced by chronic fluorosis, the research subjects were chosen from the areas of coal-burning type of endemic fluorosis in Guizhou of China. Synchronous lead ECG machine was adopted to examine the subjects selected. The results discovered that ECG abnormal detection rates were obviously increased in population from endemic fluorosis areas (Liao et al. 2013b), which mainly lied in sinus bradycardia, sinus tachycardia, LBBB, T-wave change, left ventricular high voltage, premature systole, ST-T change, low voltage, electrical axis abnormal and short PR syndrome. Since high accumulation of fluoride can affect the tricarboxylic acid cycle and reduce the production of ATP, it may lead to the energy metabolism disorder and cause cardiac muscle damage, especially cardiomyocyte was sensitive to energy deficiency. By combining with calcium or magnesium in blood, fluoride can cause calcium or magnesium metabolic disorder and trigger ventricular myoconductivity and autonomic abnormalities. However, the study discovered that population with similar conditions showed no obvious differences in ECG abnormal detection rates among groups received different time of comprehensive treatments. Such results indicated that the chosen duration of comprehensive measure was not long enough to improve the cardiac function of endemic population.

Changes of Oxidative Stress Related Indexes

The hypothesis of oxidative stress in pathogenesis of endemic fluorosis has been widely accepted and serves as a hotspot in mechanic study of the disease (Guan et al. 1989, 2017). Substantial researches, based on the investigations by patients with endemic fluorosis, animal or cell models exposed to high amount of fluoride, have proven that the declined activities of antioxidant enzymes and contents of antioxidant substances, and the increased level of free radicals and the contents of lipid peroxide, protein and DNA oxidations induced by fluorosis result in the extensive damages to body (Zhang et al. 2015; Gao et al. 2008).

Activities of Enzymes

In our previous survey to the residents in the areas of coal-burning of endemic fluorosis after comprehensive treatments, the activities of superoxide dismutase (SOD) and glutathione peroxidase (GSH-px) were obviously raised.

Recently, our team has also determined the activities and expressions of catalase (CAT), glutathione-S-transferase (GSTs), and myeloperoxidase (MPO) to better understand the prevention and treatment effect of comprehensive measure. CAT is the major enzyme for regulating hydrogen peroxide metabolism in the body, which can clean hydrogen peroxide, produce water and oxygen molecule, and reduce the generation of hydroxyl radical. GST fell into the category of isodynamic enzyme with multiple physiological functions. The major function of GST is to catalyze sulfhydryl-group coupling between glutathione (GSH) and certain intrinsic or extrinsic electrophilic group from noxious substance; then enhances its hydrophobic nature to penetrate membrane easily and to be excreted after degradation. Thus, the process could achieve detoxication purpose. The substrates of GST catalytic reaction include carcinogen, chemotherapeutics, and free radicals. GST is not only an important detoxification enzyme, but also has the activity of GSH-px. Relatively higher activity of GST plays an active role in reducing tissue oxidation and cell damage, as it can clean lipid free radicals and prevent lipid peroxidation impairment. MPO, a member of peroxidase super family, is a key iron-containing lysosome which is mainly found in azurophilic granule of myeloid cells (mainly neutrophil granulocyte and monocyte). MPO can use hydrogen peroxide and chloride ion to produce chlorate and form free radicals with oxidizing ability. The main function of MPO is to eliminate microorganisms in phagocyte. However, as the antioxidation defense reaction was lowered, MPO can increase oxidative stress level and lead to oxidative damage of tissues.

The study revealed that the activities of CAT and GST in the blood of these residents in the areas of coal-burning type of endemic fluorosis were obviously lowered, while activity of MPO was significantly increased as compared with control group from non-endemic area (Zhang et al. 2013). After comprehensive treatments in the endemic fluorosis areas, the activities of plasma CAT and GST of the local people in the areas were obviously increased, while the activity of MPO was moderately decreased compared with the group from endemic area received no treatment (Zhang et al. 2013). The results in the study also indicate that the comprehensive measure play a significant protection by increasing activity of antioxidant enzymes and lowering damage induced by oxidative stress.

Expressions of GST and MPO mRNAs

Peripheral blood samples were collected from people living in the areas of coal-burning type of endemic fluorosis and non-endemic areas. After extracting the total RNA from blood cells, real time fluorescent quantitative PCR assay was adopted, with SYBR green I as fluorescent agent, to test expressions of GST and MPO mRNAs (Zhang et al. 2013). The results showed that coal-burning type of endemic fluorosis could lower the mRNA expression of GST in peripheral blood but enhancing MPO. The lowered mRNA expression of GST could trigger lowered activity of GST and further affect its ability of antioxidative damage, which ultimately lead to oxidative damage in tissues. On the contrary, the enhanced mRNA expression of MPO could improve the activity of MPO and enhance oxidative stress. After receiving comprehensive treatment, mRNA expression of GST in blood of these residents in the areas began to increase while that of MPO decreased. The implementation of the comprehensive measure in the areas of coal-burning type of endemic fluorosis indeed detained the spread of endemic fluorosis. The fluoride intake through digestive and respiratory tracts was greatly reduced, thus, the inhibition effect on mRNA expression of GST and stimulating MPO mRNA were both reduced. This result would flatten the increasing trend of free radicals and decline the effect of fluorosis on the body. All these results indicate that the comprehensive measure contributes to the relieving effect on the damage over patients with coal-burning type of endemic fluorosis.

Changes of Gene Polymorphism of Antioxidant Enzymes

It has been proven that gene polymorphism causes individual differences of susceptibility towards certain diseases and is closely related to the occurrence of multiple diseases and their progression, which has been gained more and more attention in the academia. Gene promoter polymorphisms of antioxidant enzymes do not only affect expressions of genes and change the activities of enzymes, but also cause the abnormality in functions of antioxidant enzymes and then aggravate oxidative stress.

A study concerned the analysis of *GSTP1* Ilel05Val site polymorphism by using genome DNA extracted from venous blood of people from areas of coal-burning type of endemic fluorosis (Zhang et al. 2009). In the results, the comparison of *GSTP1* Ilel05Val site genotype frequency revealed a significant difference, i.e., the genotype frequency of Ilel05Val site of *GSTP1* AA gene in fluorosis group was lower than that of the control group. GST belongs to a supergene family with eight types. *GSTP1* is the coding gene of acid isozyme of *GST-pi* in GST family and has the functions of detoxification and antioxidation as well as regulating cellular cell kinase to affect cell proliferation. The transformation of *GSTPl* gene encoding site +313 to A-G results in the change of isoleucine (Ile) at amino acid 105 to valine (Val). It is assumed that excessive fluoride intake may cause genotype change of Ilel05Val A/G and ultimately cause lowered antioxidative ability and enhanced lipid peroxidation. However, the comparison of this genotype frequency between comprehensive measures treated group and untreated group was not statistically significant.

Five single nucleotide polymorphism (SNP) genotypes of MPO were detected by using genome DNA extracted from venous blood of people living in the areas of coal-burning type of endemic fluorosis and that of the control group (Zhang et al. 2013). Results indicated that the different genotype frequencies of MPO (including -1228 g/A, -585 T/C, -463 g/A, and -163 C/T) and 3 haplotypes with higher frequency (including -163C-463G-585T-1228G-1276T, -163C-463G-585T-1228G-1276C, and -163C-463G-585T-1228A-1276T) were obviously correlated with fluorosis, but not obviously related with comprehensive treatment time.

To assess the distribution of CAT C-262T and GPX1 Pro198Leu genotypic variants in a Turkish population, genotyping analyses of CAT and GPX1 were conducted in 250 unrelated, healthy volunteers by the PCR-RFLP assay (Suzen et al. 2010). The allele frequencies were 0.784 (C) and 0.216 (T) for CAT and 0.636 (C) and 0.364 (T) for GPX1 Pro198Leu. The genotype frequencies were 0.632 (CC), 0.304 (CT), and 0.064 (TT) for CAT and 0.416 (CC), 0.44 (CT), and 0.144 (TT) for GPX1 Pro198Leu. The genotype frequencies did not deviate from Hardy–Weinberg equilibrium. The results as compared with those of other reported populations showed marked ethnic group differences. In addition, genotypes of CAT-262 C/T and -21 A/T polymorphisms of people from the areas of coal-burning type of endemic fluorosis and controls were determined (Tu et al. 2011). During the study, CC and CT polymorphism genotypes of CAT-262C/T were tested, but no TT genotype. It was discovered that CAT gene -262 C/T and -21A/T polymorphisms showed no significant correlation with coal-burning type of endemic fluorosis.

Quality of Life and Its Influence Factors of the Residents Living in the Areas of Coal-Burning Fluorosis with Comprehensive Interference

Quality of life can be used as the evaluation index for the effect of treatment and prevention to endemic fluorosis and can give a comprehensive estimate on the physical and mental health of the population. This type of study used to evaluate the effect of prevention to drinking-water type of endemic fluorosis. Recently, based on the investigation over coal-burning type of endemic fluorosis, the health condition and quality of life with its influence factors were collected concerning the residents living in such endemic areas received comprehensive treatments (Cai et al. 2013). Questionnaires were employed for the residents of endemic fluorosis areas to evaluate the correlation of comprehensive treatments and quality of life with the help of multiple linear regression analysis. The analysis showed that skeletal fluorosis had obvious influence on quality of life. Quality of life was negatively correlated with age, disease suffering condition, dental and skeletal fluorosis, and the food dried by burning coal, and while positively correlated with higher education background, higher family average income, and improved stoves. The analysis also indicated that as the condition of dental or skeletal fluorosis aggravated, the quality of life score would be lower. Such results stressed the necessity of the comprehensive measure for the endemic fluorosis areas. The implementation of the comprehensive measure and long-term management mechanism would rise people's awareness of health care. Hence, unhealthy lifestyle of people in the endemic fluorosis areas would be changed and ultimately lend support to eradication of endemic fluorosis. These would all contribute to improve the quality of life for people living in endemic fluorosis areas.

Conclusion

A series of survey and research have been taken for the effect of comprehensive prevention and control measures on medical health of the residents living in the areas of coal-burning type of endemic fluorosis. It was discovered that in the endemic fluorosis areas received comprehensive treatment, prevalence rate and severity of dental fluorosis in primary school students were lowered, mental retardation condition was improved in children of younger age and skeletal fluorosis in adults was alleviated. Many biochemical parameters, including Hb, HDL, the activities of GTP, GOP, CAT, GST, and MPO, had tended to normalize. The inhibition effect of fluorosis on mRNA expression of GST and the stimulation on MPO were receded. In addition, quality of life score was significantly increased. All these results indicated that the comprehensive measure, including improved stoves, health education, boosted economy, and improved health care competence, made favorable achievements. However, it must be noted that of all the tested parameters, TC, TG, gene polymorphism of related enzymes and cardiac functions have not been met such evaluation criteria yet. The recovery of elder patients is still not optimistic. With the improvement of life quality of patients suffered from chronic fluorosis, certain testing parameters could be chosen as evaluation criteria of patients' health condition after comprehensive treatments. Sustainably comprehensive measure and long-term management mechanism must be established to ultimately eradicate the harm of coal-burning type of endemic fluorosis.

References

Cai YY, Huang WY, Yang JY, Guan ZZ, Yu YL. Effects of health education in integrated program for controlling endemic fluorosis in Bijie of Guizhou Province. Mod Prevent Med. 2013;40:1952–4.

Cao K, Xiang J, Dong YT, Xu Y, Li Y, Song H, et al. Exposure to fluoride aggravates the impairment in learning and memory and neuropathological lesions in mice carrying the APP/PS1 double-transgenic mutation. Alzheimers Res Ther. 2019;11:35.

Duan QH, Chen XG, Hu J, Jiao J, Guan ZZ, Yu YL, et al. The effect of improved stoves on skeletal fluorosis of coal-burning type endemic fluorosis in Bijie city of Guizhou province. Chin J Endem. 2013;32:189–91.

Gao Q, Liu YJ, Guan ZZ. Oxidative stress might be a mechanism connected with the decreased alpha 7 nicotinic receptor influenced by high-concentration of fluoride in SH-SY5Y neuroblastoma cells. Toxicol In Vitro. 2008;22:837–43.

Guan ZZ, Yang PS, Pan S, Zhuang ZJ. An experimental study of blood biochemical diagnostic indices for chronic fluorosis. Fluoride. 1989;22:112–8.

Guan ZZ, Wang LH, Sun DJ. Endemic fluorosis. In: Sun DJ, editor. Endemiology. Beijing: People's Medical Publishing House of China; 2017.

Li XS, Zhi JL, Gao RO. Effect of fluoride exposure on intelligence in children. Fluoride. 1995;28:189–92.

Liao W, Guan ZZ, Liu YF, Xu SQ, Wu CX, Li Y, et al. Changes of serum lipids in coal-burning-borne endemic fluorosis areas after implementation of comprehensive control measures. Chin J Endem. 2013a;32:42–5.

Liao W, Guan ZZ, Liu YF, Xu SQ, Wu CX, Li Y, et al. Electrocardiogram observation in the populations lived in coal-burning type of endemic fluorosis areas after comprehensive treatments. Chin J Endem. 2013b;32:424–6.

OuYang K, Zhang T, Xu SQ, Zhou BZ, Xie M, Wu CX, et al. Changes in blood glucose levels of residents in the area of coal-burning type of endemic fluorosis in Guizhou. Chin J Public Health. 2013;9:895–6.

Suzen HS, Gucyener E, Sakalli O, Uckun Z, Kose G, Ustel D, et al. CAT C-262T and GPX1 Pro198Leu polymorphisms in a Turkish population. Mol Biol Rep. 2010;37:87–92.

Tu X, Shan KR, Zhang T, He Y, Xu SQ, Zhou SG, et al. Relationship between -262C/T and -21A/T polymorphism of catalase gene and coal-burning type of endemic fluorosis. Chin J Endem. 2011;30:148–51.

Wang LJZ, Zhao L, Yang GQ, Li Q. Measures designed to prevent endemic fluorosis caused by burning coal improve the health of most residents in the area of Guizhou, China, but not of the elderly. Fluoride. 2020;53:249–56.

Wei N, Li Y, Deng J, Xu SQ, Guan ZZ. The effects of comprehensive control measures intelligence of on school-age children in coal-burning borne endemic fluorosis areas. Chin J Endem. 2014;33:320–2.

Wu CX, Shan KR, Guan ZZ. Expression of glutathione S-transferase mRNA in peripheral blood of the population in coal-burning fluorosis area. Chin J Endem. 2010;29:121–3.

Zhang T, Shan KR, An D, Xu SQ, Zhou SG, He Y, et al. Analysis of the GSTP1 gene Ile105Val polymorphism in fluorosis areas in Guizhou Province. Chin J Endem. 2009;28:268–71.

Zhang T, Shan KR, Tu X, He Y, Pei JJ, Guan ZZ. Myeloperoxidase activity and its corresponding mRNA expression as well as gene polymorphism in the population living in the coal burning endemic fluorosis area in Guizhou of China. Biol Trace Elem Res. 2013;152:379–86.

Zhang KL, Lou DD, Guan ZZ. Activation of the AGE/RAGE system in the brains of rats and in SH-SY5Y cells exposed to high level of fluoride might connect to oxidative stress. Neurotoxicol Teratol. 2015;48:49–55.

List of Abbreviations

4-HNE	4-hydroxynonenal
8-OHdG	8-hydroxy-2′-deoxyguanosine
8-iso-PGF2 α	8-isoprostaglandin F2α
α-HBDH	α-hydroxybutyric acid dehydrogenase
α-MSH	α-melanocyte-stimulating hormone

A

A	Albumin
ABP	Androgen-binding protein
AC	Adenylyl cyclase
ACh	Acetylcholine
AChE	Acetylcholinesterase
ACP	Acid phosphatase
ADP	Adenosine diphosphate
AGEs	Advanced glycation end products
AIF	Apoptosis-inducing factor
AKP	Alkaline phosphatase
Al_3^+	Aluminum’s free metal cation
ALK5	Activin receptor-like kinase 5
ALM-ASA	Adapter ligation-mediated allele-specific amplification
ALP	Alkaline phosphatase
ALT	Alanine aminotransferase
AMPA	Amino-3-hydroxy-5-methyl-4-isoxazolepropionate
AOPPs	Advanced oxidation protein products
AP	Action potential
AP-1	Activator protein-1
APD	Action potential duration

Z.-Z. Guan (ed.), *Coal-burning Type of Endemic Fluorosis*,
https://doi.org/10.1007/978-981-16-1498-9

APP	Amyloid precursor protein
APTs	Anesthetic pre-evaluative temporaries
ARE	Antioxidant response element
ASD	Autism spectrum disorder
AST	Aspartate aminotransferase
ATF4	Activated transcription factor 4
ATF6	Activating transcription factor 6
ATG	Autophagy-related
ATP	Adenosine triphosphate
AUC	Area under the curve

B

BACH1	BTB domain and CNC homolog 1
BALP	Bone alkaline phosphatase
Bax	Bcl-2 associated X protein
BBB	Blood–brain barrier
Bcl-2	B cell lymphoma-2
bFGF	Basic fibroblast growth factor
BGPs	Bone gamma carboxyglutamic acid containing proteins
Bim	Bcl-2 interacting mediator
Bip	Binding immunoglobulin protein
BMI	Body mass index
BMP	Bone morphogenetic protein
BMSCs	Bone mesenchymal stem cells
BMU	Basic multicellular units
BPK	Bone-parathyroid-kidney
BSP	Bone sialoprotein
BuChE	Butyrylcholinesterase
BUN	Blood urea nitrogen
bZIP	Basic leucine zipper
CO_3^{2-}	Bicarbonate

C

$Ca_{10}(PO_4)_6(OH)_2$	Crystalline hydroxyapatite
Ca^{2+}	Calcium
Ca^{2+}-ATPase	Ca^{2+} pump
CAD	Chronic actinic dermatitis
CaM	Calmodulin
CaMK	Calcium/calmodulin-dependent protein kinase
CaMKIIq	Calcium/calmodulin-dependent protein kinase II

cAMP	Cyclic adenosine monophosphate
Casp	Cysteine aspartic acid protease
caspase-3	Cysteine aspartate specific protease-3
caspase-8	Cysteine aspartate specific protease-8
CaSR	Calcium-sensing receptor
CAT	Catalase
Cath K	Cathepsin K
Cbf1	Core-binding factor 1
CDC	Centers for Disease Control and Prevention
CDK4	Cyclin-dependent kinase
CFI	Community Fluorosis Index
c-Fos	Cellular oncogene fos
CG	Control group
CHIP	Chromatin immunoprecipitation
CHOP	CCAAT/enhancer-binding protein homologous protein
CNS	Central nervous system
COL1A1	Collagen type I α1 chain
COL1A2	Collagen type I α2
COLIXA3	Collagen type I α3
COMT	Catechol-*O*-methyltransferase
COX-2	Cyclooxygenase-2
Cr	Creatinine
CREA	Creatinine
CRP	C-reactive protein
CSFL	Cerebrospinal fluid leakage
CT	Calcitonin
CT	Computer tomography
CTR	CT receptor
Cu/Zn-SOD	Copper/zinc SOD
CysC	Cystatin C
Cyt C	Cytochrome C
JNK	c-Jun N-terminal kinase
LC3-I	Cytoplasmic LC3
PI3KC3	Class III phosphatidylinositol 3-kinase
PKA	cAMP and protein kinase A

D

DAG	Diacylglycerol
DHPs	Dihydropyridines
DI	Dean index
DMD	Duchenne muscular dystrophy

DNA	Deoxyribonucleic acid
DNMT	DNA methyltransferases
DNMT1	DNA methyltransferase 1
DPP	Dentin phosphoprotein
DR	Digital radiography
Drp1	Dynamin-related protein 1
Fas	Death receptor
FasL	Death receptor ligand
SMAD	Drosophila mothers against decapentaplegic protein

E

E	Estrogen
EAAs	Excitatory amino acid systems
ECG	Electrocardiogram
ECM	Extracellular matrix
EC-SOD	Extracellular-superoxide dismutase
EDEMs	ER degradation-enhancing alpha-mannosidase-like proteins
eDQE	Effective detective quantum efficiency
EGCG	Epigallocatechin gallate
EMG	Electromyography
ENO	Endothelial nitric oxide
eNOS	Endothelial NOS
ER	Endoplasmic reticulum
ERK	Extracellular regulated protein kinases
ERK1/2	Extracellular signal-regulated protein kinases
ERKs	Extracellular signal-regulated kinases
ERR	Estrogen-related receptor
ERS	Endoplasmic reticulum stress
ESR	Estrogen receptor
ESR1	Estrogen receptor 1
ESRα	Estrogen receptor α
ET-1	Endothelin-1
RNase	Endoribonuclease

F

BF^{4-}	Fluoroborate
F^-	Fluoride ion
FADD	Fas-associated death domain protein
Fas-L	Fas ligand
FB	Fibroblasts

FGFs	Fibroblast growth factors
Fis1	Fission 1
FSH	Follicle-stimulating hormone
FT_4	Free T_4
Fzo	Fuzzy onions

G

G	Globulin
GAGs	Glycosaminoglycans
GATA4	GATA-binding protein 4
GBHA	Glyoxal-bis-(2-hydroxyanil)
GC	Glucocorticoid
G-domain	GTPase domain
GDP	Guanosine diphosphate
CHOP	GADD153/C/EBP homologous protein
Glc	Glucose
GM-CSF	Granulocyte-macrophage colony-stimulating factor
Gn	Glycogen
GnRH	Gonadotropin-releasing hormone
GPCR	G-protein-coupled receptor
GPT	Glutamic-pyruvic transaminase
GPx3	Glutathione peroxidase 3
GRP78	Glucose-regulated protein 78
GSH	Glutathione
GSH-Px	Glutathione peroxidase
GSK-3β	Glycogen synthase kinase-3β
GSTP1	Glutathione S-transferase pi 1
GST	Glutathione-S-transferase
GTP	Guanosine triphosphate

H

•OH	Hydroxyl radical
$CaHPO_4$	Hydrated calcium hydrogen phosphate
H_2O_2	Hydrogen peroxide
Hb	Hemoglobin
hCG	Human chorionic gonadotropin
HCT	Hematocrit
HDAC1	Histone deacetylase 1
HDL	High-density lipoprotein
HDL-C	High-density lipoprotein cholesterol

Hh Hedgehog
HO-1 Heme oxygenase-1
HOS Human osteosarcoma
HPO Hypothalamic-pituitary-ovary
HPT Hypothalamic-pituitary-testicular
HRE Hormone response element
HRRI Heart rate recovery index
HRTEM High-resolution transmission electron microscopy

I

IFN-γ Interferon-γ
IGF-I Insulin-like growth factor-I
Ihh Indian hedgehog protein
IL Interleukin
IL-1β Interleukin-1β
IL-6 Interleukin-6
IL-8 Interleukin-8
Ile Isoleucine
Inhb Inhibin b
iNOS Inducible nitric oxide synthase
INS Insulin
InsR Insulin receptor
IP3 Inositol 1,4,5-triphosphate
iPTH Intact parathyroid hormone
IQ Intelligence quotient
IRE1 Inositol-requiring enzyme 1
IκB Inhibitor of NF-κB

K

KA Kainite

L

GSSG L-Glutathione oxidized
iNOS Leukocyte/macrophage NOS
LC3 Light chain 3
LC3-II/I Light chain 3-II/I
LDH Lactate dehydrogenase
LDL-C Low-density lipoprotein cholesterol

LH	Luteinizing hormone
LINE-1	Long interspersed nuclear elements-1
LPO	Lipid peroxides
LPS	Lipopolysaccharide
LRP	Lipoprotein receptor-related protein
LTCC	L-type calcium channels
LTP	Long-term potentiation
LVH	Left ventricular hypertrophy
$RO^{\bullet}$	Lipid oxygen free radicals

M

mAChRs	Muscarinic acetylcholine receptors
MAPK	Mitogen-activated protein kinase
MAPKK	Mitogen-activated protein kinase kinase
MAPKKK	Mitogen-activated protein kinase kinase kinase
MAPKs	Mitogen-activated protein kinases
MC3T3-E1	Mouse embryonic osteoblast precursor cells
MCHC	Mean corpuscular hemoglobin concentration
M-CSF	Macrophage colony-stimulating factor
MCV	Mean corpuscular volume
MDA	Malondialdehyde
Met	Methionine
METAP2	Methionyl aminopeptidase 2
Mfn1	Mitofusion1
Mfn2	Mitofusion2
MgF_2	Magnesium fluoride
miRNAs	micro-ribonucleic acids
MLH1	MutL homolog 1
MMP13	Matrix metalloproteinases-13
MMP-9	Matrix metalloproteinase-9
Mn-SOD	Manganese SOD
MPO	Myeloperoxidase
Mr	Molecular mass
MRI	Magnetic resonance imaging
MSCs	Mesenchymal stem cells
mtDNA	mitochondrial DNA
mTOR	Mammalian/mechanistic target of rapamycin
p38 MAPK	Mitogen-activated protein kinase p38
Vmax	Maximal shortening velocity

N

nAChR	Nicotinic acetylcholine receptor
NADH	Nicotinamide adenine dinucleotide
NF-κB	Nuclear factor-kappa B
NIS	Sodium/iodide symporter
NMDARs	*N*-methyl-D-aspartate receptors
nNOS	Neuronal nitric oxide synthase
NO	Nitric oxide
NOS	Nitric oxide synthase
NOX2	NADPH oxidase-2
NQO1	NADPH quinone oxidoreductase 1
NRF2	Nuclear factor erythroid 2-related factor 2

O

MGMT	O^6-methylguanine-DNA-methyltransferase
OA	Osteoarthritis
OB	Osteoblasts
OC	Osteoclasts
OCN	Osteocalcin
OPA1	Optic atrophy 1
OPG	Osteoprotegerin
OPN	Osteopontin
OS	Oxidative stress
OSI	Oxidative stress index

P

eIF2α	PERK/eukaryotic initiation factor 2α
p38 MAPK	p38 mitogen-activated protein kinase
PAI-1	Plasminogen activator inhibitor-1
PARP	Poly adenosine diphosphate-ribose polymerase
PC	Phosphatidylcholine
PCOS	Polycystic ovary syndrome
PCR	Polymerase chain reaction
PD	Parkinson disease
PDI	Protein disulfide isomerase
PE	Phosphatidylethanolamine
Peg3	Paternally expressed 3
PERK	Protein kinase R (PKR)-like endoplasmic reticulum kinase
PG	Prostaglandin

PGC-1α	Peroxisome proliferator-activated receptors-gamma coactivator-1α
P-gp	P-glycoprotein
PGs	Proteoglycans
PI	Propidium iodide
PI3K	Phosphatidylinositol-3-kinase
PI3K/AKT	Phosphatidylinositol 3-kinase/protein kinase B
PI3KC3	Phosphatidylinositol 3-kinase catalytic subunit type 3
PINK1	PTEN-induced putative kinase 1
PIP2	Phosphatidylinositol-4,5-bisphosphate
PKB/Akt	Protein kinase B
PLC	Phospholipase C
PLT	Platelets
PPARs	Peroxisome proliferators-activated receptors
PRL	Prolactin
PS	Phosphatidylserine
PSD	Postsynaptic density
PTH	Parathyroid hormone
PTH1R	Parathyroid hormone 1 receptor (PTH1R)/PTH/PTHrP type I receptor
PTHrP	PTH-related peptide

R

RAGE	Receptor for advanced glycation end products
RANK	Receptor activator of nuclear factor kappa B
RANKL	Receptor activator of nuclear factor kappa B ligand
RBC	Red blood cells
RBP	Retinol-binding protein
RDW-CV	Red blood cell distribution width
rER	Rough endoplasmic reticulum
RNS	Reactive nitrogen species
ROM	Range of motion
ROS	Reactive oxygen species
RP	Resting potential
RSV	Resveratrol (trans-3,4',5-trihydroxystilbene)
RUNX2	Runt-related transcription factor 2
RXR	Retinoid X receptor

S

CuZn-SOD	SOD with copper and zinc
Na^+, K^+-ATPase	Sodium- and potassium-activated adenosine triphosphatase

NaF	Sodium fluoride
NIS	Sodium/iodine symporter
SAA	Serum amyloid A
SCs	Sertoli cells
SD	Sprague-Dawley
Se	Selenium
SHBG	Sex hormone-binding globulin
Shh	Sonic hedgehog protein
siRNA	Small interfering RNA
SIRTs	Sirtuins
sirtuin 1, Sirt1	Silent information regulator factor 2-related enzyme 1
SMAD3	SMAD family member 3
SMO	Smoothened
SNP	Single nucleotide polymorphism
SOD	Superoxide dismutase
SOST	Sclerosteosis

T

Col-I	Type I collagen
T_3	Tri-iodothyronine
T_4	Thyroxin
TAC	Total antioxidant capacity
TBARS	Thiobarbituric acid reaction substances
T-Bil	Total bilirubin
TC	Total cholesterol
TCM	Traditional Chinese medicine
TDFI	Total daily fluoride intake
TFI	Thylstrup-Fejerskov Index
Tg	Thyroglobulin
TG	Triglyceride
TGFBR2	Transforming growth factor beta receptor 2
TGF-β	Transforming growth factor β
TGFβ1	Transforming growth factor-β1
TH	Thyroid hormones
THA	Total hip arthroplasty
THs	Thyroid hormones
TKA	Total knee arthroplasty
TNF	Tumor necrosis factor
TNFR	Tumor necrosis factor receptor
TNF-R1	Tumor necrosis factor receptor-1
TOS	Total oxidative status
TPA	Terephthalic acid

TPO Thyroid peroxidase
Tpx/Prx Thioredoxin peroxidase
TRACP Tartrate-resistant acid phosphatase
TRAF TNF recruited-receptor associated factor
TRAIL TNF-related apoptosis-inducing ligand
TRH Thyrotropin-releasing hormone
TrxR1 Thioredoxin reductase 1
TSH Thyroid-stimulating hormone
TSIF Tooth Surface Index of Fluorosis
TT_4 Total T_4
TxA2 Thromboxane A2
TβRs TGF-β receptors

U

UA Uric acid
ucOCN Uncarboxylated osteocalcin
UDPG Uridine diphosphate glucose
UFE Urinary fluoride excretion
UN Urea nitrogen
UPR Unfolded protein response

V

Val Valine
VDR Vitamin D receptor
Vit D Vitamin D
Vit E Vitamin E

W

WBC White blood cells

X

Xbp1 X-box-binding protein 1
XIAP X-linked inhibitor of apoptosis protein

The manufacturer's authorised representative in the EU is Springer Nature Customer Service Centre GmbH, Europaplatz 3, 69115 Heidelberg, Germany. If you have any concerns regarding our products, please contact ProductSafety@springernature.com

Printed and bound by CPI Group (UK) Ltd, Croydon, CR0 4YY
30/09/2026
02225240-0011